7th EMBRIO International Symposium: "Ocean for the Future: Integrating Marine Biodiversity, Sustainability, and Resilience" (EIS 2025)

Online
2-3 September 2025

Editors:

S. Ferse **T. Bodur**

ISBN: 979-8-3313-3268-6

Printed from e-media with permission by:

Curran Associates, Inc.
57 Morehouse Lane
Red Hook, NY 12571

CURRAN ASSOCIATES INC. proceedings.com

Some format issues inherent in the e-media version may also appear in this print version.

For additional information, please contact EDP Sciences – Web of Conferences at the address below.

EDP Sciences – Web of Conferences
17, Avenue du Hoggar
Parc d'Activité de Courtabœuf
BP 112
F-91944 Les Ulis Cedex A
France

Phone: +33 (0) 1 69 18 75 75
Fax: +33 (0) 1 69 28 84 91

contact-edps@webofconferences.org

Additional copies of this publication are available from:

Curran Associates, Inc.
57 Morehouse Lane
Red Hook, NY 12571 USA
Phone: 845-758-0400
Fax: 845-758-2633
Email: curran@proceedings.com
Web: www.proceedings.com

TABLE OF CONTENTS

AQUACULTURE

TECHNOLOGY AND INNOVATION

MARINE BIODIVERSITY

BLUE ECONOMY, SUSTAINABILITY, AND RESILIENCE

https://doi.org/10.1051/bioconf/202622100001

PREFACE

The 7th EMBRIO International Symposium (EIS 2025), entitled "Oceans for the Future: Integrating Marine Biodiversity, Sustainability, and Resilience," was organized by EMBRIO (Enhancing Marine Biodiversity Research in Indonesia), a Center of Excellence (CoE) at IPB University coordinated by the Faculty of Fisheries and Marine Sciences. The symposium was held online on September 2nd–3rd, 2025, and served as a scientific forum to strengthen international collaboration and disseminate recent research on marine and aquatic sciences.

Established in 2013, EMBRIO has been dedicated to promoting science and innovation-based utilization of marine biodiversity resources while strengthening efforts to understand, protect, rehabilitate, and sustainably manage Indonesia's marine ecosystems. EMBRIO aims to become a central gateway for international scientists and research institutions seeking collaboration in Indonesia through multidisciplinary research, training, and academic exchange.

The theme *"Oceans for the Future"* highlighted the importance of integrated approaches in managing marine biodiversity, advancing sustainability, and enhancing ecosystem resilience in the face of global challenges such as climate change, pollution, and overexploitation. The symposium covered key thematic areas including marine biodiversity conservation, sustainable ocean practices, ecosystem resilience, integrated coastal management, technological innovation, blue economy development, and international cooperation.

The symposium featured keynote and invited speakers from national and international institutions. The keynote speakers were Prof. Dr. Rachmat Pambudy (Minister of National Development Planning/ Bappenas RI) with the topic was Navigating Indonesia's Ocean Future and Dr. Arif Havas Oegroseno (Vise Minister of Foreign Affairs RI) with the topic was International Collaboration. The invited speakers were Prof. Fang Qinhua (Xiamen University), Prof. Dr. Indra Jaya (IPB University), Dr. Muhamad Naimullah (Universiti Malaysia Terengganu), and Pilar Velasquez Jofre (ICAAD Guatemala).

The event was attended by 140 participants from various domestic and international institutions, resulting in 32 manuscripts to be published in the proceedings following a peer-review process. All accepted manuscripts underwent a rigorous review process, and selected papers were published in Bio Web of Conferences to ensure international visibility and scientific impact.

The organizing committee sincerely appreciated the contributions of all keynote and invited speakers, authors, reviewers, guest editors, and supporting institutions. It was hoped that the 7th EMBRIO International Symposium would foster meaningful scientific dialogue, strengthen research networks, and contribute to sustainable ocean management for present and future generations. This symposium and publication is funded by The Indonesia Endowment Fund for Education (LPDP) on behalf of Indonesian Ministry of Higher Education, Science, and Technology and managed under EQUITY Program (Contract No. 4297/B3/DT.03.08/2025-42011/IT3/HK.07.00-4/P/B/2025.

OPENING ADDRESS

Distinguished keynote and invited speakers, honorable guests, colleagues, and participants, and welcome to the 7th EMBRIO International Symposium (EIS 2025).

On behalf of EMBRIO and the organizing committee, it was a great honor to welcome all participants to this international scientific forum, held under the theme "Oceans for the Future: Integrating Marine Biodiversity, Sustainability, and Resilience." The symposium was conducted online on September 2nd–3rd, 2025, and brought together researchers, academics, students, and professionals from various countries.

EMBRIO, established in 2013 as a Center of Excellence at IPB University, has been committed to advancing research, innovation, and international collaboration based on Indonesia's rich marine biodiversity. Through activities such as international symposia, workshops, research collaboration, and academic exchange, EMBRIO has continuously strengthened Indonesia's role in global marine science.

The theme of this symposium reflected the urgent need for integrated and collaborative approaches to address complex ocean challenges. Marine biodiversity conservation, sustainable resource use, ecosystem resilience, technological innovation, and the development of a responsible blue economy were recognized as essential components for ensuring healthy oceans that support human well-being and future generations.

The symposium was honored by the participation of distinguished keynote and invited speakers representing government, academia, and international institutions, whose contributions provided valuable insights into ocean governance, marine science, sustainability, and resilience.

The organizing committee encouraged all participants to actively engage in scientific discussions, exchange ideas, and explore opportunities for future collaboration. It was sincerely hoped that the 7th EMBRIO International Symposium would not only enrich scientific understanding but also strengthen partnerships and inspire innovative solutions for sustainable ocean management.

Finally, the organizing committee expressed its deepest appreciation to the Rector of IPB University, the Dean of the Faculty of Fisheries and Marine Sciences, the guest editors, reviewers, speakers, and all committee members for their dedication and support in making this symposium a success.

 https://doi.org/10.1051/bioconf/202622100002

Dr. Roza Yusfiandayani

Chief Committee of The 7th EMBRIO International Symposium 2025

THE EVENT

Figure 1. Poster of EMBRIO International Symposium 2025

Figure 2. The Keynote Speakers Prof. Dr. Rachmat Pambudy (Minister of National Development Planning/ Bappenas RI)

Figure 3. The Keynote Speaker Dr. Arif Havas Oegroseno (Vise Minister of Foreign Affairs RI)

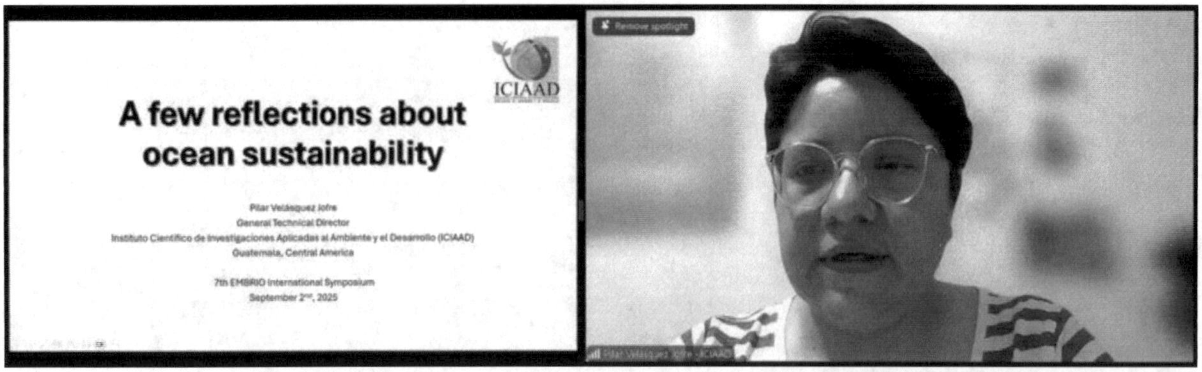

Figure 4. Invited Speaker Pilar Velasquez Jofre (ICAAD Guatemala)

Figure 5. Invited Speakers Prof. Dr. Indra Jaya (IPB University, Indonesia)

Figure 6. Invited Speakers Dr. Lisa Becking (Wageningen University and Research, Netherlands)

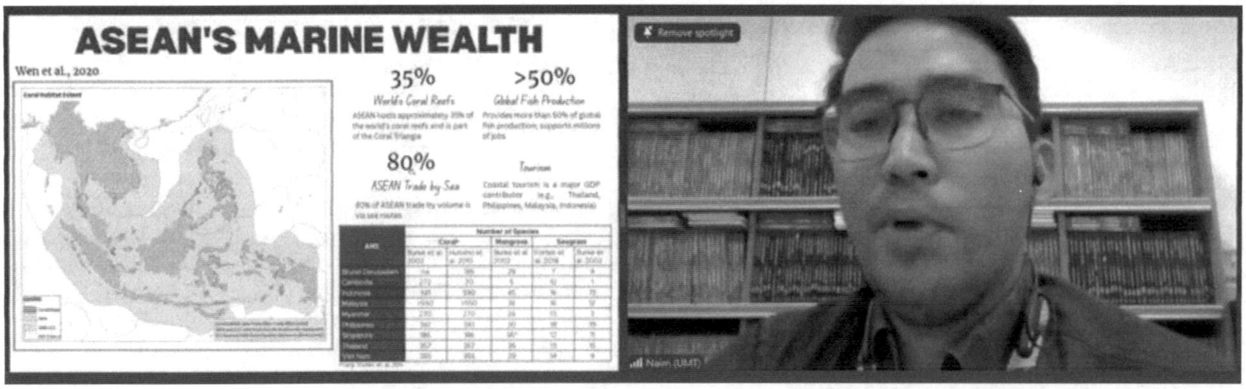

Figure 7. Invited Speaker Dr. Muhamad Naimullah (Universiti Malaysia Terengganu, Malaysia)

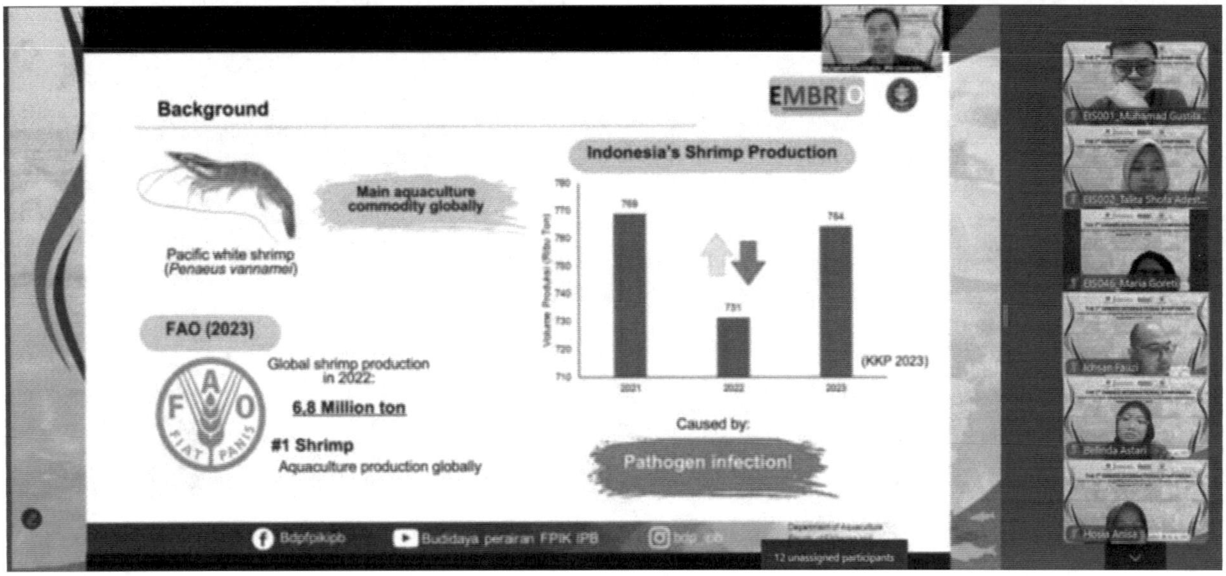
Background
Main aquaculture commodity globally
Pacific white shrimp
(Penaeus vannamei)
FAO (2023)
Global shrimp production in 2022:
6,8 Million ton
#1 Shrimp
Aquaculture production globally
Indonesia's Shrimp Production
(KKP 2023)
Caused by:
Pathogen infection!

(a)

EVALUATION OF FLOATING NET CAGE AQUACULTURE IMPACTS ON TROPHIC STATUS AND ORGANIC POLLUTION USING DIATOMS AS BIOINDICATORS IN SITU CIPONDOH, TANGERANG
PANDU GURUH GUMELAR
Dr. Muh. Herjayanto, S.Pi., M.Si.
Desy Aryani, S.Si., M.Si
FISHERIES SCIENCE DEPARTMENT
FACULTY OF AGRICULTURE
UNIVERSITY OF SULTAN AGENG TIRTAYASA

(b)

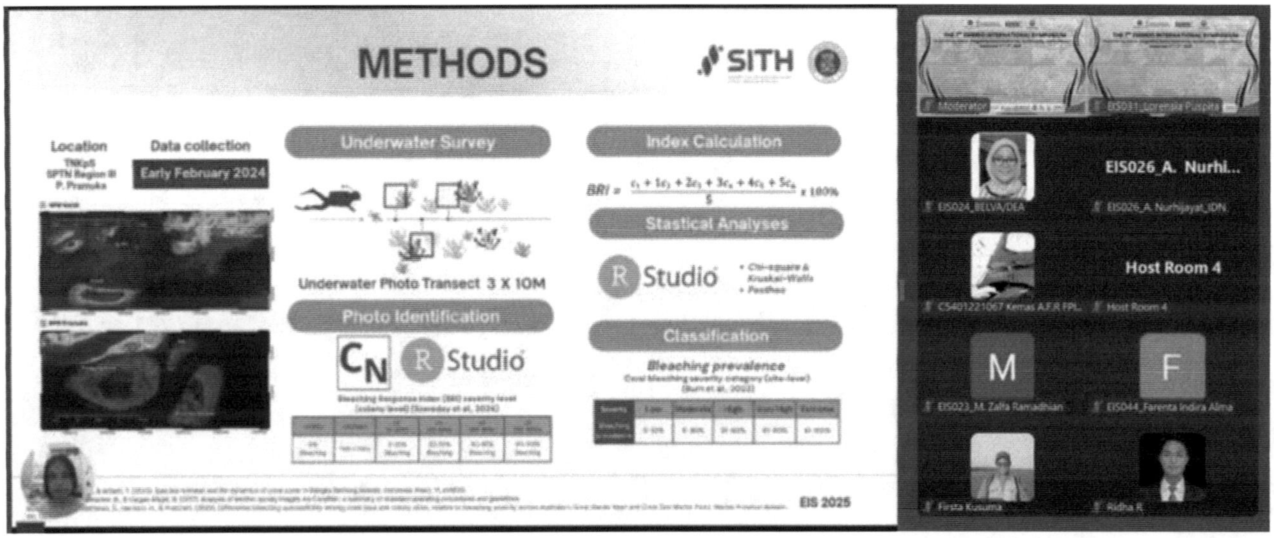
METHODS
Location
Data collection
Early February 2024
Underwater Survey
Underwater Photo Transect 3 X 10M
Photo Identification
Index Calculation
Stastical Analyses
Classification
Bleaching prevalence
EIS 2025
Host Room 4

(c)

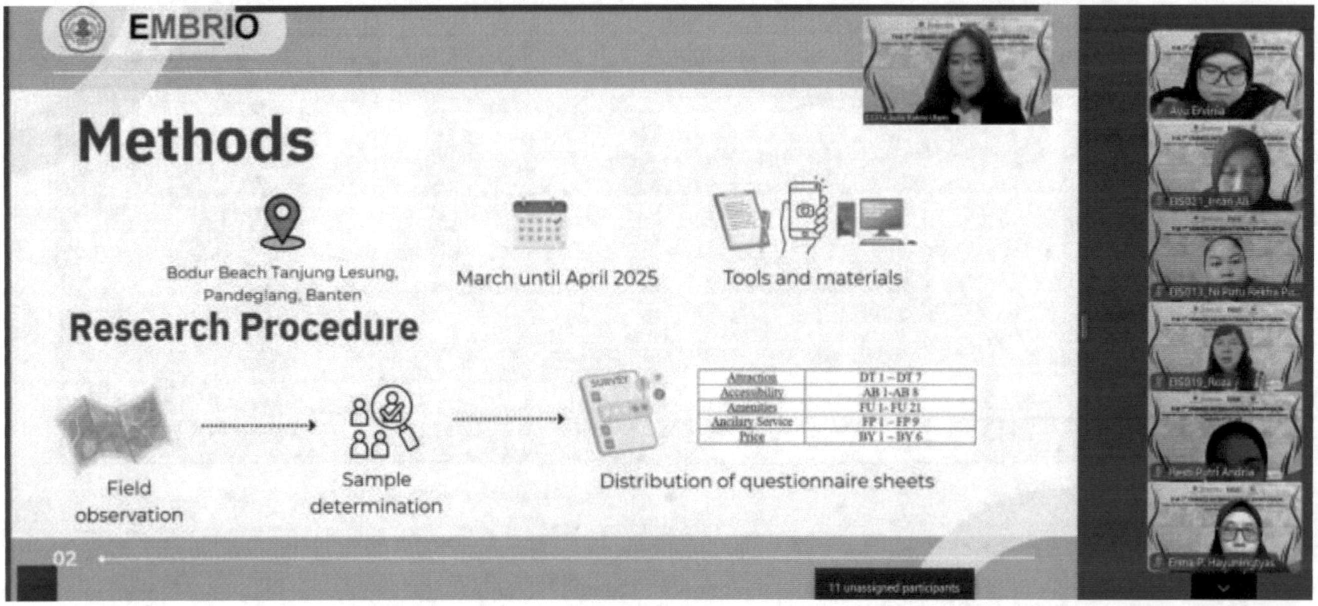

(d)

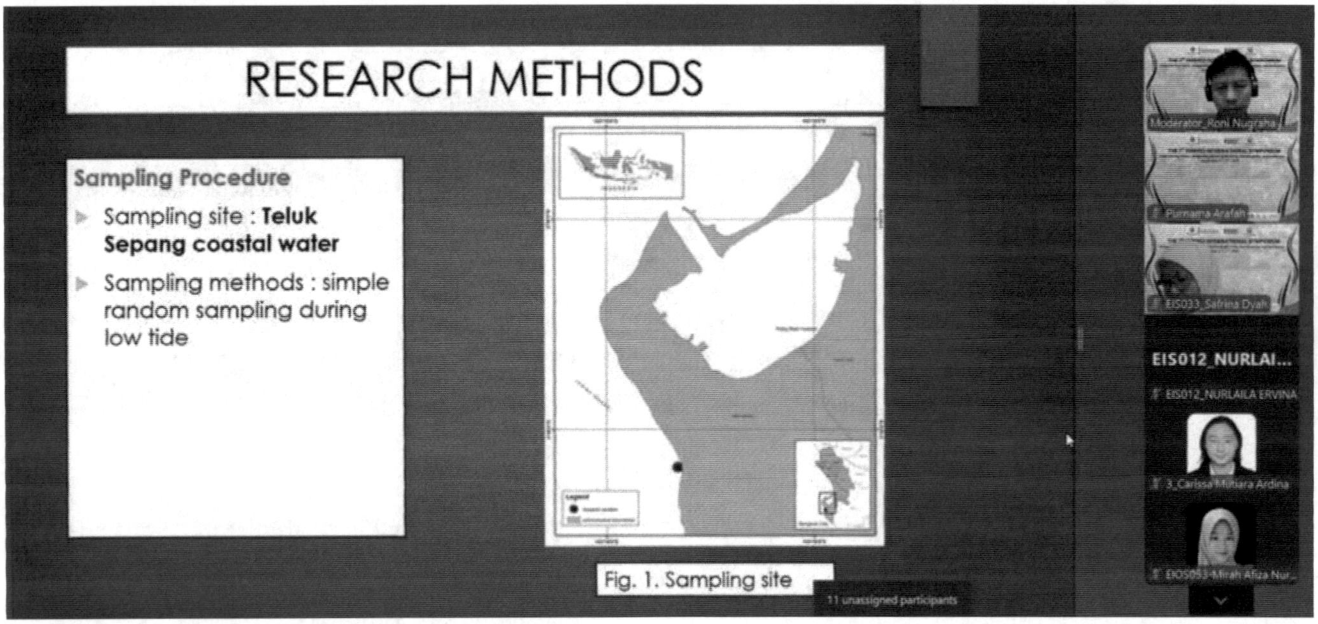

(e)

Figure 8. Participant Presentation (a-e)

Statement of Peer review

In submitting conference proceedings to *Web of Conferences*, the editors of the proceedings certify to the Publisher that

1. They adhere to its **Policy on Publishing Integrity** in order to safeguard good scientific practice in publishing.
2. All articles have been subjected to peer review administered by the proceedings editors.
3. Reviews have been conducted by expert referees, who have been requested to provide unbiased and constructive comments aimed, whenever possible, at improving the work.
4. Proceedings editors have taken all reasonable steps to ensure the quality of the materials they publish and their decision to accept or reject a paper for publication has been based only on the merits of the work and the relevance to the journal.

Title, date and place of the conference

7th EMBRIO International Symposium (EIS 2025), Bogor / Indonesia

Proceedings editor(s):

Assoc. Prof. Dr. Türker BODUR
Akdeniz University, Türkiye

Date and editor's signature

14.01.2026

Statement of Peer review

In submitting conference proceedings to *Web of Conferences*, the editors of the proceedings certify to the Publisher that

1. They adhere to its **Policy on Publishing Integrity** in order to safeguard good scientific practice in publishing.
2. All articles have been subjected to peer review administered by the proceedings editors.
3. Reviews have been conducted by expert referees, who have been requested to provide unbiased and constructive comments aimed, whenever possible, at improving the work.
4. Proceedings editors have taken all reasonable steps to ensure the quality of the materials they publish and their decision to accept or reject a paper for publication has been based only on the merits of the work and the relevance to the journal.

Title, date and place of the conference

The 7th EMBRIO International Symposium, September 2-3 2025, Online

Proceedings editor(s):

Dr. Beginer Subhan, S.Pi., M.Si.
Institut Pertanian Bogor

Date and editor's signature

02.02.2026

edp sciences

Web of Conferences

Statement of Peer review

In submitting conference proceedings to *Web of Conferences*, the editors of the proceedings certify to the Publisher that

1. They adhere to its **Policy on Publishing Integrity** in order to safeguard good scientific practice in publishing.
2. All articles have been subjected to peer review administered by the proceedings editors.
3. Reviews have been conducted by expert referees, who have been requested to provide unbiased and constructive comments aimed, whenever possible, at improving the work.
4. Proceedings editors have taken all reasonable steps to ensure the quality of the materials they publish and their decision to accept or reject a paper for publication has been based only on the merits of the work and the relevance to the journal.

Title, date and place of the conference

The 7th EMBRIO International Symposium, September 2-3 2025, Online

Proceedings editor(s): Dr. Sebastian Ferse

Date and editor's signature: 21.01.2026

edp sciences

Web of Conferences

Statement of Peer review

In submitting conference proceedings to *Web of Conferences*, the editors of the proceedings certify to the Publisher that

1. They adhere to its **Policy on Publishing Integrity** in order to safeguard good scientific practice in publishing.
2. All articles have been subjected to peer review administered by the proceedings editors.
3. Reviews have been conducted by expert referees, who have been requested to provide unbiased and constructive comments aimed, whenever possible, at improving the work.
4. Proceedings editors have taken all reasonable steps to ensure the quality of the materials they publish and their decision to accept or reject a paper for publication has been based only on the merits of the work and the relevance to the journal.

Title, date and place of the conference

The 7th EMBRIO International Symposium, September 2-3 2025, Online

Proceedings editor(s): Dr Vita Rumanti Kurniawati, SPi MT

Date and editor's signature 29 January 2026

Statement of Peer review

In submitting conference proceedings to *Web of Conferences*, the editors of the proceedings certify to the Publisher that

1. They adhere to its **Policy on Publishing Integrity** in order to safeguard good scientific practice in publishing.
2. All articles have been subjected to peer review administered by the proceedings editors.
3. Reviews have been conducted by expert referees, who have been requested to provide unbiased and constructive comments aimed, whenever possible, at improving the work.
4. Proceedings editors have taken all reasonable steps to ensure the quality of the materials they publish and their decision to accept or reject a paper for publication has been based only on the merits of the work and the relevance to the journal.

Title, date and place of the conference

The 7th EMBRIO International Symposium 2025; Bogor, Indonesia; September 2-3, 2025

Proceedings editor(s):

Dr. Muhamad Gustilatov

Date and editor's signature

January 12th, 2026

Statement of Peer review

In submitting conference proceedings to *Web of Conferences*, the editors of the proceedings certify to the Publisher that

1. They adhere to its **Policy on Publishing Integrity** in order to safeguard good scientific practice in publishing.
2. All articles have been subjected to peer review administered by the proceedings editors.
3. Reviews have been conducted by expert referees, who have been requested to provide unbiased and constructive comments aimed, whenever possible, at improving the work.
4. Proceedings editors have taken all reasonable steps to ensure the quality of the materials they publish and their decision to accept or reject a paper for publication has been based only on the merits of the work and the relevance to the journal.

Title, date and place of the conference

The 7th EMBRIO International Symposium, September 2-3 2025, Online

Proceedings editor(s): Roni Nugraha, Ph.D

3 February 2026

Date and editor's signature

17, avenue du Hoggar - PA de Courtabœuf – BP 112 - 91944 Les Ulis Cedex A (France)
Tél. : 33 (0)1 69 18 75 75 - Fax : 33(0)1 69 07 45 17 – www.edpsciences.org

Biofloc system enhances growth, immunity, and stress tolerance in high-density cultured Pacific White Shrimp

Muhamad Gustilatov[1*], *Widanarni* Widanarni[1*], *Sukenda* Sukenda[1], *Munti* Yuhana[1], *Indira* Faramudhita[1], and *Mohammad* Rakha Syahla Dewangga[1]

[1]Departement of Aquaculture, Faculty of Fisheries and Marine Sciences, IPB University, Dramaga, Bogor, Indonesia

Abstract. The main challenge in Pacific white shrimp farming is disease. One of the technologies applied to address this issue is the biofloc system, which enhances growth performance, survival rate (SR), feed conversion ratio (FCR), water quality, suppresses pathogen activity, and boosts immune response. This study aimed to evaluate the effects of the biofloc system on growth performance, health status, and stress response of high-density cultured vannamei shrimp at an intermediate scale. Postlarvae (PL-10) were reared for 56 days in biofloc and control systems at a density of 500 shrimp m^{-3} using Intermediate Bulk Containers (IBC) with a volume of 1000 L (1.2 × 1.0 × 1.15 m^3). A completely randomized design (CRD) was used with two treatments: control (C) and biofloc (BF) with three replicates. Parameters observed included growth rate, weight gain, biomass gain, FCR, SR, bacterial abundance, total hemocyte count (THC), respiratory burst (RB), phenoloxidase (PO) activity, and water quality. Results showed that the BF treatment yielded superior outcomes, indicated by SR (81.24±0.02%), FCR (1.67±0.08), reduced *Vibrio* sp. and *V. parahaemolythicus* abundance, and improved immune and stress responses.

1 Introduction

The Pacific white shrimp (*Litopenaeus vannamei*) remains one of Indonesia's principal aquaculture species. Despite its strong contribution to national aquaculture output, overall productivity has not yet achieved levels required to compete effectively in global markets. In 2023, national production reached approximately 764 thousand tons—a modest 4.5% increase from 2022 yet still slightly below the 2021 output of 768 thousand tons [1]. To enhance yields, farmers have widely adopted intensification strategies; however, higher stocking densities frequently lead to greater susceptibility to infectious diseases. Major health problems commonly encountered in shrimp farming include acute hepatopancreatic necrosis disease (AHPND), infectious myonecrosis, white feces disease (WFD), white spot disease (WSD), and yellow head disease (YHD) [2].

One of the key approaches implemented to mitigate disease risks is the use of biofloc technology. This system promotes the proliferation of beneficial microbial communities—comprising bacteria, microalgae, protozoa, and suspended organic aggregates—through the addition of carbon sources. Biofloc-based culture has been reported to improve survival rate (SR), feed conversion ratio (FCR), and growth performance. These improvements arise because microbes within the system recycle organic residues into nutritionally valuable flocs

* Corresponding author: widanarni@apps.ipb.ac.id

containing microbial biomass, amino acids, vitamins, and essential fatty acids that serve as natural supplemental feed for shrimp [3].

Beyond nutritional contributions, biofloc technology offers health-related benefits. Microbial interactions within biofloc systems can suppress pathogen activity, partly through quorum sensing (QS) interference. QS allows bacteria to regulate gene expression based on cell density, including virulence factors. Findings by Widanarni et al. (2024) [4] demonstrated that biofloc systems can downregulate QS-associated mechanisms in *Vibrio parahaemolyticus*, reducing the expression of *pirA* and *pirB*—key toxin genes responsible for AHPND. Consequently, toxin production and overall virulence can be diminished. Biofloc-derived microbial compounds have also been shown to enhance innate immune responses in *L. vannamei* [3].

Although biofloc technology has been widely studied, the performance of this system under high-density culture at an intermediate operational scale remains insufficiently explored. An intermediate-scale assessment is essential as a transitional step before implementation at commercial farm levels. Therefore, this study evaluates the growth, health condition, and stress responses of *L. vannamei* reared in a high-density biofloc system compared with a conventional control system.

2 Materials and methods

2.1 Experimental design

This study employed a completely randomized design (CRD) consisting of two treatments: a control treatment (without the biofloc system) and a biofloc treatment (with the biofloc system). Each treatment was carried out with three replicates. The stocking density of Pacific white shrimp was 500 individuals m^3 across all treatments. The culture was conducted indoors at PT. Suri Tani Pemuka, Purwakarta, West Java, for 56 days.

2.2 Biofloc preparation

The biofloc inoculum was sourced from an existing *L. vannamei* biofloc culture maintained in a 1000 L intermediate bulk container (IBC). Molasses served as the primary carbon source and was added once daily to maintain an estimated C/N ratio of 10. The application occurred approximately two hours after the morning feeding. The amount of molasses was calculated using the carbon supplementation formula described by based on assumptions of 36% feed protein, 16% nitrogen content in protein, 85% uneaten feed, and 38% carbon content in molasses. For instance, an estimated daily ration of 50 g required 64.42 g of molasses to achieve the desired C/N ratio. Prior to stocking, the prepared biofloc inoculum was added to the culture tanks at a 4:1 ratio (seawater : inoculum) and allowed to stabilize for 24 hours.

2.3 Shrimp maintenance

The test animals used in this study were Pacific white shrimp (*Litopenaeus vannamei*) postlarvae (PL10) with a total of 2,250 individuals obtained from the hatchery of PT. Suri Tani Pemuka, Indramayu, West Java. The shrimp postlarvae were first acclimatized in plastic container boxes for 24 hours. The acclimatization process was conducted to assess the health condition of the shrimp postlarvae to ensure their suitability for the experiment. The monitoring indicators included responsiveness to disturbance and good physical condition.

After acclimatization, the postlarvae were stocked into treatment tanks consisting of six Intermediate Bulk Containers (IBC) measuring $1.2 \times 1.0 \times 1.15$ m^3, each with a capacity of

1,000 L and filled with 750 L of water. Feeding was carried out using the blind feeding method for the first 30 days (DOC 1–30) and the demand feeding method for the following 26 days (DOC 31–56). Blind feeding followed standard pond operation guidelines at a rate of 2 kg per 100,000 shrimp, while demand feeding was adjusted according to the feeding rate (FR). Feed was provided four times daily at 08:00, 11:00, 14:00, and 17:00 WIB.

Water quality monitoring was conducted daily and weekly. Daily measurements included dissolved oxygen (DO), temperature, pH, and salinity, while weekly measurements included total ammonia nitrogen (TAN), nitrite, nitrate, and alkalinity. In the control treatment (K), water exchange was performed when water quality parameters (TAN and nitrite) reached maximum threshold levels.

2.4 Parameters of observation

2.4.1 Growth performance

The parameters measured in this study included survival rate (SR), average daily growth (ADG), specific growth rate (SGR), weight gain (ΔW), biomass gain (ΔB), and feed conversion ratio (FCR). ADG was calculated as $(W_t-W_0)/t(W_t-W_0)/t$, while SGR was calculated as $100\times[(\ln W_t-\ln W_0)/t]100\times[(\ln W_t-\ln W_0)/t]$, where W_t represents the final shrimp weight, W_0 the initial shrimp weight, and tt the duration of the culture period in days.

2.4.2 Bacterial abundance

Bacterial enumeration followed the total plate count (TPC) approach. Samples from rearing water, hepatopancreas, and intestine were plated on three types of agar media: thiosulfate citrate bile salt sucrose (TCBS) agar for total *Vibrio* count (TVC), seawater complete (SWC) agar for total bacterial count (TBC), and HiChrome agar for quantifying *Vibrio parahaemolyticus* (TVP). Approximately 0.1 g of organ tissue or 0.1 mL of water sample was homogenized in phosphate-buffered saline (PBS), followed by serial dilutions. A 50 µL aliquot of each dilution was then evenly spread across the respective media. All plates were incubated at 27°C for 24 hours, after which visible colonies were enumerated using standard TPC calculations.

$$Total\ bacterial\ count = bacterial\ colonies \times \frac{1}{dilution\ factor} \times \frac{1}{sample\ volume}$$

2.4.3 Immune response

To quantify total hemocyte count (THC), hemolymph was mixed with anticoagulant at a 1:3 ratio and examined using a hemacytometer under a light microscope (100×). Respiratory burst (RB) activity was assessed spectrophotometrically at 630 nm using nitroblue tetrazolium (NBT) reduction as an indicator. Phenoloxidase (PO) activity was measured by converting L-DOPA into dopachrome, with absorbance read at 490 nm. Results were expressed as dopachrome formation per 100 µL of hemolymph.

2.4.4 Water quality

Daily monitoring included dissolved oxygen (DO), pH, temperature, and salinity using standard portable meters, while weekly analyses involved total ammonia nitrogen (TAN), nitrite, nitrate, and alkalinity. Biofloc volume was measured using an Imhoff cone. When

TAN and nitrite levels in the control treatment exceeded threshold limits, partial water exchange was performed.

2.4.5. Statistical analysis

The data obtained were analyzed using Microsoft Excel 2016 and IBM SPSS Statistics version 26. Differences between the two treatments were assessed using an Independent Sample T-Test with a significance level of 0.05 ($\alpha = 5\%$).

3 Result

3.2 Growth performance

Based on the observations, the growth performance of Pacific white shrimp cultured in the biofloc system and the control over 56 days is presented in Table 1. Survival rate (SR) and feed conversion ratio (FCR) showed significant differences ($p < 0.05$) between the two treatments. In contrast, average daily growth (ADG), specific growth rate (SGR), weight gain (Wt, ΔW), and biomass gain (Bt, ΔB) of the shrimp did not differ significantly between the treatments according to the statistical analysis.

Table 1. Growth performance of Pacific white shrimp cultured in the biofloc system and control.

Parameter	K	BF
W_0 (g)	$0{,}01 \pm 0{,}00^a$	$0{,}01 \pm 0{,}00^a$
W_t (g)	$3{,}46 \pm 0{,}12^a$	$3{,}27 \pm 0{,}42^a$
ΔW (g)	$3{,}45 \pm 0{,}12^a$	$3{,}26 \pm 0{,}42^a$
ADG (g day^{-1})	$0{,}06 \pm 0{,}00^a$	$0{,}06 \pm 0{,}01^a$
SGR (% day^{-1})	$10{,}44 \pm 0{,}06^a$	$0{,}33 \pm 0{,}22^a$
B_0 (g)	$3{,}75 \pm 0{,}00^a$	$3{,}75 \pm 0{,}00^a$
Bt (g)	868 ± 8^a	997 ± 142^a
ΔB (g)	865 ± 8^a	994 ± 142^a
FCR	$1{,}67 \pm 0{,}08^a$	$1{,}48 \pm 0{,}15^b$
SR (%)	$67{,}02 \pm 0{,}02^a$	$81{,}24 \pm 0{,}02^b$

Different superscript letters within the same row indicate significant differences ($p < 0.05$). K = Pacific white shrimp cultured without the biofloc system (control); BF = Pacific white shrimp cultured with the biofloc system.

3.3 Bacterial abundance

On day 56, total bacterial count (TBC) showed a significant decrease ($p < 0.05$) in the hepatopancreas and intestine samples, with the biofloc (BF) treatment exhibiting better results, ranging from 10^5 to 10^6 CFU mL^{-1} (Fig. 1). In contrast, no significant differences were observed in the water samples on day 56.

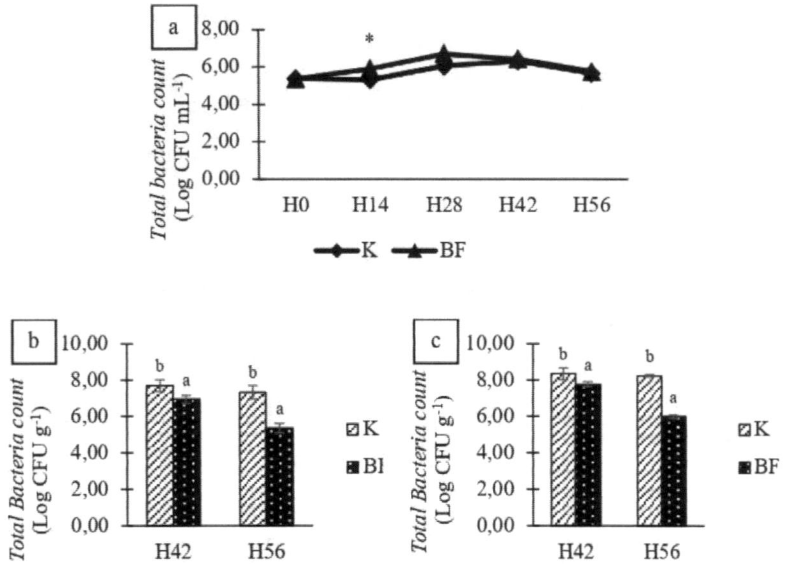

Fig. 1. Total bacterial count in the rearing water (a), hepatopancreas (b), and intestine (c) of Pacific white shrimp cultured in the biofloc system and control over 56 days. Different letters between treatments or * indicate significant differences ($p < 0.05$). There were two treatments with three replicates. K = control; BF = biofloc.

At the end of the culture period, presumptive *Vibrio* count (PVC) showed significant differences ($p < 0.05$) in the water, hepatopancreas, and intestine samples, with the biofloc (BF) treatment exhibiting a decrease, resulting in better outcomes ranging from 10^3 to 10^4 CFU mL^{-1} (Fig. 2). In contrast, the control (K) treatment tended to increase, ranging from 10^4 to 10^7 CFU mL^{-1}.

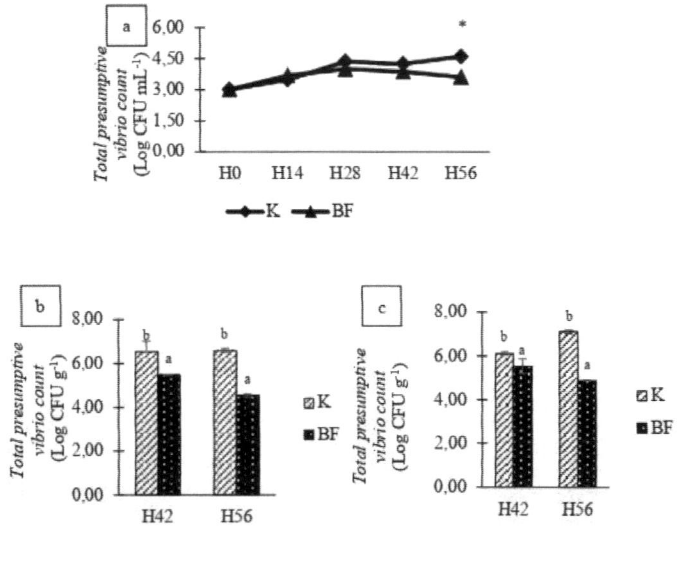

Fig. 2. Presumptive *Vibrio* count in the rearing water (a), hepatopancreas (b), and intestine (c) of Pacific white shrimp cultured in the biofloc system and control over 56 days. Different letters between treatments or * indicate significant differences ($p < 0.05$). There were two treatments with three replicates. K = control; BF = biofloc.

At the end of the culture period, total *Vibrio parahaemolyticus* (TVP) showed significant differences ($p < 0.05$) in the water, hepatopancreas, and intestine samples. The biofloc (BF) treatment exhibited a decrease, resulting in better outcomes of approximately 10^3 CFU mL^{-1} (Fig. 3), whereas the control (K) treatment tended to increase, ranging from 10^4 to 10^6 CFU mL^{-1}.

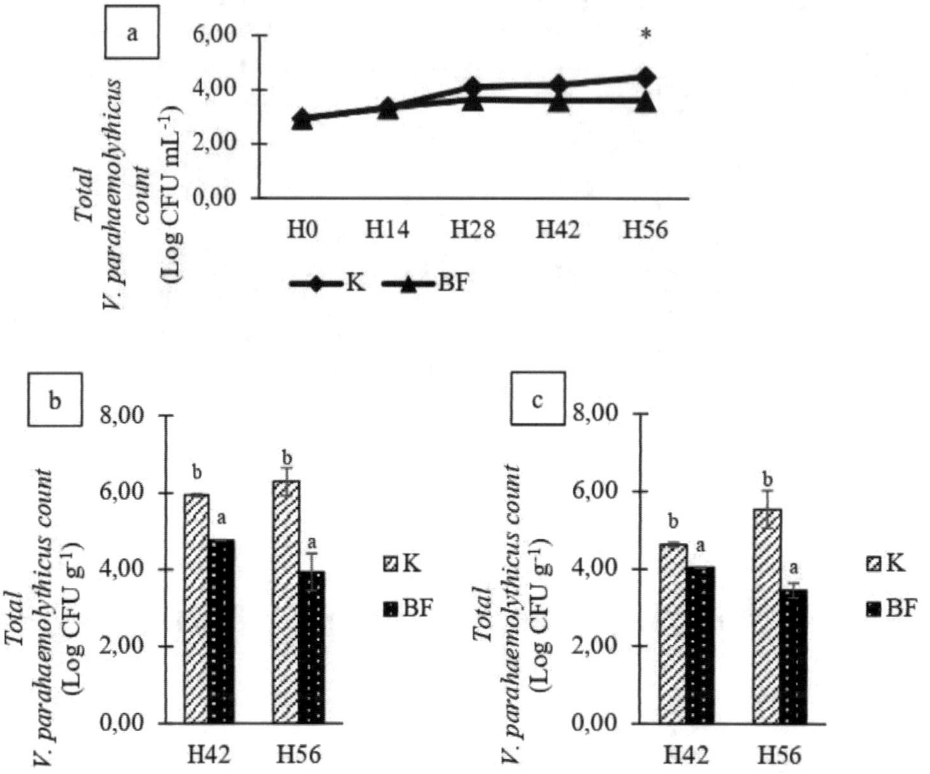

Fig. 3. Total *Vibrio parahaemolyticus* abundance in the rearing water (a), hepatopancreas (b), and intestine (c) of Pacific white shrimp cultured in the biofloc system and control over 56 days. Sampling was performed at day 0 (H0), day 14 (H14), day 28 (H28), day 42 (H42), and day 56 (H56) of the rearing period. Different letters between treatments or *indicate significant differences ($p < 0.05$). There were two treatments with three replicates each. K = control; BF = biofloc.

3.4 Immune response

Total hemocyte count (THC) on days 14, 28, and 42 showed an increase, with the control (K) treatment exhibiting significant differences ($p < 0.05$), followed by a decrease on day 56. However, on day 56, the biofloc (BF) treatment showed a significant difference ($p < 0.05$), with a higher value of $4.50 \pm 0.35 \times 10^6$ cells mL^{-1} (Fig. 4). Respiratory burst (RB) activity in the BF treatment increased on day 14, then decreased until day 42, with a slight increase on day 56 (Fig. 5). On day 56, BF showed a significant difference ($p < 0.05$), with a higher value of 0.08 ± 0.00 per 10 μL hemolymph. In contrast, phenoloxidase (PO) activity tended to increase until day 56 (Fig. 6); however, no significant differences were observed between the two treatments.

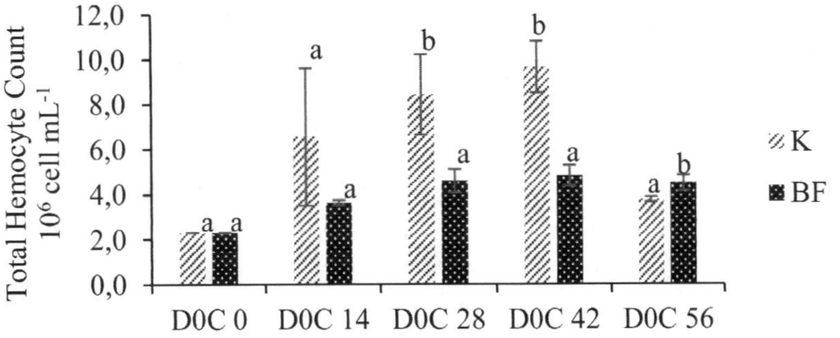

Fig. 4. Total hemocyte count of Pacific white shrimp cultured in the biofloc system and control over 56 days. Sampling was conducted at different culture periods, namely day 0 (DOC 0), day 14 (DOC 14), day 28 (DOC 28), day 42 (DOC 42), and day 56 (DOC 56), where DOC refers to *days of culture*. Different letters between treatments indicate significant differences ($p < 0.05$). There were two treatments with three replicates each. K = control; BF = biofloc.

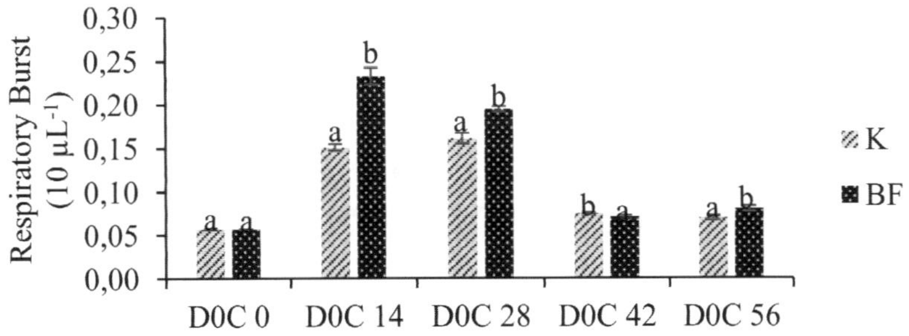

Fig. 5. Respiratory burst of Pacific white shrimp cultured in the biofloc system and control over 56 days. Sampling was conducted at different culture periods, namely day 0 (DOC 0), day 14 (DOC 14), day 28 (DOC 28), day 42 (DOC 42), and day 56 (DOC 56), where DOC refers to *days of culture*. Different letters between treatments indicate significant differences ($p < 0.05$). There were two treatments with three replicates each. K = control; BF = biofloc.

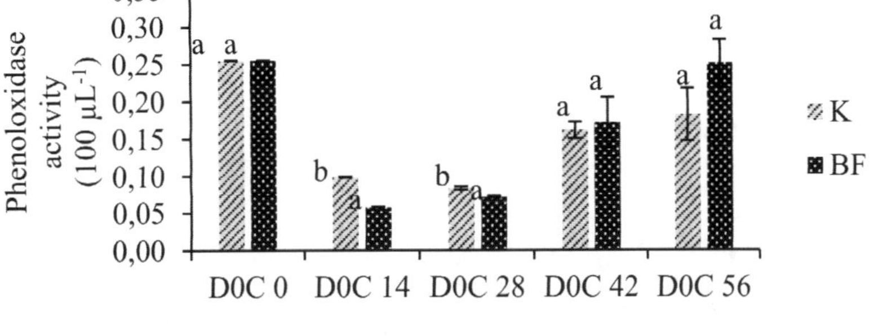

Fig. 6. Phenoloxidase activity of Pacific white shrimp cultured in the biofloc system and control over 56 days. Sampling was conducted at different culture periods, namely day 0 (DOC 0), day 14 (DOC 14), day 28 (DOC 28), day 42 (DOC 42), and day 56 (DOC 56), where DOC refers to *days of culture*. Different letters between treatments indicate significant differences ($p < 0.05$). There were two treatments with three replicates each. K = control; BF = biofloc.

3.5 Water quality

Daily water quality parameters, including dissolved oxygen (DO), temperature, pH, and salinity, remained within the optimal ranges according to the Indonesian National Standard (*Standar Nasional Indonesia*, SNI 9267-5:2024) for Pacific white shrimp (*Litopenaeus vannamei*) culture . Based on this standard, the optimal ranges are as follows: DO $\geq$ 3.5 mg L^{-1}, temperature 25–32 °C, pH 7.5–8.5, and salinity 15–35 g L^{-1}. In the present study, daily water quality values for the control (K) and biofloc (BF) treatments were: DO = 3.58–6.93 mg L^{-1}, temperature = 25.50–28.00 °C, pH = 6.83–8.28, and salinity = 29.20–36.00 g L^{-1}, which were generally within or close to the optimal ranges. Weekly parameters, including total ammonia nitrogen (TAN), nitrite (NO_2^-), nitrate (NO_3^-), alkalinity, and biofloc volume, were also monitored (Fig. 7). These parameters generally increased over the 56-day culture period, except for alkalinity in the control treatment, which decreased. Significant differences ($p < 0.05$) were observed between the two treatments on day 56 for TAN, nitrite, nitrate, and alkalinity, with the BF treatment showing better values compared to K. In addition, biofloc volume increased throughout the culture period (Fig. 8), with significant increases observed from DOC 28 to DOC 56. The highest biofloc volume was recorded on day 56, reaching 50.25 mL L^{-1}.

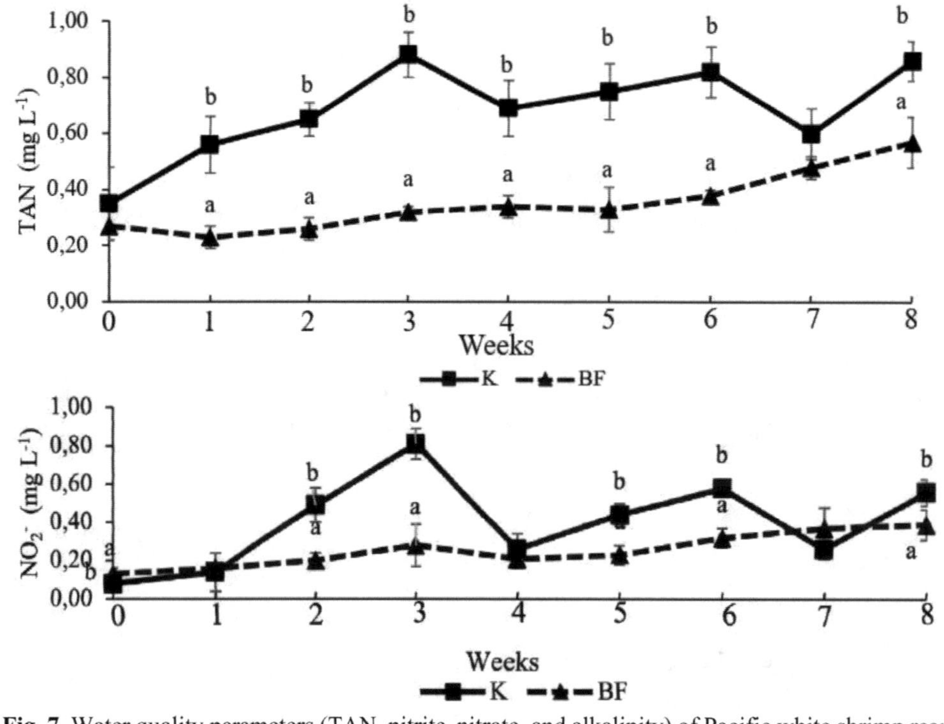

Fig. 7. Water quality parameters (TAN, nitrite, nitrate, and alkalinity) of Pacific white shrimp reared in the biofloc and control systems for 56 days. Different superscript letters between treatments indicate significant differences ($p<0.05$). Two treatments were conducted with three replicates each. K = control, BF = biofloc.

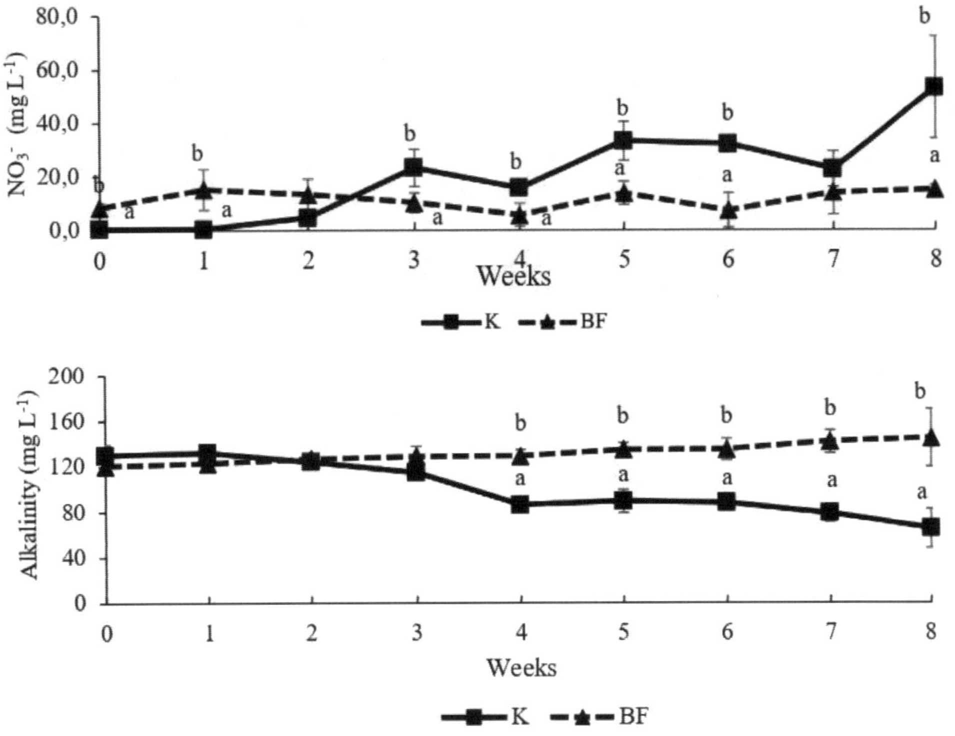

Fig. 7 (continued).

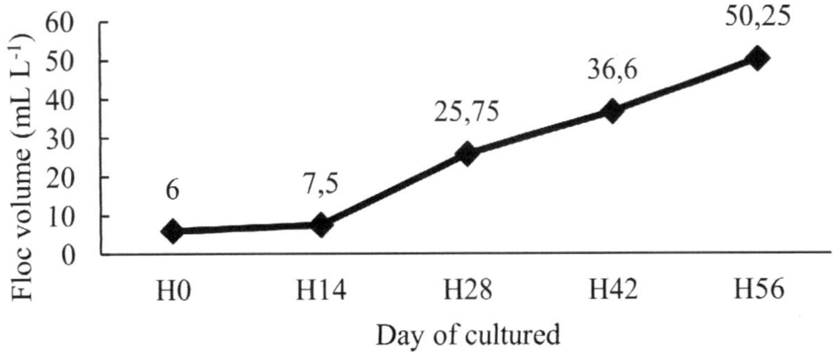

Fig. 8. Floc volume of Pacific white shrimp reared in the biofloc and control systems for 56 days. Different superscript letters between treatments indicate significant differences ($p<0.05$). Two treatments were conducted with three replicates each. K = control, BF = biofloc.

4 Discussion

Survival rate (SR) of *L. vannamei* after 56 days differed significantly between treatments, with the biofloc (BF) system yielding notably higher SR than the control (K). These findings agree with reports by El-Sayed (2021) [5], stating that biofloc systems enhance shrimp survival. The improved SR in BF was likely driven by more stable water quality, particularly lower TAN concentrations resulting from active heterotrophic bacterial assimilation. Additional pathogen-suppressing mechanisms within biofloc including extracellular enzymatic degradation of autoinducers, production of antimicrobial metabolites such as bromophenols, and phenotypic interference in *V. parahaemolyticus* further contributed to better survival outcomes [6]. Improvements in FCR under biofloc conditions were attributed to the availability of microbial biomass rich in nutrients such as proteins, lipids, minerals, and functional compounds (e.g., carotenoids and chlorophenols) that enhance digestion and nutrient uptake [6].

Despite improvements in health-related parameters, growth performance metrics such as ADG, SGR, final weight, and biomass gain showed no significant differences between treatments. High stocking density likely constrained these variables by intensifying competition for feed, reducing space availability, and increasing susceptibility to oxidative stress factors previously recognized as limiting growth under crowded conditions.

Fluctuations in bacterial abundance were observed throughout the culture period. Mid-cycle increases in TBC and presumptive *Vibrio* counts were consistent with elevated nutrient availability from carbon supplementation, feed residues, and organic detritus, which favor rapid proliferation of heterotrophic bacteria and *Vibrio* spp. [7]. Although PVC levels temporarily approached 10^4 CFU mL^{-1} the upper threshold considered safe the counts remained within non-pathogenic limits. Elevated TVP levels detected during days 28–56 were still below concentrations known to initiate infections, typically 10^5–10^7 CFU mL^{-1} within 24–48 hours. Importantly, only certain *V. parahaemolyticus* strains harboring virulence plasmids induce AHPND. Competition for nutrients and QS suppression within the biofloc environment likely played major roles in reducing pathogenicity and TVP abundance [4].

Within the digestive tract, biofloc exposure influenced microbiota composition and immune activation. Although some *Vibrio* strains proliferate in biofloc systems, most are non-pathogenic and may stimulate immune responses without causing disease. Previous studies have shown that biofloc enhances phagocytic activity and suppresses virulence gene expression in pathogenic *Vibrio* spp. [8], consistent with the lower bacterial loads found in BF shrimp organs.

Shrimp under high-density conditions often experience suppressed immune functions, including reduced THC, RB, and PO activities, due to increased metabolic demands and stress [9]. Decreases in THC may also indicate active hemocyte mobilization to infected tissues [10]. RB, representing the production of reactive oxygen intermediates during phagocytosis, is sensitive to environmental stressors such as poor water quality and crowding [9, 11]. Higher RB values indicate robust immune defense, enabling more efficient elimination of pathogens. PO activity, a key component of the crustacean melanization response, was enhanced in the presence of biofloc, likely due to continuous exposure to microbial-associated molecular patterns such as β-glucans and peptidoglycans [12].

Water quality stability is essential under intensive culture. Increasing floc volume in BF systems supports microbial nutrient recycling but must remain below the recommended threshold of 50 mL L^{-1} to avoid oxygen depletion. TAN reductions in BF were attributed to the activities of ammonia-oxidizing and nitrite-oxidizing bacteria involved in nitrification and denitrification pathways [13-14]. Alkalinity trends further indicated active nitrogen cycling, with BF systems maintaining levels above the recommended 100 mg L^{-1}, preventing pH declines that could inhibit nitrification [15].

5 Conclusion

The cultivation of Pacific white shrimp using the biofloc system at high density on an intermediate scale was proven to increase total hemocyte count (THC), improve feed conversion ratio (FCR), enhance immune responses, and reduce the abundance of *Vibrio* sp. and *Vibrio parahaemolyticus*. Furthermore, the high-density biofloc system is more recommended for application in the nursery phase. The addition of commercial probiotics is also suggested for further evaluation to determine their effects on the productivity of Pacific white shrimp in the biofloc system at the field scale.

References

1. Ministry of Maritime Affairs and Fisheries. Fishery Production. Jakarta (ID): Ministry of Maritime Affairs and Fisheries. https://portaldata.kkp.go.id/portals/data-statistik/prod-ikan/summary. 2023.
2. S. Thitamadee, A. Prachumwat, J. Srisala, P. Jaroenlak, P.V. Salachan, K. Sritunyalucksana, T.W. Flegel, O. Itsathitphaisarn. Review of current disease threats for cultivated *penaeid shrimp* in Asia. Aquaculture. **452**, 69–87 (2016). https://doi.org/10.1016/j.aquaculture.2015.10.028
3. S.K. Kim, Z. Pang, H.C. Seo, Y.R. Cho, T. Samocha, I.K. Jang. Effect of bioflocs on growth and immune activity of Pacific white shrimp, *Litopenaeus vannamei* postlarvae. Aq. Rsch. **45**, 362–371 (2014). https://doi.org/10.1111/are.12319
4. W. Widanarni, M. Gustilatov, J. Ekasari, P.G. Julyantoro, D.E. Waturangi, S. Sukenda. Unveiling the positive impact of biofloc culture on *Vibrio parahaemolyticus* infection of pacific white shrimp by reducing quorum sensing and virulence gene expression and enhancing immunity. J. Fish. Diss. **47**, e13932 (2024). https://doi.org/10.1111/jfd.13932
5. A.F. El-Sayed. Use of biofloc technology in shrimp aquaculture: a comprehensive review, with emphasis on the last decade. Rev. Aquacul. **13**, 676–705 (2021). https://doi.org/10.1111/raq.12494
6. M. Gustilatov, J. Ekasari, P.G. Julyantoro, D.E. Waturangi, F.L. Silalahi. The effect of biofloc density and *Bacillus* sp. NP5 supplementation on bacterial inhibition, antibiofilm activity, and the immunity of the pacific white shrimp (*Penaeus vannamei*) against *Vibrio parahaemolyticus*. Comp. Immunol. Rep. **8**, 200228 (2025). https://doi.org/10.1016/j.cirep.2025.200228.
7. Y. Deng, X. Xu, X. Yin, H. Lu, G. Chen, J. Yu, Y. Ruan. Effect of stock density on the microbial community in biofloc water and Pacific white shrimp (*Litopenaeus vannamei*) gut microbiota. App. Microbio. Biotech. **103**, 4241–4252 (2019). https://doi.org/10.1007/s00253-019-09773-4
8. M. Gustilatov, J. Ekasari, G.S. Pande. Protective effects of the biofloc system in Pacific white shrimp (*Penaeus vannamei*) culture against pathogenic *Vibrio parahaemolyticus* infection. Fish Shellfish Immunol. **124**, 66–73 (2022). https://doi.org/10.1016/j.fsi.2022.03.037
9. Y.C. Lin, J.C. Chen, Y.Y. Chen, S.T. Yeh, L.L. Chen, C.L. Huang, J.F. Hsieh, C.C. Li. Crowding of white shrimp *Litopenaeus vananmei* depresses their immunity to and resistance against *Vibrio alginolyticus* and white spot syndrome virus. Fish Shellfish Immunol. **45**, 104–111 (2015). https://doi.org/10.1016/j.fsi.2015.02.012
10. H. Hamsah, W. Widanarni, A. Alimuddin, M. Yuhana, M.Z Junior, D. Hidayatullah. Immune response and resistance of Pacific white shrimp larvae administered probiotic, prebiotic, and synbiotic through the bio-encapsulation of *Artemia* sp. Aquacul. Intern. **27**, 567–580 (2019). https://doi.org/10.1007/s10499-019-00346-w
11. Y.Y. Chen, J.C. Chen, K.C. Tseng, Y.C. Lin, C.L. Huang. Activation of immunity, immune response, antioxidant ability, and resistance against *Vibrio alginolyticus* in white shrimp *Litopenaeus vannamei* decrease under long-term culture at low pH. Fish Shellfish Immunol. **46**, 192–199 (2015). https://doi.org/10.1016/j.fsi.2015.05.055

12. A. Panigrahi, M.R. Sivakumar, M. Sundaram, A. Saravanan, R.R. Das, V.K. Katneni, K. Ambasankar, J.S. Dayal, G. Gopikrishna. Comparative study on phenoloxidase activity of biofloc-reared pacific white shrimp *Penaeus vannamei* and Indian white shrimp *Penaeus indicus* on graded protein diet. Aquaculture. **518**, 734654 (2020). https://doi.org/10.1016/j.aquaculture.2019.734654
13. G.R. Robles-Porchas, T. Gollas-Galvan, M. Martinez-Porchas, L.R. Martinez-Cordova, A. Miranda-Baeza, F. Vargas-Albores. 2020. The nitrification process for nitrogen removal in biofloc system aquaculture. Rev. in Aquaculture. **12**, 2228–2249 (2020). https://doi.org/10.1111/raq.12431
14. J.C. Marin, A.H. Caravelli, N.E. Zaritzky. Nitrification and aerobic denitrification in anoxic–aerobic sequencing batch reactor. Biores. Tech. **200**, 380–387 (2016). https://doi.org/10.1016/j.biortech.2015.10.024
15. J.M. Ebeling, M.B. Timmons, J.J. Bisogni. Engineering analysis of the stoichiometry of photoautotrophic, autotrophic, and heterotrophic removal of ammonia–nitrogen in aquaculture systems. Aquaculture. **257**, 346–358 (2006). https://doi.org/10.1016/j.aquaculture.2006.03.019

Effectiveness of various commercial feeds on the growth performance of common carp (*Cyprinus carpio*) in floating net cages in Cirata Reservoir, West Java

Muhammad Agus Suprayudi[1], *Julie* Ekasari[1], *Muhammad* Fakhri Muslim[1] and *Talita* Shofa Adestia[1*]

[1]Departement of Aquaculture, Faculty of Fisheries and Marine Sciences, Bogor Agricultural University, Bogor, West Java 16680, Indonesia

Abstract. This study aims to compare the growth performance of common carp (*Cyprinus carpio*) reared in floating net cages (FNC) in Cirata Reservoir, West Java, with different commercial feeds. This study utilized four different commercial feeds, namely: commercial feed A (protein content: 28.10%), treatment B (protein content: 28.80%), treatment C (protein content: 28.90%), and treatment D (protein content: 29.12%). The average weight of the fish used at the beginning of stocking was 23 g. The FNC plot used was 3 × 3 × 3 m in size. The frequency of feeding was three times a day (08:00, 12:00, 17:00) with feeding at satiation for 100 days of rearing period. The results showed that treatment D (29.12%) provided the best final biomass performance, as well as the highest average weight, specific growth rate, and feed cost per gain and feed conversion value, compared to the other three treatments. Based on the research results obtained, it can be concluded that the highest growth performance was achieved in commercial feed treatment D, which had a protein content of 29.12%.

1 Introduction

The common carp (*Cyprinus carpio* L., 1758) is one of the aquaculture commodities that has been cultivated for a long time in Indonesia. It is considered a popular commodity across various provinces, such as West Java, North Sumatra, South Sulawesi, and North Sulawesi [8]. Among these regions, West Java has become a central hub for carp aquaculture, with Cirata Reservoir being one of the prominent cultivation sites. Feed expenses represent the largest proportion of the total production cost in aquaculture operations, accounting for approximately 85% to 98% of overall production expenses [1]. Commercial feeds, driven by the rising demand for higher fish production outputs [2]. For example, in Cirata Reservoir alone, the annual demand for commercial fish feed has reached 149,321 tons.

An ideal commercial fish feed must have a balanced and optimal composition that includes essential nutrients such as proteins, fats, ash, crude fiber, moisture content, vitamins, and minerals [3]. A feed with an appropriate nutritional formulation can significantly enhance fish growth performance. The more digestible the nutrients are for the fish, the more efficient the feed becomes in promoting growth and health. However, feed efficiency is also influenced by the balance of nutrients—when the composition exceeds the fish's actual

* Corresponding talitashofa@apps.ipb.ac.id

nutritional needs, the surplus cannot be digested and is ultimately wasted, thus lowering overall efficiency [4]. According to the Indonesian National Standard (SNI) 01-4266-2006, during the grow-out phase, common carp (*Cyprinus carpio*) require feed that contains at least 25% protein and a minimum of 5% fat, with a maximum of 12% ash, a maximum moisture content of 12%, and a maximum crude fiber content of 8%.

The availability of various types of commercial fish feed in the market—differing in both quality and price—makes the selection of appropriate feed a critical decision in aquaculture operations. Practical considerations and economic factors play a decisive role in determining the feasibility and continued use of commercial fish feeds [5]. A good commercial feed is characterized by a nutritionally balanced formulation that aligns with the dietary needs of the fish, while also being economically viable—meaning it is affordable for fish farmers and fits within their purchasing capacity [6]. With many types of commercial feed available on the market, it becomes essential to evaluate their effectiveness. To ensure the efficient use of feed in supporting aquaculture production, it is necessary to conduct thorough evaluations of both the biological performance and economic outcomes within the targeted aquaculture business units, such as floating net cages.

2 Materials and Method

2.1 Test materials

The materials used in this research included four units of floating net cages measuring 3 meters × 3 meters × 3 meters each. Four types of commercial sinking feed with similar protein levels (ranging from 27.90% to 29.12%), each with a pellet size of 3 mm. Additional supporting materials included a quarantine pond, a feed storage warehouse, and daily records for monitoring feed quantity. The net cages were equipped with repair materials and tools, as net inspections and stitching repairs were carried out prior to the rearing period.

2.2 Experimental design

This research was conducted at the Denox floating net cage in the Cirata Reservoir, West Java, as shown in Figure 1. The Denox floating net cages used in this study are aquaculture facilities that have been operationally implemented at the research site. The floating net cages used consisted of 4 compartments, with supporting facilities including a quarantine pond and a feed warehouse, as presented in Figure 1. This study employed a quasi-experimental design [14], using four treatment feeds with relatively similar protein levels and one replication, as follows:

Treatment A: Commercial feed A with a protein content of 28.10%
Treatment B: Commercial feed B with a protein content of 28.80%
Treatment C: Commercial feed C with a protein content of 27.90%
Treatment D: Commercial feed D with a protein content of 29.12%

https://doi.org/10.1051/bioconf/202622101002

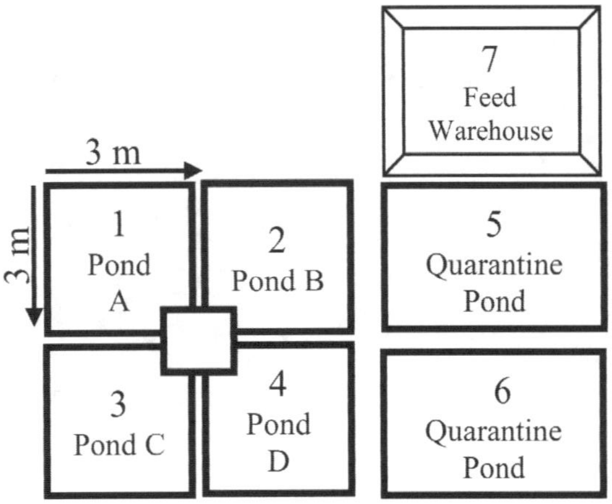

Fig. 1. FNC layout in Cirata Reservoir.

Fig. 2. Location map of Denox floating net cages.

2.3 Fish maintenance and feed experiment

The test fish were reared in four floating net cage compartments, each measuring 3 m × 3 m × 3 m. The nets of the floating cages were inspected prior to the rearing period; if any damage was found, repairs were made to the damaged sections using stitching methods. The test fish in this study were common carp (*Cyprinus carpio*) with an average weight of 23.61 g and a stocking density of 211 fish/m^3 (4.98 kg/m^3). The rearing period lasted for 100 days.

Feed was administered using the *at satiation* method, with a feeding frequency of three times a day at 08:00–09:00, 12:00–13:00, and 16:00–17:00. The feed used was a commercial sinking feed with a pellet size of 3 mm. The amount of feed was recorded daily. The proximate composition of the test feeds is presented in Table 1.

Table 1. Feed proximate analysis.

Component	Feed Brand			
	A	B	C	D
Moisture (%)	10.87	9.80	10.12	9.76
Crude Fiber (%)	4.70	5.06	5.10	5.23
Protein (%)	28.10	28.80	28.90	29.12
Ash (%)	7.13	9.24	9.45	9.22

Fat (%)	7.04	6.90	6.89	6.99
NFE(%)*	42.16	40.21	39.54	39.68
GE (kcal kg^{-1})**	4447	4334	4324	4338

* Nitrogen Free Extract (NFE) = 100% – (moisture + protein + fat + ash + fiber)
** Gross Energy Composition (GE) was calculated based on protein = 5.64 kcal/g protein, fat = 9.44 kcal/g fat, and carbohydrate/NFE = 4.11 kcal/g carbohydrate [7].

2.4 Analysis

Evaluating growth performance in aquatic species typically involves assessing parameters that track changes throughout their life cycle [8]. Emphasis is often placed on early growth stages to better understand feed efficiency and overall development. This principle informed the choice of growth indicators in this study, which included Survival Rate (SR), Specific Growth Rate (SGR), Feed Intake (FI) [9], Feed Conversion Ratio (FCR) [10], Protein Efficiency Ratio (PER) [11], and Feed Cost Gain (FCG).

Proximate analysis was conducted following the Association of Official Analytical Chemists (AOAC) standard procedures. This analysis was performed on both the experimental feed and the test fish at the beginning and end of the rearing period. The proximate components analyzed included crude protein, moisture content, crude fiber, crude fat, ash content, and nitrogen-free extract (NFE). Protein content was determined using the Kjeldahl method, which involves three main steps: digestion, distillation, and titration. Fat content was assessed using two methods—Folch method for wet fat content and Soxhlet extraction for dry fat content. Moisture content was measured by drying samples in an oven at 110°C, while ash content was determined by incinerating samples in a muffle furnace at 600°C. Crude fiber was analyzed through a process involving strong acid and alkali digestion followed by heating. The nitrogen-free extract (NFE) was calculated by subtracting the sum of other proximate components from 100%.

Water quality monitoring included daily measurements of temperature, dissolved oxygen (DO), and pH at 06:00, 12:00, and 17:00. In addition, total ammonia nitrogen (TAN), nitrite, and nitrate levels were assessed three times during the culture period—on days 7, 40, and 90. The collected data were tabulated using Microsoft Excel 365 and analyzed descriptively, following the approach used in previous research [12].

3 Result and discussion

Water quality is a crucial factor in aquaculture, significantly influencing fish health, growth rates, and feed efficiency. Throughout the 100-day rearing period, several key water quality parameters were monitored regularly on days 7, 40, and 90 across all four treatment groups (A, B, C, and D). The results indicate consistent and suitable environmental conditions for common carp (*Cyprinus carpio*). The water quality is shown on Table 2.

Table 2. Water quality in floating net cage.

Parameter	Net Cage			
	A	B	C	D
DO (mg L^{-1})	2.7-5.9	2.7-5.9	2.7-5.9	2.7-5.9
Temperature (°C)	26.9–30.8	26.9–30.8	26.9–30.8	26.9–30.8
pH	6.72–7.79	6.72–7.79	6.72–7.79	6.72–7.79

TAN*	0.12	0.10	0.11	0.08
Nitrite (mg L^{-1})	0.02	0.02	0.02	0.02
Nitrate (mg L^{-1})	0.04	0.2	0.4	0.3

*Total Ammonia Nitrogen

The dissolved oxygen levels ranged from 2.7 to 5.9 mg L^{-1} across all treatments, showing no variation between groups. These DO concentrations are generally within the acceptable range for common carp culture. However, the lower end of 2.7 mg L^{-1} approaches the minimum threshold at which fish may experience stress. Ideally, DO levels should be maintained above 3 mg L^{-1} to support optimal metabolic activity and growth.

Water temperature remained stable across all treatments, ranging between 26.9°C and 30.8°C. This temperature range is considered optimal for carp culture as it promotes healthy feed intake and metabolic functions. The absence of sudden temperature fluctuations ensured a stable environment conducive to consistent fish growth. The pH values observed ranged from 6.72 to 7.79, indicating a slightly acidic to neutral environment, which is appropriate for common carp. The minimal fluctuations in pH suggest effective buffering capacity in the water and low accumulation of organic waste, contributing to a healthy aquatic environment.

TAN concentrations varied slightly among treatments, with values ranging from 0.08 mg L^{-1} in treatment D to 0.12 mg L^{-1} in treatment A. All values remained well below the commonly accepted safe limit of 0.5 mg L^{-1} for carp, indicating that ammonia toxicity was unlikely during the study. The small differences in TAN may reflect variations in feed utilization efficiency or waste production. Nitrite levels were consistent across all treatments, measured at 0.02 mg L^{-1}. This low concentration is considered safe, as elevated nitrite can interfere with oxygen transport in fish, causing conditions such as brown blood disease. No adverse effects were reported, indicating that nitrite levels were well controlled. Nitrate concentrations ranged from 0.04 mg L^{-1} in treatment A to 0.4 mg L^{-1} in treatment C. While nitrate is less toxic compared to ammonia and nitrite, it serves as an important indicator of the nitrogen cycle and overall water quality stability. The slightly higher nitrate level in treatment C may indicate more efficient nitrification processes or differences in organic matter decomposition linked to feed breakdown.

All water quality parameters remained within safe and acceptable ranges for common carp culture throughout the duration of the experiment. The uniformity of these parameters across treatments suggests that any observed differences in fish growth performance and feed efficiency are likely due to the feed treatments themselves rather than variations in environmental conditions.

Feed plays a crucial role in supporting fish growth, and its rates can vary significantly between fish species due to differences in individual nutritional requirements. In aquaculture, the success of fish growth largely depends on how well the feed meets these nutritional needs. Fish growth is also influenced by factors such as sex, environmental conditions, larval development, and overall health status [13].

Efficient protein absorption from feed contributes significantly to growth by providing amino acids necessary for repairing damaged muscle cells, thereby supporting tissue development rather than being diverted for energy use [14]. Proteins also serve as the primary building blocks of muscles, ligaments, tendons, and internal organs [15]. The growth performance of common carp reared for 100 days in the Cirata Reservoir, West Java, and fed with different commercial feeds is presented in Table 3.

Table 3. The growth performance of common carp reared for 100 days in the Cirata Reservoir, West Java.

Parameter	Feed Brand			
	A	B	C	D
SR (%)	83	95	83	87
FI (kg)	123.5	156.4	157.3	160.6
B0 (kg)	15	15	15	15
Bt (kg)	70	100	99	104
ADG (g)	133.53	165.33	188.46	187.50
FCR	2.09	1.84	1.87	1.80
SGR (%)	3.67	5.67	5.69	5.93
PER	1.58	1.89	1.85	1.90
FCG (Rp)	22,188	19,320	19,943	18,947

*(SR) Survival Rate, (FI) Feed Intake, (B0) Initial Biomass, (Bt) Final Biomass, (ADG) Average Daily Gain, (FCR) Feed Conversion Ratio, (SGR) Specific Growth Rate, (PER) Protein Efficiency Ratio, (FCG) Feed Cost Gain

Table 3 presents the growth performance of common carp (*Cyprinus carpio*) reared for 100 days in the Cirata Reservoir, West Java, under four different commercial feed treatments (A–D). Although all feeds had comparable crude protein levels (approximately 28–29%), notable differences were observed in growth parameters, indicating that factors beyond crude protein content—such as feed formulation, digestibility, and palatability—play a significant role in influencing fish performance.

Survival Rate (SR) ranged from 83% in Treatments A and C to 95% in Treatment B, suggesting relatively consistent fish health across treatments, with Treatment B demonstrating the highest survivability. Feed Intake (FI) was highest in Treatment D (160.6 kg), slightly above the other groups, potentially reflecting better feed palatability or higher appetite due to improved feed quality. Despite all treatments starting with the same initial biomass (15 kg), final biomass (Bt) varied significantly, with Treatment D achieving the highest at 104 kg and Treatment A the lowest at 70 kg, demonstrating superior overall growth in Treatment D.

Although the crude protein contents of the experimental feeds were relatively similar (28.10 – 29.12%), significant differences in growth performance and feed utilization were observed among treatments. This indicates that fish growth was not determined solely by total protein level, but rather by overall feed quality, including protein source, digestibility, amino acid balance, fiber content, and processing characteristics. Previous studies have consistently reported that diets with comparable crude protein levels can results in different growth and feed efficiency due to variations in protein origin and nutrient availability [15].

Average Daily Gain (ADG) values were also highest in Treatments C and D, at 188.46 g and 187.50 g respectively, indicating more efficient nutrient conversion into body mass. In terms of feed efficiency, Treatment D recorded the lowest Feed Conversion Ratio (FCR) at 1.80, meaning less feed was required to produce each kilogram of fish, while Treatment A showed the least efficient conversion with an FCR of 2.09. Similarly, Specific Growth Rate (SGR) was highest in Treatments D (5.93%) and C (5.69%), further highlighting their superior daily growth performance.

Protein Efficiency Ratio (PER), which reflects the efficiency of protein utilization, was again highest in Treatment D (1.90), suggesting better assimilation of dietary protein into fish biomass. This may be attributed to differences in protein sources used in the feeds, such as varying proportions of fish meal, soybean meal, or other plant – or animal – based ingredients, which influence amino acids profiles and digestibility. Diets with better-balanced essential amino acids and lower anti-nutritional factors have been shown to enhance protein retention and growth efficiency in common carp. Furthermore, Feed Cost per Gain (FCG)—a critical economic indicator—was lowest in Treatment D at Rp 18,947, making it the most cost-effective option despite its slightly higher feed intake. Collectively, these results suggest that Treatment D provided the best combination of biological performance and economic return, underscoring the importance of overall feed quality in aquaculture production.

4 Conclusion

Although the commercial feeds used had similar protein levels, growth performance varied significantly. The best growth and feed utilization were achieved with Treatment D (protein 29.12%), which showed the lowest FCR (1.80) and the most economical feed cost per gain (Rp 18,947 or $1.13). This suggests that not only protein percentage but overall feed quality and composition significantly influence production outcomes.

References

1. M.A. Suprayudi, D. Jusadi, M. Setiawati, J. Ekasari, I.A. Fauzi, Peningkatan efisiensi protein dan pakan organisme akuatik. IPB Press, Bogor (2023)
2. H. Ali, B. Belton, M.M. Haque, K. Murshed-e-Jahan, Transformation of the feed supply segment of the aquaculture value chain in Bangladesh. Aquaculture **576**, 739–897 (2023)
3. E. Prabu, S. Felix, N. Felix, B. Ahilan, P. Ruby, An overview on significance of fish nutrition in aquaculture industry. Int. J. Fish. Aquat. Stud. **5**(6), 349–355 (2017)
4. G.P. Putranti, Subandiyono, Pinandoyo, Pengaruh protein dan energi yang berbeda pada pakan buatan terhadap efisiensi pemanfaatan pakan dan pertumbuhan ikan mas (Cyprinus carpio). J. Aquac. Manag. Technol. **4**(3), 38–45 (2015)
5. J.E. Halver, R.W. Hardy, Fish nutrition, 3rd ed. Academic Press, California (2002)
6. S.N.N. Hamid, M.F. Abdullah, Z. Zakaria, S.J. Hanim, M. Yusof, R. Abdullah, Formulation of fish feed with optimum protein-bound lysine for African catfish (Clarias gariepinus) fingerlings. Procedia Eng. **148**, 361–369 (2016)
7. T. Watanabe, Fish nutrition and marine culture. Kanagawa Fisheries Training Center–JICA, Tokyo (1998)
8. R. Stickney, D. Gatlin, Aquaculture: An introductory text, 4th ed. CAB International (2022)
9. S. Goddard, Feed management in intensive aquaculture. Chapman and Hall (1996)
10. E. Huisman, Principles of fish production. Department of Fish Culture and Fisheries, Wageningen Agriculture University, Wageningen (1987)
11. A. Tacon, The nutrition and feeding of farmed fish and shrimp: A training manual. FAO of the United Nations, Brazil, pp. 106–109 (1987)
12. M. Akbarurrasyid, S. Nurazizah, R.S. Furqon, Manajemen pembenihan ikan mas marwana (Cyprinus carpio) di Satuan Pelayanan Konservasi Perairan Daerah Wanayasa, Purwakarta, Jawa Barat. J. Aquac. Fish Health **9**(1), 30–37 (2020)

13. V.K. Manam, Fish feed nutrition and its management in aquaculture. Int. J. Fish Aquat. Stud. **11**(2), 58–61 (2023)
14. S. Wulanningrum, Subandiyono, Pinandoyo, Pengaruh kadar protein pakan yang berbeda dengan rasio E/P 8.5 kkal/g protein terhadap pertumbuhan ikan nila (Oreochromis niloticus). J. Sains Akuakult. Trop. **3**(2), 1–10 (2019)
15. NRC (National Research Council), Nutrient requirements of fish and shrimp. The National Academies Press, Washington (2011)

Effect of heat-shock duration and zygote age on tetraploid induction for triploid broodstock production in platinum koi (*Cyprinus rubrofuscus*)

Odang Carman[1], *Andreas Brehme* Sitanggang[2], *Noval Dandy* Putra[2], Alimuddin[1], *Dinar Tri* Soelistyowati[1] *and Kurnia Anggraini* Rahmi[1*]

[1]Department of Aquaculture, Faculty of Fisheries and Marine Sciences, IPB University, Dramaga, Bogor, Indonesia

[2]Undergraduate Study Program, Department of Aquaculture, Faculty of Fisheries and Marine Sciences, IPB University, Dramaga, Bogor, Indonesia

Abstract. The koi fish (*Cyprinus rubrofuscus*) is an ornamental freshwater species valued for its vibrant body patterns, colors, and high market price. Despite extensive breeding efforts, the demand for koi fry remains unmet. Advances in biotechnology, such as chromosome manipulation, offer a promising solution to increase production yields. This study evaluated the effectiveness of tetraploid induction in koi through heat shock treatment, testing different zygote ages and shock durations. The highest proportion of tetraploids (86.6%) was achieved with a heat shock of 39 °C for 2.5 minutes. When considering zygote age, heat shock applied at 31 minutes post fertilization (mpf) produced 66.7% tetraploids, while treatment at 27 mpf yielded a comparatively higher proportion than other age groups. Tetraploidy was confirmed through cytogenetic analysis, ensuring accurate identification of induced individuals. These findings provide a practical protocol for koi breeders to produce tetraploid broodstock, which can be used for triploid production or to enhance genetic resources in ornamental fish aquaculture.

1 Introduction

Koi fish (*Cyprinus rubrofuscus*) is one of the introduced species of freshwater ornamental fish that is highly favored by ornamental fish enthusiasts and has high economic value. The beautiful coloration of koi fish makes it one of the most popular freshwater ornamental fish. From 2015 to 2018, the average production of koi fish reached 11.6%, in line with the increasing demand each year. In fact, in 2018, koi fish ranked first in ornamental fish production, with a total of 476,345 individuals produced [2]. In 2022, Japan's ornamental fish exports reached around 48.6 million USD more than doubling over the past 14 years. Other leading exporters included Indonesia, with a trade value of 36.4 million USD, and Singapore, at 35.1 million USD. Among these, Koi fish stands out as one of the most traded species. The high trade volume demonstrates the global popularity of Koi among hobbyists, highlighting its potential as a thriving and expanding business sector. Although koi fish breeding efforts have been ongoing for a long time, the demand for seedlings has still not been fully met.

* Corresponding author: kurniaang@apps.ipb.ac.id

A significant increase in koi fish productivity needs to be achieved through chromosome manipulation. Chromosome manipulation is a method used to modify genetic material in order to produce individuals with superior qualities. This technique can be applied at the level of genes, chromosome sets, or populations. Genetic modification represents another biotechnological method that can be applied to enhance fish breeding [8].

Several models of chromosome manipulation that have been tested include androgenesis, gynogenesis, triploidization, and tetraploidization. These chromosome manipulation techniques can be applied in the production of tetraploid fish within the teleostei group [11]. The sterility of triploid fish results in the redirection of energy that would normally support gonadal development and gametogenesis toward somatic growth instead. During gonadal formation, triploid fish are unable to properly align their chromosome sets, preventing normal gonad development and leading to sterility. Typically, the testes and ovaries of triploid fish are smaller and show abnormal structures compared to those of diploid fish. In many cases of triploid induction, chromosomes appear damaged or incomplete. Sterility in triploid fish arises from the presence of an odd number of chromosome sets, which disrupts meiosis and leads to abnormal gonadal development or the formation of aneuploid gametes aneuploidy being the primary cause of their sterility. Triploid fish or sterile fish have faster growth compared to diploid fish [10].

Polyploidization can be carried out using shock methods, such as cold or heat treatment, hydrostatic pressure, or chemical treatments, aimed at preventing the release of the second polar body or the first mitotic division in fertilized eggs [5]. Polyploidy induction in fish can be achieved through the application of physical shock treatments that disrupt specific cellular events during early embryogenesis. Inhibition of the second polar body extrusion results in the formation of triploid individuals, whereas suppression of the first mitotic cleavage leads to the development of tetraploid embryos. Induction of polyploidy through heat shock is a method applied immediately after the first mitotic division (for tetraploid induction) at a lethal temperature. This technique has several advantages, including relatively low cost and effectiveness on a large scale. Temperature shock treatment, with variations in temperature, is the most commonly used physical method for producing polyploid fish.

One of the physical methods used in tetraploidization techniques is heat shock. Tetraploid organisms typically exhibit greater heterozygosity owing to the presence of four chromosome sets, which can contribute to the development of larger body size [3]. Moreover, tetraploids play a significant role in evolutionary processes because of their elevated genetic variability [8]. In artificially induced tetraploids, all chromosome sets are derived from diploid parental individuals [15]. The success of tetraploid induction is influenced by several factors, such as the age of the zygote at the time of treatment (initial time), the duration of the shock, and the intensity of the shock temperature. These factors need to be adjusted according to the characteristics of the specific fish species, especially when determining the age of the zygote, because each species has a different timing for the first mitotic division. To optimize the determination of zygote age before tetraploid induction, observations of the cell division process can be conducted beforehand.

Ploidy induction is highly time-sensitive, with even minute differences affecting outcomes, and the precise developmental timing specific to koi fish has not been thoroughly investigated. This study aims to determine the optimal zygote age and heat shock duration for inducing tetraploidy in platinum koi (*C. rubrofuscus*), providing a basis for developing tetraploid broodstock for mass production.

 https://doi.org/10.1051/bioconf/202622101003

2 Materials and methods

2.1 Test materials

The platinum variety of koi fish (*C. rubrofuscus*) has an average weight of 0.72±0.04 kg for males and 0.9±0.15 kg for females. The mature gonad broodstock used came from breeders in the Bogor area, consisting of two pairs. Broodstock were acclimatized before the experiment and under what water quality conditions. The material used consisted of koi fish eggs resulting from artificial spawning of broodstock using ovaprim (GnRH + Anti Dopamine). Other materials used included ethanol solution, Carnoy's solution, 50% acetic acid, glacial acetic acid, gelatin, glycerin, and formic acid as fixatives and stains for observing the number of nucleoli.

2.2 Experimental design

The study was designed using an experimental method in the form of a completely randomized design. The treatments in the study related to heat shock duration included: K1 (control for duration), A1 (heat shock at 39°C for 1.5 minutes with zygote age of 31 minutes after fertilization), A2 (heat shock at 39°C for 2 minutes with zygote age of 31 minutes after fertilization), and A3 (heat shock at 39°C for 2.5 minutes with zygote age of 31 minutes after fertilization). Meanwhile, the treatments in the study comparing zygote age were: K2 (control for zygote age), B1 (heat shock at 39°C for 1.5 minutes with zygote age of 27 minutes after fertilization), B2 (heat shock at 39°C for 1.5 minutes with zygote age of 29 minutes after fertilization), and B3 (heat shock at 39°C for 1.5 minutes with zygote age of 31 minutes after fertilization). This study applied a heat shock treatment at 39 °C, based on conditions determined during preliminary trials

2.3 Spawning

Spawning of the broodstock was carried out in a concrete tank measuring 3×1.2×0.7 m. The broodstock was induced using ovaprim (GnRH + Antidopamine). The ovaprim dosage administered was 0.2 ml/kg of body weight for males and 0.4 ml/kg for females. Eight hours after hormone induction, the broodstock were checked for ovulation and then stripped qith dry method. The eggs and sperm obtained through stripping were fertilized artificially by mixing them in a container and adding sufficient water.

2.4 Heatshock treatments

Before determining the appropriate zygote age for the heat shock treatment, observations of the first mitotic process were conducted using a microscope. This observation was carried out prior to the heat shock treatment by fertilizing the eggs with sperm, then recording the time required from fertilization to the first cell division from one cell to two cells. In platinum koi fish, the first cell division was observed to occur at 31 minutes after fertilization.

The fertilized eggs were then spread onto a glass which served as a substrate for the eggs to adhere to. Each treatment used approximately 656 fertilized eggs. After that, the eggs w plates measuring 10 cm × 23 cm, ere placed into an incubation container in the form of a styrofoam box measuring 75 cm × 42 cm × 32 cm, with a water volume of 78 liters. During the first cell division phase, the water temperature in the medium was maintained at 27°C. After this stage, the eggs were transferred to a styrofoam container filled with water at 39°C to receive the heat shock treatment. The duration of the heat shock was precisely controlled using a stopwatch.

2.5 Hatching and larval rearing

Egg hatching was carried out in 12 tray-shaped containers measuring 33 × 25 × 10 cm, with a water depth of 8 cm, each equipped with aeration. The hatching process lasted for 24–48 hours at a water temperature ranging from 26–30°C. To prevent fungal infections, methylene blue was added to each hatching container at a dose of 0.3 mg/L. The stocking density of each treatment varies because it is adjusted to the eggs produced by the parents during spawning in each treatment.

Observations were conducted at several stages after fertilization. At the 6th hour, the fertilization rate was calculated by comparing the number of fertilized eggs to the total number of eggs observed. At the 24th hour, the embryo survival rate was determined by counting the number of surviving embryos relative to the total number of embryos observed. Then, at the 48th hour or after all the eggs had hatched, the hatching rate was calculated by comparing the number of hatched larvae to the total number of fertilized eggs.

The hatched larvae were then reared for 14 days. After the yolk sac was depleted on the third day, the larvae were fed Artemia nauplii for six days, with a feeding frequency of once every three hours. At nine days old, the larvae began receiving powdered feed (Feng-Li 0) from PT. Matahari Sakti company with ad libitum method. Siphoning was carried out every two days to remove 20–25% of the water, cleaning waste and uneaten feed.

2.6 Tetraploid identification

Tetraploid identification was carried out by counting the maximum number of nucleoli per cell. The process began with fixation of 2-week-old larval samples using Carnoy's solution at a volume ten times that of the sample, for a duration of 2 hours, with the solution being replaced with a fresh one at the 30-minute mark. After fixation, the larval samples were placed on tissue paper to remove any remaining fixative solution before being transferred to a new microtube. Cell dissociation was then performed by adding 1–2 drops of 50% acetic acid solution, followed by homogenization with a dropper pipette until a cell suspension was formed. The suspension was then dropped onto microscope slides that had been soaked in 70% alcohol for two hours and placed on a hotplate at 45–50 °C. Once a ring with a diameter of 1–1.5 cm formed, the suspension was aspirated back, and 2–3 rings were made on each slide, which were then air-dried.

The slide was then stained using solution A (10 g of $AgNO_3$ dissolved in 20 mL of distilled water) and solution B (2 g of gelatin dissolved in 50 mL of water mixed with 50 mL of glycerin solution and added with 2 drops of formic acid per 10 mL). Two drops of solution A and one drop of solution B were placed on the slide, mixed, and spread using a toothpick. The slide was then placed in a staining box at a temperature of 40–45 °C for 16 minutes until it turned yellowish-brown. Afterward, the slide was rinsed with distilled water, air-dried, and observed under a microscope at 100× magnification to identify 2n and 4n chromosomes in the cells.

2.7 Analysis

All test data were collected and initially processed using Microsoft Excel 2016. Statistical analysis was performed using IBM SPSS Statistics 25.0, with differences among treatments evaluated by one-way ANOVA.

3 Result

3.2 Growth performance

Based on the observations, the parameters of total eggs, fertilization rate, 24-hour embryo survival, hatching rate and 2-weeks larval survival are presented in Table 1 and Table 2.

Table 1. Parameters of total eggs, fertilization rate, 24-hour embryo survival, hatching rate and 2-weeks larval survival in the heat shock duration treatment

Treatments	Eggs	Survival rate (24 hours) (%)	Hatching rate (%)	Survival rate (2 weeks) (%)
K1	385	61.80	8.27	80
A1	656	71,63	17.69	86
A2	762	70.19	10,90	80
A3	747	71.67	11.47	81

Control (K1): treatment without heat shock, A1: heat shock at 39°C for 1.5 minutes with zygote age of 31 minutes after fertilization, A2: heat shock at 39°C for 2 minutes with zygote age of 31 minutes after fertilization, A3: heat shock at 39°C for 2.5 minutes with zygote age of 31 minutes after fertilization.

Table 2. Parameters for total eggs, fertilization rate, 24-hour embryo survival, hatching rate, and 2- week larval survival in the zygote age treatment.

Treatments	Eggs	Survival rate (24 hours) (%)	Hatching rate (%)	Survival rate (2 weeks) (%)
K2	1870	79.82	24.44	78.51
B1	2303	78.37	39.71	77.32
B2	1718	77.97	32.99	78.64
B3	1699	76.40	32.91	81.00

Control (K): treatment without heat shock. B1: heat shock at 39 °C for 1.5 minutes with a zygote age of 27 minutes after fertilization; B2: heat shock at 39 °C for 1.5 minutes with a zygote age of 29 minutes after fertilization; B3: heat shock at 39 °C for 1.5 minutes with a zygote age of 31 minutes after fertilization.

The number of nucleoli per cell in koi fish fry resulting from tetraploidization treatment using the heat shock method is presented in Table 3 and Table 4, with images of the nucleoli preparations shown in Figure 1. Observations for each treatment were conducted by examining a sample of 120 cells. Cells observed to have 3-4 nucleoli were assumed to be tetraploid cells, while those with 1-2 nucleoli were assumed to be diploid cells.

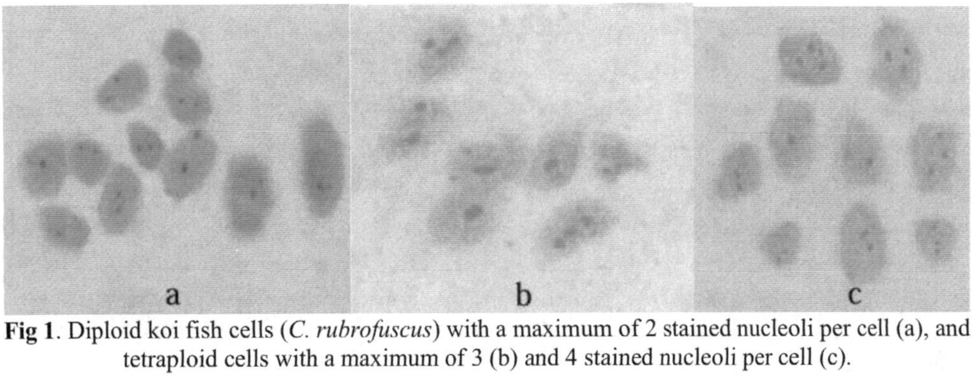

Fig 1. Diploid koi fish cells (*C. rubrofuscus*) with a maximum of 2 stained nucleoli per cell (a), and tetraploid cells with a maximum of 3 (b) and 4 stained nucleoli per cell (c).

Table 3. Distribution of the number of nucleoli per cell in diploid and tetraploid koi fish (*C. rubrofuscus*) from 30 random samples in the heat shock duration treatment

Sample	Number of Nucleoli per Cell (%)				Ploidy Level
	1	2	3	4	
1	44 (36.7)	60 (50)	16 (13.3)	0 (0)	Tetraploid
2	38 (31.7)	72 (60)	10 (8.3)	0 (0)	Tetraploid
3	45 (38))	60 (50)	13 (10.3)	2 (1.7)	Tetraploid
4	39 (32)	71 (59.7)	10(8.3)	0 (0)	Tetraploid
5	50 (41.7)	65 (55)	5 (3.3)	0 (0)	Tetraploid
6	40 (33.3)	70 (58.4)	10 (8.3)	0 (0)	Tetraploid
7	55 (45.3)	65 (54.7)	0 (0)	0 (0)	Diploid
8	46 (38.3)	62 (51.7)	12 (10)	0 (0)	Tetraploid
9	35 (29.7)	70 (58.3)	15 (12)	0 (0)	Tetraploid
10	60 (50)	60 (50)	0 (0)	0 (0)	Diploid
11	42 (35)	62 (62)	16 (13)	0 (0)	Tetraploid
12	50 (41.7)	60 (50)	10 (8.3)	0 (0)	Tetraploid
13	38 (31.7)	65 (54))	17 (14.3)	0 (0)	Tetraploid
14	50 (41.7)	70 (58.3)	0 (0)	0 (0)	Diploid
15	45 (37)	65 (54.7)	10 (8.3)	0 (0)	Tetraploid
16	35 (29.7)	60 (50)	25 (20.3)	0 (0)	Tetraploid
17	53 (44.7)	58 (48.3)	9 (7.2)	0 (0)	Tetraploid
18	42 (35)	65 (54.7)	13 (10.3)	0 (0)	Tetraploid
19	40 (33.3)	64 (53.3)	14 (11.7)	2 (1.7)	Tetraploid
20	40 (33.3)	64 (53.3)	16 (13.4)	0 (0)	Tetraploid
21	55 (45.3)	60 (50)	5 (41.7)	0 (0)	Tetraploid
22	32 (26.7)	71 (60)	16 (13)	1 (0.8)	Tetraploid
23	35 (29.7)	70 (58.3)	15 (12)	0 (0)	Tetraploid
24	54 (45)	60(50)	6 (5)	0 (0)	Tetraploid
25	50 (41.7)	55 (45.3)	15 (13)	0 (0)	Tetraploid
26	45 (38)	68 (56.7)	7 (5.3)	0 (0)	Tetraploid
27	35 (29.7)	60 (50)	25 (20.3)	0 (0)	Tetraploid
28	47 (39.7)	63(52)	10 (8.3)	0 (0)	Tetraploid
29	50 (41.7)	60 (50)	10 (8.3)	0 (0)	Tetraploid
30	39 (33)	55 (45.3)	26 (21.7)	0 (0)	Tetraploid

The percentage of diploid and tetraploid larvae originating from eggs treated with different heat shock durations is shown in Figure 2. The highest number of tetraploid larvae was obtained in treatment A3, with a value of 86.66%, followed by treatment A2 at 80%, and treatment A1 at 76.67%.

https://doi.org/10.1051/bioconf/202622101003

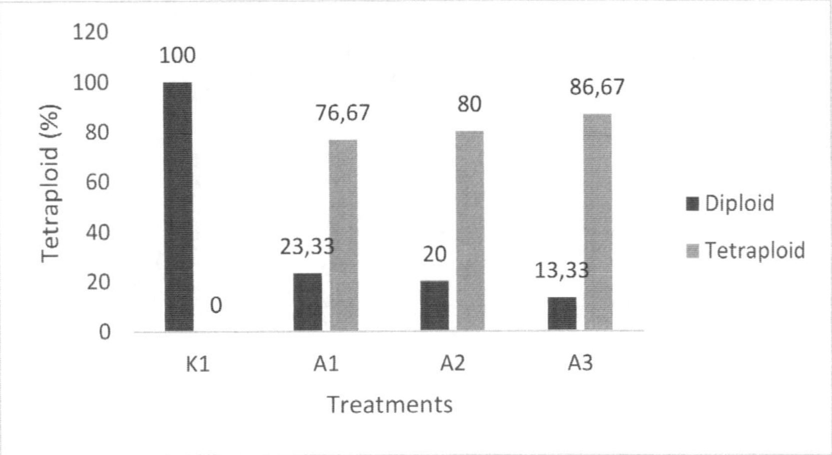

Fig. 2. Percentage of diploid and tetraploid koi larvae resulting from heat shock treatments with different durations.

Table 4. Distribution of the number of nucleoli per cell in diploid and tetraploid koi fish (*C. rubrofuscus*) from 30 random samples in the zygote age treatment

Sample	Number of Nucleoli per Cell (%)				Ploidy Level
	1	2	3	4	
1	38 (31.7)	82 (68.3)	0 (0)	0 (0)	Diploid
2	46 (38.3)	74 (61.7)	0 (0)	0 (0)	Diploid
3	12 (10)	86 (71.7)	22 (18.3)	0 (0)	Tetraploid
4	46 (38.3)	74 (61.7)	0 (0)	0 (0)	Diploid
5	13 (10.8)	78 (65)	29 (24.2)	1 (0.8)	Tetraploid
6	15 (12.5)	87 (72.5)	18 (15)	0 (0)	Tetraploid
7	26 (21.7)	82 (68.3)	10 (8.3)	2 (1.7)	Tetraploid
8	49 (40.8)	71 (59.2)	0 (0)	0 (0)	Diploid
9	41 (34.2)	69 (57.5)	10 (8.3)	0 (0)	Tetraploid
10	32 (26.7)	73 (60.8)	15 (12.5)	0 (0)	Tetraploid
11	43 (35.8)	77 (64.2)	0 (0)	0 (0)	Diploid
12	37 (30.8)	75 (62.5)	7 (5.8)	1 (0.8)	Tetraploid
13	27 (22.5)	81 (67.5)	12 (10)	0 (0)	Tetraploid
14	47 (39.2)	73 (60.8)	0 (0)	0 (0)	Diploid
15	37 (30.8)	71 (59.2)	12 (10)	0 (0)	Tetraploid
16	38 (31.7)	82 (68.3)	0 (0)	0 (0)	Diploid
17	34 (28.3)	69 (57.5)	17 (14.2)	0 (0)	Tetraploid
18	44 (36.7)	76 (63.3)	0 (0)	0 (0)	Diploid
19	50 (41.7)	70 (58.3)	0 (0)	0 (0)	Diploid
20	37 (30.8)	83 (69.2)	0 (0)	0 (0)	Diploid
21	46 (38.3)	74 (61.7)	0 (0)	0 (0)	Diploid
22	40 (33.3)	80 (66.7)	0 (0)	0 (0)	Diploid
23	37 (30.8)	65 (54.2)	18 (15)	0 (0)	Tetraploid
24	49 (40.8)	71 (59.2)	0 (0)	0 (0)	Diploid
25	45 (37.5)	75 (62.5)	0 (0)	0 (0)	Diploid
26	63 (52.5)	48 (4)	0 (0)	0 (0)	Diploid
27	39 (32.5)	81 (67.5)	0 (0)	0 (0)	Diploid
28	38 (31.7)	69 (57.5)	13 (10.8)	0 (0)	Tetraploid
29	26 (21.7)	82 (68.3)	12 (10)	0 (0)	Tetraploid
30	36 (30)	84 (70)	0 (0)	0 (0)	Diploid

The percentage of diploid and tetraploid larvae originating from eggs treated with heat shock at different zygote ages is shown in Figure 3. The highest number of tetraploid larvae was obtained in treatment B2 with a value of 6.08%, followed by treatment B3 at 5.37%, and treatment B1 at 3.69%.

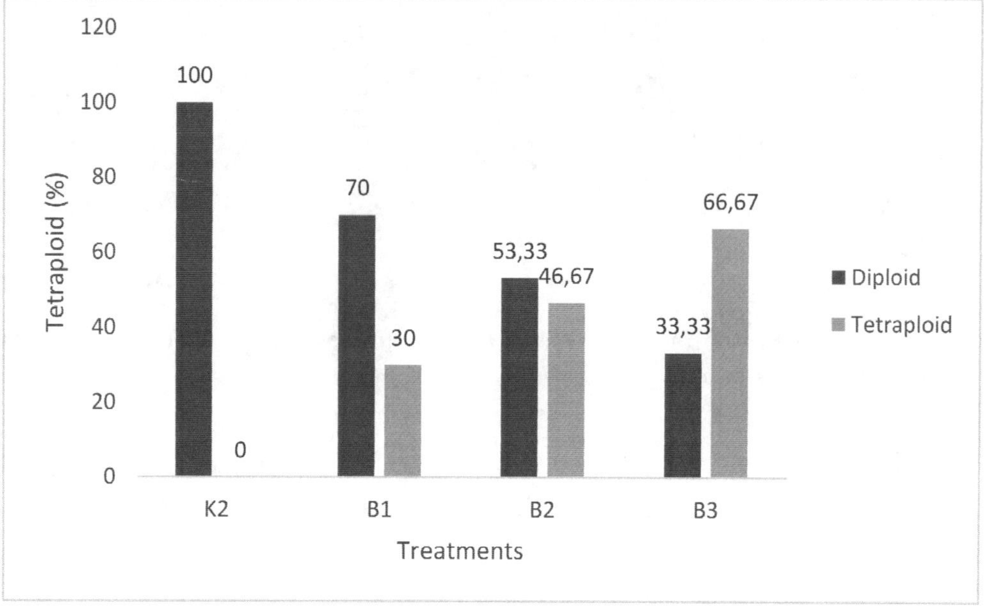

Fig 3. Percentage of diploid and tetraploid koi larvae resulting from heat shock treatment at different zygote ages

Abnormal larvae found in the heat shock duration treatment were characterized by inactive movement, smaller size compared to normal fish, a bent growth shape resembling the letter L, and abnormal swimming behavior. In the zygote age treatment, spinal deformities such as scoliosis and kyphosis, as well as bent neck bones, were observed. The percentage of abnormalities in koi fish (*C. rubrofuscus*) larvae is presented in Figures 5 and 6.

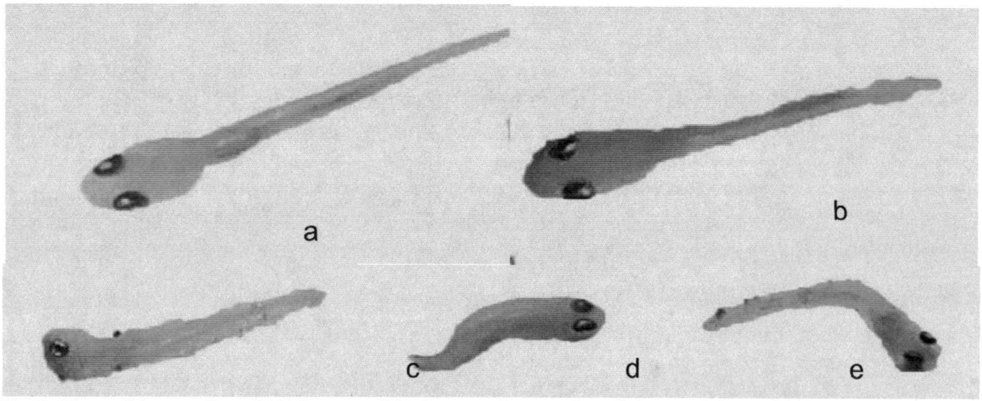

Fig 4. Normal larvae (a and b), larvae with abnormal curved neck bones (c), scoliosis (d), and kyphosis (e) resulting from the heat shock treatment.

4 Discussion

Based on Table 1, the highest total number of eggs was found in treatment A2 at 762, while the lowest was in treatment K1 at 385.33. Meanwhile, Table 2 shows that the highest total number of eggs was observed in treatment B1 with 2303 eggs, and the lowest in treatment B3 with 1699 eggs. Fecundity is related to metabolism, which responds to changes in food availability and, in turn, affects egg development, including age, size, number of eggs, and spawning cycles [10].

The highest fertilization rate in the heat shock duration treatment was observed in treatment A3, at 62.15%, while the lowest was in treatment K1, at 42.96 %. In the zygote age treatment, the fertilization rate was consistent across treatments K2 to B3, at 46.7%. The main factor influencing the fertilization success rate in these results is believed to be related to egg quality. This assumption is based on the fact that immature eggs can hinder the fertilization process, and egg maturation itself is influenced by hormonal activity. Fertilization rate can also be affected by broodstock factors such as the quality of male and female gametes, as well as handling factors like sperm mixing. Low fertilization rates may be caused by various factors, including the quality of eggs and sperm, broodstock maturity, and water quality [14].

Embryo survival during 24 hours in the heat shock duration treatments showed the highest survival rate in treatment A3 at 71.67 %, while the lowest survival rate was found in treatment K1 at 61.80 %. In the zygote age treatments, the highest survival rate was observed in treatment K2 at 79.82%, and the lowest in treatment B3 at 76.40%. The embryo mortality rate was higher in the heat shock treatments compared to the control. The high temperature during heat shock affected larval abnormalities, causing egg membrane hardening and inhibiting hatching enzyme activity [2]. The timing of cell division post-fertilization varies among fish species depending on light, yolk amount diversity, and incubation temperature [4]. Embryos that survived the heat shock process did so due to sub-lethal factors. Egg mortality was caused by various factors, including incomplete fertilization and eggs sticking or piling up during spreading, which can hinder oxygen circulation and ultimately lead to egg death [7].

The highest hatching rate for the heat shock duration treatment was found in treatment A1 at 17.69 %, while the lowest was in treatment K1 at 8.27%. For the zygote age treatment, the highest hatching rate was in treatment B1 at 39.71%, and the lowest was in treatment K2 at 24.44%. The hatching rate in the control group was lower than all treatment groups, which was suspected to be due to technical errors during egg spreading, causing the eggs to be too dense and clumped (the eggs were first spread in the control, resulting in uneven distribution on the glass substrate). Factors affecting the hatching rate include the quality of the sperm and broodstock eggs, as well as the environmental conditions of the medium. Prolonged heat shock can damage the embryo membrane, causing issues in cell division such as loss of genetic information, spindle fiber damage, and disrupted enzyme activity [6]. This is consistent with the results showing that treatment A1 had a higher hatching rate compared to treatments A2 and A3.

Larval survival data over two weeks for the heat shock duration treatment showed the highest survival rate in treatment A1 at 86 ± 1.00% and the lowest in treatment A2 at 80%. For the zygote age treatment, the highest two-week larval survival was observed in treatment B3 at 81.00%, while the lowest survival was in treatment B1 at 77.32 %. Larval survival refers to the percentage of larvae that survive until harvest out of the total larvae reared. This survival rate is influenced by the quality of the eggs produced. Additionally, environmental factors also play a role in determining larval survival success, with temperature being one of the most influential aspects.

Based on Tables 3 and 4, the highest number of four nucleoli was found in fewer tetraploid larvae (0.1%), while the highest number of three nucleoli was found more frequently in tetraploid larvae (8%). However, the maximum number of three nucleoli can be considered tetraploid rather than triploid. This is because the treatment could not have produced triploids, as tetraploid induction disrupts the first cell division, whereas triploids result from the retention of the second polar body. This is relates to anomalies observed in the variation of the maximum number of nucleoli [13].

Additionally, anomalies in the maximum number of nucleoli were also found in tetraploid catfish [11]. The probability of the maximum number of nucleoli per cell is suspected to be due to differences in ribosome synthesis activity responsible for protein production. Variation in the maximum number of nucleoli in each cell nucleus can be influenced by differences in cellular activity. The count of the maximum number of nucleoli was determined by the number of black dots (nucleoli) excluding impurities in each cell nucleus, which appeared yellowish-brown due to the staining process. Impurities can resemble nucleoli as black dots, so careful observation is needed to accurately count the number of nucleoli.

Based on the results of the tetraploid induction obtained, there is an indication that in fish resulting from tetraploid induction, embryos that received heat shock at a later zygote age tend to have lower tetraploid induction success. Each fish species has different timing for cell division after fertilization, which can be influenced by incubation temperature, species, and the maturity level of the broodstock gonads [9]. The timing of the first mitotic division is a key determinant for the success of heat shock, especially in producing tetraploid fish. The results from the heat shock duration treatment showed that the treatment with a heat shock duration of 2.5 minutes (A3) produced more tetraploids compared to durations of 1.5 and 2 minutes. Meanwhile, in the zygote age treatment, heat shock applied at 31 minutes after fertilization (B3) produced the highest number of tetraploid fish. This is consistent with observations showing that cell division in koi embryos occurs at 31 minutes post-fertilization.

Abnormalities in larvae with different heat shock durations were highest in treatment A3, followed by treatments A1, A2, and the control. The highest abnormality rate was observed in the heat shock treatment with the longest duration, at 10.33%. The lowest abnormality rate was in the control group at 4.33%. Meanwhile, in the zygote age treatment, the highest abnormal larvae rate was found in treatment B2 at 6.08%, and the lowest in treatment B1 at 3.69%. Heat shock applied during the embryonic development phase can damage the embryo membrane, cause the outer layer of the egg to harden, and inhibit the activity of hatching enzymes. This can lead to abnormalities in the larvae. Another form of larval abnormality caused by heat shock treatment is a short and curved tail.

5 Conclusion

Tetraploid induction in koi fish (*C. rubrofuscus*) was influenced by both heat shock duration and the timing of treatment relative to zygote age. The highest proportion of tetraploids (86.6%) was achieved with a heat shock of 39 °C for 2.5 minutes, while treatment at 39 °C for 90 seconds also successfully induced tetraploidy, albeit at a lower rate. Regarding zygote age, applying heat shock at 31 minutes post-fertilization (mpf) resulted in 66.7% tetraploids, which was higher than other age treatments, indicating that the timing of first cleavage plays a critical role in inhibiting cytokinesis and promoting tetraploidy.

This study demonstrates that different experimental factors specifically heat shock duration and zygote age had distinct effects on the success of tetraploid induction. While variations in heat shock duration primarily determined the overall efficiency of chromosome doubling, differences in zygote age significantly influenced the developmental stage responsiveness to treatment. Thus, each factor contributed differently to the outcome, highlighting the need to optimize both duration and timing to maximize tetraploid yield.

These findings provide a practical reference for koi breeders aiming to produce tetraploid broodstock for triploid production or to enhance genetic resources in ornamental fish. However, some treatments showed larval abnormalities and reduced hatching rates, suggesting that further optimization and confirmation are necessary to refine the protocol.

References

1. D. Chourrout, B. Chevassus, F. Krieg, A. Happe, G. Burger, P. Renard, Production of second generation triploid and tetraploid rainbow trout by mating tetraploid males and diploid females – Potential of tetraploid fish. *Theor. Appl. Genet.* 72, 193–206 (1986). https://doi.org/10.1007/BF00266992
2. Direktorat Jenderal Perikanan Budidaya (DJPB), Laporan kinerja DPIB. KKP, Jakarta (2018).
3. E. Braum, Ecological aspect of the survival of fish eggs, embryos and larvae. In: Ecology of Fresh Water Fish Production (Blackwell Scientific Publications, Oxford), pp. 102–131 (1978).
4. E. Kusrini, S. Cindelaras, A.B. Prasetio, Pengembangan budidaya ikan hias koi (Cyprinus carpio) lokal di Balai Penelitian dan Pengembangan Budidaya Ikan Hias Depok. Media Akuakultur 10(2), 71–78 (2015).
5. G.H. Thorgaard, Chromosome set manipulation and sex control in fish. In: W.S. Hoar, D.J. Randall, E.M. Donaldson (eds.) Fish Physiology, Vol. IX, Part B (Academic Press, New York), pp. 405–434 (1983).
6. H.W. Gill, H.J. Kong, C.N. An, B.S. Kim, S.G. Lim, I.S. Park, Cytogenetic study of diploid and induced tetraploid in Korean Rose bitterling, Rhodeus uyekii. SpringerPlus 5, 186 (2016).
7. Hendriana, F. Ridwansyah, A. Iskandar, A.S. Munawar, D. Lugina, Method of Hatchery for koi Cyprinus carpio to produce quality seeds at Mizumi Farm, Sukabumi, West Java. J. Penelitian Pertanian Terapan 2(1), 17–26 (2021).
8. J. Pasquier, I. Braasch, P. Batzel, C. Cabau, J. Montfort, T. Nguyen, et al., Evolution of gene expression after whole-genome duplication: new insights from the spotted gar genome. *J. Exp. Zool. Part B Mol. Dev. Evol.* 328, 709–721 (2017). https://doi.org/10.1002/jez.b.22770
9. M. Walidin, M. Admi, Effectiveness of pineapple leaf extract (Ananas cosmosus) on hatching milkfish eggs (*Chanos chanos* Forskall). Arwana: J. Ilmiah Progr. Studi Perairan 2(2), 80–89 (2020).
10. M.J. Sellars. Triploidy In: Maloy S, Hughes K (eds). Brenner's Encyclopedia of Genetics, Second Edition. Academic Press, Cambridge (2013).
11. M.S. Daryanto, O. Carman, D.T. Soelistyowati, Rahman, Ploidy level determination in genetically modified polyploid striped catfish (Pangasianodon hypophthalmus) Sauvage, 1878 based on the number of nucleoli per cell. Iktiologi Indonesia 19(1), 43–52 (2019).
12. O. Carman, A.T. Mukti, M. Zairin Jr., Alimuddin. Reproductive performances of triploid male and female Nile tilapia *Oreochromis niloticus* (Linnaeus, 1758) at different ages. Biodiversitas, 24(8) : 4235-4242 (2023).
13. O. Carman, T. Oshiro, F. Takashima, Variation in the maximum number of nucleoli in diploid and triploid common carp. Jpn. Soc. Sci. Fish. 58(12), 2303–2309 (1992).
14. P. Keshavanath, B. Gangadhara, N. Basavaraja, M.C. Nandeesha, Artificial induction of ovulation in pond-raised mahseer, Tor khudree using carp pituitary and ovaprim. Asian Fish. Sci. 19, 411–422 (2006).
15. Z. Wu, L. Wang, Y. Lu, X. Zhu, X. Yue, F. You. Artificial induction and genetic structure analysis of tetraploid turbot *Scophthalmus maximus*. Front. Mar. Sci. 6(637): 1-12 (2019).

Operational efficiency of feeding in Mud Crab fattening culture *Scylla serrata* in recirculating aquaculture system (RAS) apartment system

Iis Diatin[1], *Melani* Oktaviani Sianturi[1], *Irzal* Effendi[1], *Belinda* Astari[1*]

[1]Department of Aquaculture, Faculty of Fisheries and Marine Science, IPB University, Jl. Raya Dramaga Kampus IPB, Babakan, Dramaga, Bogor, West Java, 16680, Indonesia

Abstract. Mud crabs (*Scylla serrata*) are highly valued aquaculture commodities in the international market. However, feeding operations in mud crab culture using a Recirculating Aquaculture System (RAS) are often inefficient, increasing labor time and operational costs. This study aimed to evaluate the operational efficiency of feeding *S. serrata* once, twice, and three times per day by assessing production performance and conducting a business analysis in a RAS apartment system. A completely randomized design was used with three feeding frequency treatments (one, two, and three times daily), each with three replicates. Parameters observed included survival rate, absolute weight gain, carapace length and width growth, molting frequency and percentage, feed consumption, feed conversion ratio (FCR), feed utilization efficiency, and labor efficiency. Results showed that feeding frequency had no significant effect on survival rate, growth parameters, molting performance, or feed utilization efficiency. However, feeding twice per day produced the highest profit (IDR 5,342,008) and a labor efficiency of 1.08%. These findings indicate that feeding twice daily optimizes both biological performance and economic returns in mud crab fattening using a RAS apartment system.

1 Introduction

Mud crabs (*Scylla serrata*) are among the most valuable crustacean species in global aquaculture due to their high market demand and export potential. Indonesia, one of the major producers of mud crabs, contributes significantly to international markets such as China, the United States, Japan, and the European Union. Despite this, approximately 81% of Indonesia's crab production still relies on wild catches, indicating limited aquaculture-based supply and the need to increase cultured production [1]. To meet rising demand while promoting sustainable practices, the development of effective mud crab aquaculture systems has become increasingly important.

Conventional pond culture systems, although widely practiced, face several challenges including large land requirements, poor water quality control, and high cannibalism rates among crabs [2]. To address these limitations, the Recirculating Aquaculture System (RAS) "apartment" system has been introduced as an innovative approach for mud crab fattening. This system enables intensive culture within a controlled environment, reduces land use, and minimizes cannibalism by housing crabs individually [3–4]. However, despite these advantages, the efficiency of feeding operations in RAS systems remains a key constraint. Feeding in mud crab culture is labor-intensive because crabs must be fed individually, increasing operational costs and time requirements.

* Corresponding author: belindast@apps.ipb.ac.id

Efficient feed management is essential not only for improving growth and survival but also for enhancing labor productivity and overall economic performance in aquaculture. Previous studies have shown that feeding frequency can influence growth and feed conversion efficiency in crustaceans, yet the relationship between feeding operations and labor efficiency in mud crab culture has not been comprehensively evaluated [5–6]. Therefore, this study aims to analyze the operational efficiency of feeding mud crabs once, twice, and three times per day by evaluating production performance and conducting a business analysis within the RAS apartment system. The findings are expected to provide insights into optimizing feeding management strategies for sustainable and profitable mud crab farming.

2 Method

2.1 Site location and time

The study was carried out between October 2023 and January 2024. The study was conducted at the RAS apartment system mud crab culture business development unit of the IPB Fisheries and Marine Observation Station (IFMOS), Ancol, North Jakarta, DKI Jakarta. Collectors in Keramat Village, Pakuhaji District, Tangerang Regency, Banten, provided the seeds. Water quality assessments were conducted in the Aquaculture Environment Laboratory and the Fish Nutrition and Feed Technology Division Laboratory, Department of Aquaculture, FPIK, IPB University.

2.2 Experimental design

A Completely Randomized Design (CRD) was employed in the trial, with three replications for each of the three treatments: one, two, and three times daily feeding. In the RAS apartment system, the effects of feeding crabs once, twice, and three times a day were compared using these three treatments.

2.3 Culture techniques

2.3.1 Rearing preparation

Synthetic sponge cleaning is used to remove food residue, moss, and excrement from the apartment boxes used as mud crab rearing containers. It takes one to two days to clean the rearing containers. The rearing containers are rinsed with new water to get rid of germs and parasites. The RAS apartment system consisted of 90 individual rearing boxes (30 cm × 33 cm × 15 cm each) arranged vertically in stacked modules. Each box contained 7–8 L of rearing water, giving a total system capacity of approximately 720 L. The system operated continuously with an aeration system and a filtration unit consisting of physical (synthetic cotton, biofoam, Japanese filter mats), chemical (UV sterilizer), and biological (bioballs, crushed coral, kaldness) filters. The water flow rate was maintained at 1.5 L min^{-1}, and the water exchange rate was approximately 10% per day to maintain optimal water quality (Fig. 1).

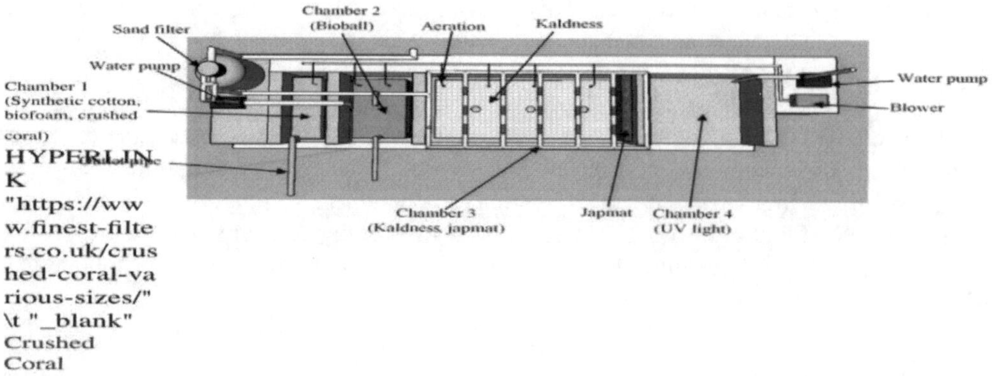

Fig. 1. Physical, chemical, and biological filter design for a recirculating aquaculture system in an apartment box for cultivating mud crabs (*Scylla serrata*).

2.3.2 Mud crab selection and acclimation

Crab selection is carried out at the crab collector's location in the village Keramat, Pakuhaji District, Tangerang Regency, Banten. Mud crab criteria which is chosen among others. Healthy mud crab with complete bodies. No spots are black or barnacles that stick to the crab's body [2]. Live crabs are also characterised by their reactions, which fasten when the foot is withdrawn and the eye, which actively move if disturbed. Gender is also observed from the external shape of the body. Male crabs are characterised by genitals that are triangular in shape and slightly pointed and attached to the part of their stomach. The crab female is characterised by an organ of his gender, which has a relatively wide triangular shape, and the front part is somewhat blunt [5] (Fig. 2). The type of mud crab was used in this research. There were 23 male mud crabs and 67 female mud crabs.

2.3.3 Mud crab rearing

Crabs are transported to the culture location using a dry transportation system. A receptacle used for transportation is a basket sized 75 cm × 42 cm × 32 cm. The time required from the transportation location to the culture location is 1-1.5 hours. The container for acclimatising mud crab is a round fibre tank. Fibre tank for acclimatisation is filled with a mixture of seawater and freshwater by adjusting the salinity of the source seed to 10-15 g/L and giving aeration. Crabs that arrive in the location are entered into the basket with a concentrated spread of 15 crabs/basket and left for 10-15 minutes. The crab is splashed with water, rearing, and left alone for 10-15 minutes. Crab acclimatised in a tub fibre round, which has been filled with rearing water to a height of 8-10 cm for 1-2 hours. The optimal salinity for mud crab is 25-30 g/L, and the temperature range is 27-28°C. Crab was acclimatised to the condition of the claw and foot road. Still bound, the bond is released after the acclimatisation process is finished. The initial size of the crab was measured using a digital scale. The length and width of the carapace were measured. Average weight of a beginning mud crab is 140.3±2.83 g, average carapace length 59.98±0.51 cm, and average carapace width 81.72±3.33 cm. Crabs were kept in apartment boxes randomly, each block (G, H, I) containing 90 crabs (Fig. 3).

Fig. 2. Morphological differences between male mud crabs (left) and female (right).

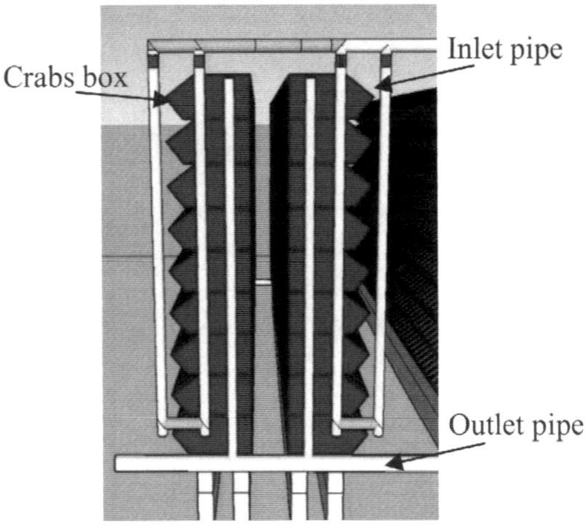

Fig. 3. RAS system apartment box design for mud crab (*Scylla serrata*) culture.

Crab given feed fish trash in a restricted way, 5-10% from the biomass crab. Type fish trash, which is fish *Selaroides leptolepis*. Fish was purchased from Muara Angke Fish Market, Jakarta North. Giving fish trash done with chopping fish so that it is more easily consumed, and weighing as much as ± 6 g/crab, then given to the mud crab with stainless steel tongs to prevent the hand from being hit by a crab's pincers. Fresh fish was stored as feed in the freezer for seven days to prevent the growth of bacteria that cause decay, which can cause damage to the feed.

2.3.4 Sampling

The crab census is carried out every 14 days with 90 crabs. The measurements taken were weight, carapace length, carapace width, and completeness of body organs. Measurements were taken using digital scales with an accuracy of 0.01 g and a vernier calliper with an accuracy of 0.1 cm. Mud crabs are observed every four hours to determine which ones are molting. Crabs immediately after molting are weighed and recorded. Harvest target crab sized 250-400 g with a rearing time range of 30 days. Water quality management is maintained by siphoning off the remaining feed. Moreover, other waste accumulates before feeding—leftover feed cleaned in the morning (07.00 WIB). Water quality parameters consisting of temperature, pH, salinity, and dissolved oxygen (DO) were measured twice a day, namely in the morning (09.00 WIB) and afternoon (17.00 WIB). Total ammonia Nitrogen (TAN), nitrite, nitrate, and alkalinity are measured every 7 days in the Laboratory Aquaculture Environment and Laboratory of Nutrition and Feed Technology Division, Aquaculture Department, IPB University.

2.4 Parameter test

Parameters observed during rearing include production performance, water quality, and business analysis. The production performance parameters measured consist of labor efficiency, survival rate, absolute weight growth, carapace length growth, carapace width growth, molting time, number and percentage, feed consumption, feed conversion ratio (FCR), and feed utilization efficiency.

$$\text{Labor efficiency (\%)} = \frac{Amount\ of\ feed\ given\ (kg)}{Amount\ of\ labor\ time\ (hours/day)} \times 100 \quad (1)$$

$$\text{Survival rate (SR, \%)} = \frac{Final\ number\ of\ crab\ (crab)}{Initial\ number\ of\ crab\ (crab)} \times 100 \quad (2)$$

$$\text{Absolute weight growth (AWG, \%)} = \frac{Final\ average\ weight\ (g) - Initial\ average\ weight\ (g)}{Time\ of\ rearing\ (Day)} \quad (3)$$

Carapace length growth is the difference between the final carapace length and the initial carapace length during the rearing period. Carapace width growth is the difference between the final carapace width and the initial carapace width during the rearing period. Molting time can be defined as the length of time it takes for a mud crab to shed its shell during the rearing period. Molting time is calculated from the first day the crab molts during the rearing period. Molting time for mud crabs is checked every four hours.

$$\text{Percentage of molting (\%)} = \frac{Final\ number\ of\ crab\ molting\ (Crab)}{Initial\ number\ of\ crab\ molti\ (Crab)} \quad (4)$$

The amount of feed consumption (g) is calculated from the total feed given, minus the total remaining feed that is not consumed during the rearing period.

$$\text{Feed conversion ratio (FCR)} = \frac{Amount\ of\ feed\ consumption\ (g)}{(Final\ biomass\ of\ crab\ (g) - Initial\ biomass\ of\ crab\ (g)) + Biomass\ of\ death\ crab\ (g)} \quad (5)$$

$$\text{Feed utilization efficiency} = \frac{Final\ average\ weight\ (g) - Initial\ average\ weight\ (g)}{Amount\ of\ feed\ consumption\ (g)} \times 100 \quad (6)$$

The water quality parameters measured consisted of temperature, pH, dissolved oxygen (DO), salinity, total ammonia nitrogen (TAN), nitrate, nitrite, and alkalinity (Table 1).

Table 1. Water quality parameters of mud crab (*Scylla serrata*) rearing media in Recirculating Aquaculture System (RAS) apartment boxes for 30 days of rearing.

Parameter	Measurement tool	Time
Salinity (g/L)	Refractometer	Every day
Temperature (°C)	Thermometer	Every day
pH	pH meter	Every day
DO (mg/L)	DO meter	Every day
TAN (mg/L)	Spectophotometer	Every 7 days
Nitrite (mg/L)	Spectophotometer	Every 7 days
Nitrate (mg/L)	Spectophotometer	Every 7 days
Alkalinity (mg/L)	Titration	Every 7 days

The business analysis consisted of investment cost, variable cost, fixed cost, total cost, total revenue, profit, cost of good sold, revenue/cost ratio, payback period, and break-even point.

$$\text{Total cost (TC, IDR)} = \text{Fixed cost (IDR)} + \text{Variable cost (IDR)} \quad (7)$$

$$\text{Total revenue (TR, IDR)} = \text{Number of production (Crab)} \times \text{Selling price (IDR)} \quad (8)$$

$$\text{Profit (IDR)} = \text{TR (IDR)} - \text{TC (IDR)} \quad (9)$$

$$\text{Cost of good sold (CGS, IDR)} = \frac{Total\ cost\ (IDR)}{Number\ of\ production\ (Crab)} \quad (10)$$

$$\text{Revenue/cost (R/C) ratio} = \frac{Total\ revenue\ (IDR)}{Total\ cost\ (IDR)} \quad (11)$$

$$\text{Payback period (Year)} = \frac{Investment\ cost\ (IDR)}{Profit\ (IDR)} \quad (12)$$

$$\text{Break-even point (BEP) unit (Crab)} = \frac{Fixed\ cost\ (IDR)}{Selling\ price\ (IDR) - \frac{Variable\ cost\ (IDR)}{Number\ of\ production\ (Crab)}} \quad (13)$$

$$\text{Break-even point (BEP) price (IDR)} = \frac{Fixed\ cost\ (IDR)}{1 - \frac{Variable\ cost\ (IDR)}{Total\ revenue\ (IDR)}} \quad (14)$$

2.5 Data analysis

Data were analyzed using Microsoft Excel 2019 and SPSS software version 26. Test parameters were analyzed using Analysis of Variance (ANOVA) with a 95% confidence interval, and if there was a significant effect, Duncan's test was continued. Business performance analysis is carried out based on assumptions obtained from the results of production performance in the research. Analysis of the water quality of the rearing media and costs was carried out descriptively through table. The number of production cycles carried out in one year is 12. The number of workers employed is one person. The business analysis was conducted on softshell and non-softshell crabs.

3 Result

3.1 Production performance

The allocation of working time workers spend cultivating mud crabs in RAS apartment boxes for 30 days of rearing, with once-daily feeding is 141 hours, twice-daily 171 hours, and three times-daily 201 hours (Table 2). More frequent feeding will affect the number of molting times, so the harvest of soft-shelled crabs will be more frequent and require more working time. In the research conducted, only one worker was used, so the workload received is heavier. Labor efficiency for feeding once, twice, and three times daily in mud crab (*Scylla serrata*) culture is presented in Table 3. Labor efficiency for once-daily feeding was 0.78%, twice-daily 1.08%, and three times-daily 1.53% (Table 3). Labor efficiency in feeding showed a significant increase with more frequent feeding. At once-daily feeding, labor efficiency only reached 0.78%. However, when fed twice daily, efficiency increased to 1.08%. The highest efficiency was achieved with three times-daily feeding, namely 1.53%. This indicates that frequent feeding allows labor to work more effectively and productively, optimising their time to achieve better productivity.

Table 2. Labor load on feeding one, two, and three times a day in the cultivation of mangrove crabs(*Scylla serrata*) in RAS apartment boxes for 30 days of rearing.

Sub-activities	**Feeding frequency (times/day)**		
	1	**2**	**3**
	Working hours (Hours)		
Container preparation (apartment box, RAS filter, acclimatization)	1	1	1
Filter system setup and water filling	2.5	2.5	2.5
Crab apartment and filter system cleaning	2.5	2.5	2.5
Mud crab seed selection	1	1	1
Mud crab seed transportation	2	2	2
Mud crab acclimatization	3	3	3
Measurement of carapace length, width, and weight	2	2	2
In-situ water quality check	30	30	30
Ex-situ water quality check	30	30	30
Siphoning	30	30	30
Mud crab feeding	30	60	90
Mud crab molting check	6	6	6
Mud crab census	4	4	4
Harvesting soft-shell crabs	1	1	1
Post-harvest	1	1	1
Total work time allocation (hours)	141	171	201
Feed consumption (kg/30 days)	1.1	1,8	3,1
Working hours (30 days)	141	171	201
Number of workers (persons/day)	1	1	1
Labor efficiency (%)	0.8	1.1	1.5

Table 3. Production performance of mud crab (*Scylla serrata*) rearing media in Recirculating Aquaculture System (RAS) apartment boxes for 30 days of rearing.

Parameter	**Feeding frequency (times/day)**		
	1	**2**	**3**
Survival rate (%)	60.00 ± 20.00^{a}	73.33 ± 20.82^{a}	66.67 ± 15.28^{a}
Absolute weight growth rate (g/day)	0.33 ± 0.21^{a}	0.64 ± 0.05^{a}	0.67 ± 0.53^{a}
Carapace length growth (cm)	1.79 ± 0.71^{a}	2.85 ± 0.38^{a}	2.41 ± 1.23^{a}
Carapace width growth (cm)	1.60 ± 0.53^{a}	2.62 ± 0.13^{a}	2.33 ± 1.23^{a}
Amount of feed consumption (g)	368.30 ± 83.51^{a}	617.67 ± 314.24^{ab}	1,026.70 ± 345.25^{b}
Molting percentage (%)	6.67 ± 5.77^{a}	16.67 ± 5.77^{a}	16.67 ± 5.77^{a}
Feed conversion ratio	3.49 ± 0.83^{a}	3.00 ± 0.33^{a}	3.56 ± 0.73^{a}
Feed utilization efficiency	2.94 ± 1.77^{a}	3.58 ± 1.57^{a}	2.25 ± 1.82

Feeding once, twice, and three times a day showed no significant differences (P>0.05) in survival rate, absolute weight growth rate, carapace length growth, carapace width growth, percentage of molting, and feed conversion ratio. The amount of feed consumption parameter showed significant differences (P<0.05). Better results in survival rate, carapace length width growth, molting percentage, feed conversion ratio, and feed utilization efficiency were obtained in the 2 times/day feeding frequency treatment, while the highest absolute weight growth rate and feed consumption were obtained in the 3 times/day feeding frequency treatment. Table 3 shows the results of mud crab production performance in the RAS apartment box for 30 days of rearing.

Table 4. Time and number of molting (crab/day) of mud crabs (*Scylla serrata*) when fed once, twice, and three times a day during culture in RAS apartment boxes for 30 days of rearing.

Feeding (times/ day)	**Day of rearing (day)**														
	1	**2**	**3**	**4**	**5**	**6**	**7**	**8**	**9**	**10**	**11**	**12**	**13**	**14**	**15**
1	0	0	0	0	0	0	0	0	0	0	0	0	0	[redacted]	0
2	0	0	[redacted]	0	0	0	0	[redacted]	0	0	0	0	0	0	0
3	0	0	[redacted]	0	0	0	0	0	0	0	0	0	0	0	0
Feeding (times/ day)	Day of rearing (day)														
	16	17	18	19	20	21	22	23	24	25	26	27	28	29	30
1	0	0	0	0	0	0	0	0	0	[redacted]	0	0	0	0	0
2	0	[redacted]	0	0	[redacted]	0	[redacted]	0	0	0	0	0	0	0	0
3	0	0	0	0	[redacted]			0	0	0	0	0	0	0	0

Molting observations conducted every four hours showed that feeding two and three times a day resulted in the same number of molts, namely 5, and feeding once a day resulted in the fewest molts, namely 2 (Table 4). More frequent feeding can increase the molting frequency in crabs. However, this also increases the workload for workers, as more frequent feeding requires extra time and attention. Furthermore, the molting period requires intensive care, as crabs are more vulnerable and require optimal environmental conditions. However, with efficient scheduling and good management, increasing the molting frequency can increase productivity and yields in the long term, although it requires more labor in the short term. Based on Table 5, the water quality of the mud crab (*Scylla serrata*) culture media in the RAS apartment box for 30 days was within the optimal range for mud crab culture.

Table 5. Water quality of the mud crab (*Scylla serrata*) rearing media during 30 days of culture in RAS apartment boxes.

Parameter	**Range of values**	**Optimum values [6]**
Temperature (°C)	27.8 – 29.4	25 – 35
pH	7.0 – 8.3	7 – 9
Dissolved oxygen (mg/L)	6.2 – 7.8	≥ 5
Salinity (g/L)	25 – 33	25
Nitrate (mg/L)	0.253 – 1.886	< 10
Nitrite (g/L)	0.045 – 1.723	< 10
Total ammonia nitrogen (g/L)	0.01 – 0.47	< 3
Alkalinity (g/L)	54.67 – 97.33	> 80

The assumptions for the mud crab business, based on feeding once, twice, and three times daily in RAS apartment boxes, are shown in Table 6. Based on the basic business assumptions, calculations of investment costs, fixed costs, variable costs, and income for mud crabs were obtained when fed one, two, and three times a day. An analysis of the mud crab business, feeding one, two, and three times a day in culture in RAS apartment boxes, can be seen in Table 7. Mud crab culture in the RAS apartment with twice-daily feeding yielded the highest profit of IDR 5,342,008, with a revenue/cost ratio of 1.09, meaning that every 1 rupiah of total costs incurred will generate a profit of 0.09 rupiah, and a payback period of 13.25 years. The 13.25-year payback period indicates that the income from the culture business is not high enough compared to the initial investment costs. Low income, high operational costs, and a significant initial investment could cause this. The payback period can be reduced by increasing income through market development and product diversification, reducing operational costs through efficiency and negotiation with suppliers, and improving production efficiency with better health and nutrition management. In addition, wiser investment management and increasing business scale to achieve economies of scale can also help accelerate returns.

Table 6. Assumptions for mud crab culture with one, two, and three feeding times daily in RAS apartment boxes during 30 days of rearing (12 production cycles/year).

Business Assumptions	**Feeding (times/day)**		
	1	**2**	**3**
Initial number of crabs (crab)	300	300	300
Initial crab weight (g)	135	135	135
Initial biomass (kg)	40.50	40.50	40.50
Number of workers (persons)	1	1	1
Working hours (hours)	141	171	201
Survival rate of soka and non-soka (%)	60	73.33	66.67
Survival rate of non-soka (%)	53.33	56.66	50.00
Survival rate of soka (%)	6.67	16.67	16.67
Mortality (%)	40	27	33
Weight increase of soft-shelled crabs (%)	46.31	62.56	64.38
Weight increase of non-soft-shelled crabs (%)	5.48	4.20	4.13
Weight increase of dead crabs (%)	1.37	1.58	7.69
Price of soft-shelled crabs (IDR/kg)	150,000	150,000	150,000
Price of non-soft-shelled crabs (IDR/kg)	120,000	120,000	120,000
Price of dead crab (IDR/kg)	70,000	70,000	70,000

Table 7. Analysis of mud crab culture using one, two, and three feeding times daily in RAS apartment boxes.

Parameter	**Feeding Frequency (times/day)**		
	1	**2**	**3**
Investment cost (IDR)	70,760,400	70,760,400	70,760,400
Fixed Cost (IDR)	14,714,433	14,714,433	14,714,433
Variable Cost (IDR)	41,887,080	43,467,120	44,715,360
Total Cost (IDR)	56,601,513	58,181,553	59,429,793
Revenue (IDR)	53,715,100	63,523,561	62,430,398
Profit (IDR)	-2,886,413	5,342,008	3,000,605
R/C ratio	0.5	1.09	1.05
Payback period (years)	-	13.25	23.58
BEP price (IDR)	66,823,294	46,604,139	51,855,824
BEP softshell crab unit (kg)	59	97	111
BEP non-softshell crab unit (kg)	340	211	210
BEP dead crab (kg)	245	98	143
Cost of Goods Sold (IDR/kg)	1,193,416	441,774	446,255
Cost of Goods Sold (IDR/kg)	207,038	202,735	344,095
Cost of dead crabs (IDR/kg)	287,225	436,493	234,867

4 Discussion

Feeding frequency is one of the most critical factors affecting production efficiency, feed utilization, and labor management in aquaculture. In this research, feeding mud crabs (*Scylla serrata*) once, twice, and three times per day did not significantly affect survival rate, growth parameters, or feed conversion ratio (FCR). However, feeding twice daily yielded the best combination of biological performance, feed utilization efficiency, and profitability. This indicates that optimizing feeding schedules can enhance overall culture performance without increasing feed costs or labor requirements unnecessarily. The number of workers is related to working hours in a business.

Labor efficiency is a concept that describes how efficiently labor is used to achieve production goals. Feeding frequencies of one, two, and three times daily can affect labor efficiency. Optimising technician work schedules appropriately and efficiently can reduce workloads and increase productivity. Optimising technician work schedules and mapping work processes aims to increase productivity and effectiveness in mud crab culture. A structured work schedule and a clear division of tasks and responsibilities allow for more efficient use of work time. The process flow for each work stage must be clear to guide technicians in minimising errors in cultivating activities. Work time spent by labor covers various stages of the culture process, from cleaning the crab apartment and filter, acclimatisation, initial sampling, feeding, maintenance, daily quality checks, to harvesting and post-harvest. Optimising work time can be achieved by increasing efficiency at each stage of the culture process, thereby increasing the productivity of soft-shelled crab culture using the apartment system. Workforce skills are crucial because they can help workers perform better and produce higher-quality output. Aquaculture activities require high skill levels and are closely related to work experience. Due to their complexity, closed-cage crab culture systems require highly skilled workers [8].

The survival rate results obtained during the study were influenced by several factors, such as the mud crabs' water quality, feed, age, and adaptability. This is also supported by research conducted by Fatihah *et al.* [9] that factors such as molting, water quality, and shelter influence the survival rate of crabs. Mud crabs were raised for 30 days using an apartment system. The apartment system has the advantage of reducing cannibalism in mud crabs, thus affecting the survival rate of mud crabs [10]. Growth is influenced by two factors: internal and external. Internal factors include heredity, sex, age, disease resistance, and the crab's ability to utilise food. External factors include food, physical and chemical conditions of the water, food quantity and quality, and movement space. One external factor that significantly impacts growth is food availability [11]. The growth of carapace length and width is related to the molting process in mud crabs. This is because the number of crabs that experience molting is minimal and the crabs used for the study were adult crabs with an average weight of 140.3 ± 2.83 g. Mud crabs (*Scylla serrata*) have a discontinuous growth pattern, namely rapid growth only occurs during molting, and growth becomes very slow after molting. Suryani *et al.* [12] stated that molting rarely occurs in adult crabs weighing above 70 g.

Molting is unavoidable in crustaceans, including mud crabs. Internal and external factors influence the molting process in mud crabs. Internal factors include age, size, and sex, while external factors include light, temperature, and food availability. One factor that causes mud crab mortality is failure in the molting process. The molting process in mud crabs requires sufficient energy from consumed food and suitable environmental conditions. The molting process in mud crabs involves several stages: premolt, ecdysis, molting, postmolt, and intermolt. The critical phase of the molting process is the ecdysis stage. At this stage, mud crabs shed their old exoskeletons and increase their water absorption. This process requires significant energy and an optimal environment, making mud crabs susceptible to molting failure [13].

Feed acts as an energy provider that drives growth. Efforts are made to provide sufficient feed for the optimal growth of mud crabs. Feed availability significantly impacts crab growth and survival. Feed consumption can be defined as the amount of food consumed by crabs during culture. Factors influencing feed consumption include the size or stage of the mud crab, appetite, feed nutritional quality, feeding frequency, environmental conditions, mortality rate during culture, and feed availability. Underfeeding can inhibit crab growth and molting, while overfeeding can cause water pollution due to the accumulation of leftover feed at the bottom of the water [11]. The feed conversion ratio (FCR) is the ability of mud crabs to convert feed into meat. The lower the FCR, the better the feed quality. The FCR reflects how efficiently the consumed feed is converted into biomass in the crab's body. A higher FCR indicates low feed quality and poor feed utilisation efficiency [14]. The FCRs of mud crabs obtained during maintenance with feeding once, twice, and three times daily were 3.49 ± 0.83, 3.00 ± 0.33, and 3.56 ± 0.73, respectively. The high FCR value occurs because the growth of mud crabs is not linear and only occurs during molting. Feed utilisation efficiency is an organism's capacity to utilise feed for optimal metabolism and growth [13]. Uneaten feed will rot, affecting the quality of the rearing environment and disrupting the crab's appetite. Rotting feed releases ammonia (NH_3), which can disrupt the survival of mud crabs. Stress factors in crabs can also reduce feed utilisation efficiency, resulting in limited energy for survival and growth and increasing the risk of mortality. Feed quality can be assessed by the feed utilisation efficiency value, which states that the lower the feed utilisation efficiency value, the lower the quality of the feed consumed [15].

The relationship between feeding efficiency and growth can also be linked to molting behavior. Since *S. serrata* exhibits discontinuous growth through molting, feed management directly influences the energy available during the premolt stage. Feeding twice per day appears sufficient to sustain metabolic demands while maintaining stable water quality, reducing stress that can disrupt molting success. This finding supports Suryani et al. [12], who emphasized that molting frequency and success rate are highly sensitive to feeding management and environmental stability. The crab will feed after 6 hours, and its body will harden in less than 24 hours. Mud crabs can molt 20 – 25 times during their lifetime. Each molt increases the size of the mud crab along with its weight [11]. Feeding too frequently can limit the digestive tract's capacity to digest food, thus affecting the rate of gastric emptying. If the feed and energy from the previous feed are not used up before new feed is provided, most of the feed is uneaten and is excreted as faeces.

Mud crab culture using the apartment system employs a closed system using recirculating aquaculture.Compared to previous studies, this research uniquely integrates biological performance, labor efficiency, and financial analysis within a controlled RAS apartment system. Previous research primarily focused on growth and FCR in open or pond systems [4; 14]. By applying a recirculating and compartmentalized system, this research demonstrates that feeding management not only affects feed utilization but also determines economic feasibility under intensive rearing conditions. The RAS apartment system offers stable water quality, reduced cannibalism, and efficient space utilization, making it a promising model for sustainable soft-shell crab production.

Filters are a crucial element in the recirculation system. The physical filters used include synthetic cotton, sand filters, biofoam, and Japanese filter mats (japmat). Dacron filters out water-soluble particles. Sand filters function to filter non-aqueous materials such as algae or other types of algae found in the maintenance water. Biofoam is a foam that filters coarse dirt. Japanese filter mats (japmat) filter dirt so that it is not carried back to the maintenance water container. The chemical filter used is UV (ultraviolet) light. UV light kills bacteria, so the water is sterile. The biological filters used are bioballs, Crushed coral, and kaldness. Bioballs function as a habitat for bacteria and filter dirt. Crushed coral functions to stabilise water pH and decompose ammonia (NH_3). Kaldness functions as a nitrification medium.

The best temperature for mud crab culture is 29 °C, because this temperature impacts feed conversion, survival, physiological response, and specific growth rate of mud crabs [4]. During culture, DO values range from 6.2 to 7.8 mg/L. This value indicates that dissolved oxygen during the culture period is in the optimum range of ≥ 5 mg/L [6]. The range of salinity values for mud crabs during the 30-day culture period ranges from 25 to 33 g/L. This range is within the optimal range for mud crab growth. This is supported by research conducted by Hastuti [7], which found that the best salinity for mud crabs is 25 g/L, because this salinity has the best impact on the growth and survival of mud crabs. The range of nitrate values during the maintenance period is 0.253 – 1.886 mg/L. This value is within the normal range of <10 mg/L [6]. The range of nitrite values during the maintenance period is 0.045 – 1.723 mg/L. This value is within the normal range of <10 mg/L [6]. Total ammonia in the mud crab culture environment is the main result of nitrogen waste from aquatic organisms [4]. The range of TAN values during the maintenance period is 0.01 - 0.47 mg/L. This value is within the safe limit of <3 mg/L [6]. The range of alkalinity values during the rearing period was 54.67 – 97.33 mg/L.

In this research, feeding twice per day produced the highest profit (IDR 5,342,008) with an R/C ratio of 1.09, indicating feasible business performance. More frequent feeding (three times daily) increased labor time and operational costs without proportionate gains in growth, resulting in lower profit margins. Labor productivity and feeding management are major determinants of profitability in mud crab aquaculture. Optimizing labor schedules through efficient feeding routines can thus enhance both technical and economic outcomes. The business is profitable and feasible if the R/C ratio is >1. Conversely, if the R/C ratio is = 1, the business is considered to have reached break-even, meaning revenue and costs are equal, with no profit or loss. If the R/C ratio is less than one, the business is experiencing a loss.

5 Conclusion

Feeding frequency did not significantly affect the survival rate, growth rate, or molting performance of Scylla serrata cultured in the RAS apartment system. However, feeding twice per day resulted in the most efficient combination of biological and economic outcomes. This treatment produced a feed conversion ratio (FCR) of 3.00 ± 0.33, a labor efficiency of 1.08%, and the highest profit of IDR 5,342,008 per production cycle. In contrast, once-daily feeding led to reduced profitability, while three-times-daily feeding increased labor costs without improving growth performance. These results demonstrate that optimizing feeding frequency to twice daily enhances operational efficiency, reduces labor workload, and improves the economic feasibility of mud crab fattening in a recirculating aquaculture system.

Acknowledgements

The authors would like to express their sincere gratitude to Fisheries and Marine Observation Station (IFMOS) Ancol Jakarta facilities, IPB University for their generous support and collaboration in this research project. The provision of essential research facilities, including mud crab seeds, seawater, and feed, has greatly contributed to the successful implementation of this study. Their valuable assistance and commitment to supporting scientific advancement in aquaculture are deeply appreciated.

Fundings

The authors thank the Directorate of Research and Innovation of IPB University for funding this research through the 2023 IPB Action Research Program.

Data availability statement

The data supporting the findings of this study are available from the corresponding author upon reasonable request.

References

1. Ministry Marine And Fishery. 2024. Data volume production mud crabs 2021-2022 in Indonesia [internet]. [downloaded 2024 April 7]. Available from https://satudata.kkp.go.id/
2. D. Mutamimah, Y. Wahyudin, Mangrove crab (*Scylla* spp.*)* enlargement techniques in Pangpang Bay, Banyuwangi Regency. *Insologi: Journal of Science and Technology.* **2(1)**, 244 – 249 (2023) https://doi.org/10.55123/insologi.v2i1.1703
3. A. Adila, Septifitri, M. Ali, Fattening of mangrove crab (*Scylla serrata)* with different feed. *Journal Sciences Fishery and Cultivation Waters.* **15(2)**, 86 – 94 (2020) https://doi.org/10.31851/jipbp.v15i2.5086
4. Y.-P. Hastuti, R. Affandi, R. Millaty, W. Nurussalam, S. Tridesianti, Temperature best to increase the growth and survival of crab seeds *Scylla serrata* mangrove in a recirculating system. *Journal of Tropical Oceanography and Science.* **11(2)**, 311 – 322 (2019) https://doi.org/10.29244/jitkt.v11i2.22727
5. E. Tiurlan, A. Djunaidi, E. Supriyantini, Reproduction aspects of mud crab (*Scylla* sp.) in Kendal waters, Central Java. *Journal of Tropical Marine Science*. **2(1)**, 29 – 36 (2019). https://doi.org/10.33019/jour.trop.mar.sci.v2i1.911
6. Food and Agriculture Organization of the United Nations. *Mud Crab Aquaculture: A Practical Manual*. Rome (IT): FAO. (2011)
7. Y.-P. Hastuti, R. Affandi, M.-D Safrina, K. Faturrohman, W. Nurussalam, Optimum salinity for growth of mangrove crab *Scylla serrata* seed in recirculation systems. *Journal Aquaculture Indonesia.* **14(1)**, 50 – 57 (2015) https://doi.org/10.19027/jai.14.50-57
8. C.-P. Tavares, U. –A. T. Silva, L. A. Pereira, A. Ostrensky, Systems and techniques used in the culture of soft-shell swimming crabs. *Reviews in Aquaculture*. **10**(4), 913-923 (2017) https://doi.org/10.1111/raq.12207
9. S.-J. Fatihah, H.-T. Julin, C.-A. Chen, Survival, growth, and molting frequency of mud crab *Scylla tranquebarica* juveniles at different shelter conditions. *AACL Bioflux*. **10**(6), 1581-1589 (2017) https://bioflux.com.ro/docs/2017.1581-1589.pdf
10. D. Mirera, P.-O. Moksnes 2015. Comparative performance of wild juvenile mud crab (*Scylla* sp.) in different culture systems in East Africa: effect of shelter, crab size and stocking density. *Aquaculture International*. **23**, 155 – 173. (2015) https://doi.org/10.1007/s10499
11. A. Hanif, S. Herlina, Different percentages of trash fish feeding to growth crab mangrove (*Scylla* spp). *Journal of Tropical Animal Science*. **10**(1), 1-5. (2021) https://unkripjournal.com/index.php/JIHT/article/view/183/178
12. N.D.P.I. Suryani, P.G.S. Julyantor, A. P. W. K. Dewi 2018. Carapace length and specific growth rate of mangrove crabs (*Scylla serrata*) fed different types of food in the crab village ecotourism area, Bali. *Journal of Marine and Aquatic Science* **4**(1), 38 – 46 (2018). https://doi.org/10.24843/jmas.2018.v4.i01.38-46
13. L. O. M. Harisud, E. Bidayani, A. F. Sharif, Growth performance and survival of mud crabs (*Scylla serrata*) with a combination of golden snail and trash fish feed. *Journal*

of Tropical Marine Science **2**(2), 43 – 50 (2019) https://doi.org/10.33019/jour.trop.mar.sci.v2i2.1378

14. S. Rahmadiah, S. Y. Lumbessy, D. P. Sustainable DP, The effect of adding fermented mulberry leaves (*Morus* spp.) in feed formulation to stimulate molting of mud crab (*Scylla serrata*). *Journal of Aquaculture Science* **8**(1), 1 – 13 (2023) https://doi.org/10.31093/joas.v8i1.201
15. Muhlis, T. Budiardi, I. Effendi, Y. Hadiroseyani, Crab production performance mangrove, *Scylla tranquebarica* at different water levels and container sizes. *Aquaculture Media* **16**(2), 79 – 86 (2021) http://dx.doi.org/10.15578/ma.16.2.2021.79-86

BIO Web of Conferences **221**, 01005 (2026) https://doi.org/10.1051/bioconf/202622101005
7th EMBRIO International Symposium (EIS 2025)

Establishing quality standards for Kohaku Koi Fish (*Cyprinus rubrofuscus*): integration of truss morphometrics, red color area, and pigmentation intensity

Agung Luthfi Fauzan[1], *Dinar Tri* Soelistyowati[2*], *Alimuddin*[2], *Agus Oman* Sudrajat[2] and *Irzal* Effendi[2]

[1]Doctoral Study Program of Aquaculture Science, Department of Aquaculture, Faculty of Fisheries and Marine Science, IPB University, Bogor, Indonesia
[2]Department of Aquaculture, Faculty of Fisheries and Marine Science, IPB University, Bogor, Indonesia

Abstract. This study aims to quantify the quality standards for Kohaku koi fish (*Cyprinus rubrofuscus*) by integrating truss morphometric measurements, red color coverage area, and pigmentation intensity analysis. Twenty-four juvenile Kohaku specimens were selected from the results of selective crossbreeding between Nidan female parents and Sandan Kohaku male parents. Morphometric analysis measured 21 truss morphometric characters of koi fish, and digital image analysis using ImageJ software measured the area of coverage and intensity of red color. Principal component analysis and hierarchical clustering were employed to classify phenotypic patterns of individual fish characteristics. Results demonstrated significant morphometric variation with coefficients ranging from 10.76% to 27.91% within the population. Red color coverage area varied extensively from 4,200 to 442,851 pixels, while intensity values ranged from 85.2 to 160.0 on a 0-255 scale. Kohaku specimens with larger red areas show higher color intensity, wider tail fins, and greater body depth. Size distribution analysis revealed 41.7% large, 37.5% medium, and 20.8% small fish, demonstrating successful selective breeding outcomes. This integrated quantitative approach provides a scientific foundation for standardized quality evaluation, replacing subjective assessment methods and supporting data-driven breeding programs to enhance the aesthetic and economic value of Kohaku koi in commercial production.

1 Introduction

Kohaku koi fish (*Cyprinus rubrofuscus*) are one of the koi fish species that have significant economic value. The beauty of Kohaku koi fish lies in their red backs with white bodies. The color of Kohaku koi fish has not yet been genetically deciphered, but it is thought to be due to complex color inheritance in koi fish [1]. The pigmentation mechanism of koi fish has been successfully elucidated in recent molecular genetic research through the identification of mRNA associated with color patterns and pigmentation in the skin [2;3]. This research has successfully complemented conventional breeding processes by providing molecular insights into the genetic basis of color inheritance in koi fish strains. Color quality and brightness are key factors determining the selling price of koi fish. The low quality of koi fish color remains

* Corresponding author: dinar@apps.ipb.ac.id

a major challenge for breeders, while the demand for high-quality koi fish continues to rise annually. A common issue in a single koi fish farming cycle is the low percentage of koi fish fry with superior quality that will be used for the next growth phase, which affects the selling value.

According to [4], body color is an indicator of beauty in ornamental fish, especially in koi fish, where it is one of the main determinants of aesthetic value and attractiveness. The brighter the color of the fish, the more attractive it is and the higher its selling price. Recent research by [5] has measured this relationship, showing that digital image analysis can objectively measure color quality parameters that are directly correlated with market value and competitive success. Factors influencing the color quality of koi fish include genetics, diet, and environment. Genetic factors influence the inheritance of color patterns in koi fish, for example, in Kohaku, the inheritance of red color patterns on the body. Diet and environmental factors influence the brightness and intensity of red color in Kohaku. Examples of environmental factors influencing the intensity of Kohaku color include water quality, temperature, and sunlight. Recent targeted metabolomics studies have revealed that seasonal variations significantly influence skin color plasticity in ornamental koi fish, providing new insights into environmental regulation of pigments [3].

In addition to color quality, the body shape of koi fish is also a consideration in determining the quality standards for good koi fish. However, to date, there are no standardized criteria for determining such quality. Therefore, information regarding morphological variations and their relationship with color quality is a fundamental aspect that requires further research. The traditional reliance on subjective assessments by expert judges is increasingly being questioned by the need for objective and repeatable quality assessment methods. Recent developments in digital image processing and computer-assisted analysis have opened up new avenues for standardized koi assessment, overcoming the high subjectivity and low efficiency of manual classification methods [2]. Therefore, information regarding morphological variations and their relationship with color quality is a fundamental aspect that requires further research.

According to [6,3], morphometric studies are one way to classify fish populations, in addition to being used for identification. Morphometric characteristics in population classification are more influenced by genetic factors. Morphometric characteristics have a genetic basis, but environmental components can modify the expression of these characteristics during larval development, which can also affect their offspring. Contemporary morphometric analyses in aquaculture have evolved beyond traditional measurements to incorporate advanced statistical techniques such as geometric morphometrics and multivariate analysis, providing more comprehensive characterization of phenotypic variation [7;8]. The benefit of this research is to produce a standard formula for determining the quality of superior Kohaku koi fish through the analysis of morphometric diversity, coverage area, and red color intensity. The integration of truss morphometric analysis with digital color quantification represents a paradigm shift from subjective to objective quality assessment in the ornamental fish industry. Recent studies have successfully applied similar integrated approaches to various fish species, demonstrating the effectiveness of combining morphological and color parameters for quality standardization [6].

Furthermore, the development of deep learning-based classification systems, such as KRS-Net, has shown promising results in automated koi variety recognition, achieving high accuracy even for varieties with high morphological similarity [2]. These technological advances support the urgent need for standardized, data-driven approaches to koi quality evaluation. The economic significance of establishing robust quality standards extends beyond individual breeding programs. The global ornamental fish market, valued at billions of dollars, increasingly demands transparent and objective quality metrics that can support fair trade practices and consumer confidence.

The purpose of this study was to determine the characterization of morphometric diversity, coverage area, and red color intensity, and their relationship with the color quality of Kohaku koi fish in order to determine the standards for high-quality Kohaku koi fish.

2 Materials and methods

2.1 Sample collection

Twenty-four Kohaku strain koi fish were obtained from the Omah Koi Farm Indonesia breeding facility in Banyuwangi, East Java. Fish were selected based on the clear phenotypic expression of the characteristics of the Kohaku strain, namely, a red color pattern with a white body color base. All test fish were juvenile fish with a total length of 15-20 cm to ensure a consistent comparison of developmental stages of the red color pattern and intensity. Fish were maintained under standardized conditions with water temperature of 25±28°C, pH of 7.0±0.5, and dissolved oxygen levels above 6 mg/L prior to analysis.

2.2 Truss morphometric analysis

Morphometric measurements of koi fish were analyzed using the truss morphometric method [9], which has been applied to koi fish. The purpose of truss morphometric measurement of koi fish is to quantitatively evaluate the body shape and morphological variation of fish. Morphometric analysis aims to control production quality and contest standards, broodstock selection, quality improvement, population classification, and species or strain identification. The Kohaku koi fish measured were 6 months old, with a total of 24 fish. Morphometric measurements were made by directly measuring the length of certain body parts using a ruler. Fish were positioned on a laminated millimeter block paper with the left side facing forward and measured by following the morphometric characteristics guide. The morphometric parameters measured can be seen in Fig 1.

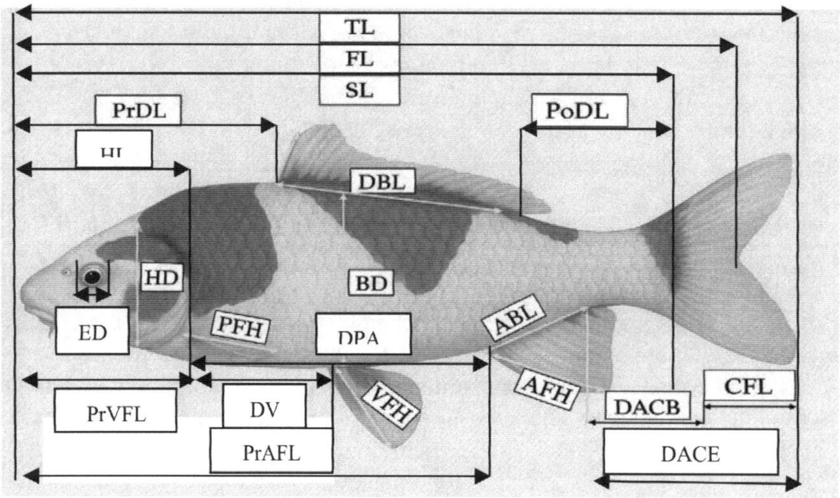

Fig. 1. Measurement of morphometric characteristics of Kohaku koi fish. TL (total length), FL (fork length, SL (standard length), HL (head length), PrDL (pre-dorsal length), PoDL (post-dorsal length), DBL (dorsal fin base length), CFL (caudal fin length), DACB (distance from anal fin to caudal base), AFH (anal fin height), ABL (anal fin base length), VFH (ventral fin height), BD (body depth), PFH (pectoral fin height), HD (head depth), PrVFL (pre-ventral fin length, PrAFL (pre-anal fin length), ED (eye diameter), DVP (distance between ventral and pectoral

fin, DPA (distance between pectoral and anal fin), DACE (distance from anal fin to caudal fin end).

2.3 Quantification of red color rea

Red color quantification was performed using digital image analysis techniques adapted from a recent study on koi color assessment [10,11]. High-resolution digital photographs were taken using standard photographic protocols under controlled lighting conditions to minimize color variations due to environmental factors.

Images were processed using ImageJ software equipped with a color analysis plugin. In ImageJ, an RGB (Red, Green, Blue) image is a type of color image that consists of three channels: red, green, and blue. Each channel has an 8-bit bit depth, which means each pixel has a color intensity value for each channel. In ImageJ, RGB images are used to represent true colors, as seen in actual specimens [11,12].

Two key color metrics were calculated: (1) red color coverage area expressed as a percentage of the total body area, and (2) average red color intensity measured on a standard scale. The calculation of red color coverage area excluded the head region in the anterior part of the operculum to focus on the body pattern, which is the main evaluation criterion in Kohaku assessment.

2.4 Pigmentation intensity scoring

The intensity of the red color is evaluated using a quantitative scoring system based on the intensity values of pixels extracted from digital images. The intensity scoring system uses a 0-255 scale corresponding to the digital pixel values, with higher values indicating a more intense red color.

Regional analysis was performed to assess the variation of red color intensity in the red color pattern in the body area. The standard body region divisions are: anterior dorsal, middle dorsal, posterior dorsal, and lateral regions. This regional approach allows the identification of pigmentation distribution patterns that may correlate with quality assessment.

2.5 Data analysis

All statistical analyses were performed using SPSS version 22.0. Data on red color coverage area, pigmentation intensity, truss morphometrics, and phenotypic characteristics were analyzed using Analysis of Variance (ANOVA) to ensure comprehensive evaluation.

3 Result and discussion

3.1 Morphometric characteristics of Kohaku Koi Fish

This study focuses on quantitative research related to the standardization of Kohaku koi fish (*Cyprinus rubrofuscus*) quality through the integration of truss morphometric analysis, red color coverage area, and red color intensity. A total of 24 juvenile Kohaku koi fish from the selective breeding of Kohaku koi fish collected from Omah Koi Farm Indonesia were analyzed using digital measurements with ImageJ software to quantify the red color coverage area and intensity. Meanwhile, morphometric measurements were taken using a ruler with a millimeter grid paper base to obtain 21 morphometric truss characteristics of Kohaku koi fish.

The objective parameters established are supported by recent research, which confirms the importance of morphological, genetic, and environmental aspects, as well as the aesthetic superiority of Kohaku koi fish, which are the stars of every international contest and market [11,12]. The development of digital-based measurement methods and the integration of multivariate statistical analysis are the latest trends in the field of quantitative genetics [11,13,14]. The results of the morphometric truss measurements of Kohaku koi fish are presented in Table 1 as follows.

Table 1. Morphometric characteristics (cm) of Kohaku koi fish.

Parameter	Description	N	Mean ± SD	Min	Max	CV (%)
TL	Total Length	24	21.39 ± 3.37	15.50	28.00	15.73
FL	Fork Length	24	18.78 ± 3.46	11.50	25.00	18.40
SL	Standard Length	24	17.25 ± 2.81	12.00	22.20	16.31
PrDL	Pre-Dorsal Length	24	8.27 ± 1.33	6.00	11.00	16.10
HL	Head Length	24	5.03 ± 1.14	3.50	9.50	22.74
DBL	Dorsal Fin Base Length	24	6.35 ± 1.18	4.50	8.00	18.54
PoDL	Post-Dorsal Length	24	3.13 ± 0.75	2.00	5.00	23.95
ED	Eye Diameter	24	1.06 ± 0.11	0.80	1.20	10.76
HD	Head Depth	24	3.83 ± 0.64	2.60	5.00	16.74
PFH	Pectoral Fin Height	24	3.02 ± 0.65	2.00	4.50	21.70
BD	Body Depth	24	5.60 ± 0.83	4.00	7.00	14.74
DPA	Dorsal to Pelvic Distance	24	8.45 ± 1.11	6.00	10.00	13.12
ABL	Anal Fin Base Length	24	1.68 ± 0.47	1.00	2.50	27.91
DACB	Distance from Anal to Caudal Base	24	2.91 ± 0.68	1.50	4.50	23.41
CFL	Caudal Fin Length	24	4.70 ± 0.72	3.30	6.00	15.30
PrVFL	Pre-Ventral Fin Length	24	8.80 ± 1.33	6.20	11.00	15.13
DVP	Distance from Dorsal to Ventral Pelvic	24	4.07 ± 0.66	3.00	5.30	16.27
PFH	Pectoral Fin Height	24	2.93 ± 0.48	2.00	4.00	16.48
AFH	Anal Fin Height	24	2.84 ± 0.51	2.00	4.00	17.79
DACE	Distance from Anal to Caudal End	24	7.34 ± 1.45	3.50	9.70	19.76
PrAFL	Pre-Anal Fin Length	24	13.14 ± 2.06	8.50	16.60	15.65

Notes: N = number of samples; SD = standard deviation; CV = coefficient of variation; Min = minimum value; Max = maximum value. All measurements are in cm unless otherwise stated.

Table 1 presents comprehensive morphometric data from 24 Kohaku koi test fish, measuring 21 different morphometric body characteristics that are important for phenotypic characterization. Morphometric parameters exhibit significant phenotypic variation within the studied population, with coefficients of variation ranging from 10.76% (eye diameter) to 27.91% (anal fin base length). The average total length measurement of 21.39 ± 3.37 cm indicates a uniform size class suitable for comparative analysis. The relatively high coefficient of variation for certain parameters, particularly anal fin base length (27.91%) and post-dorsal length (23.95%), indicates significant morphological diversity within the Kohaku strain. The standard length measurement (17.25 ± 2.81 cm) showed a strong correlation with total length, confirming the reliability of morphometric assessment. The body depth measurement (5.60 ± 0.83 cm) indicated a robust body conformation, characteristic of high-quality koi specimens. Morphometric data indicate that head length accounts for approximately 23.5% of standard length, consistent with the proportional relationships established in cyprinid fish.

Based on the research data, several key parameters were identified that can be used to characterize high-quality Kohaku koi fish, namely truss morphometry, red color coverage area, red color intensity, and the correlation between morphometry and pigmentation.

Morphometric measurements are obtained by measuring 21 standard truss distances used as geometric projections of the body. Statistical analysis shows that the average total length is 21.39±3.37 cm, body height, tail base width, and other parameters are closely related to the evaluation criteria used in koi contests. Measurement of red color coverage on the body using ImageJ resulted in an average coverage area of 141,426±112,847 pixels, with a high range of coverage between individuals. The extensive red color coverage on the back and sides is highly valued in determining the quality of superior Kohaku koi. The average red color intensity in ImageJ ranges from 113.8±19.2 (scale 0–255), with a minimum value of 85.2 and a maximum value of 160.0. The combination of high red intensity with a large red area is a characteristic of superior Kohaku. High-quality Kohaku koi have a thicker body (high body depth), wider tail fins, and a large red coverage area and intensity, which aligns with conventional evaluation criteria in international koi contests.

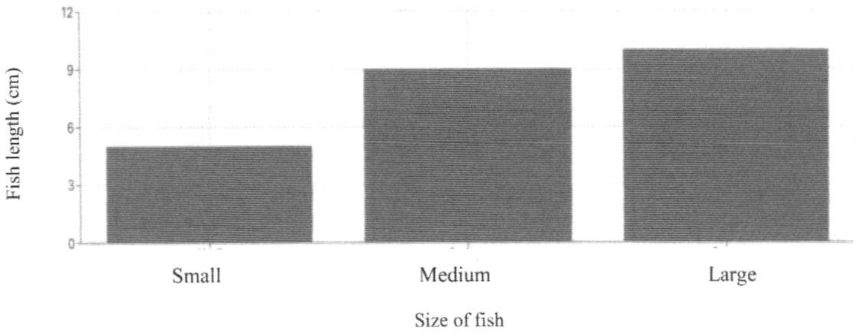

Fig. 2. Size distribution of Kohaku Koi Fish based on Total Length (TL).

Fig 2. shows the size distribution of Kohaku koi fish based on total length (TL) classification in the form of a bar chart. The size classification is divided into three categories: small (<18 cm), medium (18-23 cm), and large (>23 cm). Based on measurements from the 24 fish analyzed, the size distribution is uneven, with medium and large sizes dominating. There are 5 small fish (20.8% of the total sample), 9 medium fish (37.5% of the total sample), and 10 large fish (41.7% of the total sample). This distribution indicates a balanced population with a tendency toward medium to large sizes, suggesting the success of the selection program.

The dominance of medium (37.5%) and large (41.7%) size categories indicates that the F1 population resulting from the selective breeding of Kohaku koi carp has achieved consistent and uniform growth rates. Recent research by [10] shows that F1 Koi carp offspring from controlled breeding exhibit significant and consistent growth during the rearing period, both in terms of length and body weight. The average length of F1 after 60 days of rearing increases rapidly, and the distribution of individual sizes becomes relatively uniform, as a result of selecting superior parents and a controlled breeding environment. [6] Confirm that selection techniques and intensive breeding programs successfully produce Kohaku Koi populations with uniform growth performance and body quality, as reflected in proportional size distribution. Other studies confirm that under optimal nutritional management and density conditions, the F1 population resulting from crossbreeding will dominate in the medium-to-large size class [10].

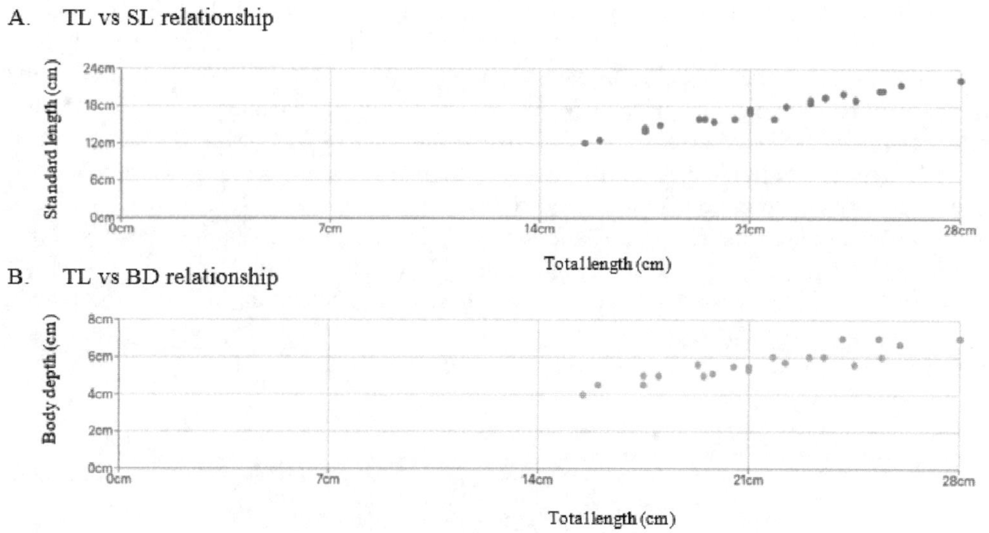

Fig. 3. Relationship of main morphometric parameters.

Fig 3. shows the relationship between the main morphometric parameters. Panel A shows the relationship between total length (TL) and standard length (SL). The analysis results indicate a positive correlation between TL and SL with a strong linear correlation coefficient ($r > 0.90$) and proportional consistency. A proportional relationship is observed between total length (TL) and body depth (BD), with an optimal TL:BD ratio of (0.26 ± 0.03). The presence of morphometric variation indicates genetic diversity within the Kohaku F1 koi fish population. This data supports a selection program for Kohaku koi fish based on morphometric characteristics.

Research related to the growth performance of juvenile koi fish by [11] shows that the parameters of total length and standard length increase over time with a highly significant statistical correlation, supporting that both are closely related as measures of growth. [6] In their evaluation of Kohaku koi fish quality using image analysis and truss morphometry noted that total length and standard length have a very high correlation ($r > 0.90$), forming the basis for morphometric classification of koi fish for breeding selection and quality.

Analysis of size distribution and morphometric correlations indicates that the Kohaku F1 fish population possesses advantageous characteristics for advanced breeding programs. The size distribution, dominated by medium-to-large categories, combined with strong and proportional morphometric correlations, suggests genetic and phenotypic stability that supports consistent production quality. The strong linear relationship between key morphometric parameters provides a scientific basis for standardizing evaluations and enables the development of more efficient selection protocols. These findings contribute to the development of a data-driven breeding program that can enhance the quality and economic value of Indonesian Kohaku koi fish in the international market.

3.2 Red Coloration Parameters

The results of measuring the coverage area and red color intensity of Kohaku koi fish are presented in Table 2 as follows:

Table 2. Red coloration coverage area and intensity measurements in fish specimens (n=24).

Specimen ID	Coverage Area (pixels)	Red Channel Intensity Values				CV (%)
		Mean ± SD	Min	Max	Range	
Fish_01	147.873	122.4 ± 27.3	0	205	205	22.3
Fish_02	442.851	137.0 ± 51.2	14	255	241	37.4
Fish_03	103.776	91.8 ± 20.5	0	194	194	22.3
Fish_04	177.150	123.4 ± 32.4	0	221	221	26.3
Fish_05	147.582	109.5 ± 27.3	3	199	196	24.9
Fish_06	141.933	118.9 ± 31.5	13	205	192	26.5
Fish_07	119.508	117.6 ± 36.2	31	216	185	30.8
Fish_08	218.625	133.7 ± 46.0	0	243	243	34.4
Fish_09	193.874	128.5 ± 45.8	15	251	236	35.6
Fish_10	12.618	94.6 ± 27.5	0	158	158	29.1
Fish_11	100.107	121.8 ± 27.6	39	221	182	22.7
Fish_12	385.209	153.2 ± 44.2	4	251	247	28.8
Fish_13	223.454	114.6 ± 35.9	22	212	190	31.3
Fish_14	61.842	109.6 ± 23.1	49	208	159	21.1
Fish_15	144.207	99.7 ± 27.2	0	204	204	27.3
Fish_16	25.299	85.2 ± 12.5	56	140	84	14.7
Fish_17	77.182	125.5 ± 35.6	15	213	198	28.4
Fish_18	179.539	101.1 ± 31.6	13	192	179	31.3
Fish_19	89.599	93.8 ± 25.3	14	186	172	27.0
Fish_20	4.200	104.4 ± 20.6	49	159	110	19.7
Fish_21	66.006	114.9 ± 20.4	50	200	150	17.8
Fish_22	100.028	110.0 ± 26.3	23	188	165	23.9
Fish_23	150.279	92.2 ± 22.5	26	146	120	24.4
Fish_24	128.877	160.0 ± 41.2	37	255	218	25.8

CV = Coefficient of Variation. Values represent mean ± standard deviation. Specimen Fish_24 showed the highest red intensity value (highlighted). Coverage area measured in pixels; intensity values range from 0-255 (8-bit scale).

Table 2 presents a quantitative analysis of the characteristics of red pigments in all test fish. Measurements of red color coverage area showed remarkable variation, ranging from 4,200 pixels (Fish_20) to 442,851 pixels (Fish_02), indicating a 105-fold difference in red pigment area. The average red intensity values showed moderate variation (85.2 to 160.0), with Fish_24 exhibiting the highest intensity (160.0 ± 41.2). The coefficient of variation for the average red intensity was 26.2%, indicating a consistent measurement protocol that captures natural biological variation. Fish_12 exhibited exceptional characteristics with a large coverage area (385,209 pixels) and high intensity (153.2 ± 44.2), representing optimal phenotypic expression. The statistical distribution shows a positive skew for coverage area (1.42), indicating that most test fish have relatively small red color coverage areas, with a few individuals exhibiting extensive red color coverage areas.

Table 3. Descriptive statistics for red coloration parameters.

Parameter	Mean ± SD	Median	Min - Max	95% CI	Skewness	Kurtosis
Coverage Area (pixels)	141.426 ± 112.847	133.405	4.200 – 442.851	93.822 – 189.030	1.42	1.89
Red Intensity	113.8 ± 19.2	114.8	85.2 - 160.0	105.6 - 122.0	0.89	0.76
Standard Deviation	30.7 ± 10.4	27.5	12.5 - 51.2	26.3 - 35.1	0.52	-0.31
Coefficient of Variation (%)	26.2 ± 5.8	26.0	14.7 - 37.4	23.7 - 28.7	0.11	-0.89

CI = Confidence Interval. Statistical analysis performed using standard descriptive methods. Normal distribution assessed using the Shapiro-Wilk test (p > 0.05 for all parameters except coverage area).

Table 3 summarizes the statistical properties of color parameters, providing important metrics for establishing quality standards. The 95% confidence interval for red intensity (105.6–122.0) sets the initial threshold for high-quality test fish. The distribution of red color coverage area shows a significant positive skewness (1.42) and high kurtosis (1.89), indicating a normal distribution. The median coverage area (133,405 pixels) differs significantly from the mean (141,426 pixels), confirming a skewed distribution. The standard deviation value indicates a normal distribution (skewness = 0.52), confirming the consistency of individual variation measurements. These statistical parameters form the basis for developing standard evaluation criteria in the commercial evaluation system for high-quality Kohaku koi.

4. Conclusion

The quantitative quality standards for Kohaku koi fish (*Cyprinus rubrofuscus*) through analysis of body morphometry, red color coverage area, and red color intensity in 24 juvenile Kohaku koi fish revealed phenotypic variation with coefficients of variation ranging from 10.76% to 27.91% in size class distribution of 41.7% large fish, 37.5% medium fish, and 20.8% small fish. A strong positive correlation was observed superior Kohaku fish, which exhibited a larger red color coverage area (up to 442,851 pixels) with higher red color intensity (maximum 160.0 on a 0-255 scale), wider tail fins, and greater body depth. The results of this study provide a measurable scientific basis as a selection criterion in koi breeding programs to replace subjective assessment methods and valid data-based determination of aesthetic value in the koi industry.

Acknowledgement

The authors gratefully acknowledge research funding through the 2024 Doctoral Dissertation Program scheme by the Directorate General of Higher Education, Research and Technology,

Ministry of Education, Culture, Research and Technology, Number: 027/E5/PG.02.00.PL/2024. We are very grateful to Omah Koi Farm Indonesia, Banyuwangi, East Java, Indonesia, for permitting us to use the place for this study.

References

1. Y. Zheng, L. Deng, Q. Lin, W. Xu, F. Wang, J. Li, KRS-Net: a classification approach based on deep learning for koi with high similarity. *Biology* **11**, 1727 (2022).
2. M. Luo, Z. Liu, H. Yang, *et al.*, Identification and characterization of skin color microRNAs in koi carp (*Cyprinus carpio* L.) by Illumina sequencing. *BMC Genomics* **19**, 779 (2018).
3. L. Liu, X. Wang, R. Zhang, H. Li, H. Zhu, Targeted metabolomics revealed the seasonal plasticity of skin color and pigment metabolites in ornamental koi carp. *Ecotoxicol. Environ. Saf.* **281**, 116595 (2024).
4. N.B.P. Utomo, O. Carman, N. Fitriyati, Pengaruh penambahan *Spirulina platensis* dengan kadar berbeda pada pakan terhadap tingkat intensitas warna merah pada ikan koi kohaku (*Cyprinus carpio* L.). *J. Akuakultur Indones.* **5**, 1–4 (2006).
5. M.A. Domasevich, H. Hasegawa, T. Yamazaki, Quality evaluation of Kohaku koi (*Cyprinus rubrofuscus*) using image analysis. *Fishes* **7**, 158 (2022).
6. C.C. Lau, S.A. Mohd Nor, M.P. Tan, *et al.*, Pigmentation enhancement techniques during ornamental fish production. *Rev. Fish Biol. Fish.* **33**, 1027–1048 (2023).
7. S.N.D. Oliveira, R.P. Ribeiro, C.A.L.D. Oliveira, *et al.*, Multivariate analysis using morphometric and ultrasound information for selection of tilapia (*Oreochromis niloticus*) breeders. *Rev. Bras. Zootec.* **48**, e20170179 (2019).
8. L.A. Jawad, J.M. Abed, A.L. Ibáñez, A. Al-Faisal, Morphometric and meristic characters of cultured and wild carp (*Cyprinus carpio* L.) populations from southern Iraq. *Fish. Aquat. Life* **30**, 95–103 (2022).
9. I. Shilman, Suparmin, F. Irmawan, Penentuan kualitas ikan arwana super red (*Scleropages formosus*) dengan metode morfometrik, meristik, dan skala warna. *J. Ruaya* **9**, 2541–3155 (2021).
10. I. Afini, D. Elfidasari, T. Kadarini, S.Z. Musthofa, Analisis morfometrik dan meristik hasil persilangan ikan pelangi Boesemani (*Melanotaenia boesemani*) dan ikan pelangi merah abnormal (*Glossolepis incisus*). *Life Sci.* **3** (2) (2014).
11. D.S. Budi, L. Davinci, A.P. Ardiyan, M. Herjayanto, Natural occurrence of hybrid koi and mahseer in artificial ponds: evidence of the dangers of releasing non-native fish into the wild. *J. Aquac. Sci.* **10**, 29–32 (2025).
12. J. Bartley, Y. Nakamura, H. Li, Quantitative evaluation of color traits in ornamental koi (*Cyprinus rubrofuscus*) using digital image analysis. *Aquac. Res.* **53**, 2801–2812 (2022).
13. Y. Mundayana, *et al.*, Evaluation of the growth performance of juvenile koi carp F1 MHC. *Egypt. J. Aquat. Biol. Fish.* **28**, 1611–1627 (2024).
14. A. Al-Aqlu, The effect of water depth on the growth and survival of koi fish seeds (*Cyprinus carpio*). *Al-Aqlu: J. Math. Eng. Sci.* **2**, 17–23 (2023).

Bioassay on horseshoe crab embryo survivability in cultures contaminated by *Bacillus thuringiensis*

Rasha Ahmed Moqbel Ghaleb [1], *Nurul* Huda Abdul Kadir[2], *Hassan* I. Sheikh[3], and *Bryan* Raveen Nelson[1*]

[1]Institut Biodiversiti Tropika dan Pembangunan Lestari, Universiti Malaysia Terengganu, 21030 Kuala Nerus, Terengganu, Malaysia.
[2]Fakulti Sains dan Sekitaran Marin, Universiti Malaysia Terengganu, 21030 Kuala Nerus, Terengganu, Malaysia.
[3]Fakulti Perikanan Dan Sains Akuakultur, Universiti Malaysia Terengganu, 21030 Kuala Nerus, Terengganu, Malaysia.

Abstract. Horseshoe crab embryos are highly sensitive to temperature and microbial stress. In this bioassay, embryos were tested under ambient, constant temperature (32 °C), dark conditions and another set with *Bacillus thuringiensis*. Six developmental stages were recorded for ambient and constant temperature settings. Survivability (100%) was maximum only in ambient temperature, it declined for 32 ºC to 76 % and it was 70.8 % together with darkness. Constant 32 °C accelerated hatching to 37 days in comparison to 40–42 days for ambient group. Exposure to *B. thuringiensis* arrested the embryogenesis in first embryonic moult stage and survivability was only 32.8%. There was abnormal coloration from degraded eggshell and these embryos were reduced in size and weight. Later, the advanced embryo had impaired limb buds. Infection rates were lowest at ambient temperature with darkness (20%) and highest under 32 °C with darkness (67.2%). Basically, thermal manipulation was stressful to horseshoe crab embryos because the hatching success rate and immunity was lowered. The presence of *B. thuringiensis* affirmed that stress increased vulnerability to pathogens. Understanding the presence of combined stressors is critical for biodiversity conservation and aquaculture management of *C. rotundicauda*.

1 Introduction

The marine environment is no longer sustainable following the expansion of human activities like industrialisation, tourism, marine traffic, and global temperature change. Furthermore, climate warming has altered the dynamics of coastal areas worldwide, leading to an increasing prevalence of vector-borne incidences in tropical regions. Typically, microorganisms develop mutual or symbiotic relationships with the ecosystem, co-supporting a network comprising bacteria, viruses, fungi, and protists. This relationship can shift for metazoans, as viruses may function as both symbionts and pathogens [1].

For example, a crayfish (*Cherax quadricarinatus*) may carry Decapod penstylhamaparvovirus 1 and show no symptoms, but upon exposure, shrimp (*Penaeus*

* Corresponding author: bryan.nelson@umt.edu.my

vannamei) could become infected with infectious hypodermal and haematopoietic necrosis. While a total of 102 agents comprising of 25 viruses, 33 bacteria, 23 protists, and 21 metazoans were responsible for notable diseases across plants, fish, corals, molluscs, crustaceans, echinoderms, turtles, and mammals, many of these diseases occur during early life stages or failed to gain widespread notice at the population level [2].

Assays were developed to understand the interactions of biological agents, from which a total of 182 half-effective concentrations involving *Penicillium* sp. and *Pseudialteromonas issachenkonii* were derived [3]. A separate assay revealed 12 bacterial extracts (*Bacillus* sp. and *Enterococcus* sp.) capable of 90 % to 100 % mosquito larvae mortality. Aside from product identification and industrial applications, bioassays are also used to evaluate growth performance, such as the exposure of *P. vannamei* post-larvae to *Vibro* sp. carrying VpPirAB genes. Although 100 % mortality was observed, histopathology revealed sloughing and detachment of the hepatopancreas, confirming the disease aetiology in contaminated hatcheries. A *Bacillus tequielensis* strain demonstrated 65 % efficacy for insecticidal activity against the mangrove pest insect (*Spodoptera frugiperda*), and these findings indicated its potential for endophytic remediation in mangrove plots. Importantly, however, not all Bacillus strains are beneficial for promoting the safe development and growth of a target species.

Meanwhile, horseshoe crabs are marine arthropods that endure a lengthy embryogenesis lasting 40 days, require 9–10 years for sexual maturity, and have a perceived lifespan of 15–18 years [4]. As an oviparous species, their embryos are left buried in the sand, enduring brief episodes of tidal exchange or otherwise experiencing periodic thermal shocks that are thought to expedite the embryonic moulting process. Essentially, with four embryonic moults, horseshoe crab embryos may pass through sensitive periods that can be identified using experimental assays. For instance, stock enhancement programmes encourage release strategies in optimal habitats, such as intertidal areas or near mangrove roots, specifically targeting low tide (when water temperatures are cooler) and preferably using individuals at the instar stages [5]. Feeding experiments indicated a blood meal containing 40 % iso-nitrogen and 14 MJ kg^{-1} as an optimal replacement for conventional feed containing 15 % to 25 % protein.

Early experiments outlined the development of horseshoe crabs, recording 16 moult cycles in 9 years for the male and 17 moult cycles over 10 years for the female of Limulus polyphemus, whereas *Tachypleus tridentatus* showed 15 and 16 moult cycles over 13 and 14 years, respectively, for the same sexual orders. While such experiments involved lengthy hatchery rearing, a simple tolerance experiment indicated optimum development below 60 ppt salinities and when water was maintained at 25–40 °C [6]. Horseshoe crab embryos were also tested with cadmium, zinc, copper, mercury, and tributyltin, with results indicating 1% embryonic malformations, which included impaired walking legs. Generally, these embryos showed higher stress protein production and accumulated these hazards in their tissue; moreover, the adults could sequester toxic compounds to their eggs, indicating a risk to predators higher up in the food chain. The growth and performance of horseshoe crabs were also evaluated through assays that used raw seawater exchange to supply organic matter, supplemented by live *Artemia* nauplii. Simple experiments such as this revealed crucial support is needed to sustain sensitive early life stages up to 10 weeks, until hatchlings reach the 3rd instar stage. These assays also provided crucial information on developmental defects as well as growth performance. Fundamentally, horseshoe crabs have persisted for 450 million years with negligible evolution, which accounts for their ancient template that has allowed them to continue existing despite climatic events and coastal developments during the course of modernisation.

Given the limited experiments investigating post-fertilisation development and embryogenesis until hatching, much has been discovered concerning cellular changes and morphological variability. This includes the confirmed presence of the amoebocyte during stage 18 of embryonic development, which accounts for its increased tolerance against abrupt environmental changes, and the formation of the book gill, similar to the scorpion book lungs [7]. A subsequent experiment evaluated ex situ spawning ground sediment characteristics in relation to nesting preferences, egg release, and spawning frequencies. Spawning grounds with 0.25–1.00 mm sediments were most preferred by 45 female crabs, which released 27,266–30,859 eggs in 3 hours [18]. However, biotic assays on this arthropod remain limited. Although the arthropod can be fed with fish, squid, clams, shrimp, fish diets, and small crabs, it is susceptible to infection by algae, fungi, cyanobacteria, Gram-negative bacteria, and parasites, and currently has few known therapeutic options aside from removal, pharmacokinetics, oxytetracycline administration, and oral dosing.

With regard to microbial infections, embryogenesis could be threatened by bacteria like *Shewanella putrefaciens*, *Bacillus cereus*, *Corynebacterium* sp., and *Enterococcus faecalis*, as well as fungi such as *Aspergillus* sp., *Aspergillus niger*, *Peniclillium* sp., *Gliocladium* sp., which cause the horseshoe crab eggshells to appear red, grey, or black. Eggs with abnormal colourations were also reported in another spawning ground, namely Johor Lama, where hatched larvae exhibited impaired or distorted body plans. The same outbreak occurred at Pantai Balok, the most studied spawning ground for horseshoe crabs in 2022–2024 surveys. Notably, this *Bacillus thuringiensis* possesses the cry1 toxin gene and is capable of digesting the proteins on the horseshoe crab egg envelope [8]. It is confirmed that *B. thuringiensis* is Gram-positive, typically discovered in soil, and may have a host-specific aetiology. A recent study indicated that this bacterium is likely to experience notable habitat gains globally by 2090 due to its use as a biopesticide. The entomopathogenic capability of *B. thuringiensis* and the presence of cry homologous toxins have made it an ideal insecticide. Crucially, however, experiments on growth performance under biotic influences have not yet been attempted. Therefore, this study aims to evaluate the embryogenesis of the horseshoe crab, *Carcinoscorpius rotundicauda*, in different culture assays alongside the administration of *B. thuringiensis* to examine host-specific developmental responses.

2 Methodology

2.1 Sampling

Fieldwork was conducted at Pantai Balok (N 03° 56.009′; E 103° 22.584′), a recognised spawning ground for horseshoe crabs on the east coast of Peninsular Malaysia. Sampling trips were scheduled in accordance with the monthly lunar cycle, when large numbers of horseshoe crabs migrate into shallow waters for amplexus during the high-tide breeding period. Potential nests were located by following the tracks left by adults on the mid- to high-beach face. A total of 15 nests were excavated at low tide between August 2022 and March 2024. Of the identified disturbances, nine nests actually contained eggs grouped in a single clutch, whilst 38 were determined to be false crawls (or abandoned nesting attempts). Freshly laid eggs were separated from the clutch using a sieve, counted, and subsequently transported to the laboratory for culture experiments.

2.2 Temperature assay

The collected eggs were subsequently raised at the hatchery. This process commenced with a two-day acclimatisation period before the initiation of the assay, which employed two distinct temperature regimes. The first regime used ambient conditions, where the temperature was permitted to fluctuate between daytime and night-time intervals (28–31 °C). The second regime was maintained at a constant 32 °C using heaters (Jager Aquarium Regler-Heizer N12350 220V-240V). Another set of eggs was maintained under the same regimes but were blacked-out to provide a consistently dark environment (Fig. 1a). For all experiments, each tank contained 250 eggs, and the water was retained at 30 ‰ salinity to accurately mimic the surface water of Pantai Balok. Finally, a portable aquarium air pump (AP-005 220-240V50/60H2) was used to oxygenate the water.

2.3 Bioassay

The soil bacteria, *Bacillus thuringiensis*, were initially cultured in nutrient broth for 24 hours. Following this, the culture underwent a ten-fold serial dilution, and the colony-forming units (CFU) were counted to quantify the inoculum. A dosage of 10^8 CFU in 1 ml of nutrient broth, consistent with the LC 50 recommendation, was subsequently introduced to the relevant conditions prepared in the temperature assay. Consequently, the experimental set-up comprised eight distinct conditions: ambient (A); ambient with bacterial exposure (A+B); ambient with permanent darkness (A+D); ambient with permanent darkness and bacterial exposure (A+D+B); constant 32 °C (C); constant 32 °C with bacterial exposure (C+B); constant 32 °C with permanent darkness (C+D); and constant 32 °C with permanent darkness and bacterial exposure (C+D+B). Conditions without bacterial addition served as control groups, whilst those with bacterial inoculation were designated as treatment groups (Fig. 1a). The embryos introduced into the assay were monitored over the maximum (more precise wording) incubation period for their development, which was 42 days post-fertilisation. We monitored the embryos daily until 100% hatching was achieved in the control groups. Recorded observations included developmental stages and duration, survivability, weight (using an analytical balance), length and width (measured with an electronic digital vernier calliper), and vulnerability, alongside any abnormal colour variation (often red colour) in the embryos.

2.4 Histology examination

Microscopic slides of all samples were prepared using the standard Haematoxylin and Eosin (H&E) protocol. Initially, samples were longitudinally sectioned (4 μm thickness) with a microtome, placed in labelled polypropylene cassettes, and fixed in 10% neutral buffered formalin (pH 7) for 24 hours. This was succeeded by a 12-hour tissue processing cycle that involved graded ethanol dehydration (70 %, 90 %, 95 %, and 100 %) and xylene clearing, carried out prior to hot paraffin wax infiltration (75 °C). Subsequently, the specimens were embedded in paraffin blocks, trimmed, and sectioned into 4 μm ribbons. These ribbons were then expanded on a 40 °C water bath, mounted onto Mayer's albumin-coated slides, fished out, and dried for 24 hours at 35 °C. The H&E stain protocol was then modified to integrate additional steps for Sudan IV and Toluidine-O staining. This primary protocol modification included integrating three washing repeats between the alcohol and the haematoxylin staining steps. Following this, further modifications involved adjusting the soak timings to three times for Sudan IV and six times dipping for Toluidine-O to accurately distinguish proteins, haemocyanin, amoebocytes, and lipids. The final stages involved a ten-second xylene dip and air drying for three minutes. Mounting was achieved using DePeX (polystyrene in xylene), followed by the placement of a cover slip. Lastly, the specimens were inspected under an advanced compound microscope (Nikon eclipse 80i, USA) using 5X, 10X, and 20X magnifications (Fig. 1b & c).

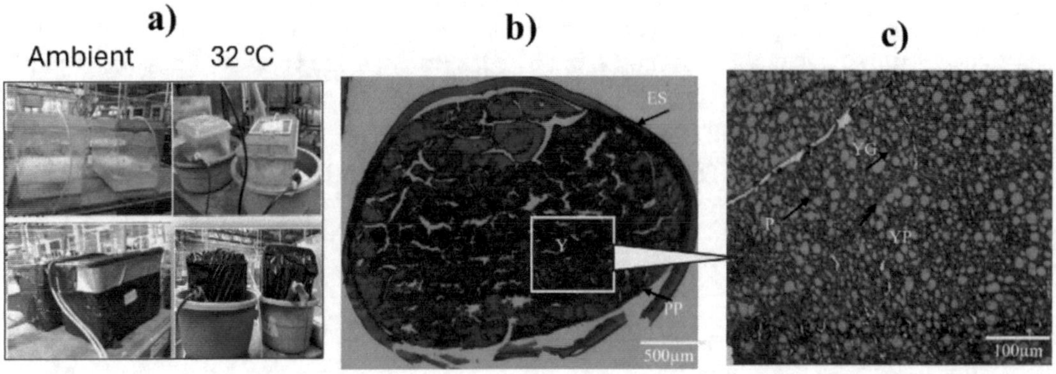

Fig. 1. A general view on the experimental setup and the organism tested. a) Bioassay setup for *Carcinoscorpius rotundicauda* embryos for all conditions including light, dark and with bacteria. b) longitudinal section of *C. rotundicauda* egg used for histological observations, c) higher magnification of egg yolk. ES= eggshell (outer chitinous layer), pp= periplasm (inner layer), Y= yolk, YG= yolk granules, YP= yolk platelets, P=pits.

2.5 Statistical analysis

Egg weight, volume, and surface area during embryogenesis were analysed using a two-way repeated measures ANOVA in GraphPad Prism software (version 5). A significance level of $P \leq 0.05$ was considered indicative of a substantial difference between means. It should be noted that the egg volume and egg surface area were calculated based on the embryo length and width measurements, employing specific equations as provided below. In addition, the developmental rate, survivability, and vulnerability were expressed as percentages and subsequently plotted using Microsoft Excel.

$$\text{Egg volume} = (0.6057 - 0.0018B)\, LB^2 \tag{1}$$

$$\text{Egg surface area} = (3.155 - 0.0136L + 0.0115B)\, LB \tag{2}$$

*L= length, B=width

3 Results

3.1 Bioassay findings

3.1.1 Life stages of embryogenesis

The developmental stages of *C. rotundicauda* eggs in control groups were divided into six categories represent in early developmental stage (ED), the first embryonic moult (M1; stage 18), the second embryonic moult (M2; stage 19), the third embryonic moult (M3; stage 20), the emergence of trilobite larvae (H; stage 21), and the first instar larvae (L) after 24 hours of hatching. During the ED stage (days 1–7, 1st week), the eggs appear olive green in colour, with cleavage nuclei emerging at the egg surface which contribute to the initiation of germ disk formation. Embryos of this stage were generally spherical and later transition into oval-like form as development progresses. In stage M1, after 7–14 days (by 2nd week), the first embryonic moult was characterized by cracking of the tough opaque chorion and the embryo appear enclosed by both outer opaque chorion and inner non-expanded extraembryonic shell. In 15–21 days or 3rd week, the embryos reached stage 19 or M2 where the second moult is characterised by chorion shedding that exposes the non-expanded extraembryonic shell (transparent membrane).

At this stage, a small undifferentiated embryo became clearly visible through the transparent membrane, with rudimentary limb buds that were initially rounded and later developed into pointed tips. This stage is regarded as the critical point for determining successful development, marking the transition from opaque to semi-translucent eggshell, swelling from perivitelline fluid and visibility of recognizable embryo. From days 22–28 (4th week or stage 20), the third embryonic moult (M3) occurs. The inner egg membrane was expanded tremendously where a moving embryo was visible. At this stage, the egg size increases by 120% due to cellular differentiation and uptake of water, constant larval movement was observed, and the prosoma and opisthosoma became distinct. By days 29–35 (5th week; stage 21), a well-developed embryo with distinguishable morphologically was ready to hatch as the trilobite larva. Finally, in days 36–42 (week 6), the trilobite larva moults into the first instar stage where the larvae resemble the adult form (Fig. 2a & Table 1). However, embryos exposed to *B. thuringiensis* failed to complete development under all tested conditions. Most of the embryos had their development arrested in the first embryonic moult, displaying abnormal coloration (often red) and gradually becoming smaller and more rounded in shape. Only a small proportion (~2%) advanced to the second embryonic moult (M2) and remained in this stage at the end of the experiment (day-35; Fig. 2b & Table 2).

3.1.2 Embryogenesis during the bioassay

In the ambient condition, early development (ED) was reached slightly earlier, on day 13. The first moult stage or M1 also lasted 13 days, beginning on day 8 and ending on day 20. Generally, M2 lasted 14 days, from day 14 to day 27, M3 lasted for 13 days, from day 21 to day 34, hatching (H) was shorter, lasting 13 days, from day 28 to day 40, and first instar (L) was reared for 6 days, beginning on day 35 and ending on day 40. Embryogenesis occurred until termination of experiment on day-40 (Fig. 3a). However, with the addition of *B. thuringiensis*, the ED stage prolonged, up to 30 days. Interestingly, embryos transitioned into the M1 stage within 11 days but unfortunately, they remained without change for 42 days. Only 2% transitioned to the M2 stage on day-35 and remained in this stage until day-42 (Fig. 3b). In dark conditions, the ED stage spanned 14 days although a majority of embryos transitioned into M1 by day-13. The M2 stage occurred from day-15 to day-28 while M3 stage occurred from day 22 to day-35 with both stages lasting 14 days apart. Hatching occurred by day-29 and lasted to day-42 and the hatched trilobite took 7 days from day day-36 to day-42 to completely transition into the first instar (Fig.3c). In dark condition, the embryos exposed to *B. thuringiensis* performed better where the ED stage lasted 28 days. The progression into stage M1 occurred by day-11 and for the remaining 97.2 %, they remained at this stage until day-42 PF. This ~2.8% took 31 days to transition into stage M2 (Fig.3d).

For the constant 32 °C exposure, stage ED lasted for 11 days despite the first transition into stage M1 by day 7. This stage lasted 7 days with the last embryo transition on day 18. Meanwhile, M2 spanned 13 days, from day 12 to day 24, M3 lasted 13 days, from day 18 to day 30, H lasted 18 days, from day 25 to day 42, L was observed for 7 days, between days 31 and 37 (Fig. 2d; Fig. 4a). However, with the addition of *B. thuringiensis*, ED lasted for 34 days. The first transition into stage M1 occurred on day 9 and the embryos remained in this stage for 34 days even until the experiment ended by day-42 (Fig. 4b). For the batch in constant 32 °C and darkness, the ED stage lasted for 11 days while the transition to stage M1 begun on day-7 and continued to day-18. About 13 days was needed for embryos to reach stage M2 which begun on day-12 to day day-24. The first transition to stage M3 occurred on day-18 and lasted to day-30. Egg hatching begun on day 25 and it continued daily until day-42. The first instar was observed by day-31 and all embryos were completely in instar stage by day-37 (Fig. 4c). However, when *B. thuringiensis* was added to this experimental design, the ED stage lasted for 34 days. The first transition to stage M1 occurred on day-34 and the embryos remained in this stage until day-42 (Fig. 4d).

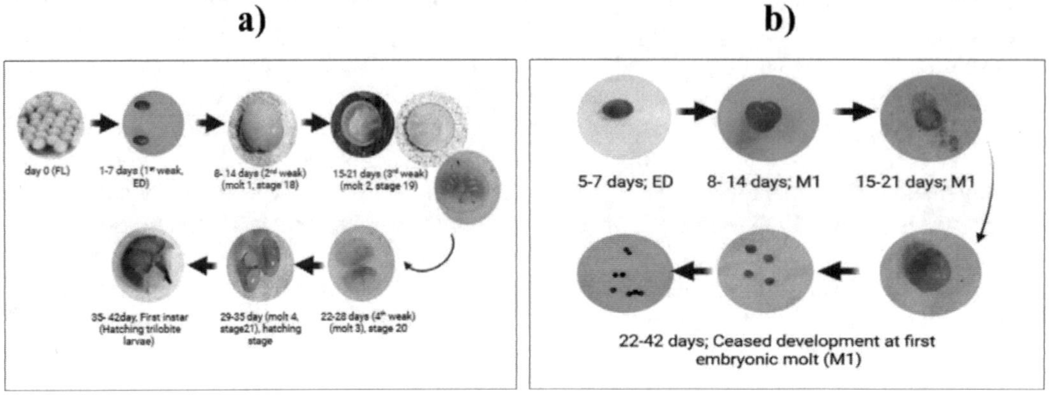

Fig. 2. Visual inspections on the embryos of *Carcinoscorpius rotundicauda*. a) The appearance of developing *C. rotundicauda* embryos, b) The appearance of *C. rotundicauda* embryos after infection. FL= freshly laid eggs, ED= early developmental stage, M1= first embryonic moult, M2= second embryonic moult, M3= third embryonic moult, H= hatching of trilobite larvae, L= first instar stage).

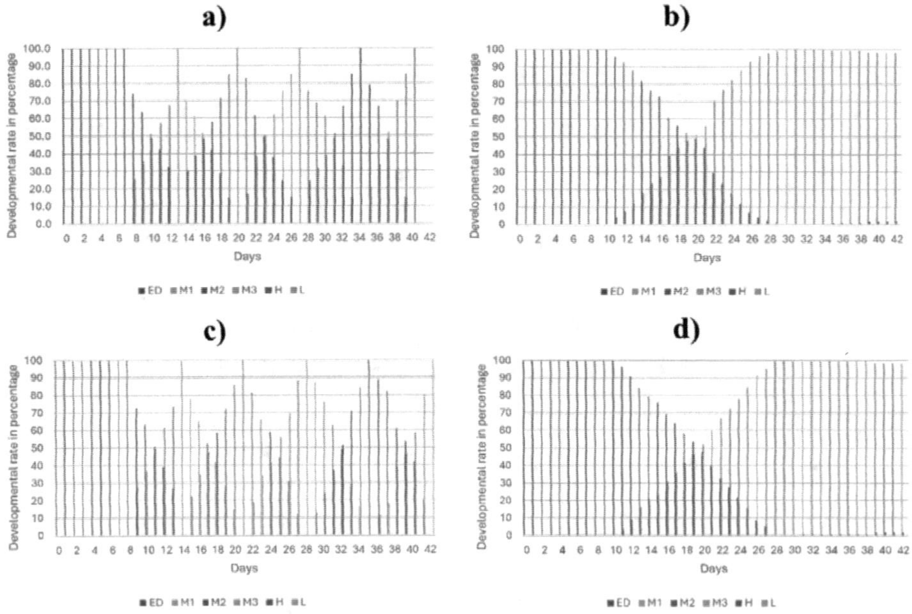

Fig. 3. Development of *Carcinoscorpius rotundicauda* embryos characterised by different embryonic moult stages. a) ambient condition, b) ambient condition with bacteria, c) ambient temperature with darkness, d) and ambient temperature with darkness and bacteria.

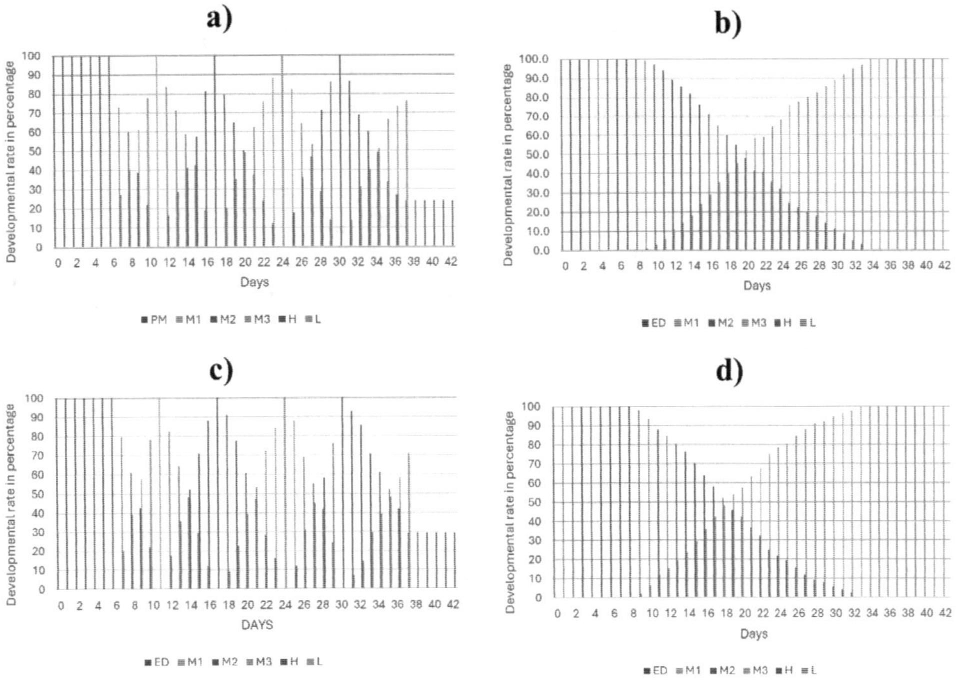

Fig. 4. Development of *Carcinoscorpius rotundicauda* embryos characterised by different embryonic moult stages. a) constant temperature (32 °C), b) constant temperature (32 °C) with bacteria, c) constant temperature (32 °C) with darkness, and d) constant temperature (32 °C) with darkness and bacteria.

3.1.3 Survival of developing embryo in bioassay

Survivability test was conducted to estimate persistence of embryos to continue developing after infection by *B. thuringiensis*. In control groups (ambient temperature, 32°C and darkness with both temperature settings) *C. rotundicauda* embryos maintained 100% survival for 42 days because trilobite larvae hatched out and moulted to the first instar. However, in constant 32 °C, the survivability was 76% because not all embryos transitioned to first. Comparatively at 32 °C and darkness, the survivability reduced to 70.8% due to prolonged hatching; after day-35 (Figure 4.a). For all temperature assays, including darkness, the addition of bacteria reduced the survivability to 70.8% (fluctuating temperature, a), 80% (a+darkness), 40.8% (c or 32°C), and 32.8% (c+darkness; Fig. 5a).

3.1.4 Vulnerability of developing embryo in bioassay

A vulnerability test was conducted to identify the most critical embryonic stages of *C. rotundicauda* for infection during the moult cycle. The assessment was based on the appearance of abnormal coloration on the eggshell, typically red, which was considered indicative of bacterial infection. In all control groups, the vulnerability score remained 0 across all embryonic moults throughout the incubation period (Table 1). In contrast, eggs exposed to *B. thuringiensis* in the treated groups exhibited varying degrees of vulnerability. The highest scores were recorded in regimes with 32 °C (59.2%), together with darkness (67.2%) and followed by ambient temperature (29 %) and together with darkness (29.2%). The vulnerability score indicated the first 14 days of the experiment as most sensitive and despite some embryos developed to stage M2, the inability to surpass this stage by 42 is an indication of ongoing vulnerability that threatened their development (Table 2).

3.1.5 Weight of developing embryo in bioassay

Embryo weight was monitored throughout the incubation period (7, 14, 21-, 28-, 35-, and 42-days) as an indicator of successful *C. rotundicauda* embryogenesis. For embryos maintained in ambient conditions and ambient temperature with darkness, these embryos showed a steady increase in weight up to day-28, reaching peak values of 133.99 mg and 148.32 mg, respectively. Thereafter, a sharp decline was observed at day-35 and day-42, with weights decreasing to 72.05 mg and 65.01 mg. In conditions like ambient and together with darkness, the embryo final weights were 78.42 mg and 72.62 mg in respective order. In 32 °C, maximum weight gain was on day-28 at 81.48 mg and with darkness it was 89.17 mg but by day-35 and day 42, the weight of embryos in the same order became 61.33 mg and 56.24 mg, and 59.83 mg and 53.97 mg, respectively (Fig. 4b). However, the presence of bacteria in this experiment led to uneven distribution of weight across all development stages. For instance, weigh gain was witness by day-14 in the order ambient (a; 41.35 mg), A +darkness (43.53 mg), 32 °C (37.64 mg) and with darkness (36.36 mg). For the same order, the final weights reduced to 28.55 mg, 28.12 mg, 22.63 mg, and 21.72 mg. Embryo weight differed significantly across all days between control and treatment groups ($P < 0.001$) (Fig. 5b).

3.1.6 Egg volume of developing embryo in bioassay

Egg volume was assessed as an additional indicator of successful growth in *C. rotundicauda* embryos, with measurements taken at regular incubation intervals (7, 14, 21-, 28-, 35-, and 42-days PF) based on embryo length and width. In the control groups maintained under both ambient and 32 °C conditions, embryos exhibited a steady increase in egg volume, reaching their maximum by day 42 (403.39 mm^3, 424.65 mm^3, 255.99 mm^3, and 200.60 mm^3) in the order ambient condition, ambient temperature with darkness, constant 32 °C, and constant 32 °C with darkness (Fig. 4c).

In contrast, embryos exposed to bacterial treatment showed only a slight increase in egg volume up to day 14 PF, with consistently lower values than those in the non-bacterial groups. Notably, embryos reared under constant temperature with bacterial exposure displayed slightly lower volumes than those at ambient conditions with bacterial exposure. Their dimensions were 32.61 mm^3, 34.48 mm^3, 22.26 mm^3, and 16.66 mm^3 in the order ambient condition with bacteria, ambient temperature with darkness and bacteria, constant 32 °C with bacteria, and constant 32 °C with darkness and bacteria. Thereafter, egg volume gradually declined toward day 42. Differences were significant on most days between control and treatment groups ($P < 0.001$), with exceptions on day 14 for C vs. C+B and C+D vs. C+D+B (no significance), day 7 (weak significance, $P < 0.05$ for C vs. C+B and C+D vs. C+D+B) as shown in Fig. 5c.

3.1.7 Egg surface area of developing embryo in bioassay

Egg surface area was calculated from egg length and width measurements as an additional indicator of successful growth in *C. rotundicauda* embryos. In the control group, egg surface area was decreased between day-7 and day-14 PF, followed by a gradual decline as development continued on day 15 onward to days 35–42 days. Among the control conditions, the steepest decline occurred at ambient temperature with permanent dark, followed by ambient condition, then 32 °C, and lastly 32 °C with permanent dark. In contrast, eggs exposed to bacterial treatment showed an early decline in surface area between day 7 and day 14 after which seems stabilized across all conditions. Most days showed significant differences between the control and treatment groups ($P < 0.001$), except on day 14 for C+D vs. C+D+B, where no significant difference was detected, and on day 7, where only weak significance was observed ($P < 0.05$ for both C vs. C+B and C+D vs. C+D+B), as illustrated in Fig. 5d.

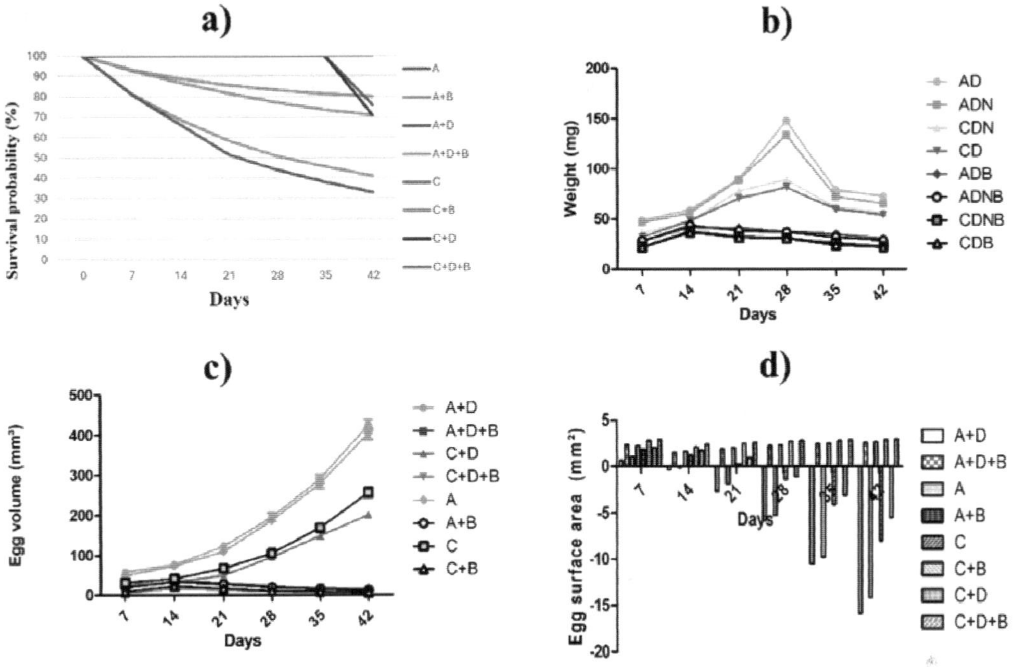

Fig. 5. Associations to *Carcinoscorpius rotundicauda* embryogenesis during the assays. a) Survival of *C. rotundicauda* embryos where n=250 eggs for each group. b) weight of *C. rotundicauda* developing embryo in bioassay, c) egg volume of *C. rotundicauda* developing embryo in bioassay, and d) egg surface area of *C. rotundicauda* developing embryo in bioassay. * A+D= Ambient temperature with permanent dark, C+D= Constant temperature 32 °C with permanent dark, A+D+B= Ambient temperature with permanent dark and bacterial exposure, C+D+B= Constant temperature 32 °C with permanent dark and bacterial exposure, A= Ambient temperature with day-night interval, C= Constant temperature 32 °C with day-night interval, A+B= Ambient temperature with day-night interval and bacterial exposure, and C+B= Constant temperature 32c° with day-night interval and bacterial exposure. * In eggs exposed to *B. thuringiensis*, embryos ceased at M1 persisted during the same period in which control group embryos advanced toward hatching.

Table 1. Developmental stages of *Carcinoscorpius rotundicauda* embryos, egg morphology and the vulnerability score. G= green, M= moult, RO= round, O= oval, T= trilobite, OP= opaque, TR= transparent, + indicates presence of feature, and - indicates absence of feature.

Condition		A+D						A						C+D						C					
Stage		ED	M1	M2	M3	H	L	ED	M1	M2	M3	H	L	ED	M1	M2	M3	H	L	ED	M1	M2	M3	H	L
General features	Colour	G	M	M	M	M	M	G	M	M	M	M	M	G	M	M	M	M	M	G	M	M	M	M	M
	Shape	Ro	O	O	O	O	T	Ro	O	O	O	O	T	Ro	O	O	O	O	T	Ro	O	O	O	O	T

Condition		A+D						A						C+D						C					
Stage		ED	M1	M2	M3	H	L	ED	M1	M2	M3	H	L	ED	M1	M2	M3	H	L	ED	M1	M2	M3	H	L
	Eggshell appearance	OP	OP	TR	TR	TR	-	OP	OP	TR	TR	TR	-	OP	OP	TR	TR	TR	-	OP	OP	TR	TR	TR	-
Specific features	Limb bud formation	-	-	+	+	+	+	-	-	+	+	+	+	-	-	+	+	+	+	-	-	+	+	+	+
	Surface cracking	-	+	-	-	-	-	-	+	-	-	-	-	-	+	-	-	-	-	-	+	-	-	-	-
	Extra-embryonic shell expansion	-	-	+	+	+	-	-	-	+	+	+	-	-	-	+	+	+	-	-	-	+	+	+	-
	Movement	-	-	-	+	+	+	-	-	-	+	+	+	-	-	-	+	+	+	-	-	-	+	+	+
	Segmentation	-	-	-	+	+	+	-	-	-	+	+	+	-	-	-	+	+	+	-	-	-	+	+	+
Vulnerability (%)		0	0	0	0	0	0	0	0	0	0	0	0	0	0	0	0	0	0	0	0	0	0	0	0

Table 2. Development stages of *Carcinoscorpius rotundicauda* embryos, egg morphology and the vulnerability score after the addition of bacteria. R= red, B= black, RO= round, O= oval, OP= opaque, + indicates presence of external feature, and - indicates absence of external feature. *Embryos that ceased at M1 were retained until embryos in control group advanced toward hatching.

Condition		A+D+B						A+B						C+D+B						C+B					
Stage		ED	M1	M2	M3	H	L	ED	M1	M2	M3	H	L	ED	M1	M2	M3	H	L	ED	M1	M2	M3	H	L
General features	Colour	R	R	R	R	R	R-B	R	R	R	R	R	R-B	R	R	R	R	R	R-B	R	R	R	R	R	R-B
	Shape	RO	O	O	O	RO	RO	RO	O	O	O	RO	RO	RO	O	O	O	RO	RO	RO	O	O	O	RO	RO
	Eggshell appearance	OP	OP	OP	OP	OP	OP	OP	OP	OP	OP	OP	OP	OP	OP	OP	OP	OP	OP	OP	OP	OP	OP	OP	OP
Specific features	Limb bud formation	-	-	-	-	-	-	-	-	-	-	-	-	-	-	-	-	-	-	-	-	-	-	-	-
	Surface cracking	-	+	-	-	-	-	-	+	-	-	-	-	-	+	-	-	-	-	-	+	-	-	-	-
	Extra-embryonic shell expansion	-	-	-	-	-	-	-	-	-	-	-	-	-	-	-	-	-	-	-	-	-	-	-	-
	Movement	-	-	-	-	-	-	-	-	-	-	-	-	-	-	-	-	-	-	-	-	-	-	-	-
	Segmentation	-	-	-	-	-	-	-	-	-	-	-	-	-	-	-	-	-	-	-	-	-	-	-	-
Vulnerability (%)		6.8	4.4	3.2	2.4	2	1.2	7.2	6	5.2	4.4	3.6	2.8	18.4	12.8	10.4	7.6	5.2	4.8	18.8	14.8	14.8	7.6	6	5.2

3.2 Histological observations of developing embryo in bioassay

After 7 days into the experiment, there was variability in periplasm thickness and yolk distribution, with pit formations (hemispherical depressions) scattered across the embryonic surface (Fig. 6a). These structures result from interactions between the periplasm and the underlying yolk layer following the cortical reaction of fertilization. At this stage, multiple nuclei were visible. It was caused by successive superficial cleavages within cytoplasmic masses that later developed into blastomeres. These blastomeres, concentrated at the embryonic periphery, contribute to blastula formation and marked the onset of germ disks. The yolk became organised into discrete masses.

After 14 days, in stage M1, the germ disk expands and develops irregular furrows extending across the entire surface (Fig. 6b). These structures initiate segmentation which then corresponds to the later formation of limb buds and future body segments. At this stage, the opaque chorion cracks to expose the non-expanded extra-embryonic shell, within which fibrous bundles are visible. By day 21, embryonic development advances into organogenesis. This is evident with the production of blood cells (amoebocytes) associated with isolated yolk masses positioned at the embryonic periphery or within connective tissue. In later stages, these cells become confined to the haemocoel cavities or the heart. Additional developmental features include the differentiation of the opisthosoma and prosomatic appendages, such as limb buds. The opisthosomal appendages, including the first and second branchial appendages, begin separating into distinct gill lamellae, while other gills remain clustered in the gill book. Furthermore, endoskeletal elements (cartilaginous structures) emerge as solid outgrowths from the ventral wall of the branchial appendages (Fig.6c&d).

In contrast, infected eggs (Fig. 6a–f) initially appeared to follow a similar developmental pattern to normal eggs during the early stage and M1, with yolk organized into distinct central and peripheral masses. However, these eggs simultaneously exhibited clear degradation of the chitinous eggshell, leading to eroding characteristics and pigmented lesions on the shell (exoskeleton). These lesions often referred to as blebs, box burns, or black spots seemed to originate from the embryo's surface and are likely the result of embryonic autolysis. Comparatively, embryos infected by bacteria were indicated with arrested development on day-14 or stage M1 and failed to progress to M2. This was a condition marked by agenesis when compared with normal embryos at the same stage (Fig. 6g).

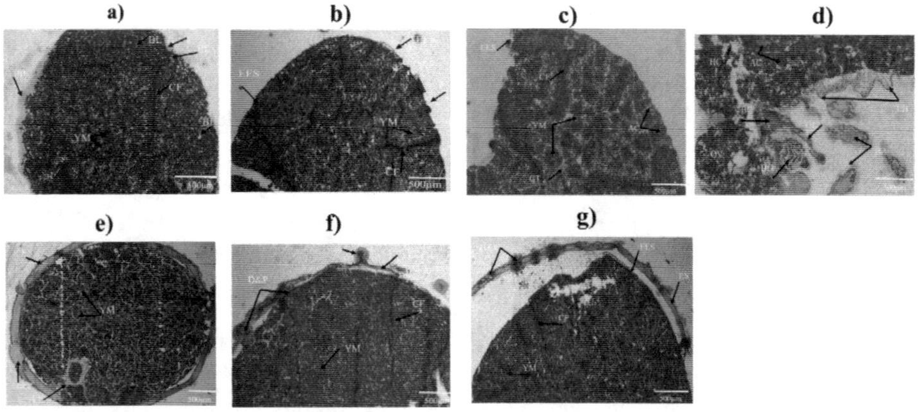

Fig. 6. Longitudinal sectioning of *Carcinoscorpius rotundicauda* embryos after experimentation. a) day 7 (early develop stage), b) day 14 (1st embryonic moult), c) day 21 (2nd embryonic moult dorsal view), d) day 21 (2nd embryonic moult, ventral view), e) after bacterial infection on day 7 (early develop stage), f) day 14 (1st embryonic moult) and g) day 21 PF (ceased develop at 1st embryonic moult or M1). PP= periplasm, P= pits, BL= blastomeres cells, YM= yolk mass, F= furrows, CF= chitinous fibrous bundles, EES= non- extraembryonic shell, CT= connective tissue, GL= gill lamella, BG =book gill, GL= gill lamella, LB= limb bud, BA= branchial appendages, C= cartilage, OV= opisthosomal vent-rum , AC=

amoebocytes, HC= hemocoel cavity, D&P= degradation and pigmentation of eggshell, B= Blebs formation.

Discussion

Carcinoscorpius rotundicauda embryos reared under ambient and ambient-dark conditions successfully completed development, progressing through four embryonic moults and hatching as trilobite larvae with ≈100% success over 40–42 days. This outcome is consistent with other reports of 41–42 days for embryogenesis and near-complete hatching under ambient conditions [9]. Early development was characterised by key milestones like shallow periplasm, uneven yolk, and pit formation marked the cortical reaction and blastoderm initiation within 7 days, followed by germ disk formations in the early blastula stages [10]. By Day 14, the first embryonic moult (M1) was evident through furrows and chorion cracking, correlating with embryo expansion after germ disk completion, yolk mobilisation, and the onset of differentiation for later limb bud formation and body segmentation [11]. By Day 21, organogenesis had produced limb buds, gills, an endoskeleton, and immune cells, consistent with circulatory and immune maturation [7]. In contrast, elevated temperature (32 °C) accelerated development, shortening the time-to-hatching by five days to ~37 days. However, not all embryos adapted well, as hatching success was reduced to 76 %, or 70.8 % in addition with constant darkness. Sustained warming can narrow the thermal safety margin, increase late-stage mortality, and ultimately restrict embryonic development, as observed in *Limulus polyphemus* and other horseshoe crabs with a safety margin generally ≥35 °C [6].

Exposure to *B. thuringiensis* produced a strikingly different outcome, marked by signs of eggshell blebs, erosion, pigmentation anomalies, and autolysis. A vast majority of embryos (98%) failed to progress beyond the M1 stage, suggesting that the bacterium disrupted the perivitelline environment, inducing developmental arrest and autolysis [6]. These pathological effects likely stemmed from *B. thuringiensis* toxins and chitinolytic enzymes that damaged epithelia, altered osmotic balance, and facilitated pathogen invasion [12]. Similar toxin-induced eggshell damage, such as emulsification and electron-dense blebs, has been reported in *Trichostrongylus colubriformis* following *B. thuringiensis* exposure. Since chitinases degrade cuticle and peritrophic membranes, this enzyme acts as a virulence factor, compromising the host defences and promoting successful bacterial establishment in the egg [12].

These 98 % of infected embryos also exhibited red colouration and shrinkage. Developmental bottlenecks are well-documented in horseshoe crabs, where the M1 stage was characterised by high metabolic demands, rupturing of the opaque chorion, and yolk reorganisation. These were observations associated with thinning of the chorion envelope which inevitably increased vulnerability [7, 13]. This vulnerability at the M1 stage may be linked to a weak immune defence, as amoebocytes are largely 'absent'. Robustness increases in progressive stages like M2 (second embryonic moult onward) due to tissue specialisation, as seen in *L. polyphemus* and *T. tridentatus* embryos during M2 (stages 15–21 of the Sekiguchi scale) [13]. The threat is further reduced with the presence of immature amoebocytes during the M3 stage. Therefore, it is not surprising that bacterial (e.g., *Bacillus* spp.) or fungal infections, alongside blebs, erosion, and arrested development, are frequently discovered in horseshoe crab nests [8].

Progressive mortality was clearly observed with *B. thuringiensis* exposure, and was exacerbated when combined with other stressors. Mortality was high at the constant temperature (32 °C; 59.2 %) and worsened under darkness (67.2 %). While temperature and light effects on metabolism and membrane properties may have acted as minor co-factors influencing the rate of embryogenesis in *C. rotundicauda*, it was similar to how light provides cues and temperature triggered the moulting process in *L. polyphemus* and *T. gigas* larvae.

In addition, the exposure to *B. thuringiensis* caused an early and persistent reduction in surface area. This pattern signifies developmental arrest before M2, which prevents organogenesis and cuticle remodelling, leaving embryos pale, undersized, and static. The synergistic effects of constant darkness may be related to the absence of photoperiodic cues that perhaps trigger embryonic moults. Previous studies noted that while *L. polyphemus* embryos showed accelerated development above 30 °C, they also exhibited increased heat shock proteins when salinity was altered. Other research linked temperatures above 32 °C, combined with factors like pH and salinity to inducing stress in embryos.

Stress often triggers organism susceptibility to pathogens. While *C. rotundicauda* embryos (control) showed steady weight and volume gains during the first 28 days, ambient temperature embryos achieved larger dimensions than those at 32 °C. This increase in weight and dimension correlated with a reduction in egg surface area during Days 7–14, confirming 'normal' morphogenesis because the rate of tissue accretion was proportional to yolk depletion. This process alters the egg geometry through compaction and fluid balance, as described for the M2 stage in *L. polyphemus* and *Tachypleus* spp. [14]. Weight gain and size increase align with late embryonic morphogenesis, where yolk depletion supports somatic growth during organogenesis; tissue and organs are relatively denser than yolk material [7, 11]. Embryos typically weigh less between Days 35–42 as the extra-embryonic membrane thins and releases perivitelline fluid to facilitate hatching [7]. Comparatively, only the 2 % of *B. thuringiensis*-exposed embryos successfully moulted to M2, while the remaining 98 % were arrested at M1. All infected embryos gained weight only until ~Day 14 before experiencing a decline due to shrinkage. Under stress, these embryos likely accelerated their metabolism through yolk utilisation, resulting in poor energy balance. Developmental arrest at M1 then prevented the mass and volume gains required for organogenesis, as energy was redirected toward managing stress and immunity. Poor energy reserves and the presence of pathogen toxins caused tissue damage and catabolism, evidenced by egg size shrinking and weight reduction. Natural selection, by forcing yolk depletion at a rate unequal to tissue synthesis, likely limited the development of the 98 % of embryos, mirroring observations during thermal-accelerated development in *L. polyphemus* [11, 15]. The ability of *B. thuringiensis* to digest embryonic envelope proteins (classes H, B1, B2, and 'rest') likely facilitated bacterial invasion, compromising osmotic balance and catalysing yolk reserves through toxin activity. This aligns with red-egg pigmentation, microbial infection, and arrested moulting seen in *T. gigas* embryos. Overall, temperature acted as an abiotic modifier by increasing metabolic demands and reducing early immune capacity, while darkness potentially disrupted the photoperiodic cues regulating moult timing. The prolonged development observed was caused by these co-factors acting together to increase vulnerability.

Conclusion

Under normal conditions, embryos of *Carcinoscorpius rotundicauda* progressed successfully through six defined developmental stages (early developmental stage, M1 to M3 moults, hatching, and the first instar larva), completing embryogenesis within 42 days. This was confirmed by high survival rates (>70%–100%) and "normal" development, evidenced by steady increases in egg volume and weight, decreasing surface area, and characteristic histological transitions like chorion shedding and organogenesis. In contrast, exposure to *Bacillus thuringiensis* severely impaired embryogenesis, causing developmental arrest at the first embryonic moult (M1). This was indicated by abnormal colouration, diminished weight and volume, increased surface area, reduced survivability (as low as 32.8% under 32 °C and darkness), and histological evidence of eggshell degradation and cellular autolysis. Only ~2%

of exposed embryos progressed beyond M1, highlighting the first 14 days as the most critical period of infection. Overall, *B. thuringiensis* disrupts normal morphogenesis and compromises embryo viability across all temperature and light regimes, suggesting that warmer climates may favour pathogens over host survivability. Further research is necessary to investigate the mechanistic pathways of *B. thuringiensis* in inducing this developmental arrest, focusing on immune responses and eggshell integrity, to inform mitigation measures for oviparous arthropods in changing environments.

Acknowledgement

This research was supported by Ocean Park Conservation Foundation Hong Kong, Project Code OT-01-2021.

References

1. Zrnčić, S., Microbial Diseases of Marine Organisms. *J. Mar. Sci. Eng.* **10**(11), 1682 (2022).
2. Byers, J. E, Marine parasites and disease in the era of global climate change. *Annu. Rev. Mar. Sci.* **13**(1), 397-420 (2021).
3. Subbaiyan, R., Ganesan, A., Dhanuskodi, S., Scientific investigation of antifouling activity from biological agents and distribution of marine foulers—Coastal Areas of Tamil Nadu. *Appl. Biochem. Biotechnol.* **196**(3), 1752-1766 (2024).
4. Smith, M. D., & Brockmann, H. J., The evolution and maintenance of sexual size dimorphism in horseshoe crabs: an evaluation of six functional hypotheses. *Anim. Behav.* **96**, 127-139 (2014).
5. Xu, P., Bai, H., Xie, X., Wang, C. C., Huang, X., Wang, X., Zhang, M., Ye, Z., Zhu, J., Zhen, W., Kwan, K. Y., Tri-spine horseshoe crab aquaculture, ranching and stock enhancement: perspectives and challenges. *Front. Mar. Sci.* **8**, 608155 (2021).
6. Ehlinger, G. S., & Tankersley, R. A., Survival and development of horseshoe crab (*Limulus polyphemus*) embryos and larvae in hypersaline conditions. *Biol. Bull.* **206**(2), 87-94 (2004).
7. Farley, R. D, Ultrastructure of book gill development in embryos and first instars of the horseshoe crab *Limulus polyphemus* L. (Chelicerata, Xiphosura). *Front. Zool.* **9**(1), 4 (2012). https://doi.org/10.1186/1742-9994-9-4
8. Moqbel, R. A. G., Nelson, B. R., Sheikh, H. I., Kadir, N. H. A., Sidique, S. N., Kumaran, J. V., Meilana, L., Parimannan, S., Raj Kumar, H.C. Wei, S. S., The risk of *Bacillus thuringiensis* contamination towards survivability of *Carcinoscorpius rotundicauda* eggs. *Available at SSRN 5358409*.
9. Zadeh, S. S., Christianus, A., Saad, C. R., Hajeb, P., Kamarudin, M. S., Comparisons in prosomal width and body weight among early instar stages of Malaysian horseshoe crabs, *Carcinoscorpius rotundicauda* and *Tachypleus gigas* in the laboratory, *Biology and Conservation of Horseshoe Crabs* (pp. 267–274), (Springer US, 2009). https://doi.org/10.1007/978-0-387-89959-6_16
10. Barnum, S. R., Clapper, D. L., Brown, G. G, Modifications of the fertilized egg surface following the cortical reaction in *Limulus polyphemus L.* as viewed with the scanning electron microscope. *Morphology*. **183**(2), 185-198 (1985).

11. Jorgensen, P., Steen, J. A. J., Steen, H., Kirschner, M. W., The mechanism and pattern of yolk consumption provide insight into embryonic nutrition in Xenopus. *Development*, **136**(9), 1539–1548 (2009). https://doi.org/10.1242/dev.032425

12. Martínez-Zavala, S. A., Barboza-Pérez, U. E., Hernández-Guzmán, G., Bideshi, D. K., Barboza-Corona, J. E, Chitinases of *Bacillus thuringiensis*: Phylogeny, Modular Structure, and Applied Potentials. *Front. Microbiol.* **10**, 3032 (2020). https://doi.org/10.3389/fmicb.2019.03032

13. Sekiguchi, K., Biology of Horseshoe Crabs. *Society*, 14(3) (1988).

14. Shuster Jr, C. N., Sekiguchi, K. Growing up takes about ten years and eighteen stages. The American horseshoe crab, 103-132 (2003).

15. Vasquez, M. C., Multiple stressor interactions and effects on embryo development of the American horseshoe crab, *Limulus polyphemus*. University of Florida, (2013).

Enriched Rotifers (*Brachionus plicatilis*) on the growth and survival rate of Barramundi (*Lates calcarifer*) larvae

Maria Goreti Eny Kristiany[1*], *Amyda Suryati* Panjaitan[1], *Nurhaidin*[1], *Ai* Setiadi[1], *DH. Guntur* Prabowo[2], *Aripudin*[2], *Taufik Ramli* Hadi[2]

[1] Aquaculture Study Program, Jakarta Technical University of Fisheries, Ministry of Marine Affairs and Fisheries, Republic of Indonesia, Jakarta, Indonesia
[2] Aquaculture Study Program, Karawang Polytechnic of Marine and Fisheries, Karawang Regency, West Java, Indonesia

Abstract. The availability of high-quality natural feed is one of the key factors in the successful cultivation of barramundi (*Lates calcarifer*) larvae. Rotifers (*Brachionus plicatilis*) are the main feed during the early larval stage; however, their nutritional content needs to be improved through an enrichment process. Rotifers, as natural feed for fish larvae, not only provide protein and other nutrients, but are also considered living capsules capable of transferring micromolecules. Another advantage of rotifers is their very slow movement, making them easy for fish larvae to catch. Several studies on the enrichment of rotifers using specific types of vitamins have been conducted, but they have not yet been applied to specific fish species. This study aimed to analyze the effect of enriched rotifers on the growth and survival of barramundi larvae. The study was conducted from day 2 to day 17. The treatments tested were: A (*Chlorella* sp.+ yeast), B (*Chlorella* sp. + yeast + Ascorbic acid), C (*Chlorella* sp. + yeast + Cobalamin), and D (*Chlorella* sp. + yeast + Ascorbic acid + Cobalamin). The results showed that treatment D produced the highest absolute length growth of 0.95 cm, while treatment B resulted in the highest survival rate of 41.7%.

1 Introduction

Several research studies on barramundi larvae have reported low survival rates. [8] Reported that a survival rate of barramundi larvae only 8.89%. Also reported that mortality barramundi larvae have rates of up to 100% [9]. Fish mortality can occur due to unsuitable low-quality feed and environmental conditions [2]. Therefore, the availability of high-quality and sufficient natural feed is very important for larval rearing.

One type of zooplankton that is commonly used is *Branchionus plicatilis*, a type of rotifer that has been used as feed for various fish and shellfish larvae. Providing rotifers at a density of 10–20 individuals/ml can increase survival rates by up to 72%. The quality and quantity of rotifers are determined by the type of nutrition they receive [10]. Rotifer nutrition can be enhanced through nutrient enrichment. Common types of feed given to rotifers include algae, yeast, vitamins, and enrichment emulsions [7].

The growth of barramundi larvae requires vitamins that serve as supplements for the juveniles [13]. A combination of ascorbic acid and cobalamin is considered ideal. Cobalamin

* coresponding author : kristiany.eny25@gmail.com

can aid in the secretion of digestive enzymes, thereby enhancing fish growth. Cobalamin deficiency can lead to symptoms such as reduced growth and decreased feed intake (anorexia) [5].

The aim of this study is to determine the effect of rotifer enrichment using various enrichment materials on the growth and survival of barramundi larvae.

2 Materials and methods

The barramundi larvae used in this study were two days old (D2), with an initial length and weight of 0.21 ± 0.04 cm and 0.49 ± 0.06 mg, respectively. A total of 300 larvae were used per container, equivalent to 10 larvae per liter. At this age, the larvae had entered the early exogenous phase, which is the stage where they begin actively searching for and consuming food from their environment, as the yolk sac reserves start to deplete. By D2, the morphological structure of the larvae's digestive tract had developed sufficiently to begin digesting external feed such as rotifers [12]. The containers used had a capacity of 40 liters, totaling 12 units, each filled with 30 liters of seawater. Each container was equipped with moderate aeration to diffuse oxygen and assist in larval movement.

This study used a Completely Randomized Design (CRD) with 4 treatments and 3 replications. The stocking of barramundi larvae into the experimental in the morning, with a stocking density of 10 fish/L for all treatments [14]. Each containers used were 30 liter water. The enrichment materials used were *Chlorella* sp. with a density of 1.5–2.0 × 10^6 cells/mL yeast = 200 µg/mL [11] ascorbic acid = 4 µg/mL [7] and and cobalamin = 1.4 µg/mL.

The treatments that have been tested in this study are:

A: Chlorella + Yeast

B: Chlorella + Yeast + Ascorbic acid

C: Chlorella + Yeast + Cobalamin

D: Chlorella + Yeast + Ascorbic acid + Cobalamin

Chlorella was cultured in 2.5-liter jars. The density of Chlorella was measured using a hemocytometer to determine the number of cells each milliliter. Harvesting was carried out by directly collecting the required amount of Chlorella as needed.

Rotifers (*Brachionus plicatilis*) used as live feed for the larvae were of SS type size and cultured under controlled conditions. Rotifers are cultured using a 25-liter container with the provision of nutrients according to the treatment being tested. Observations on individual density (ind/ml) are conducted every day for 7 days in each treatment to evaluate their growth. Nutrient enrichment of the rotifers was carried out for 4 hours prior to feeding [11]. Nutrients were administered to the rotifers twice a day, at 06:00 am and 00:00 pm, and were used as feed for the larvae at 11:00 am and 16:00 pm. The rotifers were provided to the barramundi larvae at a density of 10 individuals/ml. Rotifer density was measured before and after enrichment to ensure consistency in the number provided to the larvae.

The nutrient enrichment doses used were Chlorella at 1.5–2.0 × 10^6 cells/mL, yeast at 200 µg/ml, ascorbic acid at 4 µg/ml, and cobalamin at 1.4 µg/mL. The observed water quality parameters included temperature, pH, salinity, and dissolved oxygen (DO). Water quality parameters were measured daily in the morning and afternoon [8].

Growth monitoring was carried out by sampling every 7 days. Ten fish each container were carefully caught using a scoop net and a small container to prevent stress, then measured and recorded.

Monitoring of barramundi larval growth was conducted to assess the effects of administering rotifers enriched with various tested nutrients on both growth and survival of the barramundi larvae. The growth parameter observed was the absolute increase in length [11]. This approach aligns by [13] who stated that fish growth parameters can be assessed through changes in body length. Larval length measurements were taken every 7 days over a

15-day rearing period. These measurements were intended to monitor the growth dynamics of the larvae under each treatment.

3 Result

3.1 *Chlorella* sp. growth

The growth of *Chlorella* sp. during the culture period showed a characteristic dynamic corresponding to the growth phases of microalgae. Observations were made daily over a certain period to determine the growth rate based on cell density (cells/mL) [4]. These observations provide an overview of the adaptation ability, nutrient utilization, and the potential of *Chlorella* sp. as a live feed. The growth data of Chlorella during the culture period can be seen in Figure 1.

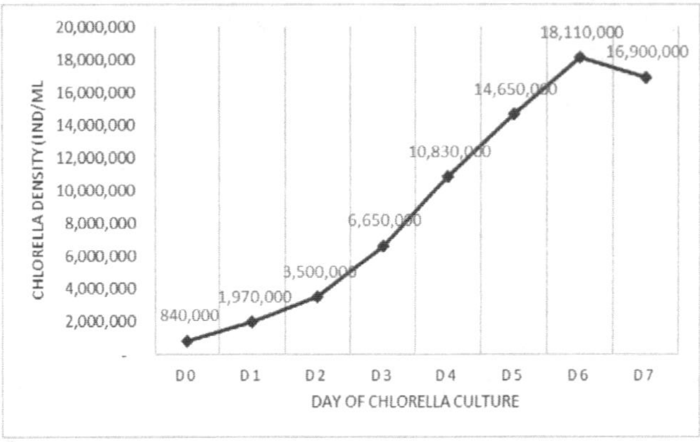

Fig. 1. *Chlorella* sp. growth graph in 7 days culture.

Based on Figure 1, the density of *Chlorella* sp. on the first day was recorded at 840,000 cells/ml. From the first day to the third day, there was a relatively slow increase in density due to the initial lag phase. According to [2], the lag phase occurs from day 0 to day 3, characterized by a slow growth rate because the algae have not yet optimally utilized nutrients for growth. Subsequently, from the fourth to the sixth day, there was a significant and steady increase in cell numbers as the culture entered the logarithmic (exponential) phase. This phase is highly optimal because metabolic processes and cell division occur actively [1] However, starting on the seventh day, the growth rate began to decline, entering the death phase. This is indicated by a decrease in cell density, which aligns with the statement by [2] that the death phase in *Chlorella* sp. occurs on day 7. *Chlorella* sp. was harvested on the sixth day, during the exponential (logarithmic) phase, because at this stage the algal cells experience their most active growth and cell density reaches its optimal peak.

3.2 Rotifers growth

The growth of Rotifers shows a dynamic pattern influenced by the type of nutrients provided. Rotifer growth can be seen in Figure 2. Rotifer density increased in all treatments until day 5, then decreased on days 6 and 7. The highest increase occurred in treatment D (yeast + Chlorella + ascorbic acid + cobalamin), followed by treatments C, B, and A. This indicates that the addition of ascorbic acid and cobalamin in the feed affects the growth rate of the rotifer population.

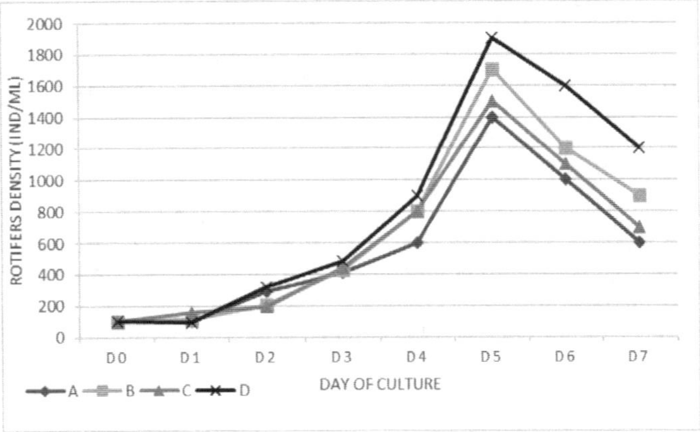

Fig. 2. Growth graph of rotifers with different nutritional treatments during 7 days of culture.

Ascorbic acid is known to act as an antioxidant that supports the immune system and energy metabolism, thus accelerating the reproduction rate of rotifers. Research by [12] shows that feeding yeast enriched with ascorbic acid significantly increases Rotifers density.

Cobalamin functions in DNA synthesis and cellular metabolism, contributing to growth of rotifer cells. Research by [6] shows that the use of yeast, ascorbic acid and cobalamin as feed enrichers has a very significant effect on the increase in rotifer populations, and the best treatment being a combination of all.

Treatment A (yeast + *Chlorella* sp) resulted in the lowest density compared to other treatments of rotifers. Although *Chlorella* sp. contains important nutrients such as proteins, lipids, and essential fatty acids, without additional vitamin enrichment, rotifers' reproductive capacity remains limited. [12] who reported that Chlorella supports only weak rotifers growth unless enriched with cobalamin.

The observed pattern of rapid growth up to day 5, followed by a decline, is likely due to diminished media quality from accumulating metabolic waste and uneaten feed. Once peak density is reached, increased competition for space and oxygen causes the rotifer population to decrease. Therefore, effective management of water quality and nutritional balance to maintaining optimal population growth.

Treatment B (yeast + *Chlorella* sp + Ascorbic acid) produced a higher rotifer density compared to treatments A and C. This indicates that ascorbic acid exerts a stronger influence on increasing the rotifer population density depend on only used cobalamin. Ascorbic acid functions as an antioxidant that protects cells from oxidative damage and enhances the efficiency of energy metabolism. According to [11] adding ascorbic acid to rotifer enrichment media significantly accelerates the reproductive rate by supporting the rotifers' resilience and physiological activity. However, treatment C, which contained only cobalamin (without ascorbic acid), the results were lower than treatments D and C but not lower than treatment A. Although is important in DNA synthesis and protein metabolism, its solitary use is not optimally effective under intensive culture conditions, which may induce oxidative stress in rotifers. Without ascorbic acid as an antioxidant protector, the effectiveness of cobalamin may diminish, so the reproductive rate does not increase maximally [14]. This aligns [4] who stated that combinations of vitamins are more effective than single vitamins, as they provide both metabolic support and cellular protection simultaneously.

Barramundi larvae reared until day 17 showed different average rotifer consumption levels across each treatment. Treatment A, the larvae consumed rotifers at a density of 1–2 ind/mL. Treatment B consumption level of 3–6 ind/mL. Treatment C ranged between 2–6 ind/mL. Treatment D highest with consumption reaching 6–8 ind/mL. Although all treatments used the same number of larvae and rotifers, the differences in consumption indicate that feeding efficiency is influenced by the feed composition provided in each treatment. This finding supports the statement by [9], that the efficiency of live feed consumption in larvae is strongly determined by physiological conditions, feed quality, and rearing management.

3.3 Absolute length growth larvae

Figure 3 presents a graph of the absolute length growth of barramundi larvae throughout the study. The values displayed represent the increase in larval length over a 15-day rearing period of rotifers enriched with different combinations of substances. The largest length increment was observed in treatment D (0.95 cm), next by treatment B (0.89), treatment C (0.83 cm), and the lowest in treatment A (0.75 cm). The highest growth occurred in larvae fed rotifers enriched with yeast + *Chlorella* sp + Ascorbic acid + Cobalamin, while the lowest growth corresponded to rotifers enriched with yeast + *Chlorella* sp. Subsequently, statistical tests were conducted to compare growth differences among treatments.

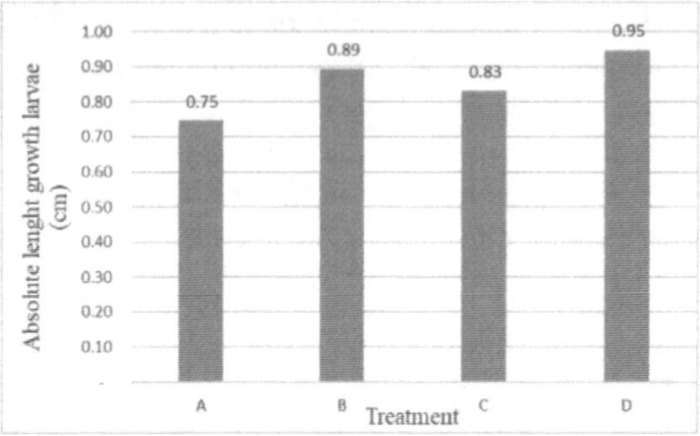

Fig. 3. Absolute length growth barramundi larvae graph.

In treatment A, rotifers enriched with yeast and *Chlorella* sp provided a relatively adequate supply of nutrients to support the growth of barramundi larvae. Yeast is known to contain protein and lipids essential for tissue synthesis and basic metabolic functions. *Chlorella* sp, is rich in unsaturated fatty acids, vitamins, and bioactive pigments, which support energy metabolism and the overall health of the larvae [6]. However, specific micronutrients such as ascorbic acid and Cobalamin crucial roles in stimulating cell division, forming new tissues, and protecting against oxidative stress, and these may not be present in sufficient quantities in yeast and *Chlorella* sp. [2].

The addition of ascorbic acid in treatment B were a significant role in increasing the absolute length growth of the larvae. Ascorbic acid functions as an antioxidant that protects cells from oxidative damage and supports the repair of damaged tissues. By enhancing resistance to stress, larvae are able to allocate more energy toward growth rather than recovery from metabolic disturbances [3]. This aligns with the view of [9], who stated that ascorbic acid is essential for larval development in fish, as most species are unable to synthesize this vitamin. ascorbic acid is involved in several biological processes, including acting as an immunostimulant and supporting stress response. Further noted that a deficiency of ascorbic acid in tissues can result in improper bone growth and may even become a limiting factor for growth. As a result, larvae fed with rotifers enriched with ascorbic acid showed better length growth compared to treatments A and C.

In treatment C, the addition of cobalamin during nutrient enrichment of rotifers contributed positively to the absolute length growth of barramundi larvae, although it was not as effective as treatments B or D. Cobalamin serves as a vital cofactor in DNA synthesis and amino acid metabolism, which are directly involved in cell division and tissue formation especially during the early larval growth phase [9]. Furthermore, cobalamin also affects the nervous system and the production of red blood cells, thereby supporting the overall health and vitality of the larvae [11]. However, without the support of antioxidants such as ascorbic acid, the larvae's ability to cope with environmental stress is more limited, resulting in suboptimal growth.

In treatment D, the combination of ascorbic acid, cobalamin and macronutrients derived from yeast and Chlorella, yielded the highest absolute length growth among the larvae. This combination created a synergistic effect that enhanced multiple metabolic and physiological pathways in the larvae [6]. Ascorbic acid functioned as an antioxidant, reducing oxidative stress and facilitating tissue repair, while vitamin B_{12} supported cell growth and energy utilization [11]. Chlorella contains of highly unsaturated fatty acids (HUFAs), carotenoids, and chlorophyll, which further support immune function and metabolism in the larvae [1]. This nutrient-rich combination enabled the barramundi larvae to maximize nutrient uptake and allocate more energy to growth rather than stress recovery. Therefore, the highest result in absolute length growth was observed in treatment D. Consequently, treatment D produced the greatest gains in absolute length growth. This combination of micronutrients provides the nutritional support critically required during the larval growth phase by accelerating tissue formation and enhancing metabolic efficiency.

3.4 Survival rate

The research results showed that rotifers given different nutrients exhibited significant effects on the survival rate of barramundi larvae (Fig. 4). Treatment A (yeast + *Chlorella* sp.) achieved a survival rate of 22.7%. This indicates that the nutrient combination was already able to support larval survival.The addition of ascorbic acid in Treatment B increased survival to 41.7%. This aligns with the statement by [9] which noted that ascorbic acid plays an important role in enhancing stress resistance and larval fish survival. Treatment C, which contained cobalamin, also showed an increase in survival to 35.3%, supporting the findings of [11] who reported that cobalamin contributes to larval metabolism and growth. In Treatment D, the addition of both ascorbic acid and cobalamin resulted in a survival rate of 39%. This indicates a potential synergistic effect between the two vitamins on the survival of barramundi larvae. If ascorbic acid and cobalamin work through different but complementary biological pathways, producing a synergistic effect that enhances larval protection and metabolism [12].

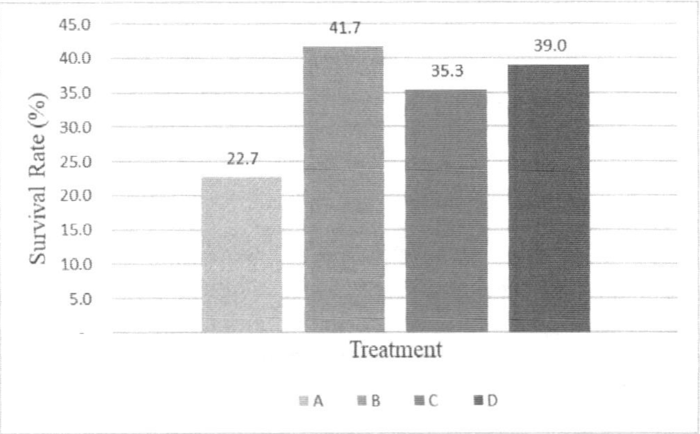

Fig. 4. Survival rate barramundi larvae graph.

According to [2,3], the minimum survival rate standard for barramundi larvae is 20%.Thus, all treatments in this study met and even exceeded the survival rate standard. This indicates that larval survival success also heavily depends on the synchronization between the larvae's nutritional needs and the availability of efficient, high-quality live feed. Therefore, vitamin supplementation particularly the combination of ascorbic acid and cobalamin with yeast and *Chlorella* sp.be recommended as a strategy to improve the survival of barramundi larvae in intensive aquaculture systems.

Physiologically, the relationship between the dosage of additional nutrients and larval survival is very close, as larvae have specific metabolic needs during the early stages of life. ascorbic acid as a powerful antioxidant, protects larval cells from oxidative stress caused by high stocking density and environmental fluctuations, while cobalamin functions as a cofactor in cellular metabolism and DNA synthesis.

The proper dosage of these two vitamins can strengthen the larvae's physiological systems and improve immune resistance, ultimately leading to increased survival rates. In addition, the combination of both vitamins in Treatment D demonstrates a potential synergistic effect that supports the stability of larval survival.

Table 1. Water quality measurement results.

Parameters	Treatment				Optimal range	Source
	A	B	C	D		
Themperature (°C)	28-32.3	28.4-32	28.3-32	28.1-32	28-32	[12]
pH	8.1-8.2	8-8.2	8.1-8.2	8.1-8.2	7.0-8.5	[12]
Salinity(g/L)	30-32	30-32	30-32	30-32	28-32	[12]
DO (mg/L)	5.3-5.6	5.3-5.6	5.3-5.6	5.3-5.6	min 4	[12]

Table 1 shows that the water quality during the rearing of barramundi larvae across all treatments remained within the optimal range to support their survival and growth. The rearing temperature ranged between 28–32°C. This range is considered appropriate and safe for rearing white snapper larvae [4]. The pH values ranged from 7.0 to 8.5, which can optimally support the growth and survival of the larvae. The salinity range of 30–32 g/L is still within the optimal range for rearing barramundi larvae (*Lates calcarifer*). [2] stated that at a salinity of 30 g/L, barramundi larvae exhibited higher growth and survival rates.

Conclusion

The use of *Chlorella* sp., yeast, ascorbic acid and cobalamin as nutrient enrichment for rotifers had an impact on the growth and survival of barramundi (*Lates calcarifer*) larvae. Treatment D resulted in the highest absolute length increment (0.95 cm), while Treatment B achieved the highest survival rate (41.7%).

Recommendation

For barramundi larval rearing, it is recommended to use rotifers enriched with *Chlorella* sp., yeast, and ascorbic acid as the initial feed. Further research is needed using lower dosage variations of ascorbic acid and cobalamin to determine the most effective optimal concentrations for larval growth and survival parameters.

References

1. B. Baptiste, A. Kermarrec, V. Beaumal, A. Riaublanc, H.-P. Ta Nguyen, Nutritional value of *Chlorella vulgaris* : implication of color. In Agriculture and Food Chemical (2025). https://Doi.Org/10.26434/Chemrxiv-2025-5ldw9
2. R. Boroh, M. Litaay, M.-R. Umar, A. Ambeng, Pertumbuhan *Chlorella* sp. pada beberapa kombinasi media kultur. BIOMA: Jurnal Biologi Makassar, 4(2), 129-137. (2019).
3. M. -R. Chaklader, J. Howieson, R. Fotedar, Growth, hepatic health, mucosal barrier status and immunity of juvenile barramundi, Lates calcarifer fed poultry by-product meal supplemented with full-fat or defatted Hermetia illucens larval meal. Aquaculture, 543, 737026. 23 (2021).
4. K. -D. Cahyani, M. Muzahar, W. -K. -A. Putra, Tingkat kelangsungan hidup dan pertumbuhan panjang larva ikan kakap putih (*Lates calcarifer*) dengan suhu pemeliharaan yang berbeda. Intek Akuakultur, 6(1), 1-9 (2022).
5. A. -C. Hansen, R. Waagbø, G. -I. Hemre, New B vitamin recommendations in fish when fed plant-based diets. Aquaculture Nutrition, 21(5), 507-527 (2015). https://Doi.Org/10.1111/Anu.12342
6. D. -R. Jerry, D.-B. Jones, M. Lillehammer, C. Massault, S. Loughnan, H. -S. Cate, N. -A. Robinson, Predicted strong genetic gains from the application of genomic selection to improve growth related traits in barramundi (Lates calcarifer). Aquaculture, 549, 737761 (2022).
7. E. -R. -Y. Kaisar, L. -G. Sumahiradewi, L. -S. Kalih, D. -Y. Sativa, Pengaruh penggunaan ragi roti dan ikan rucah sebagai pengganti (*Nanochloropsis* sp.) dalam kultur rotifer di bpbl lombok (The effect of using yeast bread and rucah fish as an replacement (*Nanochloropsis* sp.) in rotifer culture in Lombok BPBL). Al-Qalbu: Jurnal Pendidikan, Sosial dan Sains, 1(1), 23-31 (2023).
8. M. -S. -K. Khan, K.-R. Salin, A. Yakupitiyage, M. -A. -M. Siddique, Effect of stocking densities on the growth performance, cannibalism and survival of asian seabass Lates calcarifer (Bloch, 1790) fry in different nursery rearing system. Aquaculture Research, *52*(11), 5332-5339 (2021).
9. N. Kursistiyanto, S. Anggoro, S. Suminto, Penambahan vitamin C pada pakan dan pengaruhnya terhadap respon osmotik (Addition of ascorbic acid in feed and effects on osmotic responses, feed efficiency and growth of gesit tilapia (*Oreochromis* sp.) In Various Osmolarity Of Water Medium). Saintek Perikanan: Indonesian Journal Of Fisheries Science And Technology, 8(2), 66–75 (2013).
10. Nurmasyitah, D. -C. Nanda, Hasanuddin, Pengaruh pemberian pakan alami yang berbeda terhadap tingkat kelangsungan hidup larva ikan kakap putih (*Lates calcarifer*). Jurnal Ilmiah Mahasiswa Kelautan Dan Perikanan Unsyiah, *3*(1), 56–65 (2018).
11. G. Salsabila, S. Suminto, R. -A. Nugroho, Pengaruh pengkayaan Brachionus rotundiformis dengan dosis vitamin (B1, B6, B12 dan vitamin C) berbeda dalam feeding regimes terhadap kelulushidupan dan pertumbuhan larva bandeng (*Chanos*

chanos). Jurnal Sains Akuakultur Tropis : Indonesian Journal of Tropical Aquaculture, 3(2), 11-20 (2019).

12. P. -K. Singh, S. Munilkumar, J. K. Sundaray, P. Santhanam, A. Sharma, R. Haque, M. Satheesh, Effect of selenium, vitamin C and highly unsaturated fatty acids-enriched *Brachionus calyciflorus* on growth, survival, physio-metabolic and anti-oxidative responses in anabas testudineus (Bloch, 1792) larvae. Aquaculture, 568, 739293 (2023). https://doi.org/https://doi.org/10.1016/j.aquaculture.2023.739293
13. M. I. Effendie, Biologi perikanan, [Fisheries biology]. (Yayasan Pustaka Nusatama. Bogor, 1997).
14. M. Faizi, and H. Tandipayuk, Pengaruh berbagai dosis vitamin B kompleks tcerhadap laju pertumbuhan bobot spesifik larva ikan kakap putih (*Lates calcarifer*). Journal Of Aquaculture Studies And Development, 1, 31–36 (2023).

Evaluation of the survival of sand lobster seeds after transportation at different densities

Muznah Toatobun[1,2], *Eddy* Supriyono[1*], *Kukuh* Nirmala[1], *Yuni* Puji Hastuti[1], Alimuddin[1], *Irzal* Effendi[1], and *Moh* Burhanuddin Mahmud[1]

[1] Departement of Aquaculture, Faculty of Fisheries and Marine Science , IPB University.
[2] Fisheries Cultivation Technology Study Program, Department of Aquatic Resources Technology. Tual State Fisheries Polytechnic.

Abstract. This study aimed to evaluate the effect of density on the survival of sand lobster fry after transportation. The study used a completely randomized design (CRD), consisting of three density treatments: 10 individuals/container, 15 individuals/container, and 20 individuals/container. The lobster juvenile were reared for 15 days and fed twice daily. The parameters observed were total hemocytes, plasma glucose, and lobster fry survival. The results showed that the density treatment of 10 individuals/container produced the highest survival rate, namely 76.67±5.77%. The densities of 15 individuals/container and 20 individuals/container each produced survival rates of 64.44±3.85% and 43.33±5.77%, respectively.

1 Introduction

Sand lobster (*Panulirus humarus*) has become one of the leading commodities in Indonesia's aquaculture sector, especially in the southern and eastern coastal regions, which have favorable ecosystem conditions [1]. Global market demand for lobster continues to increase, whether fresh, live, or frozen, opening up significant economic opportunities. However, the development of lobster aquaculture still faces challenges, one of which is the suboptimal technical management of aquaculture, including stocking density used in the grow-out system. Stocking density is a crucial factor that directly affects the survival of farmed organisms.

Determining the ideal stocking density is crucial to avoid competition among individuals, cannibalism, water quality degradation, and physiological stress, which ultimately lead to reduced lobster survival rates. Several studies indicate that excessively high stocking densities can result in growth retardation due to increased competition for space and feed, as well as elevated concentrations of ammonia and other toxic compounds in the culture medium [2]. Conversely, while low stocking densities may improve individual performance, they are economically inefficient due to suboptimal use of space and feed [3]. Therefore, research on variations in stocking density for lobster larvae is important to identify the optimal balance between productivity and cultivation efficiency. This study was conducted to determine the differences in the effects of various stocking densities on the survival of lobster larvae in a controlled cultivation system.

* Corresponding author: eddy_supriyono@apps.ipb.ac.id

2 Methods

1.1 Time and location

This study was conducted in July 2023 at a lobster seed collector in Ujung Genteng village, Sukabumi. Test parameter analysis was carried out at the Aquatic Organism Health Laboratory, Department of Aquaculture, Faculty of Fisheries and Marine Sciences, IPB, Dramaga, Bogor.

1.2 Experimental design

This study was designed using a completely randomized design (CRD) with three treatments and three replicates. The treatments consisted of a density of 10 fish/tank, a density of 15 fish/tank, and a density of 20 fish/tank.

1.3 Experimental animal and maintenance management

Sand lobster fry used in the transportation process in the second phase of the research. The seeds weighed between 30 and 50 grams per seed. The container used was a 60 × 60 × 45 cm^3 fiber tub. Nine aquariums were equipped with aerators and a recirculation system. Each aquarium was stocked with sand lobster fry according to the treatment and replicates applied.

Lobster fry were placed in a rearing container and maintained for 30 days to monitor post-transportation survival. During the rearing period, the lobster fry were fed small fish twice daily, and the water in the container was changed daily by siphoning. Water quality measurements (temperature, DO, pH) were carried out daily throughout the rearing period, while observations (ammonia, nitrite, and nitrate) were conducted every 7 days. Hemolymph and hemolymph glucose were collected and measured at the beginning and end of the rearing period.

1.4 Test parameters

Parameters that are survival rate, total hemocytes, and hemolymph glucose levels

1.4.1 Survival rate

The survival rate based on the equation proposed by Zonneveld formula [4] is as follows:

$$SR = \frac{Nt}{N0} \times 100 \qquad (1)$$

where: *SR* = Survival Rate (%), *Nt* = Number of lobsters at the end of the cultivation period, *No* = Number of lobsters at the beginning of the cultivation period.

1.4.2 Total hemocytes

Hemolysate was collected in 0.1 mL using a 1 mL syringe containing 0.1 mL of sodium citrate anticoagulant, then homogenized for 5 minutes. The first drop of hemolysate from the syringe was discarded, then the hemolysate was dropped onto a hemocytometer and the number of cells per mL was counted using a light microscope with 40x magnification. The cell count results are calculated using the formula according to Blaxhall and Daishley [4], namely:

Total hemocytes = [(average total cells) × (1/large box volume) × (dilution factor)] (2)

1.4.3 Glucose level

Hemolymph samples were centrifuged for 10 minutes at a speed of 1000 rpm to separate the hemolymph plasma. Next, 0.5 μL of hemolymph plasma was added to 3.5 mL of ortho-toluidine color reagent in glacial acetic acid. The mixture was placed in boiling water for 10 minutes. After cooling to room temperature, the hemolysate glucose concentration was measured using a spectrophotometer at a wavelength of 635 nm. The absorbance value was then converted into hemolysate glucose concentration in mg/100 mL. The hemolysate glucose concentration was calculated using the formula from Wedemeyer and Yasutake [5], which is:

$$GD = \frac{AbsSp}{AbsSt} \times GSt \tag{3}$$

where: *GD* = Hemolym glucose concentration (mg/100 mL), *AbsSp* = Absorbance of samples, *AbsSt* = Absorbance of standart, *GSt* = Standard glucose concentration (mg/100 mL).

1.5 Data analysis

The data on diversity coefficients and physical and chemical properties of the media were tabulated and analyzed descriptively. Other data were processed and analyzed using SPSS Ver.22 software at a 95% confidence level. If there were significant differences between the mean values of the treatment data, a least significant difference test was performed [6].

2 Result

2.1 Survival rate

The survival rate for each treatment can be seen in (Figure 1). The treatment with 10 fish per container resulted in the highest survival rate (76.67±5.77%), which was significantly different from the other treatments ($p<0.05$). The treatments with densities of 15 and 20 fish per tank had survival rates of 64.44±3.85% and 43.33±5.77%, respectively.

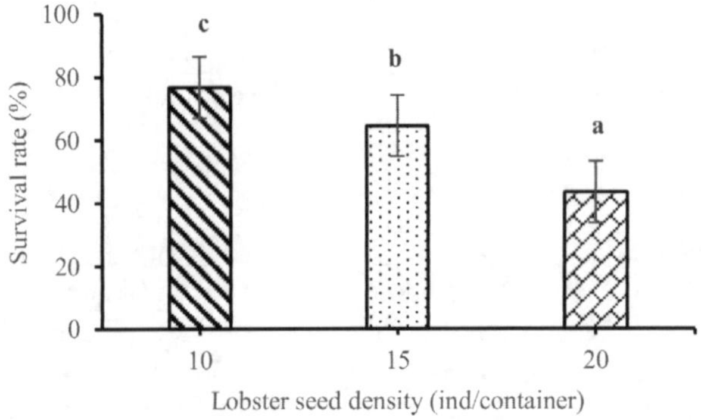

Fig.1. Survival of sand lobsters at different densities. Different letters in each bar (mean value ± standard deviation) indicate statistically significant differences (P<0.05).

2.2 Total hemocytes

The total hemocytes from each treatment can be seen in (Figure 2). The treatment of 10 individuals/container had the lowest total hemocytes (0.36±0.04 ×10^4 cells/mm^3), which was significantly different from the other treatments ($p<0.05$). The treatments with densities of 15 and 20 fish per container had total hemocytes of 0.49±0.03 ×10^4 cells/mm^3 and 0.76±0.06 ×10^4 cells/mm^3, respectively.

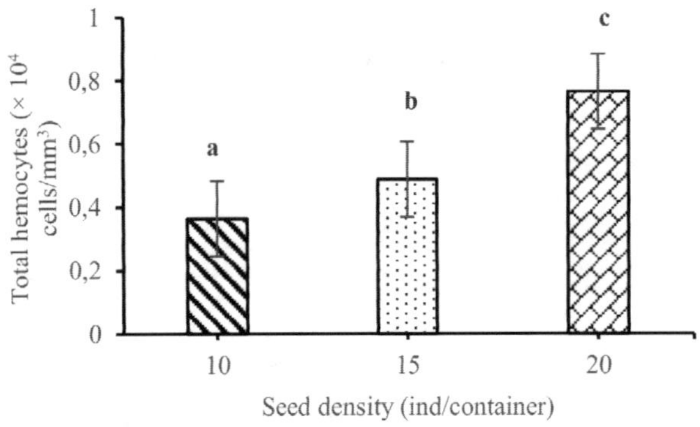

Fig. 2. Total sand lobster hemocytes at different densities. Different letters in each bar (mean value ± standard deviation) indicate statistically significant differences (P<0.05).

2.3 Glucose level

Hemolymph glucose levels for each treatment can be seen in (Figure 3). Treatment 10 fish/container had a hemolymph glucose level of 64.63±2.88 mg/100 mL, which was not significantly different from treatment 15 fish/container (63.33±5.84 mg/100 mL). The treatment with 20 individuals per container resulted in the highest hemolymph glucose level of 128.61 ± 3.06 mg/100 mL.

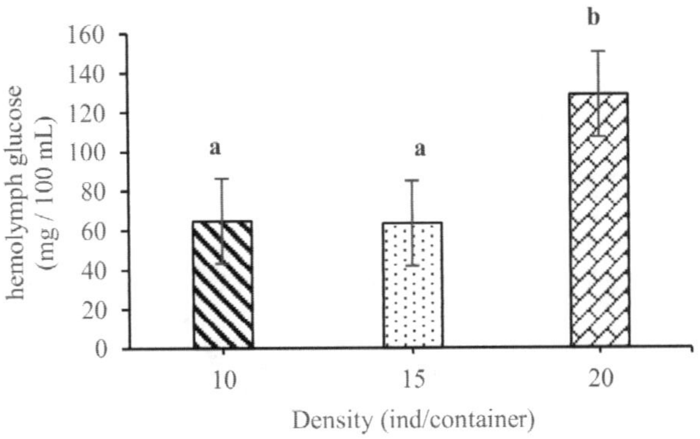

Fig. 3. Hemolymph glucose of sand lobsters at different densities. Different letters in each bar (mean value ± standard deviation) indicate statistically significant differences ($P<0.05$).

2.4 Water quality

The water quality range during lobster fry maintenance is still within the range tolerated by lobster fry (Table 1).

Tabel. 1. Range of water quality for lobster seed maintenance.

Parameter	Unit	Range
Dissolve oxygen	mg L^{-1}	5,00 – 5,13
pH	-	8,51 – 8,61
Salinity	g L^{-1}	28,0 – 30,0
Temprature	°C	27,0 – 29,0
Ammonia total (TAN)	mg L^{-1}	0,17 – 2,77
Nitrite (NO_2)	mg L^{-1}	0,03 – 1,48
Nitrate (NO_3)	mg L^{-1}	0,01 – 1,67

3 Discussion

Stress is defined as a state in which homeostasis is disturbed beyond normal limits and the recovery process continues. through metabolic pathways and affects the fish immune system [7]. In the crustacean immune system, hemocytes act against pathogens or foreign particles through phagocytosis, encapsulation, degranulation, and synodular aggregation [8].

Lobster fry housed in confined spaces at high densities can cause physical and environmental stress. Physical stress can include limited space for movement, while environmental stress can include increased ammonia concentrations and decreased water quality [9]. This stress can affect the lobster's immune system, which is directly related to the total number of hemocytes in the hemolymph.

High stocking dens can cause an increase in lobster metabolism in response to stress. When lobsters are stressed by high stocking dens, their bodies respond by increasing their metabolism, which impacts their energy needs. This energy demand can affect glucose levels in the hemolymph [10]. The stress caused by high stocking densities causes the release of stress hormones such as adrenaline and cortisol, which stimulate the breakdown of glycogen into glucose (glycogenolysis) to provide rapid energy. Glucose levels in the hemolymph increase in response to the greater energy needs, adapting to the stress [11].

A dissolved oxygen value range of 5.00 – 5.13 mg L^{-1} indicates adequate oxygen levels in the water to support normal lobster seed metabolism. In general, higher oxygen levels reduce oxidative stress in lobsters. A pH range of 8.51 – 8.61 indicates slightly alkaline water.

This pH range is considered ideal for many lobster species, including sand lobsters, which typically prefer a slightly alkaline pH to support their physiological balance [12].

A salinity between 28.0 – 30.0 g L^{-1} indicates adequate water conditions for the needs of sand lobsters. Salinities higher or lower than this range can cause osmotic stress and affect the lobster's ion balance. A temperature range of 27.00 – 29.00 °C is within the ideal temperature range for the growth of sand lobster fry. Temperatures that are too high or too low can cause a decrease in metabolic rate, weaken the immune system, or increase the likelihood of thermal stress, which can damage the lobster's body tissue [13].

A total ammonia value between 0.17 – 2.77 mg L^{-1} reflects ammonia levels that can vary and are potentially harmful at higher concentrations. Ammonia is a waste product produced by lobsters and other organisms in aquaculture systems, and its accumulation in water can cause toxicity, disrupt their respiratory system, and damage body tissues [14]. A nitrite value range of 0.03 – 1.48 mg L^{-1} indicates relatively low nitrite concentrations. Nitrite is an intermediate product in the nitrogen cycle and can be highly toxic to lobsters at high concentrations. A nitrate value range of 0.01 – 1.67 mg L^{-1} ppm indicates relatively low nitrate concentrations, which are generally harmless [15].

References

1 A. Rizka, N. Muhammad, FR. Aeko, Strategic Management for the Development of Sand Lobster (Panulirus Homarus) Aquaculture in East Lombok Regency. FSJT. **4** (2025) doi: 3793-3808. 10.55927/fjst.v4i12.359.

2 MDR. Awal, AR. Islam, A. Jaman, MD. Rashid, A. Bhadra, Effects of Stocking Densities on Growth and Production Performance of Mystus cavasius in Bottom Clean Aquaculture System. Aquaculture, Fish and Fisheries. (2025) doi: 10.1002/aff2.70063.

3 M. Fadjar, S. Andayani, D.R. Andriani, I.G.E.P.S. Sentanu, A.M. Amrillah, D. Aisyah, Cultivating lobster seeds (Panulirus homarus) using a recirculation aquaculture system (RAS) in Banyuwangi, Indonesia. JIAT. **8**, 1 (2022)

4 P.C. Blaxhall, K.W. Daishley, Routine haematological methods for use with fish blood. J. of Fish Biology. **5**, 577-581 (1973).

5 G.A. Wedemeyer, W.T. Yasutake, Clinical Methods for the Assessment of the Effects of Environmental Stress on Fish Health, (Department of the Interior Fish and Wildlife Service, Washington DC, 1977)

6 P.G.D. Steel, J. H. Torrie, Principle and Procedures of Statistics, A Biometrical Approach, (PT Gramedia, Jakarta, 1991)

7 M. Zare, E. Heidari, CSM. Hosseini Choupani, SR. Akhavan, A. Rombenso, N. Esmaeili, The Recovery Time between Early Mild Stress and Final Acute Stress Affects Survival Rate, Immunity, Health, and Physiology of Oscar (Astronotus ocellatus). Animals. 13, 1606 (2023). Doi:10.3390/ani13101606

8 M. Celi, F. Filiciotto, M. Vazzana, V. Arizza, V. Maccarrone, M. Ceraulo, S. Mazzola, G. Buscaino, Shipping noise affecting immune responses of European spiny lobster (Palinurus elephas). Canadian Journal of Zoology. **93**, 2 (2015)

9 E. Supriyono, R.W.. Prihardianto, K. Nirmala, The stress and growth responses of spiny lobster *Panulirus homarus* reared in a recirculating aquaculture system. AACL Bioflux. **10**, 2 (2017)

10 EA. Conneely, C. Coates, Meta-analytic assessment of physiological markers for decapod crustacean welfare. Fish and Fisheries. 25. 134-150 (2024). Doi: 10.1111/faf.12798.

11 MC. Soares, F. Banha, SC. Cardoso, M. Gama, R. Xavier, L. Ribeiro, P. Anastácio, Hemolymph Glycemia as an Environmental Stress Biomarker in the Invasive Red Swamp Crayfish (Procambarus clarkii). Physiol Biochem Zool. 95(3):265-277 (2022) doi: 10.1086/719857

12 Q. Ge, J. Wang, J. Li, J. Li. Effect of high alkalinity on shrimp gills: Histopathological alternations and cell specific responses. Ecotoxicol Environ Saf. 256:114902 (2023). doi: 10.1016/j.ecoenv.2023.114902

13 X. Ren, Q. Wang, H. Shao, Y. Xu, P. Liu, J. Li, Effects of low temperature on shrimp and crab physiology, behavior, and growth: a review. Front Mar Sci. **8** (2021)

14 W. Lin, H. Luo, J. Wu, TC. Hung, B. Cao, X. Liu, J. Yang, P. Yang, A Review of the Emerging Risks of Acute Ammonia Nitrogen Toxicity to Aquatic Decapod Crustaceans. Water. 15(1):27 (2023). Doi: 10.3390/w15010027

15 Q. P Fitzgibbon, C. Caraguel, S.B. Pyecroft, C. Gardner, R. Day, K. Patel, J.J. Mahadevan, Improving Post-Harvest Survivability of Southern Rock Lobster in a Changing Environment, (FRDC, Tasmania, 2019)

The Effect of guava extract on water quality and survival of tilapia (*Oreochromis niloticus*) during closed transportation

Annisa Rizka Ramadhani[1], *Eddy* Supriyono[1*], *Yuni* Puji Hastuti[1], *Wildan* Nurussalam[1], *Moh* Burhanuddin Mahmud[1]

[1] Department of Aquaculture, Faculty of Fisheries and Marine Sciences, IPB University, 16680 Bogor, Indonesia

Abstract. Transportation activities are a major cause of tilapia mortality due to stress induced by long travel, decreased dissolved oxygen, and high levels of ammonia. The addition of guava leaf extract expected to be an alternative solution. This study aimed to analyze the effect of guava leaf extract addition during transport and post-transport rearing on survival rate, water quality, blood glucose, and specific growth rate of tilapia. This study used a completely randomized design of two treatments and one control all in triplicate. Transportation 24 hours and maintenance after transportation 21 days. The results of the study showed more optimal water quality and changes in blood glucose after transport in control 117.33 mg dL^{-1}, while 0.50% was 87.33 mg dL^{-1}, which was relatively lower than the control. A 0.50% dose resulted in an increase in survival rate (SR) from 80% in the control group to 90%. After maintenance treatment with extract resulted in an SR of 100%. Specific growth rate in treatments did not have a significant effect. Conclusion is that the addition of guava leaf extract as an environmentally friendly additive at 0.50% of the total water volume used in tilapia significantly affects SR, water quality during transportation, and blood glucose levels in fish after transportation.

1 Introduction

Tilapia (*Oreochromis niloticus*) cultivation is a freshwater fishery product businesses that has significant economic value if managed and developed properly. Stated that based on the composition of production in the fourth quarter, the fish commodity with the highest production in aquaculture is tilapia of 482.25 thousand tons, which experienced an increase in growth of 43.71%. The largest number of tilapia hatchery production in 2022 according to KKP statistical data [1]. Transportation is crucial when fish fry are distributed for further shipment to nursery or grow-out sites. Generally, transportation activities are divided into two methods, namely wet transportation and dry transportation. The transportation of tilapia fry using wet transportation uses a closed system. Usually, open transportation is used for fish for consumption and relatively close distances, while closed transportation systems are used for long-distance shipping to save space [2].

Water quality during transportation must be considered as a determinant of fish survival. Fish stress is caused by the fact that during transportation there is a reduction in dissolved oxygen content in the container, accumulation of carbon dioxide and total ammonia nitrogen, as well as the effect of temperature during the journey. In addition to fish metabolic activity, other factors that cause fish stress are density and shock during transportation. The fry quality

*Corresponding author: eddy_supriyono@apps.ipb.ac.id

is highly dependent on effective transportation methods, which are an important phase in ensuring production sustainability [3].

Efforts to reduce water quality problems during transportation can be done in several ways. For example, modification of carrier media and materials that can reduce metabolic activity during fish delivery. One of the widely used methods is the chemical MS-222/Tricaine methanesulfonate as an anesthetic [4]. The use of chemicals is feared to have a negative impact on fish and also in its use requires relatively high costs. In addition, in aquaculture products, food safety is prioritized by replacing chemicals with materials that are easily degradable in nature. One alternative natural ingredient that can be used is guava leaf extract (*P. guajava*) mixed into water medium during transportation [5].

According to [5], guava leaf extract contains secondary metabolites of flavonoids, tannins, alkaloids, steroids and saponins. Guava leaves contain flavonoids in which there is quercetin which functions to inhibit the movement of the intestinal wall in fish or antimetabolites in fish. Quercetin is directly related to suppressing the metabolic and respiration processes of fish. Efforts to suppress fish metabolism can be done to maintain water quality during the transportation process. In addition, the extract of guava leaves can be used for stress management, which can reduce the rate of fish mortality [6]. The flavonoid and saponin content is beneficial in the prevention of stress during fish transport. Saponins are also secondary metabolites used in anesthesia. However, research on the effectiveness of guava leaf extract in maintaining water quality, reducing stress, and increasing the survival rate of tilapia during 24 hours of transportation and after transportation is still limited. Therefore, this study aims to analyze the effect of adding guava leaf extract during closed-system transportation and post-transportation on survival rates, water quality, blood glucose levels, and specific growth rate.

2 Methods

2.1 Time and place

This study in November – December 2023. This study was conducted at the Aquaculture Environmental Laboratory 3 and water quality at the Aquaculture Environmental Laboratory 1, Department of Aquaculture, Faculty of Fisheries and Marine Sciences, IPB University.

2.2 Test materials

The tilapia used is obtained from cultivators in Tegal Waru, Bogor Regency. The fish used had an average length of 6.76±0.06 cm and an average weight of 5.70±0.08 g. The guava leaves used are located around the IPB Dramaga campus with no physical damage taken from the same guava tree. When making guava extract solution, use freshly picked leaves. The extract was then prepared using the boiling method. Guava leaves were blended, then aqua destillata was added at a ratio of 1:16. The extraction process continued by boiling until only ¼ of the initial volume remained. The boiled mixture was then filtered using filter paper [7]. The filtered result was made into a guava leaf extract solution, which was then diluted to the desired concentrations of 0%, 0.25%, and 0.50%.

2.3 Research design

This method is Completely randomized design (CRD) with two treatments and one control. Each of these treatments has three repeats. Each plastic bag contains 20 fish L^{-1}. Each bag contains 2 L^{-1}. The dosage refers to the previous lethal concentration (LC_{50}) in Table 1.

Table 1. Dosage of guava leaf extract used during transportation.

Treatment (Dosage)	Information
K (Control)	0% of the total volume of transportation water used
JB1 (Guava 0.25%)	0.25% of the total volume of transportation water used
JB2 (Guava 0.50%)	0.50% of the total volume of transportation water used

2.4 Test parameters

2.4.1 Survival rate

Fish survival is observed during transport treatment and after transport. In addition, during maintenance keep an eye on the survival of the fish. The formula to find out SR is:

$$SR\ (\%) = Nt/No \times 100 \quad (1)$$

where: *SR* = Survival Rate (%), *Nt* = Number of fish at the end of the cultivation period (individuals), *No* = Number of fish at the beginning of the cultivation period (individuals).

2.4.2 Blood glucose

Blood glucose levels of tilapia fry were measured before, after transportation, and one week after transportation with extract concentrations of 0%, 0.25%, and 0.50% were performed. Three fish in each treatment were taken as samples for blood collection from the caudal vein using a 1 ml syringe [8]. Three samples because to minimize stress and over mortality after transport since the fish are not anesthetized due to their small size, blood sampling is performed quickly and carefully to minimize stress and to avoid the effects of anesthetics when observed during maintenance. The blood is then tested with a glucose level test kit (*Gluco Easy Touch code: 2692*).

2.4.3 Specific growth rate

The specific growth rate represents the percentage increase in fish body weight per day. It is determined using the following formula:

$$SGR(\%/day) = [(ln(Wt) - ln(Wo)/t] \times 100 \quad (2)$$

where: *SGR*=Specific Growth Rate (%/day), *Wt*=Average weight of fish at the end of the study (g), *Wo*=Average weight of fish at the beginning of the study (g), *t*=Time period (days), *ln*=Natural logarithm.

2.4.4 Water quality

Water quality measurements are carried out during transportation and maintenance. The parameters tested at the time of transportation are DO, pH, temperature, nitrite, TAN, and during maintenance DO, pH, temperature, nitrite, TAN, alkalinity, and hardnees.

Table 2. Water quality parameters observed.

Parameters	Unit	Method	Time	
			Transportation	Maintenance
DO	mg L^{-1}	SNI 06-6989.14:2004	6 hours once	Once a day
pH	-	SNI 6989.11:2019	6 hours once	Once a day
Temprature	°C	SNI 6989.23-2005	6 hours once	Once a day
TAN	mg L^{-1}	APHA, ed. 21, 2005,4500-NH_3-F	6 hours once	once a week
Nitrite	mg L^{-1}	APHA, ed. 21, 2005,4500-NO_2-B	6 hours once	once a week
CO_2	mg L^{-1}	IK Titimetri	6 hours once	once a week
Alkalinity	mg L^{-1} CaCO_3	IK Titimetri	–	once a week
Hardnees	mg L^{-1} CaCO_3	SNI 06-6989.12-2004	–	once a week

3 Results

3.1 Survival rate transportation

The following is the percentage of tilapia survival rate transportation which can be seen in Figure 1.

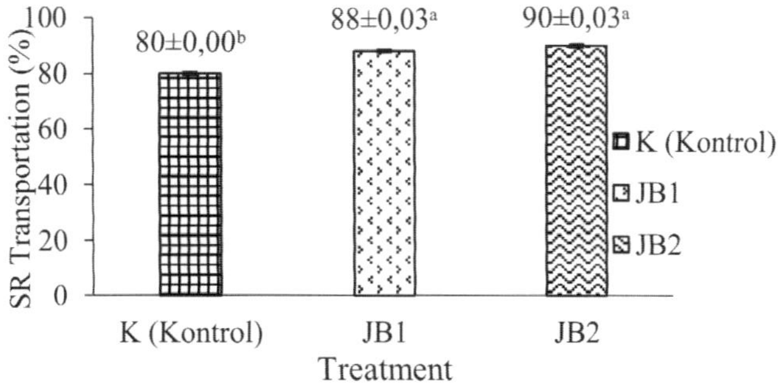

*Different superscript letters indicate significant differences (p<0.05).

Fig. 1. The survival in tilapia transport water with different guava leaf extract concentrations for 24 h (K: control, JB1: 0.25%, JB2: 0.50%).

Based on the ANOVA test, SR transportation in the K treatment was significantly different (P<0.05) from the JB1 and JB 2 treatments, but JB1 and JB2 were not significantly different (P>0.05).

3.2 Dissolved oxygen (DO) transportation

Figure 2 shows the dissolved oxygen (DO) levels in the tilapia transportation medium.

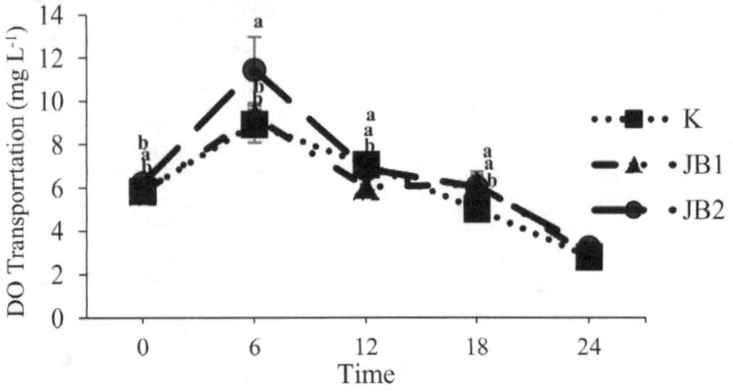

*Different superscript letters indicate significant differences (P<0.05).

Fig. 2. Dissolved oxygen in tilapia transport water with different guava leaf extract concentrations for 24 h (K: control, JB1: 0.25%, JB2: 0.50%) .

Dissolved oxygen (DO) levels in all treatments increased at 6 hours and then decreased from 12 to 24 hours. Based on the ANOVA test, the transport DO values at hours 0, 6, 12, and 18 were significantly different ($P<0.05$), but at the 24 hour there was no significant difference ($P>0.05$).

3.3 Carbon dioxide (CO_2) transportation

Figure 3 shows the characteristics of the tilapia transportation medium.

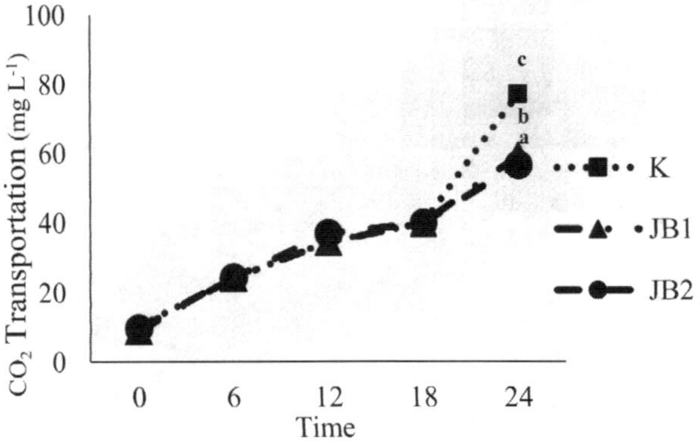

*Different superscript letters indicate significant differences (P<0.05).

Fig. 3. Carbon dioxide in tilapia transport water with different guava leaf extract concentrations for 24 h (K: control, JB1: 0.25%, JB2: 0.50%).

The CO_2 value in each treatment continues to increase as the transportation time increases. Based on the ANOVA test, the transportation CO_2 value at the 0, 6, 12, 18th hour was not significantly different ($P>0.05$) and the 24th hour was significantly different ($P<0.05$) for each treatment.

3.4 pH transportation

The following is the pH of tilapia transportation medium transportation which can be seen in Figure 4.

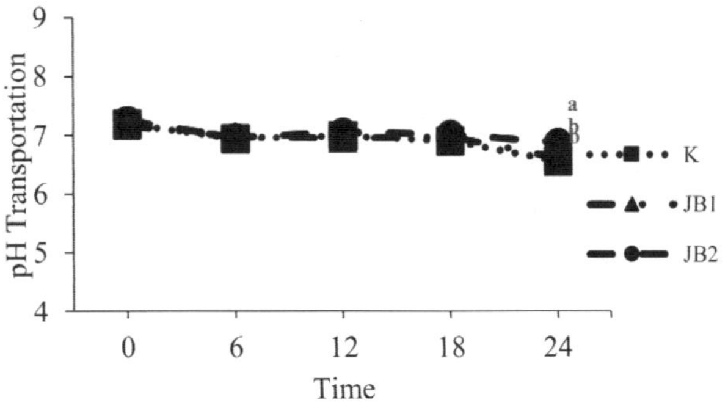

*Different superscript letters indicate significant differences (P<0.05).

Fig. 4. pH in tilapia transport water with different guava leaf extract concentrations for 24 h (K: control, JB1: 0.25%, JB2: 0.50%.

Based on the ANOVA test, the pH of transportation at the 0, 6, 12, 18th hour was not significantly different (P>0.05) and the 24th hour was significantly different (P<0.05) for each treatment.

3.5 Temperature transportation

The following is the temperature of tilapia transportation media transportation which can be seen in Figure 5.

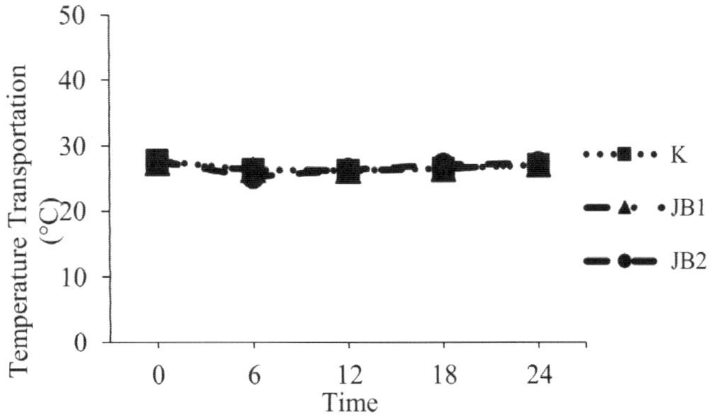

Fig. 5. Temperature in tilapia transport water with different guava leaf extract concentrations for 24 h (K: control, JB1: 0.25%, JB2: 0.50%).

Based on the ANOVA test, the transportation temperature at hours 0, 6, 12, 18, and 24 were not significantly different (P>0.05) for each treatment. The temperature range transportation is 25.3-27.7 °C.

3.6 TAN Transportation

Figure 6 shows the total ammonia nitrogen (TAN) concentration in the tilapia transportation medium.

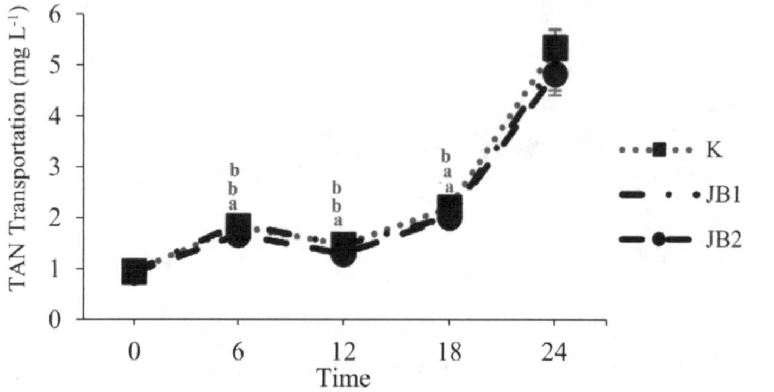

*Different superscript letters indicate significant differences (P<0.05).

Fig. 6. TAN in tilapia transport water with different guava leaf extract concentrations for 24 h (K: control, JB1: 0.25%, JB2: 0.50%).

Based on the ANOVA test, the of transportation TAN at the 6, 12 and 18 hours was significantly different (P<0.05) and at 0 and 24 hours there was no significant difference (P>0.05).

3.7 Nitrite transportation

Figure 7 shows the nitrite concentration in the tilapia transportation medium.

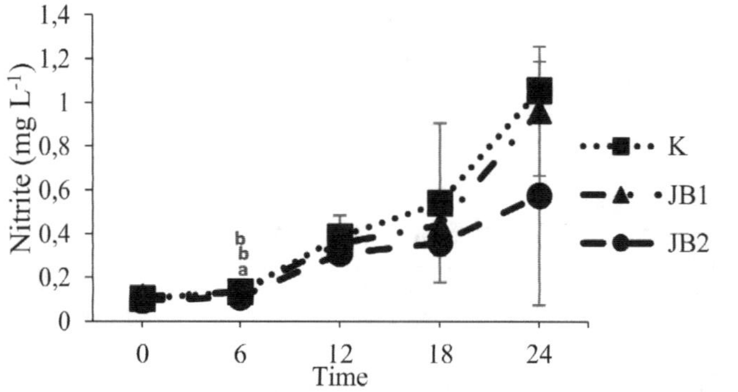

*Different superscript letters indicate significant differences (P<0.05).

Fig. 7. Nitrite in tilapia transport water with different guava leaf extract concentrations for 24 h (K: control, JB1: 0.25%, JB2: 0.50%).

Based on the ANOVA test, the of transport nitrite at hours 0, 12, 18, and 24 was not significantly different (P>0.05), but at the 6th hour it was significantly different (P<0.05).

3.8 Survival rate after maintenance

The following is the percentage value of tilapia survival rate after maintenance which can be seen in Figure 8.

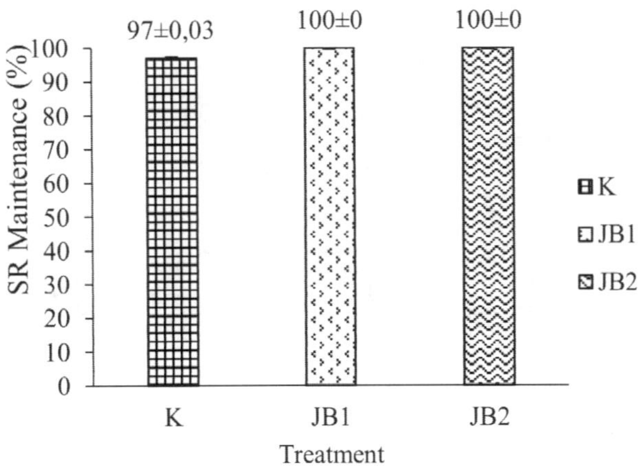

Fig. 8. Survival rate of tilapia fry after 21 days of rearing.

The maintenance SR of K, JB1, JB2 treatment was not significantly different (P>0.05). The JB1 and JB2 treatments did not result in death, while controls resulted in 3% of the initial stocking after transport.

3.9 Water quality after maintenance

The values of water quality parameters in situ (DO, pH, temperature) and ex situ (Nitrite, TAN, CO_2, alkalinity, and salinity) during 21 days of maintenance after transportation are shown in Table 3.

Table 3. Value of maintenance water quality parameters after transportation.

Parameters	**Range values**			**Optimal value**	**Source**
	K	**JB1**	**JB2**		
DO	6.63 – 9.17	6.80 – 9.17	6.83 – 9.23	>5 mg L^{-1}	[9]
CO_2	5.33 – 7.99	4.00 – 7.99	4.00 – 7.99	<15 mg L^{-1}	[9]
pH	6.21 – 8.48	6.44 – 8.47	6.12 – 8.51	6.5 – 8.5	[9]
Temperature	25.90 – 27.03	26.03 – 27.23	25.97 – 27.07	23 – 30 ^{oC}	[9]
TAN	0.13 – 0.17	0.12 – 0.14	0.11 – 0.14	<0.02 mg L^{-1}	[9]
Nitrite	0.10 – 0.12	0.09 – 0.11	0.09 – 0.11	<1 mg L^{-1}	[9]
Alkalinity	41.33 – 46.67	42.67 – 48.00	41.33 – 49.33	20 – 300 mg L^{-1}	[9]
Hardness	41-37 – 44.04	44.04 – 48.05	44.04 – 52.05	20 – 300 mg L^{-1}	[9]

3.10 Blood glucose

The following is the of blood glucose measured before transportation, after transportation, and after a week of maintenance in tilapia is shown in Figure 9.

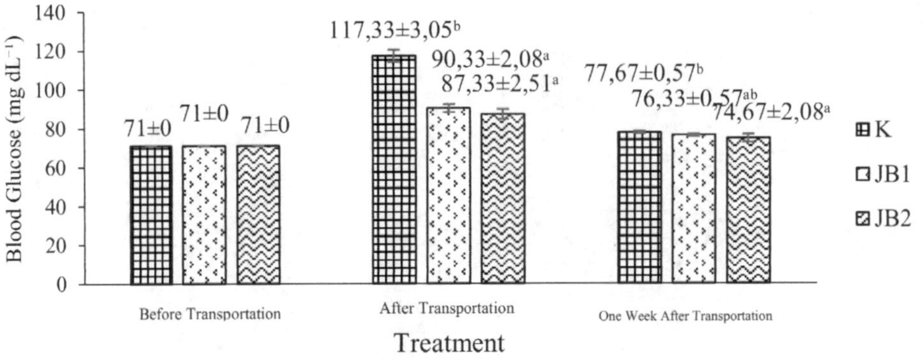

*Different superscript letters indicate significant differences (P<0.05).

Fig. 9. Blood glucose of tilapia fry in three treatments before transportation, after transportation, and after a week of maintenance.

Based on the results of the ANOVA test, the blood glucose after transport of treatment K were significantly different (P<0.05) with JB1 and JB2, but JB1 and JB2 were not significantly different (P>0.05), and after a week of maintenance treatment K was significantly different (P<0.05) with JB2 but JB1 was not significantly different (P>0.05) from K and JB2. JB2 reduced glucose levels significantly vs. control, indicating lower stress.

3.11 Specific growth rate

The following is the percentage of the specific growth rate of tilapia during maintenance is shown in Figure 10.

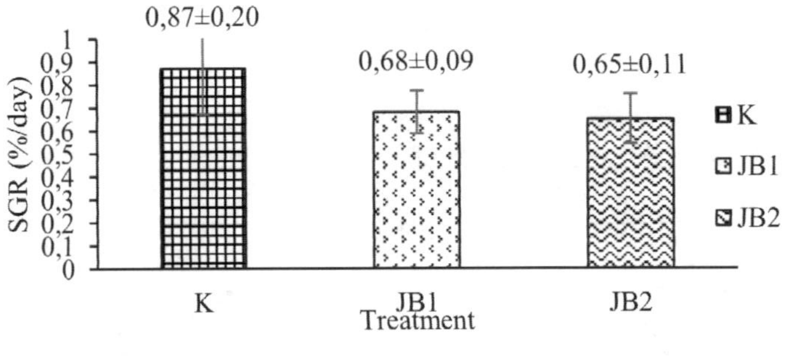

Fig. 10. Value of the specific growth rate of tilapia after 21 days of rearing.

Based on the results of the ANOVA SGR test, K, JB1 and JB2 treatment were not significantly different (P>0.05).

4 Discussion

Indicators of stress levels in fish can be correlated with increased blood glucose levels that occur from measurements at the beginning of transport and after simulations are conducted. The optimal level for fish blood glucose ranges from 40 – 90 mg dL^{-1}. Indicate the level of stress of fish, the results of blood sugar content measurement can be carried out. Increased blood glucose levels are a secondary effect of stress, namely through the release of corticosteroids and catecholamines. Stressful conditions can cause an increase in glucocorticoids which results in an increase in blood glucose levels used to overcome high energy needs [10]. Based on the results of blood glucose measurements, the administration of guava leaf extract resulted in blood glucose levels during transport compared to controls without being given extract. The addition of guava leaf extract to the JB1 and JB2 treatment is considered to reduce the level of glucose use in the blood for homeostasis activities against environmental stress. Supported by the opinion of [5], guava leaves contain secondary metabolites that are able to suppress stress levels and as an antihyperglycemic in fish.

The Stocking density in the container is a crucial factor that influences fish stress levels and subsequently affects survival during transportation. The application of unsuitable transportation methods can negatively affect post-transport conditions, thereby increasing mortality rates at arrival and during the post-arrival period, it was carried out in a 24-hour transportation simulation [11].

Water quality during transportation can have an effect on the survival rate of fry. Dissolved oxygen concentrations in the water during this transportation will decrease along with the length of transportation [12]. DO and CO_2 play an important role in the survival of transported fish. A decrease in DO levels will increase CO_2 during the transportation process. Fish, as poikilothermic organisms, fish are highly affected by external stressors such as temperature, salinity, and water quality, which can cause stress levels [13]. The pH concentration indicates a decrease or is categorized as relatively acidic with a range of 6.5–7.3. This can be caused by the addition of guava leaf extract which contains 9–12% tannin compounds. The temperature of all treatments showed a range of 25.3–27.7°C. Temperature is important for fish health, especially for fish physiological processes. A similar study conducted by [14], Recent studies have shown that Nile tilapia can tolerate elevated ammonia concentrations during short-term exposure; however, prolonged exposure, even at relatively low ammonia levels, can induce significant metabolic disturbances. Therefore, maintaining TAN concentrations at low levels is crucial to support fish welfare and growth during transportation. Nitrite and nitrate compounds are the result of the decomposition of ammonia compounds, both photoautotrophs by algae and autotrophs and heterotrophs by bacteria. The results obtained from JB1 and JB2 treatment are still within optimal limits when compared to optimal levels according to [15], the optimal level for fish is <1 mg L^{-1} and nitrites will be toxic to fish if it reaches a level of 5 mg L^{-1}.

Furthermore, maintenance activities were carried out after transportation for 21 days). These results show that after transportation, tilapia can adapt to the environmental conditions where they are reared. Although there was a 3% decrease in controls, JB1 and JB2 had no significant effect or further effect of administration of guava leaf extract during post-transport maintenance on survival rates. Based on the results of measuring water quality parameters during the 21 days of overall maintenance, it can be said that the water conditions are suitable and optimal for the maintenance of tilapia fry. The SGR value from the 21 day maintenance results showed the highest growth produced in K. because the stocking density used in the study was in accordance with the TKH after transportation, which was as follows: (K: 80%, JB1: 88%, JB2: 90%). Stating that increasing stocking density can result in a decrease in fish growth. Competition will move space, affect the competition and physiology of the fish. Competition in obtaining feed evenly will be smaller. This competition of motion space makes fish unable to optimally utilize feed so that fish growth is disrupted and slow.

5 Conclusion

The addition of 0.50% guava leaf extract to the total volume of transport water used for tilapia not only serves as a cost-effective and environmentally friendly additive but also has a significant effect on survival rates, water quality during transport, and blood glucose levels in fish after transport. A 0.50% dose resulted in an increase in survival rate during transport from 80% in the control group to 90%, and after 21 days of maintenance, the 0.25% and

0.50% doses yielded a survival rate of 100% compared to the control group. Specific growth rate (SGR) in all treatments did not show a significant effect.

References

1 KKP, Rilis data kelautan dan perikanan triwulan IV tahun 2022. Jakarta: Pusat Data, Statistik dan Informasi KKP (2022).

2 M. K. Maraja, N. Salindeho, J. Pongoh. Penanganan ikan nila (*Oreochromis niloticus*) hidup dengan dengan menggunakan es sebagai pengawet. MTHP. **5**, 3 (2017). https://doi.org/10.35800/mthp.5.3.2017.16849

3 E. Supriyono, R. Anandasari, K. Adiyana, L. Thesiana, R. Rasidi, S. Hastuti, The effectiveness of eugenol in mitigating water quality, biochemical, and histopathological alterations during transportation of giant freshwater prawn juveniles. Iraqi j. vet. sci. **39**, 2 (2025). https://doi.org/10.33899/ijvs.2025.157578.4131.

4 L.A. Espinoza-Ramos, Y. Puma-Vilca, J.1. Huanacuni, R. Pepe-Victoriano, R. Evaluation of Tricaine (MS-222) and Eugenol for Sedation of Peruvian Grunt Anisotremus scapularis. *Animals.* **15**, 9 (2025). https://doi.org/10.3390/ani15091322.

5 I. Batubara, I.H. Suparto, N.S. Wulandari. The best extraction technique for kaempferol and quercetin isolation from guava leaves (*Psidium guajava*). Earth Environ. Sci. **58**, 1 (2017). https://doi.org/10.1088/1755-1315/58/1/012060

6 I.W. Sudira, I.M Merdana, S.N. Qurani, Preliminary phytochemical analysis of guava leaves (*Psidium guajava L.*) as antidiarrheal in calves. ATBES. **3**, 2 (2019). https://doi.org/10.24843/atbes.v03.i02.p01.

7 T Birdi, P Daswani, S Brijesh, P Tetali, A Natu, N Antia. Newer insights into the mechanism of action of *Psidium guajava* L. leaves in infectious diarrhea. BMC Complement. Med. Ther. **10**, 33 (2010). https://doi.org/10.1186/1472-6882-10-33.

8 T. Yuniastutik, Observasi waktu maksimal penyimpanan darah terhadap nilai hematokrit ikan lele (*Clarias* sp.). Temapela. **4**, 2 (2021).

9 SNI, Standardisasi Nasional Indonesia, Produksi Ikan nila (*Oreochromis niloticus Bleeker*) kelas pembesaran di kolam air tenang. Jakarta: SNI (2009).

10 T.N. Widianto, I. Malhani, N. Priyanto, Simulasi transportasi ikan nila hidup menggunakan sistem basah terbuka pada suhu rendah. JPBKP. **17**, 1 (2022). https://doi.org/10.15578/jpbkp.v17i1.791.

11 Y. Yang, E. Narayan, S.R. Planellas, C.Z. Phillips, L. Zheng, B. Xu, L. Wang, Y. Liu, Y. Sun, G. Sagada, H.Y. Shih. Effects of stocking density during simulated transport on physiology and behavior of largemouth bass (*Micropterus salmoides*). J. World Aquacult. Soc. **55**, 2 (2024). https://doi.org/10.1111/jwas.13054

12 D. Fang, J. Mei, J. Xie, W. Qiu. The effects of transport stress (temperature and vibration) on blood biochemical parameters, oxidative stress, and gill histomorphology of pearl gentian groupers. Fishes. **8**, 4 (2023). https://doi.org/10.3390/fishes8040218

13 H.T. Ferdinand, E. Supriyono, T. Budiardi, M. Setiawati, Ridwan, The secondary stres responses of transported glass eel (*Anguilla bicolor bicolor*) with various packing densities. AACL Bioflux. **15**, 2 (2022).

14 G. Stephen, R.E. Edziyie, K. A. Obirikorang, D. Adjei-Boateng, P. Vilhelm Skov. Nile tilapia (*Oreochromis niloticus*) show high tolerance to acute ammonia exposure

but lose metabolic scope during prolonged exposure at low concentration. Aquat. Toxicol. **27**, 1 (2022). https://doi.org/10.1016/j.aqrep.2022.101325

15 D. Deswati, E. Yani, S. Safni, O. N. Tetra, H. Pardi, Development methods in aquaponics systems using biofloc to improve water quality (ammonia, nitrite, nitrate) and growth of tilapia and samhong mustard. Int. J. Environ. Anal. Chem. **1**, 11 (2020) https://doi.org/10.1080/03067319.2020.1839437

Enhancing marine-derived functional drinks with natural antioxidants: a sustainable approach using seaweed and indigenous botanicals

Cahyuning Isnaini[1], *Wini* Trilaksani[1*], *Bambang* Riyanto[1], and *Sendy* Rindi Budiarto[1]

[1] Department of Aquatic Product Technology, Faculty of Fisheries and Marine Science, IPB University, Bogor, 16680, Indonesia

Abstract. Seaweed is a marine biomass known for its dietary fiber, bioactive compounds, and hydrocolloids. Carrageenan extracted from seaweed is suitable for the health-promoting beverages. Jelly drinks, while popular in the market, especially among children, often contain high levels of sugar and low nutritional value. This study aimed to formulate antioxidant-enriched functional jelly drinks by incorporating *Kappaphycus alvarezii* with natural colorants from roselle, curcuma, and beetroot. Formulations using different seaweed porridge -to-water ratios (1:8, 1:9, 1:10 V/V) were evaluated through hedonic sensory testing. The 1:10 ratio was found to be the most preferred based on overall liking. The selected formula was enriched with indigenous botanical at concentrations of 5% and 10% (w/w), respectively. Among these, the 10% concentration of each is better than 5% concentration. At this level, the antioxidant activity by DPPH assay IC_{50} values of 1153.08 ppm (roselle), 537.40 ppm (curcuma), and 409.04 ppm (beetroot). Meanwhile, the dietary fiber content in the control and botanical-enriched samples ranged from 1.88 to 2.82% (w/w). The selected product (1:10 seaweed porridge:water with 10% coloran) contained 83.89% water, 0.20% ash, 0.17% protein, 0.39% fat, 1.93% dietary fiber, and 15.42% carbohydrate. The addition of indigenous botanicals not only enhanced antioxidant capacity but also improved consumer acceptability at certain level.

1 Introduction

The beverage industry in Indonesia has experienced substantial growth over the past two decades, driven by evolving consumer preferences and a dynamic product at the small and medium sized enterprise level to the large scales industries [1]. Alongside the increase in product diversity, there is a growing public awareness of the importance of healthy dietary choices. Modern consumers increasingly seek functional foods and beverages that not only offer sensory appeal but also provide health benefits, particularly those supporting digestive health and overall well-being [2]. This trend is reflected in the rising market share of products enriched with probiotics, dietary fiber, and natural bioactive compounds [3].

Seaweed is one of the most promising marine-based resources for functional food development. It is rich in protein, minerals, vitamins, carbohydrates, and dietary fiber, yet low in fat [4]. Moreover, many seaweed species, including *Kappaphycus alvarezii*, contain

* Corresponding author: wtrilaksani@apps.ipb.ac.id

bioactive compounds with antioxidant and antimicrobial properties [4]. Seaweed is also a source of carrageenan, a long-chain sulfated polysaccharide widely used in the food industry as a gelling agent, stabilizer, and thickener [5]. The unique functional and nutritional attributes of carrageenan make it highly suitable for incorporation into health-oriented beverages such as jelly drinks.

Jelly drinks are semi-solid beverages characterized by a soft gel consistency, offering both textural appeal and convenience. However, most commercially available jelly drinks are high in sugar and contain synthetic additives, including artificial colorants, which may have negative health implications [6]. In response to increasing demand for clean-label and nutrient-rich products, the use of natural pigments from plant sources has gained attention. Plant-derived pigments such as anthocyanins from roselle (*Hibiscus sabdariffa*), curcuminoids from curcuma (*Curcuma longa*), and betalains from beetroot (*Beta vulgaris*) not only impart vibrant colors but also contribute antioxidant activity, enhancing the functional value of the final product [7, 8, 9].

Integrating seaweed with natural plant-based pigments represents a sustainable and innovative approach to beverage diversification. Such products align with the principles of marine biodiversity utilization and the blue bioeconomy, offering a pathway toward functional beverages that promote human health while supporting the sustainable use of marine resources. This study was conducted to develop seaweed-based jelly drinks enriched with roselle, curcuma, and beetroot extracts, evaluate their physicochemical and sensory properties, and determine their antioxidant potential.

2 Materials and methods

2.1 Materials

The materials used in the formulation of the jelly drink included fresh seaweed (*Kappaphycus alvarezii*, 45 days of cultivation), dried curcuma (*Curcuma longa*, 10 months of cultivation), dried roselle (*Hibiscus sabdariffa*) calyces (2 months of cultivation), and fresh beetroot (*Beta vulgaris*, 3 months of cultivation). For the antioxidant activity assay, the reagents employed were methanol (Merck), 1,1-diphenyl-2-picrylhydrazyl (DPPH) free radical, butylated hydroxytoluene (BHT), and ascorbic acid, which served as a reference antioxidant.

2.2 Research stages

This study was conducted in three main stages: (1) determination of the jelly drink formula by comparing different ratios of seaweed to water, (2) extraction of liquid natural colorants and formulation of jelly drinks with their addition, and (3) evaluation of antioxidant activity and dietary fiber content.

2.1.1 Determination of jelly drink formula

Seaweed was soaked for 24 hours and washed. The cleaned seaweed was blended to reach a slurry texture and cooked at 100 °C with different ratios of slurry to water, namely 1:8, 1:9, and 1:10 (modified from Trilaksani *et al.* [10]). The resulting seaweed drink was filtered to remove the residue. The filtrate was added with 10% sugar and 0.1% citric acid w/w (modified from Trilaksani *et al.* [10]). The jelly drink was cooled down to room temperature and packaged in a plastic cup. The selected formulation was chosen based on the consumer preference using hedonic sensory test.

2.1.2 Natural colorant extraction

Roselle petals (*H. sabdariffa*), curcuma (*C. longa*), and beetroot (*B. vulgaris*) was extracted to be natural colorants (modified procedure by Andarwulan and Faradillah [11]). Roselle extract was produced by sun-dried the fresh petal and boiled in water. Curcuma extract was produced by sun-dried the peeled rhizomes and boiled in water. Beetroot extract was produced by boiled the peeled tuber. All of them are boiled at ratio 1:5 and 1:10 (w/v) for 15 minutes and filtered to obtain the liquid free from residues.

2.1.3 Formulation of jelly drink with natural colorants

The best jelly drink formulation obtained from the previous step was subsequently enriched with natural colorants at 5% and 10% (w/w) of the total ingredient weight. The mixture was homogenized using a wooden stirrer to ensure uniform distribution of the pigments. The jelly drink was then cooled to 27 °C and packed in plastic cups.

2.3 Analysis

2.3.1 Sensory analysis (SNI 01-2346-2006 [12])

A total of 30 semi-trained panelists participated in this evaluation, and the assessment included five parameters: appearance, color, aroma, texture, and taste. The hedonic test was designed to measure consumer preference levels toward a product using structured scoring sheets. The scale used in this study ranged from 1 to 9, where: (1) dislike extremely; (2) dislike very much; (3) dislike moderately; (4) dislike slightly; (5) neutral; (6) like slightly; (7) like moderately; (8) like very much; and (9) like extremely. Since sensory evaluation is inherently subjective and relies on human perception, statistical analysis was applied to validate the results. The data obtained were analyzed using the non-parametric Kruskal–Wallis test, and significant differences were further examined through Multiple Comparison tests.

2.3.1 Proximate analysis

Proximate analysis (moisture, ash, protein, fat, and carbohydrate *by difference*) was conducted according to AOAC [13] methods. Moisture content was determined by oven-drying 5 g of sample at 105 °C for 5–8 h until constant weight. Ash content was measured by incinerating 5 g of sample in a muffle furnace at 600 °C for 1 h. Protein content was determined by the Kjeldahl method. Two grams of sample were digested with H_2SO_4, K_2SO_4, HgO, and H_2O_2, then distilled and titrated using 0.1 N HCl. Fat content was measured by Soxhlet extraction using hexane as solvent at 80 °C for 6 h. The solvent was evaporated, and the residue was dried at 105 °C. Carbohydrate content was calculated by difference.

2.3.2 Antioxidant DPPH analysis

Antioxidant activity was assessed using the DPPH radical scavenging method Brand-Williams [14]. Jelly drink extracts were prepared at various concentrations (200, 400, 600, and 800 ppm), mixed with 0.1 mM DPPH, incubated for 30 min at room temperature, and measured at 517 nm. The IC_{50} value was determined from linear regression of % inhibition against sample concentration, with BHT and ascorbic acid as references.

2.3.3 Dietary fiber analysis

Dietary fiber content was determined using the AOAC enzymatic–gravimetric method [13]. Samples were sequentially digested with α-amylase, protease, and amyloglucosidase, followed by ethanol precipitation and filtration. Residues were dried, weighed, and expressed as total dietary fiber (TDF), calculated as the sum of insoluble (IDF) and soluble dietary fiber (SDF).

2.3.4 Data analysis

A Completely Randomized Design (CRD) was employed for this study to evaluate the effects of different treatments on the jelly drink formulations. Sensory hedonic data were statistically analyzed using the Kruskal–Wallis test, followed by multiple comparison analysis in SPSS to determine significant differences among treatments. Additional data analyses, including descriptive and comparative statistics, were conducted using Microsoft Excel to support the interpretation of results.

3 Result and discussion

3.1 Determination of jelly drink formula

Development of the jelly drink formula was conducted with varying water-to-seaweed ratios. The produced jelly drinks were subsequently assessed using hedonic sensory evaluation. Figure 1 shown the result of hedonic organoleptic test of Jelly Drink.

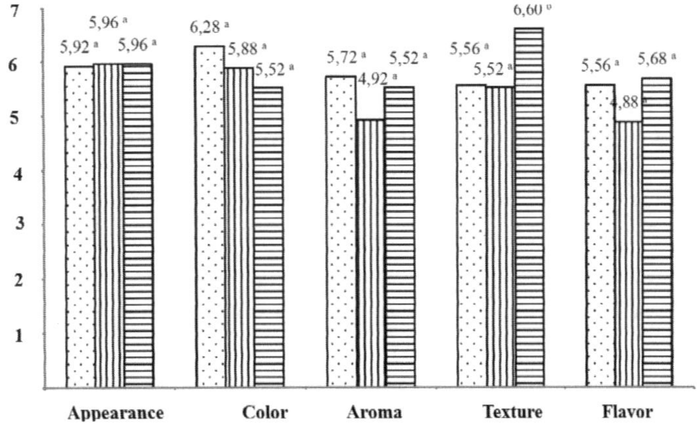

Fig. 1. Hedonic organoleptic result of Jelly Drink.

Note: ☐ 1:8 seaweed:water ratio, ▥ 1:9 seaweed:water ratio, ▤ 1:10 seaweed:water ratio

The appearance hedonic scores shown in the Figure 1 ranged from 5.92 to 5.96 or categorized as neutral but not significantly effect among the treatments as $p>0.05$. The overall appearance of the jelly drink formulations was characterized by the presence of finely ground seaweed granules and a slightly dark color and yellowish white color jelly which not significantly affected by the treatment. The use of fresh seaweed required soaking, which altered the original seaweed color, thereby influencing the brightness levels observed in the jelly drinks. The aroma likeness of jelly drinks with different seaweed-to-water ratios ranged from 4.92 to 5.72 (neutral). The aroma of the jelly drinks was primarily derived from seaweed, which was consistent across all formulations, as there were no differences in the amount of added ingredients. Importantly, the products did not exhibit a fishy odor, as increasing the seaweed proportion allowed the characteristic aroma to dissipate during the cooking process.Texture of jelly drinks ranged from 5.52 to 6.60 (slightly liked). The highest score was obtained at the 1:10 ratio, while the lowest was at the 1:9 ratio. Results from the Kruskal–Wallis test showed significant differences ($p < 0.05$) in texture preference among treatments. Further analysis using Multiple Comparison revealed that the 1:10 ratio (seaweed:water) was significantly different from the other treatments The desirable texture for jelly drinks is one that can be easily consumed through suction. Increasing the proportion of seaweed produced a softer, more fragile gel structure, which facilitated consumption. This is due to the presence of hydrocolloids, particularly carrageenan, in seaweed [15]. A higher concentration of hydrocolloids increases the ratio of solids to liquids, resulting in a stronger and stiffer gel. Furthermore, the carrageenan content of red seaweed plays a crucial role in gel formation, as it is an extracted hydrocolloid from Rhodophyta, which also contains floridean starch. This structural property contributes to the desired consistency in jelly drink formulations. Taste scores of jelly drinks with different seaweed-to-water ratios ranged from 4.88 to 5.68 (neutral). The highest preference score was observed at the 1:10 ratio, while the lowest was recorded at the 1:9 ratio. The Kruskal–Wallis test indicated that differences in seaweed-to-water ratios did not significantly affect ($p > 0.05$) the taste preference of the jelly drinks as found in Hu [15], the more concentrated seaweed to water ratio, the likeness preference is lower.

The similarity in taste among treatments was due to the standardized use of sugar and citric acid across all formulations, which provided consistent sweetness and sourness. Based on sensory evaluation, the 1:10 ratio (seaweed:water) was identified as the most preferred formulation by panelists. This optimized formula was then used in subsequent experiments to evaluate the effect of incorporating natural colorants into jelly drinks.

3.2 Sensory evaluation of jelly drinks with the addition of natural colorants

The selected jelly drink formulation (1:10 seaweed:water ratio) obtained from the previous step was subsequently enriched with natural colorants at 5% (w/w) and 10% (w/w) of the total ingredient weight and homogenized. The jelly drink was then cooled to room temperature and packaged in plastic cups. Sensory evaluation was conducted to assess the sensory attributes of the product. The parameters observed included appearance, color, aroma, texture, and taste.

3.2.1 Jelly drinks with roselle colorants

The results of sensory evaluation of jelly drinks enriched with roselle extract showed mean scores ranging from 4.60 to 6.92 (slightly liked) across the five parameters tested. Overall, the product formulated with 10% roselle extract received the highest scores for most parameters. Based on the results of the Kruskal–Wallis test, differences in the proportion of roselle extract significantly affected the appearance and color parameters ($p < 0.05$). The jelly drink with 10% roselle extract exhibited significantly higher scores for appearance and color preference compared to the 5% formulation. Further pairwise comparisons were conducted using multiple comparison procedures in SPSS confirmed that the 10% roselle formulation was significantly different from the other treatments in both appearance and color parameters. Hedonic test results are shown in the Figure 2.

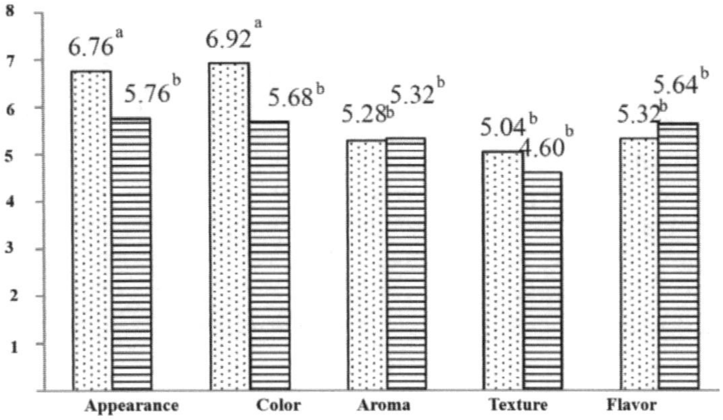

Fig. 2. Hedonic of jelly drinks with roselle colorants.

Note: ▫ 10% of roselle colorants, ⊟ 5% of roselle colorants

For roselle-enriched jelly drinks according to the Figure 2, the formulation with 10% extract achieved the highest scores for appearance, color, and texture, whereas the 5% formulation was preferred in terms of aroma and taste. The jelly drink had a sour taste with a distinctive roselle aroma [7]. Roselle petals are characterized by their high content of organic acids, including oxalic, tartaric, malic, and succinic acids [7].

The ratio of roselle to water during extraction influenced the color and appearance of the product, both of which were significantly affected. The total anthocyanin content in roselle extract decreased as the water-to-flower ratio increased, as larger water fractions dilute the concentration of anthocyanins. The red coloration of roselle is primarily derived from anthocyanin pigments, the stability or degradation of this red pigment is influenced by factors such as pH, light, temperature, and oxygen exposure [7].

3.2.2 Jelly drinks with curcuma colorants

The results of the sensory evaluation of jelly drinks enriched with curcuma extract showed in Figure 3 have mean scores ranging from 4.12 to 6.48 (slightly liked). Differences in the proportion of curcuma extract had a significant effect ($p < 0.05$) on the texture parameter, with the 10% formulation being preferred by panelists over the 5% formulation. Further analysis with the T-test confirmed that the 10% formulation was significantly different from other treatments for texture

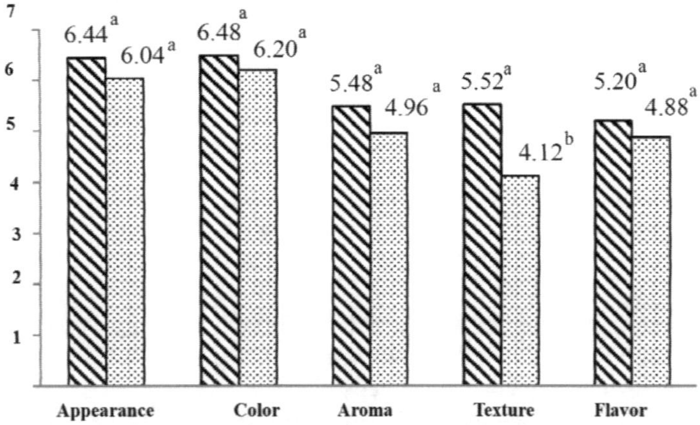

Fig. 3. Hedonic of jelly drinks with curcuma colorants.

Note: ▧ 10% of curcuma colorants, ▨ 5% of curcuma colorants

For curcuma-enriched jelly drinks, the 10% (curcuma-to-water) formulation obtained the highest scores across all sensory parameters. This is attributed to the higher concentration of natural curcumin pigments in this treatment. Increased curcumin content in food products influences appearance, color, aroma, texture, and taste. The jelly drink with a 10% curcuma extract demonstrated a deeper yellow color compared to the 5% formulation, consistent with the higher curcumin concentration. Curcumin exhibits a bright yellow hue under acidic conditions and shifts toward a yellow-brown coloration under neutral or alkaline conditions [11]. Moreover, this formulation was estimated to have a lower pH, which affected both the texture and taste of the product. A lower pH in jelly drinks weakens gel strength, resulting in a texture that is easier to sip.

3.2.3 Jelly drinks with beetroot colorants

The beetroot-enriched jelly drinks sensory evaluation results showed mean scores ranging from 4.52 to 6.64 (slightly liked). Differences in the proportion of beetroot extract had a significant effect ($p < 0.05$) on the color parameter, with the 10% formulation being preferred by panelists compared to the 5% formulation. T-test results further confirmed that the 10% beetroot formulation was significantly different from other treatments in terms of color. The hedonic result of jelly drinks with beetroot colorants is shown in Figure 4.

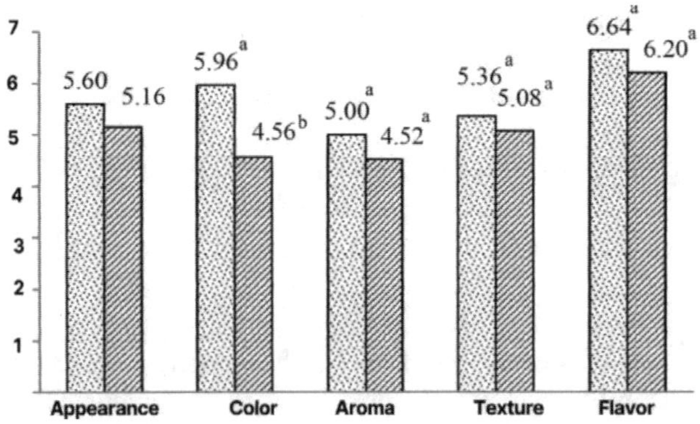

Fig. 4. Hedonic of jelly drinks with beetroot colorants

Note: ▨ 10% of beetroot colorants, ▧ 5% of beetroot colorants

The seaweed jelly drink with 10% beetroot extract was the most preferred by panelists, which can be attributed to the higher concentration of betalain pigments in this formulation. According to Trishitman [9], the stability of betalains is influenced by both internal factors—such as pigment concentration, pH, and moisture—and external factors including temperature, light, and oxygen, all of which play an important role in optimizing pigment stability and maintaining color quality in food products.

The purple color observed in beetroot-enriched jelly drinks originates from betalain pigments, which are water-soluble and therefore readily extracted. Beetroot is recognized as a natural source of red-purple coloration derived from betalains, a combination of betacyanin (purple pigment) and betaxanthin (yellow pigment). Betacyanins share similar characteristics with anthocyanins as natural food colorants, with their stability being highly affected by temperature. In addition to its vibrant color, beetroot imparts a distinctive flavor attributed to the presence of the chemical compound geosmin [9].

Fruits and vegetables, including beetroot, are valuable sources of natural colorants. However, natural pigments also present some limitations, particularly their relatively high production costs and low stability compared to synthetic colorants. The acceptance score of jelly drinks with natural colorants increased because the flavor and aroma of the spices were able to mask the fishy odor of seaweed [15].

3.3 Chemical characteristic of natural colorants jelly drink

The chemical characteristics analyzed in this study included antioxidant activity, moisture, protein, ash, fat, carbohydrate, and dietary fiber contents of the seaweed-based jelly drink.The antioxidant activity of natural colorants jelly drink is shown in Table 1.

Table 1. Antioxidant activity of natural colorants jelly drink.

Samples	**% Inhibition**				**IC_{50} (ppm)**
	2 ppm	**4 ppm**	**6 ppm**	**8 ppm**	
BHT	28.35	31.37	33.42	36.52	18.24
	200 ppm	**400 ppm**	**600 ppm**	**800 ppm**	
Ascobic Acid	25.27	50.95	86.84	94.37	380.69
Roselle 10%	26.68	30.65	34.82	41.70	1,153.08
Roselle 5%	25.04	27.26	32.27	36.25	1,550.37
Curcuma 10%	24.15	25.90	27.51	32.00	537.40
Curcuma 5%	30.46	32.59	34.58	36.52	2,223.54
Beetroot 10%	37.27	42.91	49.15	59.07	409.04
Beetroot 5%	37.43	47.31	53.77	62.26	605.94

3.3.1 Antioxidant activity

Antioxidant activity was assessed using the 2,2-diphenyl-1-picrylhydrazyl (DPPH) radical scavenging assay, a widely adopted method for evaluating free radical inhibition. The key parameter used in this method is the IC_{50} value, which represents the concentration of sample required to inhibit 50% of the DPPH radicals [14].

As presented in Table 1, the IC_{50} value for jelly drinks enriched with 10% roselle extract was 1153.08 ppm, indicating the concentration required to achieve 50% free radical inhibition. By comparison, the 5% extract formulation showed a weaker activity with an IC_{50} of 1550.37 ppm. These values were significantly higher (indicating weaker antioxidant activity) compared to reference standards, namely ascorbic acid (380.69 ppm) and BHT (18.24 ppm). The higher activity observed in the 10% formulation is attributable to greater anthocyanin content, whereas reduced extract concentration at 5% resulted in lower pigment availability and weaker antioxidant capacity. reported that anthocyanin levels decline with increased dilution ratios during extraction. Anthocyanins in roselle not only function as antioxidants but also have additional benefits, including antihypertensive and gastrointestinal protective properties. However, the relatively weak antioxidant performance in jelly drinks compared to pure extracts is likely due to thermal degradation of anthocyanins during heating. According to Lai et al [7] anthocyanins are sensitive to pH and high temperatures, being more stable under acidic and low-temperature conditions.The IC_{50} value of jelly drinks supplemented with 10% curcuma extract was 537.40 ppm, markedly stronger than the 5%

formulation (2223.54 ppm). This demonstrates that higher concentrations of curcumin significantly enhanced antioxidant activity. Despite this improvement, the values remained higher (less active) compared to ascorbic acid and BHT standards. Curcumin, the primary pigment in curcuma, not only imparts a yellow coloration but is also recognized for its antioxidative and anticancer properties [8]. The superior performance of the 10% treatment is consistent with higher curcumin availability, which correlates with enhanced radical scavenging ability. Previous studies have also confirmed that curcumin functions effectively as a free radical scavenger and lipid peroxidation inhibitor. Nonetheless, both treatments were categorized as having weak antioxidant activity when evaluated against the classification proposed by Blois, which defines strong antioxidants as those with IC_{50} values below 100 ppm. The reduced activity is most likely attributed to curcumin degradation under heat and alkaline pH during processing [8].

According to Table 1, the IC_{50} value of jelly drinks formulated with 10% beetroot extract was 605.94 ppm, whereas the 5% formulation exhibited an IC_{50} of 409.04 ppm. This indicates that the 10% beetroot extract was more effective in achieving 50% DPPH radical scavenging compared to the 5% treatment. As explained by Brand-Williams et al. [14], lower IC_{50} values correspond to higher antioxidant activity.

The stronger antioxidant performance of the 10% beetroot formulation is attributed to the higher betalain content present in the extract. A greater proportion of water during extraction reduces pigment concentration, leading to diminished bioactivity. Betalains, the principal pigments in beetroot, are water-soluble nitrogen-containing compounds, consisting of two groups: betacyanins (red-violet pigments) and betaxanthins (yellow pigments). These pigments not only provide vibrant coloration but also possess biological activities including antioxidant, anti-inflammatory, hepatoprotective, and antitumor properties [5].

The antioxidant reference standards, ascorbic acid (380.69 ppm) and BHT (18.24 ppm), demonstrated substantially stronger activity than both beetroot formulations. Thus, although the beetroot-based jelly drinks showed functional antioxidant potential, their activity was relatively weak compared to pure antioxidant compounds. Nonetheless, the incorporation of beetroot extract provided dual benefits by enhancing product color and contributing bioactive compounds with health-promoting properties.

3.3.2 *Proximate and dietary fiber of natural colorant jelly drink*

The seaweed-based jelly drink formulations selected from sensory evaluation and antioxidant activity tests were further analyzed for proximate composition and dietary fiber content. The results are presented in Table 2.

Table 2. Proximate and dietary fiber of natural colorants jelly drink.

Parameter	**Roselle**	**Curcuma**	**Beetroot**
Water (%)	83.3	83.49	83.89
Ash (%)	0.20	0.20	0.20
Protein (%)	0.88	1.51	0.17
Fat (%)0.2	0.20	0.20	0.39
Carbohydrate (%)	15.42	14.59	15.34
Dietary fiber (%)	1.93	1.88	1.93

The proximate analysis indicated that the jelly drinks contained relatively high levels of carbohydrates, second only to moisture content, across all treatments with natural pigments. This composition was primarily derived from the seaweed (*Kappaphycus alvarezii*) used as the main raw material.

The dietary fiber content of the jelly drinks was relatively similar among the three types of natural pigments (roselle, curcuma, and beet). This is expected, as the fiber component mainly originated from seaweed. Kumar [4] reported that carrageenan extracted from *Eucheuma cottonii* contained dietary fiber levels of approximately 25.05%. Seaweed-derived fiber is considered superior to that from fruits and vegetables, as it contains acidic groups such as sulfate groups, which provide distinct physicochemical and physiological effects, including high water- and oil-binding capacity, swelling ability, and the ability to bind vitamins and minerals.

According to the Indonesian National Food and Nutrition Workshop (WNPG, 2004), the recommended daily intake of dietary fiber is 25 g/day. The seaweed jelly drinks supplemented with roselle, curcuma, and beet extracts contributed 15.44%, 15.04%, and 15.44%, respectively, of the daily fiber requirement per 200 g serving. Therefore, these jelly drinks can serve as a practical alternative to help fulfill daily dietary fiber intake. Moreover, dietary fiber functions as a nutrient source for beneficial gut microbiota, thereby supporting digestive health.

4 Conclusion

Seaweed-based jelly drinks enriched with natural pigments from roselle, curcuma, and beetroot at 10% concentration were more preferred by panelists compared to those with 5%, and antioxidant analysis confirmed that the 10% formulations exhibited lower IC_{50} values, indicating higher antioxidant capacity. Furthermore, the dietary fiber analysis showed that the enriched drinks contributed 15.44% (roselle), 15.04% (curcuma), and 15.44% (beetroot) of the recommended daily fiber intake per 200 g serving size, demonstrating their potential as functional beverages to support digestive health. To advance this innovation, further studies are recommended, particularly on shelf-life stability and packaging performance, as well as comparative evaluations between natural and synthetic colorants in terms of sensory quality and antioxidant activity, to better highlight the health-promoting and sustainable advantages of marine-based functional drinks.

References

1. Ragimun, S. Widodo. Strategy of Strengthening Food and Beverage Industry in Indonesia. Journal of Economics and Behavioral Studies 11(4), 102-110. (2019).
2. T. Estiasih, J. M. Maligan, J. E. Witoyo, A. A. H. Mu'alim, K. Ahmadi, T. Mahatmanto, E. Zubaidah. Indonesian traditional herbal drinks: diversity, processing, and health benefits. Journal of Ethnic Foods 12(7), 1-24. (2025).
3. M. Jannah, N. Wakiah, O. M. Sandi, P. A. Islamawan, N. L. N. Wulan. Microorganism in prebiotic beverage fermentation: Health benefits and mechanismin dairy and non-dairy products. Jurnal Inovasi Pangan dan Gizi 2(1): 1-18. (2025).
4. Y. Kumar, A. Tarafdar, P. C. Badgujar. Seaweed as a source of natural antioxidants: therapeutic activity and food applications. Journal of Food Quality, 5753391, 1-17.(2021).
5. N. Rochmawati, D. E. Ermawati. The effect of carrageenan concentration on the physical and chemical properties of gummy turmeric acid jamu. Majalah Farmaseutik 17(2), 230-237. (2021).

6. H. Han, Y. Kim, M. Gim, H. Shin, H. Jang, W. J. Yoon, G. H. Lee, Y. K. Park. Effect of Sugar-Free Jelly on Glycemic Metabolism and Its Potential Health Benefits in Non-Diabetic Adults. Foods 13(920), 1-12. (2024).
7. Lai, Y.-J.; Chiang, Y.-C.; Jhan, Y.-S.; Song, T.-Y.; Cheng, M.-C. Extraction Effects on Roselle Functionalities: Antioxidant, Antiglycation, and Antibacterial Capacities. Foods 2024, 13, 2172. https://doi.org/10.3390/ foods13142172.
8. Coqueiro A, BrumEDS, et al. Endogenous antioxidant properties of curcuminoids from *Curcuma longa* L. obtained by a single-step extraction/nanoencapsulation approach. J Food Biochem.2020;44:e13531. https://doi.org/10.1111/jfbc.13531help
9. D. Trishitman. Effect of storage conditions on quality, microbial stability, and shelf life kinetics of beetroot juice concentrate. Sustainable Food Technology. (2025).
10. W. Trilaksani, B. Riyanto, W. Ramadhan. Diversifikasi dan Pengembangan Produk Hasil Perikanan/Perairan. Bogor (ID): Departemen Teknologi Hasil Perikanan. ISBN 978-602-19460-1-5. (2013).
11. Andarwulan N, Faradilla RHF. 2012. Pewarna Alami untuk Pangan. Bogor (ID): South East Asian Food and Agricultural Science and Technology (SEAFAST) Center.
12. [BSN] Badan Standardisasi Nasional. 2006. SNI-01-2346-2006. Hedonic test (Uji Hedonik). Jakarta (ID): BSN.
13. [AOAC] Association of Analytical Communities. 2007. Official Method of Analysis. Gaithersburg (USA): AOAC.
14. Brand-Williams W, Cuveller ME, Berset C. 1995. Use of a free radical method to evaluate antioxidant activity. J Food Scien Tech. 28: 25-30.
15. A. Hu, S. Wu. Formulation of *Gracilaria lemaneiformis* jelly using fuzzy mathematics evaluation. Food Chemistry 10(28). (2025).

Effect of ultrafiltration on the antioxidant activity of alcalase-hydrolyzed catfish (*Clarias gariepinus*) protein

Nadia Latifah Widiana[1], *Purnama* Arafah[1], *Agnes* Anathasia[1], and *Roni* Nugraha[1*]

[1]Department of Aquatic Product Technology, Faculty of Fisheries and Marine Sciences, IPB University, Jl Agatis Kampus IPB Dramaga, Bogor, West Java 16680, Indonesia

Abstract. Catfish (*Clarias* sp.) is one of Indonesia's leading aquaculture commodities with potential utilization as a source of functional protein. Protein hydrolysis using alcalase enzyme can generate bioactive peptides with antioxidant activity. These antioxidant properties are typically more prominent in low-molecular-weight peptides, thus requiring separation via ultrafiltration. This study aimed to evaluate the effect of ultrafiltration on the antioxidant activity of catfish protein hydrolysate. Hydrolysis was carried out using alcalase at concentrations of 1%, 2%, and 3%, followed by fractionation using a 3 kDa molecular weight cut-off (MWCO) ultrafiltration membrane. Antioxidant activity was assessed using the DPPH and ABTS assays. The highest degree of hydrolysis (62,42%) was obtained at 3% enzyme concentration. The <3 kDa fraction exhibited the highest antioxidant activity, with IC_{50} values of 41,20 ppm (DPPH) and 2,80 (ABTS), both categorized as very strong antioxidants. These findings indicate that ultrafiltration is an effective method for enriching short peptides with high antioxidant potential.

1 Introduction

Fish protein hydrolysate is a product of the degradation of fish protein into short-chain peptide compounds through a hydrolysis process that can occur with the aid of enzymes, acids, or bases. Fish protein hydrolysate products have the advantages of high solubility, good emulsifying capacity, and swelling ability. Hydrolysis also increases the solubility, biological activity, and benefits of fish protein, either through a chemical process with inexpensive acids or enzymatically using proteases in low-value fish [1]. Protease enzymes for protein hydrolysis, with their proteolytic ability to specifically break down peptide chains, produce hydrolysates that maintain essential amino acid content. One enzyme commonly used in protein hydrolysate production is alcalase. The alcalase enzyme from the bacterium *Bacillus licheniformis* is known to have a broader spectrum and greater ability to break down peptides into simpler forms. Fish protein hydrolysates have been known to contain bioactive peptides that to display biological activities, including antimicrobial, antihypertensive, antithrombotic, antioxidant, and immunomodulatory activities with low toxicity [3].

Catfish (*Clarias* sp.) is a major fishery commodity in Indonesia, widely cultivated and distributed to various segments of society. The advantage of catfish over other animal

* Corresponding author: rnugraha@apps.ipb.ac.id

products is high content of leucine and lysine. Leucine is known for its function in growth, nitrogen balance, and muscle formation, while lysine is important for growth, tissue repair, and child development [4]. The protein content of catfish has the potential to produce bioactive peptides through hydrolysis. Putra *et al.* [5] study on catfish gill hydrolysate (CGH) showed that a 3% increase in enzyme concentration increased the antioxidant activity of ABTS to 9.57 μM TEAC.

Protein hydrolysates resulting from enzymatic hydrolysis are heterogeneous because they consist of peptides with varying molecular sizes. Advanced separation methods, such as ultrafiltration, are used to obtain peptide fractions based on their molecular size. Ultrafiltration aims to obtain peptide solutions with specific molecular weights, which can potentially influence their biological activity. Due to its ability to separate and concentrate bioactive compounds, ultrafiltration is widely utilized in purification and fractionation processes. Thus, this research was aimed to determine the effect of ultrafiltration on the antioxidant activity of the alcalase-hydrolyzed catfish protein.

2 Materials and methods

2.1 Equipment and materials

The tools used in this study consisted of knives, cutting boards, blenders, porcelain cups, volumetric pipettes, funnels, beakers, micropipettes, microtubes, 96-well microplates, vortex, test tubes (Pyrex), measuring cups (Pyrex), Whatman 41 filter paper, ultrafiltration membranes (Amicon® ultra-15 centrifugal filters 3K MWCO milipore, Ireland), , digital scales (OHAUS AX224 New Jersey, USA), pH meters (Hanna Instruments, Rhode Island, USA), SPECRO UV Vis (RS Spectrophotometer UV-2500), centrifugators (FC5178R 120 V), electrophoresis equipment (PEQ Lab, Erlangen, Germany), and water baths (B-ONE, China).

The main ingredients in this study were catfish (*Clarias gariepinus*) and alcalase with an enzyme activity of 2,4 U/g (Sigma-Aldrich, St. Louis, USA). Other additional materials used in the analysis were distilled water, Phosphate Buffer Saline (PBS) buffer pH 8, trichloroacetic acid (TCA) (Merck, Darmstadt, Germany), Bradford Reagent, Tris HCl 60 mM pH 6,8, glycerol, β-Mercaptoethanol, Bovine Serum Albumin (BSA), 2,2-diphenyl-1-picrylhydrazyl (DPPH) (Sigma, Missouri, USA), 2,2-azino-bis(3-ethylbenzthiazoline-6-sulfonic acid (ABTS) (Sigma, Missouri, USA), FeCL3.6H2O (Merck, New Jersey, USA), ascorbic acid, methanol, $K_2S_2O_8$, NaOH (Merck), SDS buffer solution (Merck, Darmstadt, Germany), marker (10-250 kDa, BioRad, UK), acetic acid, gallic acid, follin, and Na_2CO_3.

2.2 Work procedure

The research consisted of two stages. The first stage involved the preparation and characterization of catfish meat, which would then be used to produce fish protein hydrolysate (FPH). The catfish protein was hydrolyzed using the alcalase enzyme to measure the degree of hydrolysis. The selected catfish protein hydrolysate was characterized by its degree of hydrolysis. In the second stage, the selected fish protein hydrolysate was subjected to ultrafiltration testing using a membrane filter to characterize its molecular weight and antioxidant activity.

2.2.1 Preparation of catfish meat

The raw materials were purchased at the local fish market as fresh fish. Preparation of the catfish began with cleaning it under running water. The skin and internal organs were thoroughly removed. The catfish meat was separated from the head, skin, bones, and internal organs. The separated meat was then ground using a grinder or blender. The softened meat was then tested to determine its contents, such as moisture, ash, fat, and protein.

2.2.2 Protein hydrolysis with alcalase enzyme

The protein breakdown process was carried out by adding the alcalase enzyme to the previously prepared catfish meat. The fish meat was mixed with PBS buffer solution at a ratio of 1:4, then homogenized for 2 minutes and weighed. Alcalase enzyme was added at concentrations of 1%, 2%, and 3%, then the mixture was heated at 55°C for 2 hours to break down the protein. Enzyme inactivation by heating the mixture at 90°C for 15 minutes. The mixture was then cooled and spun with a centrifuge to separate the liquid containing the protein breakdown products. The supernatant was filtered again using Whatman 41 filter paper. The best hydrolysate was selected for further analysis, such as how far the protein was broken down, its molecular size, and how strong its antioxidant activity was (DPPH and ABTS methods). 12 mL of the best hydrolysate was taken from the best results to be filtered again using a special filter tool, namely an ultrafiltration membrane using a centrifuge. The results of this filtration were divided into two parts, namely molecules larger than 3 kDa and smaller than 3 kDa.

2.3 Analysis procedure

2.3.1 Moisture content

Moisture content analysis was conducted to determine the amount of water contained in the sample. The initial step involved drying the crucible in an oven at 105°C for 1 hour. The crucible was then placed in a desiccator for 15 minutes until it reached room temperature, followed by weighing. Approximately 3 g of sample was placed in the crucible and dried in the oven at 105°C for 5 hours. After drying, the crucible containing the sample was placed back into the desiccator until it reached room temperature and then weighed. The moisture content was calculated using the following formula:

$$\text{Moisture content (\%)}: \frac{B-C}{B-A} \times 100\% \quad (1)$$

Description:
A: mass of empty crucible (g)
B: mass of crucible with sample before drying (g)
C: mass of crucible with sample after drying (g)

2.3.2 Ash content

Ash content analysis was performed to determine the total mineral content in the sample. The procedure began with drying the crucible in an oven at 105°C for 1 hour. The crucible was cooled in a desiccator for 15 minutes and weighed. One gram of sample was placed in the crucible and charred over a Bunsen flame until no smoke appeared. The crucible containing the sample was then placed in a muffle furnace at 600°C for 6 hours. After ashing, the crucible was cooled in a desiccator to room temperature and weighed. The ash content was calculated using the following formula:

$$\text{Ash content (\%)}: \frac{B-A}{C} \times 100\% \quad (2)$$

Description:
A: Mass of empty porcelain crucible (g)
B: Mass of crucible with sample after ashing (g)
C: Mass of sample (g)

2.3.3 Fat content

Fat content analysis was carried out to determine the amount of fat in the sample by extracting fat or oil using a non-polar organic solvent. The procedure began with cleaning, drying, and weighing an empty flask. Approximately 5 g of sample was wrapped in filter paper and placed in a Soxhlet extraction apparatus, fitted with a condenser on top and a fat flask at the bottom. Hexane solvent was added in sufficient volume, and reflux extraction was carried out until the solvent returned to the flask. Distillation was then performed to evaporate and recover the solvent. The fat flask containing the extract was dried in an oven at 105°C to remove residual solvent, as the boiling point of fat is higher. The flask was cooled in a desiccator for 20–30 minutes and then weighed. The fat content was calculated using the following formula:

$$\text{Fat content (\%)}: \frac{W3-W1}{W2} \times 100\% \quad (3)$$

Description:
W1: Mass of empty fat flask (g)
W2: Mass of sample (g)
W3: Mass of flask with fat extract (g)

2.3.4 Protein content

Protein content analysis was performed to determine the crude protein in the sample using the Kjeldahl method.In the digestion stage, a 0.3 g sample was placed into a 100 mL Kjeldahl flask, followed by the addition of 0.3 g selenium catalyst and 20 mL concentrated H_2SO_4, and heated at 410°C until a clear greenish-yellow solution was obtained, then cooled for 15 minutes. The distillation stage was conducted by adding 300 mL of distilled water and 100 mL of 40% NaOH to the digest, after which the distillate was collected in a 125 mL Erlenmeyer flask containing 10 mL of 0.1 N H_2SO_4 with mixed indicators (0.1% methylene blue and 0.1% methylene red in a 2:1 ratio), resulting in a bluish-green solution. Finally, in the titration stage, the distillate was titrated with 0.1 N HCl until the solution changed to pink, and the volume of titrant used was recorded for protein content determination. The protein percentage was then calculated using the appropriate formulas.

$$\text{Nitrogen content (\%)}: \frac{volume\ HCl\ (mL) - volume\ blank\ (mL) \times N\ HCl \times 14}{Sample\ weight\ (mg)} \times 100\% \quad (4)$$

$$\text{Protein content (\%wb)} = \%N \times 6.25 \quad (5)$$

$$\text{Protein content (\%db)}: \frac{Protein\ content\ (\%bb)}{100 - moisture\ content\ (\%bb)} \times 100\% \quad (6)$$

Description:
wb: Wet basis
db: Dry basis
N: Normality of HCl solution

2.3.5 Degree of hydrolysis

The degree of hydrolysis was measured to determine the ability of protease to hydrolyze protein. A 20 mL sample of FPH was taken, and 20 mL of 20% (b/v) TCA was added. The mixture was allowed to stand for 30 minutes to facilitate precipitation, followed by centrifugation at 7,800 × g for 15 minutes. The obtained supernatant was analyzed for nitrogen content using the Kjeldahl method. The degree of hydrolysis was calculated using the following formula:

$$\text{Degree of Hydrolysis (\%)} = \frac{\textit{Nitrogen content in supernatant}}{\textit{Total nitrogen content in sample}}\ x\ 100\% \qquad (7)$$

2.3.6 Protein concentration by Bradford method

Protein concentration was determined using the Bradford method in a 96-well plate. A series of standard Bovine Serum Albumin (BSA) solutions were prepared at concentrations of 0; 0.1; 0.3; 0.5; 0.7; and 0.9 mg/mL. Protein hydrolysate samples of catfish were also prepared. A volume of 5 µL of either standard or sample solution was added to each well, followed by 250 µL of Bradford reagent. The mixture was gently mixed with a pipette or shaker for 15–30 seconds to ensure homogenization, then incubated at room temperature for 5–45 minutes. Absorbance was measured at 595 nm using a microplate reader. The standard absorbance values were used to construct a calibration curve (Y-axis: absorbance, X-axis: BSA concentration), and protein concentration in the samples was determined by interpolating their absorbance values on this curve.

2.3.7 Molecular weight determination by SDS-PAGE

Samples were dissolved in 5% SDS solution and heated in a water bath at 85°C for one hour followed by centrifugation at 8,500 × g for 5 minutes at room temperature. Sample buffer solution containing 60 mM Tris-HCl (pH 6.8), 2% SDS, and 25% glycerol was added at a ratio of 1:10, with additional 10% β-mercaptoethanol. The mixture was heated in boiling water for 2 minutes. Subsequently, 15 µL of the sample was loaded into polyacrylamide gels consisting of 15% separating gel and 5% stacking gel. Electrophoresis was carried out at a constant current of 50 mA per gel for 2 hours. Gels were stained with 0.05% Coomassie Brilliant Blue R-250 in 15% methanol and 5% acetic acid for 1 hour, then destained with 30% methanol and 10% acetic acid for 1 hour, followed by a de-staining process for 30 minutes.

2.3.8 Determination of peptides containing phenolic groups

. A total of 125 µL of the sample was mixed with 125 µL methanol and 625 µL distilled water in a test tube. Then, 62.5 µL of 50% (v/v) Folin–Ciocalteu reagent was added, and the mixture was homogenized using a vortex. After 5 minutes of incubation, 125 µL of 5% Na_2CO_3 solution was added and allowed to react for 60 minutes at room temperature. Absorbance was measured at 725 nm using a UV-Vis spectrophotometer. Total phenolic content was calculated from a gallic acid standard curve and expressed as mg gallic acid equivalents per gram of fresh sample (mg GAE/g).

2.3.9 DPPH antioxidant activity

The analysis measured the ability of samples to reduce the free radical DPPH (2,2-diphenyl-1-picrylhydrazyl). Samples were prepared at concentrations of 2, 5, 10, 20, 30, and 40 ppm. A total of 150 µL of each sample was placed into microplate wells, followed by 50 µL of 0.1 mM DPPH solution. Stock solutions were prepared at 100 ppm, while the blank consisted of 150 µL methanol p.a. and 50 µL DPPH solution. Vitamin C (1–5 ppm) was used as a positive control. The microplate was incubated for 30 minutes at room temperature in the dark, and absorbance was measured at 517 nm using a UV-Vis spectrophotometer.

2.3.10 ABTS antioxidant activity

The ABTS radical cation ($ABTS^{+}$) was prepared by mixing 7 mM ABTS solution with 2.45 mM potassium persulfate ($K_2S_2O_8$) in a 1:1 ratio (v/v), followed by incubation in the dark for 18 hours at room temperature. The resulting solution was diluted with methanol to an absorbance of 0.700 ± 0.02 at 734 nm. Samples were prepared at concentrations of 0.2, 0.4, 0.8, 1.2, 1.6, and 2 ppm. A total of 500 µL of each sample solution (1 mg/mL) was mixed with 500 µL of $ABTS^{+}$ solution and incubated for 10 minutes at room temperature. Then, 150 µL of each mixture was transferred into microplate wells, and absorbance was measured at 734 nm using a UV-Vis spectrophotometer. Antioxidant activity was expressed as IC_{50} values obtained from linear regression of percent inhibition versus sample concentration.

3 Result and discussion

3.1 Proximate analysis

This analysis aimed to assess the suitability of catfish for the production of protein hydrolysates. The results of the analysis were then used as the basis for evaluating the nutritional quality and functional potential of the raw material. The chemical composition of catfish flesh (*C. gariepinus*) is presented in Table 1.

Table 1. The proximate composition of catfish flesh (*C. gariepinus*).

Parameter (%)	Catfish (*C. gariepinus*)
Moisture	82.04 ± 0.07
Ash	0.84 ± 0.11
Fat	1.20 ± 0.02
Protein	15.51 ± 0.53

Catfish (*C. gariepinus*) flesh contained 15.51% protein, placing it within the 15–20% range typical of high-protein fish. Catfish can be considered a suitable raw material for producing protein hydrolysates with high nutritional value.

3.2 Characteristics of catfish protein hydrolysate

Catfish (*C. gariepinus*) protein hydrolysate was produced through enzymatic hydrolysis using Alcalase, an enzyme that acts as a biocatalyst to accelerate the hydrolysis reaction and thereby enhance the efficiency of protein breakdown. Hydrolysis is a chemical reaction that degrades complex protein molecules into simpler forms such as peptides and amino acids. In this study, Alcalase was applied at concentrations of 1%, 2%, and 3% (b/v). Prepared catfish flesh was mixed with Phosphate Buffered Saline (PBS) at a ratio of 1:4 (b/v), and the hydrolysis process was carried out at 55 °C for 2 hours. The resulting supernatant represented the catfish protein hydrolysate, as shown in Figure 1.

Fig 1. Catfish (*Clarias gariepinus*) protein hydrolysates obtained with Alcalase enzyme concentrations of (a) 1%, (b) 2%, and (c) 3%.

The hydrolysis of catfish (*C. gariepinus*) protein using alcalase produced hydrolysates with distinct visual characteristics at different enzyme concentrations. At 1% concentration, the hydrolysate appeared yellow and highly turbid; at 2%, it became yellow, more transparent, with slight turbidity; and at 3%, it showed a clearer and more transparent yellow solution. The decreasing turbidity with increasing enzyme concentration reflects more efficient hydrolysis, as proteins are broken down into smaller soluble fractions, leading to higher clarity. Turbidity is inversely related to protein solubility and pH, where lower turbidity indicates higher solubility and optimal enzymatic activity [6]. Increasing enzyme concentration provides more active sites for substrate interaction, thus accelerating hydrolysis and increasing product yield, whereas low enzyme concentration slows the reaction due to substrate excess.

3.2.1 Degree of hydrolysis

Enzyme concentration is a critical factor influencing the effectiveness of hydrolysis, as it determines how optimally peptide bonds are cleaved. Higher enzyme concentrations result in greater enzymatic activity, which in turn increases the degree of hydrolysis (DH). DH serves as an indicator of the success of protein hydrolysis, calculated as the percentage of cleaved peptide bonds relative to the total peptide bonds in the substrate. A higher DH value reflects a more efficient hydrolysis process, producing greater amounts of peptides and free amino acids. DH is commonly determined using the SN-TCA (Soluble Nitrogen After Trichloroacetic Acid Precipitation) method, which measures the amount of soluble nitrogen after precipitation of undigested proteins. The degree of hydrolysis of catfish protein hydrolysates is presented in Figure 2.

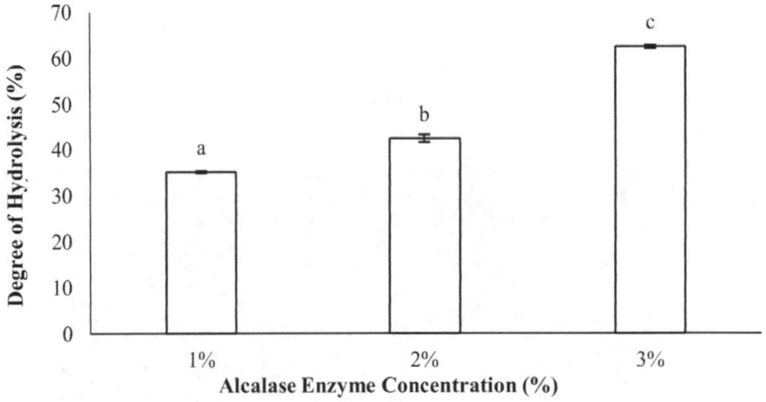

Fig. 2. Degree of hydrolysis of catfish (*C. gariepinus us*) protein hydrolysates.

The degree of hydrolysis (DH) of catfish (*C. gariepinus*) protein hydrolysates increased significantly with higher alcalase concentrations, with the highest DH value of 62.42% at 3% enzyme concentration ($p<0.05$), as confirmed by ANOVA and Duncan's test. Similar findings were reported by Aprilia *et al* [7], who obtained DH values of 58.19% using 3% calotropin, and Nurhayati *et al* [8], who reported 49% DH with crude papain compared to 25% with commercial papain. A higher DH reflects fewer high-molecular-weight proteins remaining and more small peptides and free amino acids formed, including essential amino acids such as lysine, leucine, and methionine [9].

3.2.2 Soluble protein concentration

The measurement of soluble protein concentration in catfish protein hydrolysates was carried out using the Bradford method, a rapid and practical colorimetric technique for determining total protein concentration in solution. The intensity of the blue color formed is proportional to the protein concentration and can be quantified using a spectrophotometer at a wavelength of 595 nm. The soluble protein concentration of catfish protein hydrolysates can be observed in Figure 3.

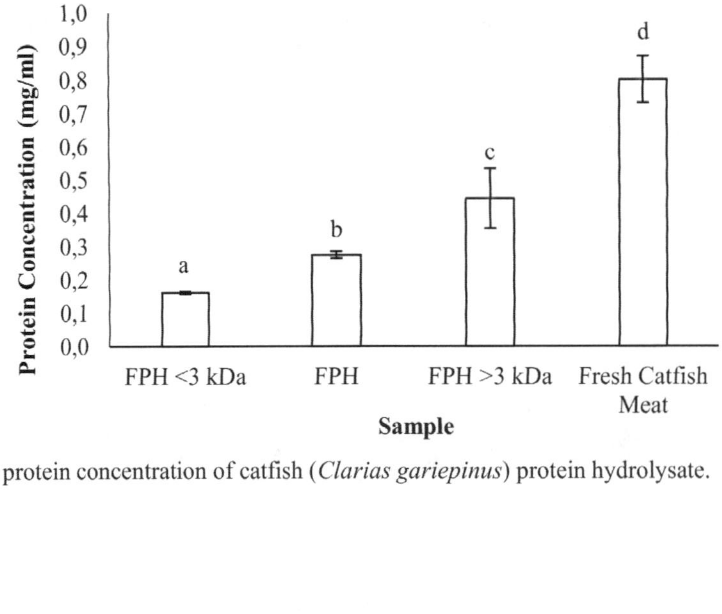

Fig. 3. Soluble protein concentration of catfish (*Clarias gariepinus*) protein hydrolysate.

The measurement of soluble protein concentration using the Bradford method showed that fresh catfish meat had the highest value (0.75 ± 0.072 mg/mL), while the <3 kDa fraction had the lowest (0.16 ± 0.004 mg/mL). The low concentration in the <3 kDa fraction indicates that most proteins were hydrolyzed into small peptides and free amino acids, which are unable to form complexes with Coomassie Brilliant Blue (CBB), making them undetectable. The Bradford method relies on the color shift of CBB G-250 from red to blue upon binding with amino acid residues, with the intensity proportional to protein concentration. The proteolytic activity of alkalase accelerates peptide bond cleavage, promoting further protein degradation and reducing the soluble protein concentration detected, particularly in the <3 kDa fraction.

3.2.3 Catfish protein hydrolysates after ultrafiltration

Ultrafiltration is a pressure-driven membrane separation technique that employs semipermeable membranes with pore sizes ranging from 1 to 200 kDa to separate solutes based on molecular size. Compared to microfiltration, ultrafiltration membranes have smaller pores but remain larger than those used in nanofiltration and reverse osmosis. In this study, ultrafiltration of catfish protein hydrolysates was performed using a 3 kDa membrane (Amicon® Ultra-15 centrifugal filters, 3K MWCO Millipore), producing two main fractions: <3 kDa and >3 kDa. The hydrolysates subjected to ultrafiltration were obtained from treatment with 3% Alkalase, which yielded the highest degree of hydrolysis (62.42%). The results of the ultrafiltration process of catfish protein hydrolysates are shown in Figure 4.

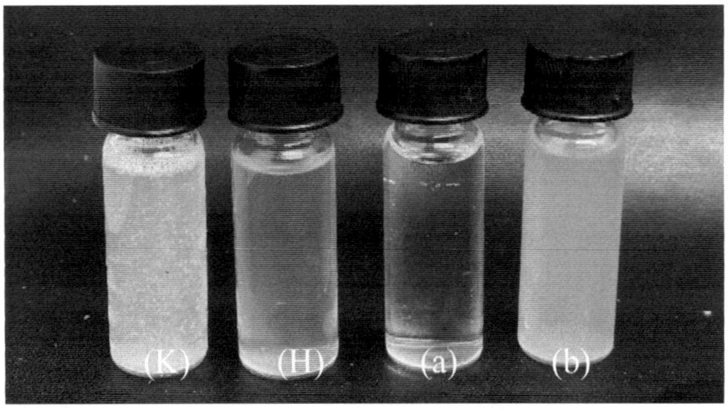

Fig. 4. Fractions of catfish protein hydrolysates: (K) fresh catfish meat, (H) protein hydrolysate, (a) FPH <3 kDa fraction, (b) FPH >3 kDa fraction.

The visual characteristics of catfish protein hydrolysates separated using a 3 kDa ultrafiltration membrane showed clear differences between fractions. The >3 kDa fraction appeared more turbid and yellowish compared to the <3 kDa fraction, which was clearer. This difference reflects the principle of ultrafiltration, which separates molecules by size: molecules larger than the membrane's cut-off are retained on the surface, where their accumulation forms a "cake layer" that scatters light and increases turbidity [10]. The efficiency of peptide separation depends on hydrolysate properties such as molecular weight distribution, concentration, hydrophobicity, charge, pH, and ionic strength,. They further reported that not only pore size but also membrane arrangement in multilayer systems affects filtrate yield. For example, using a 10 kDa membrane layered above a 5 kDa membrane (10/5) produced more filtrate than two 5 kDa membranes in sequence (5/5), since the upper 10 kDa membrane allowed smoother flow and prevented excessive accumulation on the surface, thereby improving the filtration efficiency of the smaller membrane.

3.2.4 Molecular weight of catfish protein hydrolysates

In this study, SDS-PAGE analysis was performed on fresh catfish meat, catfish protein hydrolysates, and ultrafiltration fractions (>3 kDa and <3 kDa). The SDS-PAGE profiles of catfish meat and hydrolysates are presented in Figure 5. The molecular weight of proteins was determined by comparing the protein bands obtained from SDS-PAGE with standard protein markers (10–250 kDa). The electrophoresis results showed that fresh catfish muscle exhibited thicker protein bands than the hydrolysate, indicating higher protein concentration in intact tissue with a molecular weight range of 10–216 kDa. In contrast, the catfish protein hydrolysate showed degradation into smaller peptides with bands at 6–10 kDa, while the >3 kDa fraction appeared at 7–9 kDa. Similarly, Astiana *et al* [11] reported catfish viscera hydrolysates in the range of 15–16 kDa. These findings indicate that hydrolysis with alcalase effectively degraded proteins into smaller peptides. In the <3 kDa fraction, protein bands were not detected, suggesting that the fragments were either too small (<10 kDa) or present at concentrations too low to be visualized by SDS-PAGE.

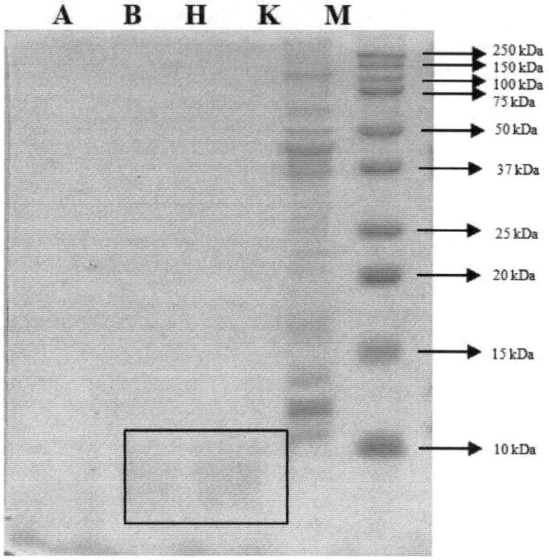

Fig. 5. Molecular weight profile of catfish (*Clarias gariepinus*) protein hydrolysates: (A) <3 kDa fraction, (B) >3 kDa fraction, (H) catfish protein hydrolysate, (K) fresh catfish meat, and (M) protein marker.

3.2.5 Phenolic-containing peptide content

Phenolic compounds characterized by the presence of one or more hydroxyl (-OH) groups attached to an aromatic ring.are recognized as one of the most effective groups of natural antioxidants, In this study, the total phenolic content of catfish protein hydrolysates was analyzed to evaluate their contribution to antioxidant activity. The results of the phenolic content analysis of catfish protein hydrolysates are presented in Figure 6.

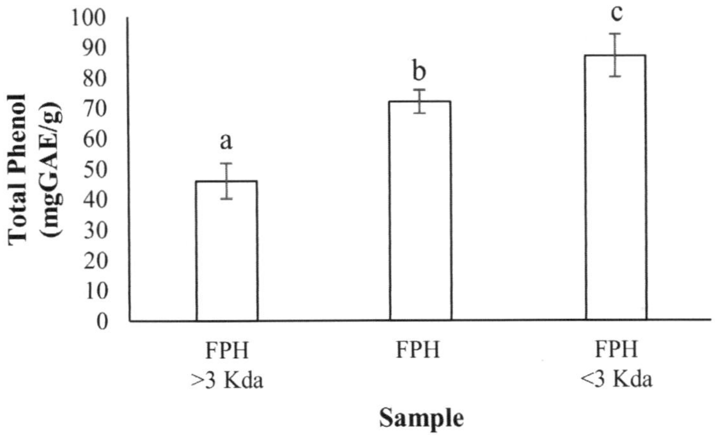

Fig. 6. Phenolic group–containing peptides in Catfish (*Clarias gariepinus*) protein hydrolysates.

Figure 7 shows that catfish protein hydrolysates phenolic compound varied from 46.49 to 87.24 mg of gallic acid equivalents/100 g of hydrlysate. . Ultrafiltration significantly affected ($p<0.05$) total phenolic content, with the <3 kDa fraction showing the highest value (87.24 mg GAE/g), followed by whole HPI (71.97 mg GAE/g) and the >3 kDa fraction (46.49 mg GAE/g). This indicates strong antioxidant potential, as low-molecular-weight peptides in the <3 kDa fraction likely contain aromatic amino acids such as tyrosine, tryptophan, and phenylalanine that enhance radical scavenging. Aromatic structure can serve as an active hydrogen donor and gives stability to neighboring amino acid residues. Smaller peptides (<3 kDa) allow free phenolics to interact more effectively, while in the >3 kDa fraction phenolics are bound to larger peptides, reducing their reactivity [12].

3.2.6 Antioxidant activity of catfish protein hydrolysate

a. DPPH antioxidant activity

Antioxidant activity was quantitatively assessed using the DPPH method, based on the reaction between DPPH radicals and antioxidant compounds capable of donating hydrogen atoms. Antioxidant activity was measured using a spectrophotometer at a wavelength of 517 nm. The DPPH antioxidant activity of catfish protein hydrolysates is shown in Figure 7.

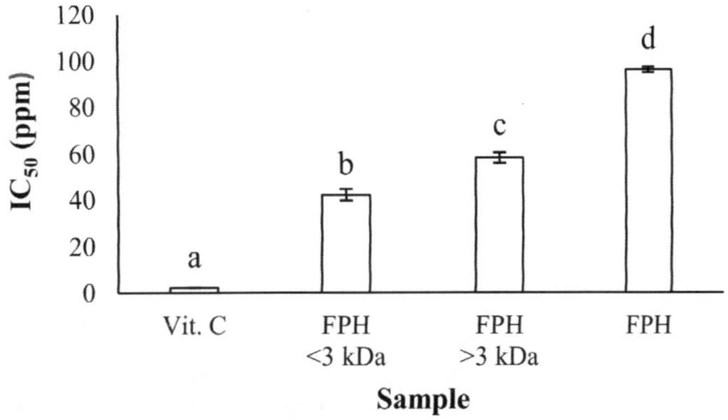

Fig. 7. DPPH antioxidant activity of catfish (*Clarias gariepinus*) protein hydrolysate.

The DPPH antioxidant activity of catfish protein hydrolysates showed that differences in molecular weight fractions obtained after ultrafiltration resulted in significantly different IC_{50} values. Analysis of variance indicated that molecular weight differences significantly affected the antioxidant activity of the hydrolysates ($P<0.05$), and Duncan's post hoc test confirmed that each fraction had a distinct effect on antioxidant values. Smaller peptides (<3 kDa) exhibited higher solubility, mobility, and proton- or electron-donating capacity, allowing more efficient reduction of DPPH radicals, as evidenced by the color change from purple to yellow. The <3 kDa fraction (IC_{50} = 41.70 ppm) was classified as a very strong antioxidant, while the >3 kDa fraction (58.20 ppm) and whole hydrolysate (95.87 ppm) were considered strong antioxidants. Lower IC_{50} values in the >3 kDa fraction compared to whole hydrolysate were attributed to turbidity and larger particles scattering light, which required additional centrifugation for accurate spectrophotometric readings. Similar trends have been reported in silver carp protein hydrolysates [13], where low-molecular-weight peptides exhibited stronger antioxidant activity due to higher concentrations of free amino acids and small peptides.

Peptide fractions with molecular weight <3 kDa consistently demonstrated higher DPPH radical scavenging activity and reduction potential than larger fractions because smaller peptides are more soluble, diffuse more readily, and interact efficiently with free radicals. Antioxidant activity is also strongly influenced by amino acid composition; the presence of hydrophobic residues such as leucine, valine, and phenylalanine, as well as aromatic and sulfur-containing amino acids like tyrosine, tryptophan, and cysteine, enhances electron transfer and radical stabilization. While small peptides excel in radical scavenging, larger fractions may perform better in metal-chelating activity due to a higher number of functional groups [13].

b. ABTS antioxidant activity

The ABTS (2,2′-Azinobis(3-ethylbenzothiazoline)-6-sulfonic acid) method was used to evaluate the antioxidant capacity of a compound by measuring its ability to scavenge free radicals. The ABTS antioxidant activity of catfish protein hydrolysates is presented in Figure 8.

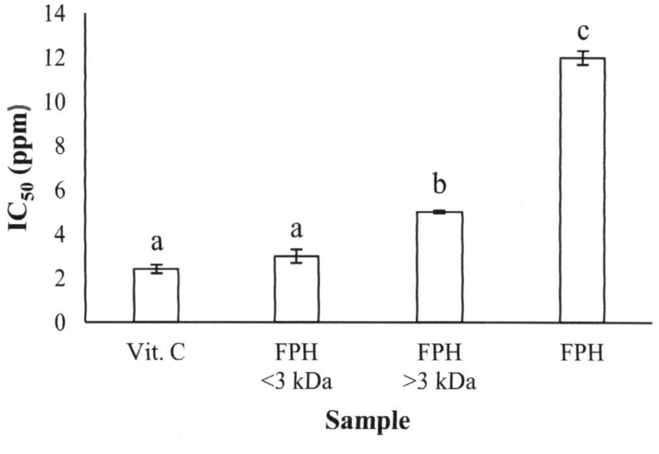

Fig. 8. ABTS antioxidant activity of catfish (*C. gariepinus*) protein hydrolysate.

The antioxidant activity of catfish protein hydrolysates (*C. gariepinus*) measured using the ABTS method showed that molecular weight differences after ultrafiltration resulted in varying IC_{50} values. Analysis of variance indicated that molecular weight significantly affected ABTS antioxidant activity ($P<0.05$), and Duncan's multiple range test confirmed significant differences among fractions. Lower IC_{50} values indicate higher antioxidant activity, as smaller peptides can more efficiently donate electrons to reduce $ABTS^{\cdot+}$ radicals, causing a decrease in absorbance [14]. The IC_{50} of unfiltered hydrolysates was higher than the <3 kDa fraction (2.80 ± 0.366 ppm), consistent with studies showing that ultrafiltration produces small peptides and bioactive amino acids that enhance radical-scavenging efficiency [13].

Peptide fractions with lower molecular weight (<3 kDa) exhibited higher antioxidant activity than larger fractions (>3 kDa), as smaller peptides diffuse faster and interact more readily with ABTS radicals. The high activity of the <3 kDa fraction is attributed to the abundance of small peptides containing functional amino acids such as tyrosine, tryptophan, cysteine, leucine, and valine, which donate electrons or hydrogen atoms to neutralize free radicals [12]. ABTS assay results also highlight the high hydrophilic antioxidant content in catfish protein hydrolysates. The ABTS method is considered superior to DPPH in detecting both hydrophilic and lipophilic antioxidants, providing faster reaction times and stable absorbance. While both DPPH and ABTS involve hydrogen atom transfer (HAT) and electron transfer (ET) mechanisms, ABTS is more sensitive to hydrophilic antioxidants and predominantly operates via HAT. In contrast, DPPH is better suited for lipophilic antioxidants and works mainly through ET, sometimes combined with HAT [15].

4 Conclusion

The protein hydrolysate produced using 3% alcalase represents the optimum condition in this current research, yielding the highest degree of hydrolysis. The ultrafiltration process was proven effective in separating peptides smaller than 3 kDa and had a significant impact on enhancing the antioxidant activity of the carp protein hydrolysate. The <3 kDa fraction exhibited higher antioxidant activity compared to the hydrolysate prior to ultrafiltration and the >3 kDa fraction.

References

1. T. Yuniarti, N.Y. Arrahmi, N. Dharmayanti, S. Sugiwati, M. Mulyono, T. Hidayat, P. Martosuyono, A. Maulani, A. Alghani, Karakteristik fisiokimia dan aktivitas antioksidan hidrolisat protein ikan kakap (*Lutjanus* sp.) skala pilot. JPB Kelautan Perikanan **19**, 61–72 (2024). https://doi.org/10.15578/jpbkp.v19i1.98
2. R. Nugraha, B. Lilyantika, T. Nurhayati, Antioxidant and antigenic capacity of pepsin-hydrolyzed carp (Cyprinus carpio) protein. BIO Web of Conferences 112, 02003 (2024). https://doi.org/10.1051/bioconf/202411202003
3. R. Nugraha, F. Kurniawan, A. Abdullah, A.L. Lopata, T. Ruethers, Antihypertensive and antidiabetic drug candidates from milkfish (*Chanos chanos*) -Identification and characterization through an integrated bioinformatic approach. Foods 13 (16), 2594, (2024). https://doi.org/10.3390/foods13162594
4. D.R. Yuniarti, D. Ratnasari, M.J. Lababan, Pemanfaatan pangan lokal lele untuk pembuatan nugget. JAMU **1**, 10–16 (2021). https://doi.org/10.46772/jamu.v1i02.343
5. M.M. Putra, W. Wisnumurti, N.H. Athallah, A.T.A. Zaen, A.R. Sahura, Ustadi, Influence of enzyme concentration, hydrolysis duration, and drying temperature on the production of antioxidant-containing peptide from catfish (*Clarias* sp.) gills. J. Appl. Food Technol. **12**, 57–64 (2025). https://doi.org/10.17728/jaft.26160
6. A.A. Safiera, E. Purpitasari, T.A. Siswoyo, Aktivitas antioksidan dan kemampuan proteksi terhadap kerusakan DNA dari protein isolat biji melinjo (*Gnetum Gnemon* L.) terhidrolisis menggunakan alkalase terimobilisasi. e-Jurnal Pustaka Kesehatan **5**, 382–388 (2017). https://doi.org/10.19184/pk.v5i2.5870

7. V. Aprilia, F.S. Hati, Formulasi bubur bayi MPASI yang diperkaya hidrolisat protein ikan lele dumbo (*Clarias gariepinus*). J. Gizi Diet. Indones. **4**, 88–96 (2016). https://doi.org/10.21927/ijnd.2016.4(2).88-96
8. T. Nurhayati, E. Salamah, Hidayat, Karakteristik hidrolisat protein ikan selar (*Caranx leptolepis*) yang diproses secara enzimatis. Buletin Teknol. Hasil Perikanan **10**, 23–34 (2007). https://doi.org/10.17844/jphpi.v10i1.966
9. A.R. Nirmala, L. Permatasari, H. Muliasari, R.F. Decatti, Review: analisis kondisi optimal metode penghambatan denaturasi protein bovine serum albumin (BSA) pada pengujian aktivitas antiinflamasi berbagai ekstrak daun tanaman. J. Agritech. Food Process. **3**, 101–112 (2023). https://doi.org/10.31764/jafp.v3i2.20953
10. A. Chabeaud, L. Vandanjon, P. Bourseau, P. Jaouen, M. Chaplain-Derouiniot, F. Guerarda, Performances of ultrafiltration membranes for fractionating a fish protein hydrolysate: Application to the refining of bioactive peptidic fractions. Sep. Purif. Technol. **66**, 463–471 (2009). https://doi.org/10.1016/j.seppur.2009.02.012
11. I. Astiana, Nurjanah, T. Nurhayati, Karakteristik kolagen larut asam dari kulit ikan ekor kuning. J. Pengolahan Hasil Perikanan Indonesia **19**, 79–93 (2016). https://doi.org/10.17844/jphpi.2016.19.1.79
12. T. B. Zou, T. P. He, H. B. Li, H. W. Tang, E. Q. Xia, The structure-activity relationship of the antioxidant peptides from natural proteins. Molecules, 21(1), 72. (2016). https://doi.org/10.3390/molecules21010072
13. S. Dong, M. Zeng, D. Wang, Z. Liu, Y. Zhao, Antioxidant and biochemical properties of protein hydrolysates prepared from silver carp (*Hypophthalmichthys molitrix*). Food Chem. **107**, 1485–1493 (2008). https://doi.org/10.1016/j.foodchem.2007.10.011
14. M.I. Tangkau, Fatimawali, E.J. Suoth, Uji aktivitas antioksidan ekstrak etanol batang lengkuas putih (*Alpinia galanga*) dengan metode ABTS. Pharmacon **12**, 358–366 (2023). https://doi.org/10.35799/pha.12.2023.49216
15. Z. Theafelicia, S.N. Wulan, Perbandingan berbagai metode pengujian aktivitas antioksidan (DPPH, ABTS dan FRAP) pada teh hitam (*Camellia sinensis*). J. Teknol. Pertanian **24**, 35–44 (2023).

Antioxidant activity of catfish (*Pangasius* sp.) protein hydrolyzed by alcalase enzyme

Roni Nugraha[1*], *Agnes* Anathasia[1], *Nadia* Latifah Widiana[1], and *Mala* Nurilmala[1]

[1]Department of Aquatic Product Technology, Faculty of Fisheries and Marine Sciences, IPB University, Jl Agatis Kampus IPB Dramaga, Bogor, West Java 16680, Indonesia

Abstract. Protein hydrolysis using alcalase enzyme yields a high degree of hydrolysis, generating bioactive peptides with potential applications as nutraceuticals and functional food ingredients. Striped catfish (*Pangasius* sp.), characterized by its high protein and moderate fat content, represents a promising source of antioxidant peptides. This study aimed to evaluate the antioxidant activity of striped catfish protein hydrolysates produced by alcalase hydrolysis and subsequently fractionated using ultrafiltration. Hydrolysis was conducted with alcalase at concentrations of 1%, 2%, and 3%, followed by ultrafiltration through a 3 kDa MWCO membrane. Antioxidant activity was assessed using DPPH and ABTS assays. The highest degree of hydrolysis (43.37%) was observed at 3% enzyme concentration. The <3 kDa fraction exhibited strong antioxidant activity with IC50 values of 21.97 ppm (DPPH) and 3.28 ppm (ABTS). These findings demonstrate that ultrafiltration effectively produces smaller peptides with enhanced antioxidant activity.

1 Introduction

Protein hydrolysates are small peptides produced from the breakdown of protein polypeptides. Fish protein hydrolysates are the result of protein degradation into smaller molecules through hydrolysis by enzymes, acids, or bases. Bioactive peptides contained in protein hydrolysates function in the human body as nutraceuticals and functional foods. The advantages of producing fish protein hydrolysates include antioxidants and antibacterial activities. They also contain a complete profile of amino acids, have a protein digestibility of 87.03%, and possess good functional properties with solubility ranging from 99.84% to 99.99%, making them suitable for use as functional food ingredients. Bioactive peptides are protein fragments and amino acid derivatives with various health benefits. They act as antioxidants, anti-inflammatory, anti-obesity, anticancer, antihypertensive, antidiabetic, antimicrobial, and antithrombotic agents [1]. Antioxidants are highly beneficial in health and food sectors. In health, they prevent cancer, diabetes, cardiovascular disease, andpremature aging. In the food industry, antioxidants prevent lipid oxidation and inhibit bacterial growth in food products. The mechanisms of antioxidant activity can be categorized into five types: (1) free radical scavenging, such as 2,2′-azino-bis(3- ethylbenzothiazoline-6-sulfonic acid) (ABTS) and 2,2-diphenyl-1-picrylhydrazyl (DPPH), (2) reducing power such as Ferric Reducing Antioxidant Power (FRAP), (3) lipid peroxidation inhibition, (4) metal ion chelation, and (5) synergistic effects [2].

* Corresponding author: rnugraha@apps.ipb.ac.id

Enzymatic hydrolysis is considered the most efficient method for producing protein hydrolysates. Enzymes promote hydrolytic changes and degradation of proteins while generating peptides that are easily degraded. A study on catfish protein hydrolysates using papain enzyme reported inhibition percentages ranging from 37.85% to 67.62% [3]. In this research, the enzyme used is alcalase. Alcalase is widely applied to obtain bioactive peptides with activities such as antioxidant, antihypertensive, and anti-inflammatory effects. The advantages of alcalase include high catalytic activity, specificity, water solubility, and eco-friendliness. Alcalase has been reported to achieve a high degree of hydrolysis, up to 78% [4].

The main source of bioactive peptides is protein, and catfish (*Pangasius* sp.) is a fish species with high protein content. Catfish contain 16.1% protein and 5.7% fat [5]. It has high demand in both domestic and international markets due to its wide availability, relatively low cost, and widespread distribution. However, its utilization is still limited. Research on fish protein hydrolysates has been widely conducted using different enzymes and fish species, yielding various bioactive peptides with potential functions, particularly antioxidants. Shorter peptides tend to exhibit higher antioxidant activity. Therefore, this study aims to determine the optimal concentration of alcalase for hydrolyzing catfish protein and to evaluate the effect of ultrafiltration on the antioxidant activity of catfish protein hydrolysates in order to enhance their bioactive potential.

2 Materials and methods

2.1 Equipment and materials

The equipment used in this study included a knife, ruler, cutting board, blender, porcelain crucible, volumetric pipette, funnel, beaker glass, micropipette, microtube, vortex, test tube (Pyrex), measuring cylinder (Pyrex), Whatman No. 41 filter paper, ultrafiltration membrane (Amicon® Ultra-15 centrifugal filters 3K MWCO, Millipore, Ireland), , digital balance (OHAUS AX224, New Jersey, USA), pH meter (Hanna Instruments, Rhode Island, USA), UV-Vis spectrophotometer (RS Spectrophotometer UV-2500), centrifuge (FC5178R 120 V), electrophoresis apparatus (PEQ Lab, Erlangen, Germany), and water bath (B-ONE, China). The main materials used in this study were fresh catfish (Pangasius sp.) fillet obtained from Dramaga Market, Bogor, and Alcalase 2.4 L enzyme from *Bacillus licheniformis* (Sigma-Aldrich, St. Louis, USA) (activity 2.4 Anson Unit/mL), as well as Phosphate Buffer Saline (PBS) pH 8. Other materials used for analysis included distilled water, trichloroacetic acid (TCA) (Merck, Darmstadt, Germany), Bradford reagent, Tris-HCl 60 mM pH 6.8, glycerol, β-mercaptoethanol, Bovine Serum Albumin (BSA), DPPH (Sigma, Missouri, USA), ABTS (Sigma, Missouri, USA), $FeCl_3{\cdot}6H_2O$ (Merck, New Jersey, USA), ascorbic acid, ethanol, $K_2S_2O_8$, NaOH (Merck), SDS buffer solution (Merck, Darmstadt, Germany), protein marker (10–250 kDa, Bio-Rad, UK), acetic acid, gallic acid, Folin reagent, and Na_2CO_3.

2.2 Work procedure

This research was carried out in two stages. The first stage was the preparation of samples and characterization of catfish (*Pangasius sp.*) meat as the main raw material for producing fish protein hydrolysates. The second stage was the hydrolysis of catfish protein and the measurement of the degree of hydrolysis. The selected hydrolysate with the highest degree of hydrolysis was then subjected to ultrafiltration, followed by characterization of molecular weight, protein concentration, and antioxidant activity. In the final step, the selected catfish protein hydrolysate was further fractionated using a 3 kDa ultrafiltration membrane filter and subsequently re-characterized in terms of molecular weight, protein concentration, and antioxidant activity.

2.2.1 Preparation of catfish meat

Catfish (*Pangasius* sp.) was obtained from the Fish Market, Dramaga. Sample preparation was carried out by taking the fish meat. The catfish was first cleaned thoroughly using running water. The filleted meat was minced using a chopper or blender. The minced meat was then subjected to chemical composition analysis, including determination of moisture, ash, fat, and protein content.

2.2.2 Protein hydrolysis with alcalase enzyme

Protein hydrolysis was carried out by mixing the prepared catfish meat with buffer and alcalase enzyme. The catfish meat was homogenized with PBS buffer pH 8 (1:4) using a shaker for 2 minutes, then weighed, followed by the addition of alcalase enzyme at concentrations of 1%, 2%, and 3% (v/v). Hydrolysis was performed at 55 °C for 2 hours using a water bath shaker. Enzyme inactivation was then conducted by heating the sample at 90 °C for 15 minutes. The sample was cooled to room temperature and subsequently centrifuged at 10,000 rpm at 4 °C for 15 minutes to obtain the supernatant as the catfish protein hydrolysate. The selected protein hydrolysate was then analyzed for degree of hydrolysis, molecular weight, phenolic compounds, protein concentration, and antioxidant activity, followed by fractionation using ultrafiltration with a 3 kDa MWCO membrane.

2.3 Analysis procedure

2.3.1 Moisture Content [6]

Moisture content analysis was conducted to measure the amount of water contained in the sample. The initial step was drying the crucible in an oven at 105 °C for 1 hour. The crucible was then placed in a desiccator for 15 minutes until it reached room temperature, followed by weighing. A 3 g sample was placed into the crucible and dried in an oven at 105 °C for 5 hours. After drying, the sample was again placed in the desiccator until it reached room temperature and then weighed. The moisture content was calculated using the following formula:

$$Moisture\ content\ (\%) = \frac{B-C}{B-A} \times 100\% \qquad (1)$$

Description:
A: Mass of empty crucible (g)
B: Mass of crucible with sample before drying (g)
C: Mass of crucible with sample after drying (g)

2.3.2 Ash content [6]

Ash content analysis was conducted to measure the amount of minerals contained in the sample. One gram of the sample was placed into the crucible and ignited over a Bunsen burner flame until no smoke was observed. The crucible containing the sample was then placed in a muffle furnace at 600 °C for 6 hours. After ashing, the crucible with the sample was cooled in a desiccator to room temperature and weighed. The ash content was calculated using the following formula:

$$Ash\ Content\ (\%): \frac{B-A}{C} \times 100\% \quad (2)$$

Description:
A: Mass of empty porcelain crucible (g)
B: Mass of crucible with sample after ashing (g)
C: Mass of sample (g)

2.3.3 Fat content [6]

Fat content analysis was performed to determine the amount of fat in the sample by extracting lipids or oils from the food material using a non-polar organic solvent. The first step was cleaning, drying, and weighing an empty flask. Five grams sample was wrapped in filter paper and placed in the Soxhlet extractor, which was mounted under a condenser with the fat flask positioned below. Hexane solvent was added in sufficient volume, and reflux was carried out until the solvent returned to the fat flask. The distillation process was conducted to evaporate the solvent and collect it. The fat flask containing the lipid extract was then heated in an oven at 105 °C to evaporate the remaining solvent, while the fat remained due to its higher boiling point. After heating, the flask was cooled in a desiccator for 20–30 minutes and weighed. The weight of the extracted fat was used to calculate the fat content in the sample. The fat content was calculated using the following formula:

$$Fat\ content\ (\%): \frac{W3-W1}{W2} \times 100\% \quad (3)$$

Description:
W1: Mass of empty fat flask (g)
W2: Mass of sample (g)
W3: Mass of flask with fat extract (g)

2.3.4 Protein content [6]

Protein content analysis is carried out to determine the crude protein in a material. The Kjeldahl method is used for protein analysis, which involves three main steps: destruction, distillation, and titration. In the destruction step, a 0.3 g sample is placed into a 100 mL Kjeldahl flask, followed by the addition of 0.3 g selenium and 20 mL concentrated H2SO4. The sample is then heated at 410°C until the solution turns clear greenish-yellow, and subsequently cooled for 15 minutes. In the distillation step, 300 mL of distilled water and 100 mL of 40% NaOH are added to the clear solution, followed by distillation. The distillate is collected in a 125 mL Erlenmeyer flask containing a mixture of 10 mL 0.1 N H2SO4 with 0.1% methylene blue and 0.1% methylene red indicators in a 2:1 ratio. The resulting distillate solution appears bluish-green. The final step is titration using 0.1 N HCl until the solution in the Erlenmeyer flask changes to pink. The volume of titrant used is recorded to calculate the protein content of the sample. Protein content can then be determined using the following formulas :

$$Nitrogen\ Content\ (\%): \frac{volume\ HCl\ (mL) - volume\ blank\ (mL) \times N\ HCl \times 14}{Sample\ weight\ (mg)} \times 100\%$$

$$Protein\ content\ (\%wb) = \%N\ \times 6{,}25$$

$$Protein\ Content\ (\%wb): \frac{Protein\ content\ (\%bb)}{100 - moisture\ content\ (\%bb)} \times 100\% \quad (4)$$

Description:
wb: Wetbasis
db: Dry basis
N: Normality of HCl solution

2.3.5 Degree of hydrolysis

Degree of hydrolysis (DH) is determined to measure the ability of protease to hydrolyze a protein. First, 20 mL of protein hydrolysate is taken and mixed with 20 mL of 20% (v/v) TCA solution. The mixture is stored for 30 minutes to allow precipitation, followed by centrifugation at 7,800 × g for 15 minutes at 4°C. The obtained supernatant is then analyzed to determine the nitrogen content using the Kjeldahl method. The degree of hydrolysis can be calculated using the following formula:

$$Degree\ of\ Hydrolysis\ (\%) = \frac{Nitrogen\ content\ in\ supernatant}{Total\ nitrogen\ cntent\ in\ sample} \times 100\% \quad (5)$$

2.3.6 Protein concentration by Bradford method

Protein concentration assay using the Bradford method was performed with a 96-well plate. A series of Bovine Serum Albumin (BSA) standard solutions with graded concentrations (e.g., 0; 0.2; 0.4; 0.6; 0.8; 1.0; 1.2; and 1.4 mg/mL) was prepared. Protein hydrolysate samples from catfish were also prepared. A volume of 5 μL of either standard or sample solution was added into each well of the 96-well plate. Then, 250 μL of Bradford reagent was added to each well. The mixtures were gently mixed with a pipette or a shaker for 15–30 seconds to ensure homogenization. Samples were incubated at room temperature for 5 to 45 minutes. After incubation, absorbance was measured at 595 nm using a microplate reader. The absorbance values of the standards were used to construct a standard curve with absorbance (Y-axis) versus BSA concentration (X-axis). The protein concentration of the samples was determined by matching their absorbance values to the standard curve.

2.3.7 Molecular weight determination by SDS-PAGE

The sample was dissolved in 5% SDS solution and heated in a water bath at 85 °C for one hour. The mixture was then centrifuged at 8,500 × g for five minutes at room temperature. The separated supernatant was mixed with buffer containing 60 mM Tris-HCl (pH 6.8), 2% SDS, 25% glycerol (v/v), and 10% β-mercaptoethanol at a ratio of 1:10. The mixture was subsequently heated in boiling water for two minutes. A volume of 15 μL of each sample was loaded onto a polyacrylamide gel consisting of 15% separating gel and 5% stacking gel. Electrophoresis was carried out for two hours at a constant current of 50 mA per gel. Following electrophoresis, the gel was stained with 0.05% Coomassie Brilliant Blue R-250 prepared in 15% methanol and 5% acetic acid for one hour. The gel was then rinsed with 30% methanol and 10% acetic acid for one hour, followed by a de-staining step for 30 minutes.

2.3.8 Determination of peptides containing phenolic groups

The determination of phenolic content was conducted using a modified method of Apostolidis and Lee (2010). One milliliter of sample was mixed with 1 mL of 95% ethanol and 5 mL of distilled water. Subsequently, 0.5 mL of 50% (v/v) Folin–Ciocalteu reagent was added, and the mixture was homogenized using a vortex. After incubation for 5 minutes, 1 mL of 5% Na_2CO_3 solution was added and allowed to react for 60 minutes at room temperature. The absorbance of the solution was measured at 725 nm using a UV-Vis spectrophotometer. The total phenolic content was calculated based on a gallic acid standard curve and expressed as milligrams of gallic acid equivalent per gram of fresh sample (mg GAE/g).

2.3.9 DPPH antioxidant activity

The antioxidant activity was determined using the DPPH method. A 50 ppm DPPH solution was prepared by dissolving 5 mg of DPPH in 50 mL of methanol. The stock solution of the sample was prepared by dissolving 10 mg of the sample in 10 mL of methanol. Working solutions with concentrations of 2, 5, 20, 30, 60, and 100 ppm were prepared by pipetting the appropriate volume of the stock solution into test tubes and diluting with methanol to a final volume of 10 mL. The blank solution was prepared by mixing 2 mL of methanol with 2 mL of DPPH solution, and its absorbance was measured at 510 nm using a UV-Vis spectrophotometer. For the sample analysis, 2 mL of each concentration was pipetted into vials, followed by the addition of 2 mL of DPPH solution, with additions carried out at 1- minute intervals. After the addition of DPPH, the samples were incubated for 30 minutes at room temperature. The absorbance of each concentration was then measured using a cuvette. The IC_{50} value was calculated using linear regression analysis, where the regression equation ($y = ax + b$) derived from the plot of sample concentration versus percentage of antioxidant inhibition was used to determine IC_{50}.

2.3.10 ABTS antioxidant activity

ABTS radicals were generated by mixing 7.4 mM ABTS stock solution with 2.6 mM potassium persulfate, followed by incubation for 18 hours in the dark at room temperature. After incubation, the solution was diluted with deionized water until the absorbance reached 1.1 ± 0.05 at 405 nm. A total of 200 μL of ABTS solution was mixed with 100 μL of collagen hydrolysate sample and incubated for 15 minutes at room temperature. The absorbance of the mixture was measured at 405 nm using a microplate reader. Deionized water was used as the control to replace ABTS in the sample (sample control), and for the ABTS control, water was used to replace the hydrolysate. A standard curve was prepared using vitamin C solutions at concentrations ranging from 2.5 to 15.5 μg/mL.

3 Result and discussion

3.1 Proximate analysis

The chemical composition of *Pangasius* sp. fillet was analyzed to determine its suitability as raw material for protein hydrolysate production. The chemical composition of *Pangasius* sp. is presented in Table 1.

Table 1. The proximate composition of catfish flesh (*C. gariepinus*).

Parameter (%)	Catfish (*Pangasius sp.*)
Moisture	76.10 ± 0.01
Ash	0.91± 0.04
Fat	6.03 ± 0.01
Protein	16.49 ± 0.50

The striped catfish (*Pangasius hypophthalmus*), a freshwater species introduced to Indonesia in 1972, is widely cultivated due to its economic value and nutritional quality. The fish used in this study was in fillet form, containing 16.49% protein.

3.2 Characteristics of Catfish protein hydrolysate

The protein hydrolysate of striped catfish was produced through enzymatic hydrolysis using alcalase. Hydrolysis is the process of breaking chemical bonds in large molecules into smaller ones, where in protein hydrolysis, peptide bonds are cleaved into smaller peptides. The resulting hydrolysates typically exhibit a light brown to yellowish color, depending on the type of enzyme applied. Enzymes act as biocatalysts, accelerating the hydrolysis reaction and promoting more specific cleavage of peptide bonds. In this study, alcalase was used at concentrations of 1%, 2%, and 3%. The resulting supernatant represented the catfish protein hydrolysate, as shown in Figure 1.

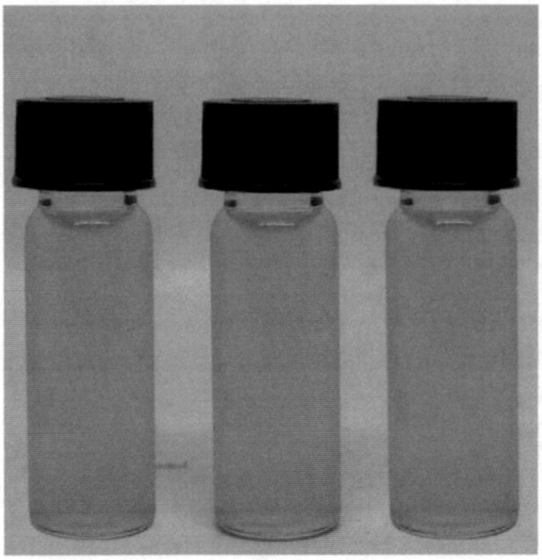

(a) **(b)** **(c)**

Fig. 1. Catfish (*Pangasius* sp) protein hydrolysates obtained with Alcalase enzyme concentrations of (a) 1%, (b) 2%, and (c) 3%.

The catfish protein hydrolysate (HPI) exhibited a turbid yellowish color across enzyme concentrations of 1%, 2%, and 3%, with only minor differences observed. The darker color observed in the 3% hydrolysate can be attributed to oxidation processes and peptide bond cleavage during hydrolysis, which release amino groups that participate in Maillard reactions with reducing sugars. This reaction, together with lipid oxidation typical of medium-fat fish such as catfish, contributes to the yellowish-brown color. Extended hydrolysis time further enhances this non-enzymatic browning, leading to a more intense brownish appearance [7]. Consistently, higher enzyme concentrations have been reported to yield darker hydrolysates, which matches the trend observed in this study.

3.2.1 Degree of hydrolysis

In this study, the DH of striped catfish protein hydrolysates was determined using the SN-TCA method, which measures the total soluble nitrogen content in 20% trichloroacetic acid (TCA) solution. The results of the DH measurement of striped catfish protein hydrolysates are presented in Figure 2.

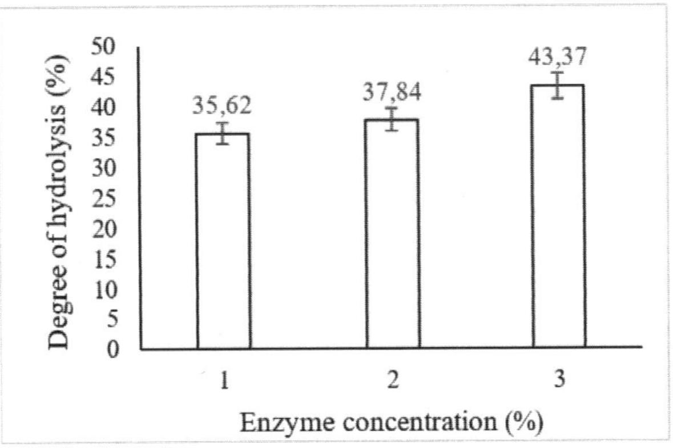

Fig. 2. Degree of hydrolysis of catfish (*Pangasius* sp) protein hydrolysates.

Figure 2 shows that the degree of hydrolysis (DH) values obtained using the SN-TCA method varied across treatments. The DH value at 1% enzyme concentration was 35.62%, which increased to 37.84% at 2% and showed a significant increase to 43.37% at 3%. This indicates that higher enzyme concentrations effectively break down proteins into peptides. A previous study reported that the DH value of shrimp head protein hydrolysates using alcalase was 30.11% [8]. The DH of striped catfish myofibrillar protein hydrolysates using alcalase at 3 mg/mL ranged from 40.60–83.6%, which was lower compared to papain hydrolysis (46.16–89.17%) [9]. Higher DH values lead to lower molecular weights, which are correlated with stronger antioxidant activity [10].

3.2.2 Soluble protein concentration

The protein hydrolysate sample of striped catfish with the highest degree of hydrolysis (DH) was further analyzed to determine protein concentration using the Bradford method. The Bradford assay is a colorimetric analysis used to quantify total protein concentration, employing Coomassie Brilliant Blue (CBB) dye, which binds to proteins and produces a blue color. The absorbance of the resulting solution is then measured spectrophotometrically at 595 nm. The measurement results of the soluble protein concentration are presented in Figure 3.

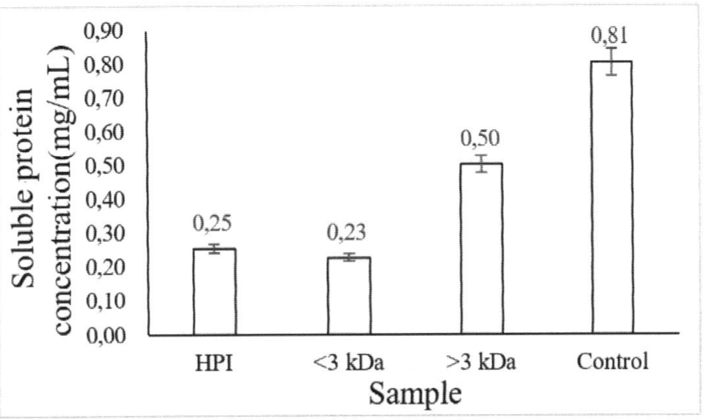

Fig. 3. Soluble protein concentration of catfish (*Pangasius* sp.) protein hydrolysate.

Figure 3 shows that the concentration of soluble protein varied across treatments. The highest protein concentration was observed in the control (0.81 ± 0.03 mg/mL), while the lowest was found in the <3 kDa hydrolysate fraction (0.23 ± 0.05 mg/mL). Analysis of variance indicated that enzyme activity and the applied method significantly affected soluble protein concentration ($P<0.05$), while Duncan's test revealed no significant difference between hydrolysates and the <3 kDa fraction, but a significant difference was observed between the >3 kDa fraction and the control. Results also showed that the <3 kDa fraction contained lower protein concentrations compared to the >3 kDa fraction.

3.2.3 Catfish protein hydrolysates after ultrafiltration

Ultrafiltration is a separation technique based on molecular weight using porous membranes, producing both filtrate and retentate fractions. In this study, crude extracts were filtered through a molecular weight cut-off (MWCO) membrane of 3 kDa (Amicon® Ultra- 15 centrifugal filters, 3K MWCO, Millipore). Ultrafiltration separates liquid components according to the size and structure of dissolved molecules under pressure. The principle of ultrafiltration is the movement of molecules across the membrane driven by concentration differences between the feed and the permeate. Smaller molecules that pass through the porous membrane form the permeate, while larger molecules retained by the membrane form the retentate [11]. The results of striped catfish protein hydrolysates fractionated by ultrafiltration are presented in Figure 4.

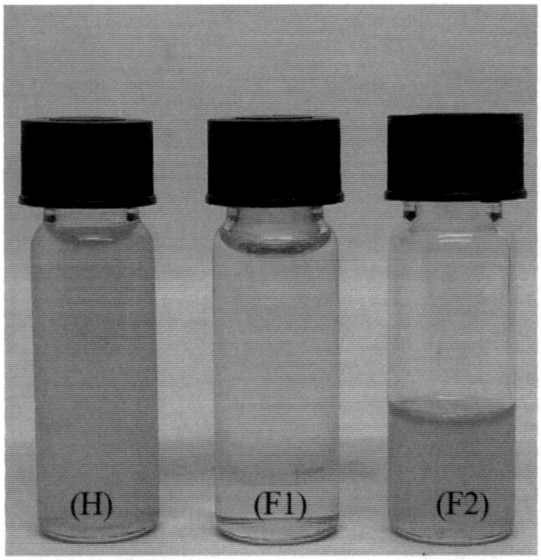

Fig. 4. Fractions of (*Pangasius* sp) protein hydrolysates: (H) protein hydrolysate, (F1) FPH <3 kDa fraction, (F2) FPH >3 kDa fraction.

Figure 4 shows that the striped catfish protein hydrolysate displayed a cloudy yellow appearance, with the <3 kDa fraction exhibiting a clearer golden-yellow color, while the >3 kDa fraction appeared darker, cloudier, and contained fine particles. This difference is attributed to the higher peptide concentration in the >3 kDa fraction compared to both the crude hydrolysate and the <3 kDa fraction. The distinct visual characteristics of each hydrolysate fraction were caused by ultrafiltration, which separates proteins based on molecular size. Larger peptides generally contain more aromatic residues, impurities, chromophores, and complex proteins, leading to a more turbid and yellowish appearance due to light scattering and higher absorbance. Ultrafiltration has been shown to enhance antioxidant activity, with lower molecular weight fractions (<5 kDa) exhibiting stronger antioxidant potential compared to larger fractions (>5 kDa) [12].

3.2.4 Molecular Weight of Catfish Protein Hydrolysates

Molecular weight represents the sum of atomic masses in a molecule. Electrophoresis is the most commonly used method to analyze molecular weight. According to Liu et al. [12], peptide fractions with molecular weight <3 kDa generally exhibit higher antioxidant activity due to their smaller size, which facilitates interactions with free radicals. The SDS-PAGE method uses discontinuous polyacrylamide gel as a supporting matrix and sodium dodecyl sulfate (SDS) to denature proteins. Protein markers with molecular weights ranging from 10–250 kDa were used. The molecular weight analysis of striped catfish fillet and protein hydrolysate can be seen in Figure 5. SDS-PAGE analysis of striped catfish fillet and protein hydrolysate revealed differences in molecular weight. Fresh catfish fillet showed molecular weights ranging from 10–227 kDa, while protein hydrolysate ranged from 5.9–9.5 kDa, and the >3 kDa fraction ranged from 6.8–9.5 kDa. In contrast, hydrolysates, particularly <3 kDa and >3 kDa fractions, displayed fewer and less visible bands.

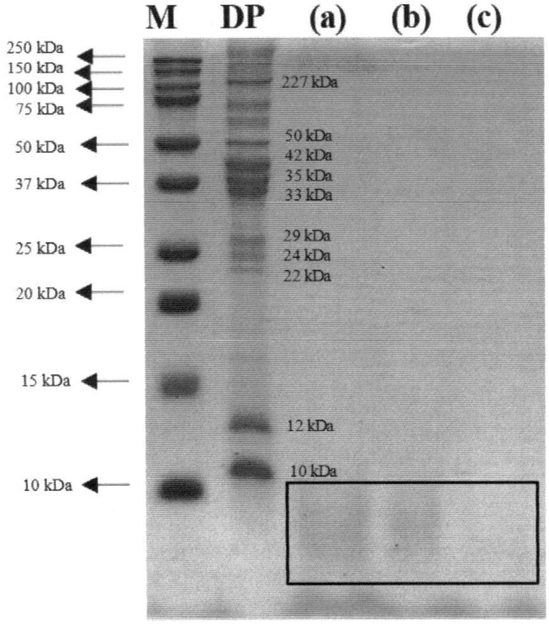

Fig. 5. Molecular weight of catfish protein hydrolysate. (M) Marker, (DP) catfish meat, (a) fish protein hydrolysate, (b) >3 kDa fraction, (c) <3 kDa fraction.

3.2.5 Phenolic-Containing Peptide Content

Phenolic compounds are characterized by hydroxyl groups attached to aromatic rings and can interact with proteins to form protein-phenolic complexes via covalent and non-covalent bonds. The total phenolic content of striped catfish fillet, protein hydrolysate, <3 kDa fraction, and >3 kDa fraction was measured to assess the contribution of phenolic compounds to the antioxidant activity of the protein hydrolysates. The results of the phenolic-containing peptides in protein hydrolysates can be seen in Figure 6.

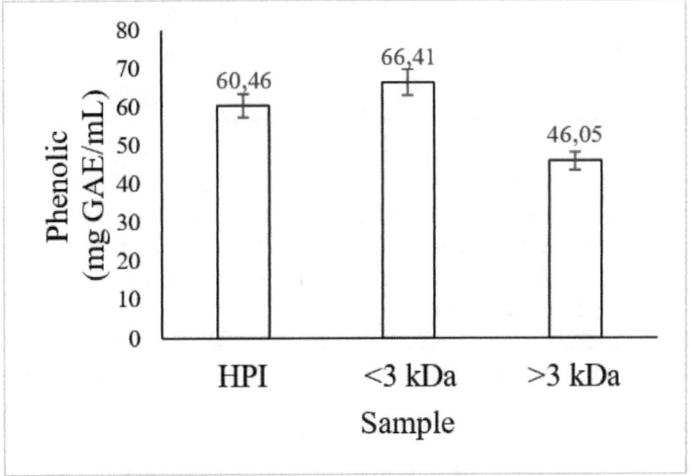

Fig. 6. Phenolic Group–Containing Peptides in Catfish (*Pangasius* sp) Protein Hydrolysates.

The determination of phenolic-containing peptides showed that striped catfish protein hydrolysates contain phenolic groups, with the highest content observed in the <3 kDa fraction. Analysis of variance indicated that molecular weight differences significantly affected the ultrafiltration results ($P<0.05$). The highest phenolic content was observed in the <3 kDa fraction (66.41 ± 6.47 mg GAE/L) and the lowest in the >3 kDa fraction (46.05 ± 9.57 mg GAE/L).

3.2.6 Antioxidant Activity of Catfish Protein Hydrolysate

a. DPPH Antioxidant Activity

The results of the DPPH antioxidant activity for protein hydrolysates can be seen in Figure 7. Duncan's multiple range test revealed that the <3 kDa fraction exhibited the lowest IC_{50} (21.97 ppm), significantly different from the >3 kDa fraction (62.42 ppm) and unfiltered hydrolysate (96.93 ppm). Previous studies reported that protein hydrolysates with 0–6% papain had DPPH inhibition values of 37.85–67.62% [3]. Factors such as enzyme conditions, pH, temperature, and time influence radical scavenging; extreme pH can reduce antioxidant activity by affecting protein structure [7]. Additionally, amino acid composition and peptide sequence are key determinants of hydrolysate antioxidant activity [13]. Several studies have reported that fish protein hydrolysates (FPH) exhibit varying antioxidant activities depending on the fish species, the type of enzyme used, and the hydrolysis conditions. For example, Kotson *et al.* [14] found that FPH derived from a mixture of rainbow trout (*Oncorhynchus mykiss*) and Atlantic salmon (*Salmo salar*) rest raw material showed antioxidant activity as well as DPP-IV inhibitory effect; the molecular weight distribution and hydrolysis pretreatment (high-pressure processing) influenced the antioxidant capacity significantly.

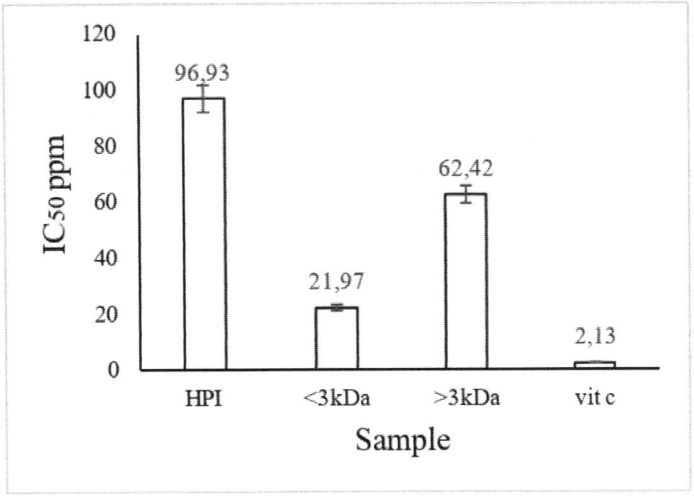

Fig. 7. DPPH antioxidant activity of catfish (*Pangasius* sp) protein hydrolysate.

b. ABTS Antioxidant Activity

The ABTS radical scavenging activity of the protein hydrolysates is shown in Figure 8. The results showed significant differences among the samples. Duncan's multiple range test revealed that the <3 kDa fraction had the lowest IC_{50} value of 3.28 ppm, indicating very strong antioxidant activity, whereas the >3 kDa fraction (5.70 ppm) and the unfiltered hydrolysate (8.35 ppm) had higher IC_{50} values. The <3 kDa fraction displayed stronger activity than the >3 kDa fraction due to its low molecular weight peptides containing hydrophobic amino acids with simpler structures that can donate hydrogen and interact effectively with free radicals [15].

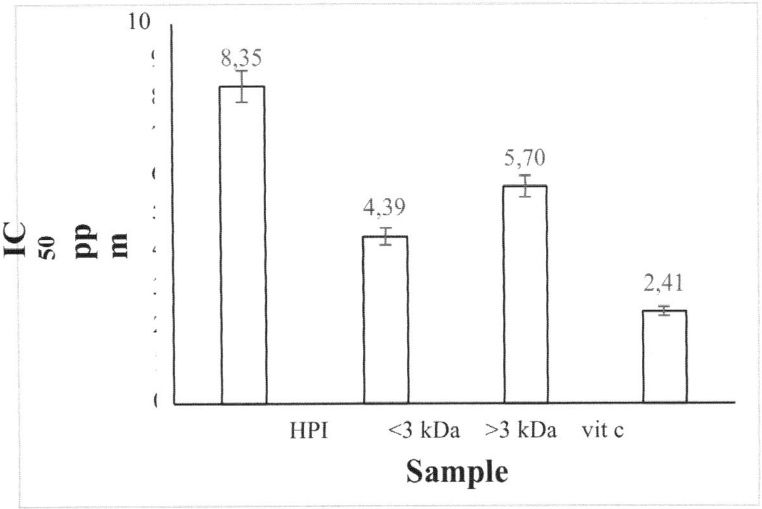

Fig. 8. ABTS antioxidant activity of catfish (*Pangasius* sp.) protein hydrolysate.

4 Conclusion

Pangasius protein hydrolysate treated with 3% alcalase enzyme concentration was the best treatment with the highest degree of hydrolysis. Ultrafiltration was able to separate small peptides <3 kDa and worked optimally on the antioxidant activity of the pangasius protein hydrolysate. The ultrafiltration results of the <3 kDa fraction had the strongest antioxidant activity compared to the *Pangasius* protein hydrolysate and the >3 kDa fraction.

References

1. A. Honrado, M. Miguel, P. Ardila, J.A. Beltrán, J.B. Calanche, From Waste to Value: Fish Protein Hydrolysates as a Technological and Functional Ingredient in Human Nutrition. *Foods 13*, 3120. (2024) https://doi.org/10.3390/foods13193120
2. N. Bibi Sadeer, D. Montesano, S. Albrizio, G. Zengin, M.F. Mahomoodally, The Versatility of Antioxidant Assays in Food Science and Safety—Chemistry, Applications, Strengths, and Limitations. *Antioxidants*, *9*, 709. (2020) https://doi.org/10.3390/antiox9080709
3. A. Baehaki, S.D. Lestari, A.R. Romadhoni, Hidrolisis protein ikan patin menggunakan enzim papain dan aktivitas antioksidan hidrolisatnya. J. Pengolah. Has. Perikan. Indones. **18**, 230– 239 (2015). https://doi.org/10.17844/jphpi.2015.18.3.230
4. V.G. Tacias-Pascacio, R. Morellon-Sterling, Siar H, Tavano O, Berenguer-Murcia Á, Use of Alcalase in the production of bioactive peptides: A review. Int. J. Biol. Macromol. 164, 1–14 (2020). https://doi.org/10.1016/j.ijbiomac.2020.10.060
5. A.E. Yunianto, I.S. Aisyah, D.S. Fatimah, A.D. Aulia, D. Salsabila, S. Khairinisa, S.R. Ningsih, V.P. Nuriyah, Z.T. Hadi, Makanan alternatif peningkatan kecukupan protein. J. Masy. Mandiri **7**, 1–2 (2023).
6. AOAC, Official Methods of Analysis of the Association of Official Analytical Chemists, 18th edn. (AOAC, Washington, 2005).
7. H.G. Kristinsson, B.A. Rasco, Fish protein hydrolysates: Production, biochemical, and functional properties. Crit. Rev. Food Sci. Nutr. **40**, 43–81 (2000).
8. T. Yuniarti, A. Prayudi, L. Supenti, H. Suhrawardan, H.M.P., Produksi dan profil kimia hidrolisat protein dari hasil samping pengolahan udang segar. J. Fish. Sci. **50**, 50–66 (2021). https://doi.org/10.22146/jfs.59906
9. M. Nurilmala, T. Nurhayati, R. Roskananda, Limbah industri filet ikan patin untuk hidrolisat protein. J. Pengol. Has. Perikan. Indones. **21**, 288–296 (2018). https://doi.org/10.17844/jphpi.v21i2.23083
10. C.P. Iduantoro, I. Zuraida, S. Sulistiawati, A. Mismawati, B.F. Pamungkas, Potensi peptida bioaktif dari hasil samping perikanan sebagai antihipertensi dan antioksidan – review. Media Teknol. Hasil Perikanan **12**, 19–26 (2024). https://doi.org/10.35800/mthp.12.1.2024.52202
11. M. Mahreni, Separation of mineral from brown algae extract using micro filtration (MF) and ultra filtration (UF) membranes – review. Eksergi **19**, 20–30 (2022). https://doi.org/10.31315/e.v19i1.6653
12. R. Liu, L. Xing, Q. Fu, G.H. Zhou, W.G. Zhang, A review of antioxidant peptides derived from meat muscle and by-products. Antioxidants **5**, 1–20 (2016). https://doi.org/10.3390/antiox5030032

13. F. Shahidi, Y. Zhong, Measurement of antioxidant activity. J. Funct. Foods **18**, 757–781 (2015). https://doi.org/10.1016/j.jff.2015.01.047

14. Kotson E, Daukšas E, Hansen Aas G, Rustad T, Tiwari BK, Lammi C, Cropotova J. 2024. Antioxidant activity and DPP-IV inhibitory effect of fish protein hydrolysates obtained from high-pressure pretreated mixture of rainbow trout (*Oncorhynchus mykiss*) and Atlantic Salmon (Salmo salar) rest raw material. *Marine Drugs*. 22(12): 568.

15. T. B. Zou, T. P. He, H. B. Li, H. W. Tang, E. Q. Xia, The structure-activity relationship of the antioxidant peptides from natural proteins. Molecules, 21(1), 72. (2016). https://doi.org/10.3390/molecules21010072

Effect of cinnamon masking and nanoencapsulation on fatty acid profile and health lipid indices (AI, TI) of tuna eye oil

Azizah Septiyani Irawan[1], *Wini* Trilaksani[1*], and *Joko* Santoso[1]

[1]Department of Aquatic Product Technology, Faculty of Fisheries and Marine Sciences, IPB University, Jl. Agatis Campus of IPB Dramaga, Bogor 16680 West Java, Indonesia

Abstract. Tuna eye oil is rich in omega-3 fatty acids, particularly eicosapentaenoic acid (EPA) and docosahexaenoic acid (DHA), yet its utilization is limited by a strong fishy odor and low oxidative stability. This study evaluated the effects of cinnamon masking and nanoencapsulation using gum arabic and Tween 80 on the fatty acid profile, atherogenicity index (AI), and thrombogenicity index (TI) of tuna eye oil. Pure tuna eye oil contained saturated fatty acids (SFA) 30.23%, monounsaturated fatty acids (MUFA) 26.93%, polyunsaturated fatty acids (PUFA) 42.76%, with EPA 6.94%, DHA 30.27%, AI 0.43, and TI 0.20. Cinnamon masking increased SFA and PUFA while decreasing MUFA, maintaining low AI and TI values due to the protective effect of cinnamon antioxidants. Nanoencapsulation reduced PUFA, particularly DHA, likely due to oxidation during processing, yet still resulted in low AI and TI values. These findings demonstrate that both treatments can modify the lipid profile and maintain favorable cardiovascular health indices, supporting the development of stable, consumer acceptable omega-3 rich products

1 Introduction

Tuna eye is a by-product of tuna processing that contains a high proportion of polyunsaturated fatty acids (PUFA), notably the long-chain omega-3 such as eicosapentaenoic acid (EPA) and docosahexaenoic acid (DHA). The PUFA content in oil obtained from tuna eye ranges from 36–48%, in wich EPA generally occurs at 4–6% and DHA can reach 25–37% of total fatty acids. [1]. PUFAs, especially those belonging to the n-3 group, are known to have positive health effects, including lowering the possibility of cerebrovascular and coronary complications, helping to decrease circulating triglycerides, contributing to blood pressure regulation, improving insulin sensitivity, supporting neural growth and maintaining visual system performance [2].

Despite its high nutritional value, tuna eye oil has limitations, such as a strong fishy odor and low oxidative stability. The fishy odor can reduce consumer acceptance, while its susceptibility to oxidation accelerates quality deterioration. One approach to overcome these issues is through masking with natural ingredients and nanoencapsulation. Masking is a method of adding strong-flavored ingredients to cover the fishy odor of fish oil. Previous studies have shown that cinnamon as a masking agent can effectively reduce fishy odor and improve overall sensory acceptance [3]. Nanoencapsulation is an encapsulation technique with a particle size of <100 nm, which serves to protect bioactive compounds from

* Corresponding author: wtrilaksani@apps.ipb.ac.id

degradation caused by exposure to oxygen, light, and temperature [4]. In addition to enhancing oxidative stability, this processing technique can also influence the composition of fatty acids in the oil [5].

Changes in in the proportion of fatty acids may affect health lipid indices such as the atherogenic index (AI) and thrombogenic index (TI). These indices help determine the potential risk or benefit of lipids to cardiovascular health. PUFAs including linoleic acid (LA), α-linolenic acid (LNA), EPA, and DHA have antithrombogenic effects, while saturated fatty acids (SFAs) such as lauric, myristic, and palmitic acids contribute to an increased risk of atherosclerosis and thrombosis [6]. This study aims to seeks the leverage of cinnamon masking and nanoencapsulation treatments on the fatty acid profile, AI, and TI of tuna eye oil. The findings are expected to add value to tuna by-products while enhancing their economic potential as a more stable and consumer-acceptable source of functional lipids.

2 Methods

2.1 Extraction

Tuna eye samples (*Thunnus albacares*) were obtained from a tuna processing company in Denpasar, Bali. The samples were collected fresh, directly stored at –20 °C, and kept under frozen conditions prior to processing. The extraction method followed the procedure described by [7]. The preparation stage involved thawing the tuna eyes, followed by cutting each eye into three parts to remove the sclera and lens. The remaining tuna eye tissue was then homogenized into a paste. The paste was subjected to ultrasonic treatment and subsequently centrifuged at a low temperature (4°C) at 10,000 rpm for 30 minutes. Centrifugation was conducted to separate the virgin fish oil (VFO) from other components (eye tissue, blood, water, and others). The extracted tuna eye oil was stored in was placed in dark bottles at –20°C.

2.2 Masking

The masking method was modified from [8]. The process was carried out by adding cinnamon powder at a concentration of 15% (w/w) of oil. Specifically, 7.5 g of cinnamon powder was added to 50 g of tuna eye oil, and the mixture was homogenized using a magnetic stirrer at 750 rpm for 15 minutes. The concentration of 15% was selected based on a previous study [3].

2.3 Nanoencapsulation

Nanoencapsulation of tuna eye oil followed the method described by [9], using gum arabic (25%) and Tween 80 (4%) as wall materials. Gum arabic was dissolved in distilled water using a magnetic stirrer at 500 rpm and 60°C for 1 hour, then stored in a refrigerator for 24 hours to achieve maximum hydration. Tween 80 was then added as an emulsifier, followed by the gradual addition of fish oil at a concentration of 6%. The mixture was immediately stirred for 30 minutes at 500 rpm using a magnetic stirrer, and subsequently subjected to size reduction using a homogenizer at 25,000 rpm for 3 minutes. The resulting emulsion was then sonicated at 24 kHz, 5 minutes.

2.4 Fatty acid profile

The analytical method was based on transforming fatty acids into their derivative forms, namely methyl esters, to enable detection by chromatography. Separation of components was performed by detecting peaks on the chromatogram corresponding to specific retention times, which were then compared with standards. The procedure began by weighing 25 mg of fish oil sample and mixing it with 1 mL of NaOH solution in methanol, followed by heating for 20 minutes. Subsequently, 2 mL of 20% BF_3 solution along with an internal standard was added, and the mixture was reheated for another 20 minutes. After cooling, 2 mL of saturated NaCl solution and 1 mL of isooctane were added, mixed, and the isooctane layer was separated and dried over anhydrous sodium sulfate prior to injection into the instrument. A 1 μL aliquot of the fatty acid methyl esters (FAME) was injected in split mode into an HP-88 capillary column (100 m × 0.25 mm × 0.20 μm). Helium was used as the carrier gas at a flow rate of 1.0 mL/min. The injector and detector temperatures were set at 250°C and 280°C, respectively. The oven temperature program was set at 120°C (held for 2 min), increased by 10°C/min to 180°C (maintained for 10 min), and further increased by 5°C/min to 220°C (held for 5 min). Identification of fatty acids was achieved by matching the chromatographic retention times of sample peaks with certified FAME references

2.5 Atherogenicity index (AI) and thrombogenicity index (TI)

The atherogenicity index (AI) and thrombogenicity index (TI) derived from the proportion of principal saturated fatty acids relative to key unsaturated fatty acid groups. The indices were computed using equations previously outlined by [6] the AI and TI values:

$$AI = \frac{C12{:}0+(4xC12{:}0)+C16{:}0}{\sum MUFA+\sum n6\ PUFA+\sum n3\ PUFA} \quad (1)$$

$$TI = \frac{C14{:}0+C16{:}0+C18{:}0}{(0.5x\sum MUFA)+(0.5x\sum n6\ PUFA)+(3x\sum n3\ PUFA)+(\frac{\sum n3\ PUFA}{\sum n6\ PUFA})} \quad (2)$$

2.6 Data analysis

The dataset was processed through one-way analysis of variance (ANOVA) to evaluate differences among the three types of fish oils: virgin fish oil (VFO), masked fish oil (MFO), and nanoencapsulated fish oil (NFO). Although these samples represent different forms of fish oil rather than treatment levels, ANOVA was applied to statistically compare their physicochemical characteristics. Prior to analysis, data were tested for normality and homogeneity of variances at a 95% confidence level ($\alpha = 0.05$). When significant differences were found ($p < 0.05$), Duncan's multiple range test was conducted to identify pairwise differences among sample types. Statistical analyses were performed using SPSS version 22.0.

3 Result and discussion

The results of the analysis indicate that processing methods, such as masking and nanoencapsulation, exert a notable influence on the fatty acid profile of tuna eye oil (Table 1). The fatty acid profile data for the masking treatment were obtained from a previous report [3]. In the masking treatment, an increase was observed in SFA, EPA, DHA, and PUFA values, while MUFA values decreased. This reduction in MUFA may be attributed to the heating step during the masking process, conducted at 50 °C. Such a temperature has the potential to cause degradation of monounsaturated fatty acids (MUFA), as MUFA contain double bonds that are susceptible to heat-induced damage.

Different heat treatments can affect both the type and quantity of fatty acids produced. High temperatures are likely to break certain double bonds in fatty acids and replace them with single bonds, thereby altering the fatty acid composition. The increase in SFA, EPA, DHA, and PUFA in the masking treatment may be related to the decrease in MUFA, given that the percentage composition is recalculated based on the total fatty acids after processing. SFA are more stable due to their saturated carbon chains, making them more likely to remain unchanged or increase proportionally during processing. The rise in PUFA levels could be attributed to higher proportion of EPA and DHA, which in turn may be linked to the presence of antioxidants in cinnamon used as a masking agent, protecting these fatty acids. [10] noted that antioxidants can remove prooxidant impurities. This finding is consistent with [11] who reported that fish oil supplemented with tocopherol antioxidants contained higher EPA, DHA, and PUFA levels compared to oil without antioxidant addition.

Table 1. Fatty acid profile.

Fatty acid (%)	**Samples**		
	VFO	**MFO**	**NFO**
C12:0	0.01	0.15	0.03
C13:0	0.03	0.03	0.02
C14:0	3.02	3.05	2.65
C15:0	0.86	0.94	0.79
C16:0	19.47	18.60	17.44
C17:0	1.26	0.99	1.09
C18:0	4.92	4.74	4.34
C20:0	0.40	0.35	1.00
SFA	30.20 ± 0.07^{b}	31.60 ± 0.00^{c}	26.95 ± 0.01^{a}
C14:1	0.08	0.08	0.04
C16:1	6.56	5.67	5.03
C17:1	1.13	1.08	0.93
C18:1	17.81	16.55	28.39
C20:1	1.23	1.35	1.20
MUFA	26.94 ± 0.05^{b}	24.88 ± 0.03^{a}	36.21 ± 0.04^{c}
C18:2	1.29	1.21	1.32
C18:3	0.45	0.39	0.38
C20:2	0.31	0.36	0.26
C20:3	0.17	0.14	1.52
C20:4	3.08	3.05	2.58
EPA, C20:5n-3	6.95 ± 0.04^{b}	7.31 ± 0.01^{c}	6.31 ± 0.01^{a}
DHA, C22:6n-3	30.24 ± 0.11^{b}	30.42 ± 0.01^{c}	25.56 ± 0.01^{a}
PUFA	42.76 ± 0.13^{b}	43.46 ± 0.04^{c}	36.84 ± 0.05^{a}

Notes: Numbers followed by different superscript letters within columns (a–c) indicate significant differences ($p < 0.05$). VFO: Virgin Fish Oil; MFO: Masking Fish Oil; NFO: Nanoencapsultion Fish Oil.

In the nanoencapsulation treatment, fatty acid contents decreased compared to the masking sample. The DHA value in this treatment was 25%, lower than the 42–43% observed in VFO and MFO. This reduction is most likely associated with the encapsulation stages, such as homogenization, ultrasonication, and freeze-drying. During freeze-drying, unsaturated fatty acids may come into contact with oxygen, triggering oxidation and lipid degradation [12]. Homogenization may also reduce DHA content due to increased exposure to prooxidant compounds and oxygen, induced by cavitation and localized heating during processing [13]. Ultrasonication has a similar impact, reducing fatty acid levels, including DHA, in nanocapsules.

The predominant MUFAs in tuna eye oil are oleic acid and palmitoleic acid, whereas the dominant SFAs are palmitate and stearate In the PUFA group, the most abundant fatty acids are EPA and DHA. Variations in the percentage of fatty acids in fish oil influence health-related lipid indices, which are represented by the atherogenic (AI) and thrombogenic index (TI). The AI values are presented in Figure 1.

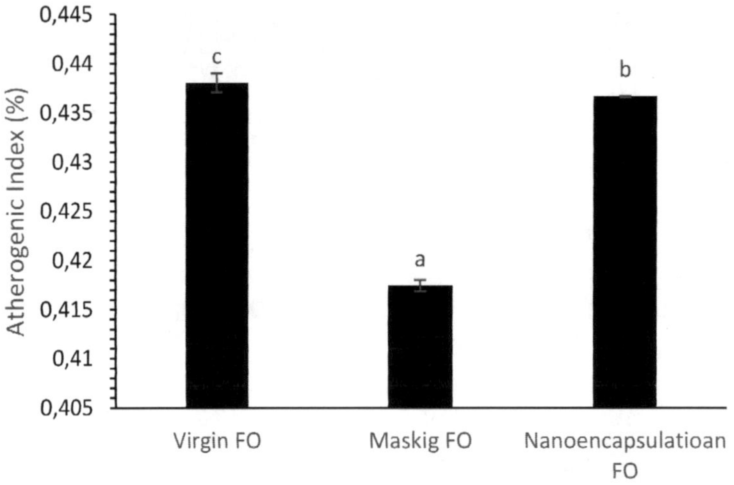

Fig. 1. Atheroghenecity index tuna eye oil. Different letters in the graph indicate significantly different results ($p < 0.05$).

Indices relating to atherogenesis (AI) and thrombogenesis (TI) can be calculated to describe the effects of fatty acid variability on human health. AI and TI indicate the potential for atherosclerotic plaque formation, thrombus development, and blood clotting. The TI values are presented in Figure 2.

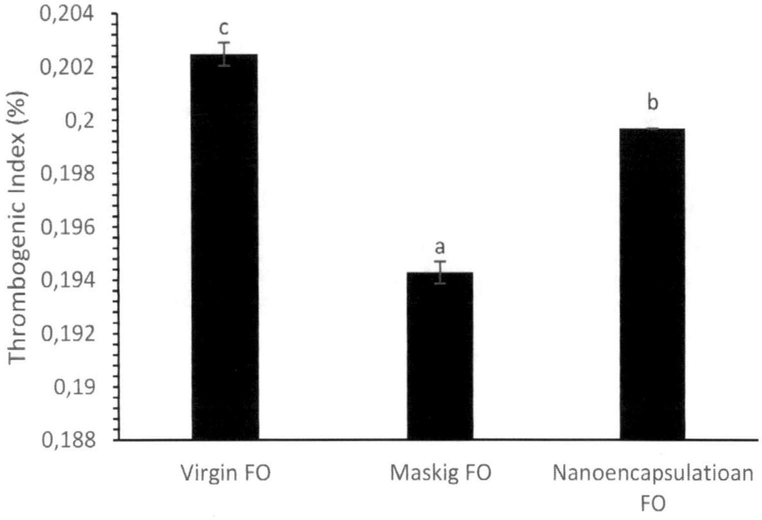

Fig. 2. Thrombogenicity index. Different letters in the graph indicate significantly different results ($p < 0.05$).

The AI and TI values for each type of tuna eye oil showed significant differences among treatments ($P < 0.05$). Variations in AI and TI values in tuna eye oil may be influenced by the extraction method, processing temperature, pretreatment, and the degree of interaction between the material and the oil [14]. Moreover, the lipid composition of the oil further influences AI and TI indices. These indices are determined by the balance between pro-thrombogenic fatty acids (saturated fatty acids) and anti-thrombogenic fatty acids (MUFA and PUFA). A high unsaturated fatty acid content in the oil reduces AI and TI values. Low index values in tuna eye oil reflect its high PUFA content, where a high proportion of PUFA indicates good nutritional quality of the studied fish oils.

In this study, AI values were recorded between 0.41 and 0.43 (Fig. 1), whereas TI values were observed between 0.19 and 0.21 (Fig. 2). These results indicate that tuna eye oil has a low potential risk for cardiovascular disease. This finding is consistent with [1], who reported an AI value of 0.47 and a TI value of 0.20 for medium-sized tuna eye oil. High AI and TI values are associated with a heightened likelihood of cardiovascular complications due to the consumption of certain fatty acids. Lipid sources exhibiting AI scores under 1 and TI levels below 0.5 are regarded as highly suitable for dietary use [15].

4 Conclusion

Cinnamon masking and nanoencapsulation significantly affected the fatty acid profile, AI, and TI of tuna eye oil. Masking increased SFA, EPA, DHA, and PUFA levels, while reducing MUFA, and maintained low AI and TI values due to the protective effect of cinnamon antioxidants. Nanoencapsulation reduced PUFA, particularly DHA, likely due to processing-induced oxidation. All treatments produced tuna eye oil with low AI (0.41–0.4) and TI (0.19–0.21) values.

References

1 W. Trilaksani, B. Riyanto, W. Ramadhan, F. Sinulingga, S. Fauziah, The characteristics of PUFAs-rich virgin fish oil as affected by size of tuna eye. Biodiversitas, **24**(12), 6545–6556 (2023).

2 Li, O. Bode, H. Drummond, A. J. Sinclair, Lipid for functional food and nutriaceuticals, omega-3 (n-3) fatty acid. The Oily Press, Bridgewater, U.K. (2003).

3 A.S. Irawan, W. Trilaksani, J. Santoso, Improving oxidative stability and sensory acceptance of DHA-rich tuna eye oil using cinnamon and mastic as natural masking agents. J. Pengolahan Hasil Perikanan Indonesia, **28(**7), 591–602 (2025). https://doi.org/10.17844/jphpi.v28i7.64337

4 A. Ye, J. Flanagan, H. Singh, Formation of stable nanoparticles via electrostatic complexation between sodium caseinate and gum arabic. J. Food Sci., **71**(2), R70–R77 (2006). https://doi.org/10.1111/j.1365-2621.2006.tb08905.

5 Y. Soleimanian, M.A. Sahari, M. Barzegar, Influence of processing parameters on physicochemical properties of fractionated fish oil at low temperature crystallization. Nutr. Food Sci., **45**(1), 2–15 (2015). https://doi.org/10.1108/NFS-05-2014-0042

6 T.L.V. Ulbricht, D.A.T. Southgate, Coronary heart disease: Seven dietary factors. Lancet. **338**(8773), 985–992 (1991). https://doi.org/10.1016/0140-6736(91)91846-M

7 W. Trilaksani, B. Riyanto, F. Azzahra, J. Santoso, K. Tarman, Recovery of tuna virgin fish oil and formulation as a product model of emulsion food supplement. IOP Conf. Ser.: Earth Environ. Sci., 1–8 (2020).

8 G. Song, M. Zhang, X. Peng, X. Yu, Z. Dai, Q. Shen, Effect of deodorization method on the chemical and nutritional properties of fish oil during refining. LWT - Food Sci. Technol., **96,** 560–567 (2018).

9 F.V. Moghadam, A. Pourahmad, D. Mortazavi, R. Davoodi, R. Azizinezhad, Use of fish oil nanoencapsulated with gum arabic carrier in low fat probiotic fermented milk. Food Sci. Anim. Resour., **39**(2), 309–323 (2019).

10 D. Cameron-Smith, B.B. Albert, W.S. Cutfield, Fishing for answers: Is oxidation of fish oil supplements a problem? J. Nutr. Sci., 4, **36** (2015). https://doi.org/10.1017/jns.2015.3

11 V. Šimat, J. Vlahović, B. Soldo, D. Skroza, I. Ljubenkov, I. Generalić Mekinić, Production and refinement of omega-3 rich oils from processing by-products of farmed fish species. Foods, **8**(4), 125 (2019). https://doi.org/10.3390/foods8040125

12 M.C. Castaneda-Saucedo, et al., Effect of freeze-drying and production process on the chemical composition and fatty acids profile of avocado pulp. Rev. Chil. Nutr., **41**(4), 404–411 (2014). https://doi.org/10.4067/S0717-75182014000400009

13 D.C. Valencia-Flores, M. Hernandez-Herrero, B. Guamis, V. Ferragut, Comparing the effects of ultra-high-pressure homogenization and conventional thermal treatments on the microbiological, physical, and chemical quality of almond beverages. J. Food Sci., **78**(2), E199–E205 (2013).

14 Q. Ghazali, N.H.M. Yasin, The effect of organic solvent, temperature and mixing time on the production of oil from Moringa oleifera seeds. IOP Conf. Ser.: Earth Environ. Sci., **36**, 012053 (2016).

15 B. Abbasi, E. Ghanbarzadeh, B. Panahizadeh, Association between atherogenic, thrombogenic, and lipophilic indices and the odds of diabetic nephropathy in type 2 diabetic patients: A case–control study. Food Sci. Nutr., (2024). https://doi.org/10.1002/fsn3

https://doi.org/10.1051/bioconf/202622102005

Effect of lighting on heart rate performance of Eel (*Anguilla bicolor*)

Anindhita Azzahra Natasya[1], *Wazir* Mawardi[1], *Ari* Purbayanto[1], *Mochammad* Riyanto[1*]

[1] Marine Fisheries Technology Study Program, Department of Fisheries Resource Utilization, Faculty of Fisheries and Marine Sciences, IPB University, Bogor, Indonesia

Abstract. Eels (*Anguilla bicolor*) are catadromous fish with a complex life cycle, in which the elver stage represents a critical transitional phase from marine to freshwater environments. Environmental factors such as light significantly influence eel physiology, migratory behavior, and metabolic activity. This study evaluated the effects of different light spectra on the cardiac performance of elver-stage eels using a non-invasive electrocardiogram (ECG) technique. The experiment was conducted at the Fish Behavior Laboratory, Faculty of Fisheries and Marine Sciences, IPB University, from January to May 2024. Three lighting treatments were applied: blue light (~450 nm), full-spectrum white light, and a control condition without additional lighting. A Completely Randomized Design (CRD) was employed with three replicates per treatment. Results indicated that white light significantly increased heart rate (69.6 ± 0.92 bpm) compared to blue light (46.6 ± 0.90 bpm) and control conditions (48.6–51.2 bpm) ($P < 0.05$), suggesting a heightened stress response under white light exposure. In contrast, blue light maintained more stable cardiac activity, indicating reduced physiological stress. These findings are relevant for the optimization of eel fishing techniques, conservation, and fish welfare management.

1 Introduction

Eels (*Anguilla* spp.) are catadromous species that reproduce in the ocean, with their larvae and juveniles migrating into coastal and freshwater habitats to grow before returning to the sea for spawning. A critical transition in this life cycle is the elver stage, when juveniles move from marine to freshwater environments and undergo major physiological adjustments, particularly in osmoregulation, morphology, and behavior, that enable freshwater residency and upstream migration [1]. Conditions at river mouths, including light intensity, temperature, and substrate type, strongly influence entry timing, migration success, and spatial distribution of arriving elvers [2, 3].

Light is a key environmental cue for fish, regulating circadian rhythms, feeding behavior, and metabolic rate. High daytime light can increase activity and therefore oxygen and nutrient demands, placing a greater workload on the heart. Conversely, abrupt or unnatural light regimes typical in captivity can induce stress responses that alter heart rate and blood pressure [4]. Organisms living under strong light fluctuations often show cardiac adaptations that reflect those conditions. Specific wavelengths also matter blue light has been reported to influence retinal signaling and melatonin pathways, improving metabolic efficiency and activity in some fish, whereas broad white light can support growth but may cause stress if it mismatches natural habitat conditions [5-9]. Eels are primarily nocturnal and prefer darker microhabitats, although artificial illumination can alter their activity patterns [10, 11].

* Corresponding author: mochammadri@apps.ipb.ac.id

Cardiac performance is a sensitive indicator of physiological response to environmental change. Electrocardiography (ECG) is widely used to measure heart rate and rhythm in fish under varying light and activity conditions [12, 13]. Despite this, studies addressing how light intensity affects cardiac function specifically in elver-stage eels are limited. Consequently, this research explored how varying light intensities influence the cardiac activity of eel elvers, offering information that can support better culture techniques and conservation efforts.

2 Materials and methods

Data collection and observation of eel cardiac performance were carried out at the Fish Behavior Laboratory, Department of Fisheries Resource Utilization, Faculty of Fisheries and Marine Sciences, IPB University, from January to May 2024. The experiment examined how different lighting conditions affect cardiac performance in elver-stage eels (*Anguilla* spp.). Given the species' nocturnal behavioral patterns, data acquisition and physiological observations were predominantly performed during nighttime to ensure ecological relevance and minimize behavioral disturbance.

2.1 Preparation of eel samples

The experimental subjects consisted of ten elver-stage *Anguilla* bicolor, with an average total length of 29.3 ± 1.1 cm and a body weight of less than 200 g. Before the commencement of the experimental trials, the specimens underwent an acclimatization period exceeding 40 days in a controlled rearing tank (dimensions: 194.5 × 101 × 62 cm), with a maintained water depth of 27 cm. The tank was continuously aerated with freshwater to ensure optimal oxygenation. During acclimatization, the eels were fed regularly using a pellet and kept under stable environmental conditions, where the water temperature was maintained around 25°C with a pH between 7 and 8.

2.2 Methods

Cardiac activity in elver-stage *Anguilla* bicolor was assessed using a non-invasive electrocardiography (ECG) technique. The instrumentation included an oscilloscope (Tektronix TDS2014C), a bio-amplifier (Nihon Kohden AB-632), graphite electrodes for signal detection [10], and auxiliary components such as multimeters, copper cables, connectors, and smartphone cameras for visual documentation. Each observation session lasted 30 minutes, with 24 experimental replicates conducted.

The trials were performed in a glass aquarium (57.5 × 31 × 29 cm) equipped with continuous aeration and an artificial lighting system utilizing a 10-watt RGB LED floodlight. Two lighting treatments were applied: blue light (~450 nm) and full-spectrum white light. Light intensity levels followed the protocol by Riyanto [10], with blue light ranging from 0 to 5.2×10^{-8} W/cm^2 and white light from 0 to 1.7×10^{-8} W/cm^2. The lamp was set at a height of 30 cm above the water so that the entire aquarium received evenly distributed light.

Electrocardiographic signals were captured via graphite electrodes placed externally on the fish's body, transmitting electrical activity to the bio-amplifier. The amplified signals were visualized as waveforms on the oscilloscope and recorded using a smartphone camera. All recordings were subsequently stored on a flash drive for post-experimental analysis.

Water quality parameters, including temperature and other relevant physicochemical properties, were monitored throughout the experiment to maintain optimal conditions. Heart rate data were statistically analyzed to determine the effects of lighting treatments on cardiac performance, employing ANOVA with additional post-hoc comparisons performed when necessary. The study used a Complete Random Design (RAL) with four sequential phases: Control 1 (dark), Blue Light, Control 2 (dark), and White Light. Each treatment lasted 30 minutes with three repetitions. A 1-hour rest interval is given to minimize residual effects.

2.3 ECG recording

Electrocardiographic (ECG) recordings were conducted to assess the cardiac activity of elver-stage *Anguilla* bicolor under different lighting conditions. Each eel elver was chosen at random from the acclimation tank and then placed into the experimental aquarium (57.5 × 31 × 29 cm). Within the aquarium, each fish was placed inside a PVC pipe (6 cm diameter, 35 cm length) designed to minimize movement and create a calm environment conductive to accurate physiological monitoring.

Two graphite electrodes were positioned at both ends of the pipe using wooden supports to ensure stability and consistent signal acquisition. These electrodes were connected via copper wires to a signal amplifier (Nihon Kohden AB-632), which amplified the bioelectrical signals generated by the fish's cardiac activity. The amplified signals were displayed on an oscilloscope (Tektronix TDS2014C). To minimize electrical noise and interference, grounding wires were attached to a metal table as a neutral reference point.

Prior to data collection, a heart rate normalization procedure was implemented to ensure that each fish's cardiac rhythm had stabilized to baseline conditions [10]. Each fish was used only once per experimental session to avoid repeated stress exposure. After the observation, the fish was transferred to a separate maintenance tank of identical dimensions to promote recovery and minimize physiological disturbance.

Heart rate waveforms were recorded via video using a smartphone camera and stored on a flash drive for subsequent analysis. Three lighting treatments were applied: blue light (~450 nm), full-spectrum white light, and dark control. Illumination was provided using an RGB LED floodlight, calibrating intensities according to established standards [10].

Every treatment was conducted in triplicate, producing 24 total trials in the experiment. For each trial, heart activity was monitored over a 30-minute recording period. The sequence of treatments within a single trial cycle was as follows: control 1 (dark), blue light, control 2 (dark), and white light. All treatments were applied sequentially to the same individual, with sufficient recovery time between exposures. The 30-minute duration was selected based on prior studies indicating its adequacy for capturing physiological responses to light stimuli. A summary of the experimental layout and its procedural steps is illustrated in Figure 1.

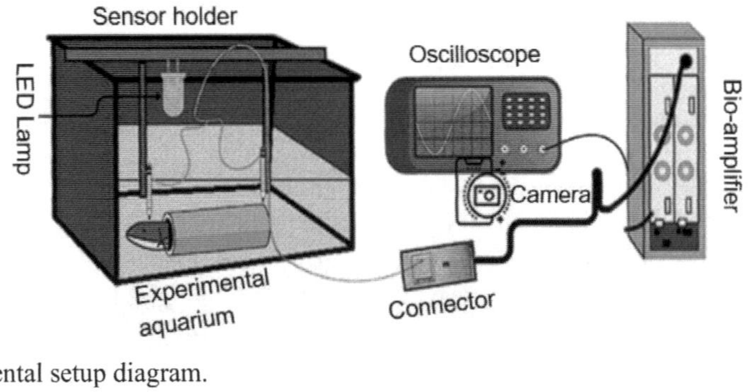

Fig. 1. Experimental setup diagram.

2.4 Data analysis

Eel heart rate data was analyzed using SAS software with the Duncan test to compare the averages between lighting treatments. The study used a complete random design (RAL) with one factor, namely light colors: blue, white, and dark as controls. The analysis was conducted to evaluate how each lighting treatment affects the physiological state of the fish. The statistical model used in this study is presented as follows:

$$Yij = \mu i + \tau i + \varepsilon ij \ atau \ Yij = \mu i + \varepsilon ij \quad (1)$$

i	:	Treatment
j	:	Deuteronomy
Yij	:	Observations on the first treatment of the j repeat
μ	:	General average
The	:	Effects of the i-treatment
€ij	:	Error of the first treatment experiment of the j test

3 Results and discussion

3.1 Results

Previous studies indicate that this lighting type has a direct influence on the cardiac rhythm of eel elvers. Under blue light, their heartbeat tends to decrease, which can be seen from the number of peak ECG waves. Figure 2 illustrates how the light treatments shaped the elvers' cardiac response.

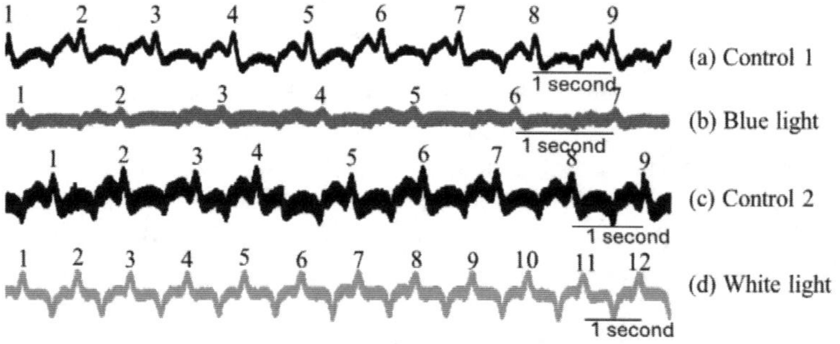

Fig. 2. Eel heart rate profile (30 cm): (a) control 1, (b) blue light, (c) control 2, (d) white light.

Overall, white light seems to have the greatest effect in increasing the heart rate of eels, which can be seen from the consistency of high heart rate points in various total lengths of fish. This suggests that no matter the length of the fish, white light causes a consistent increase in heart rate across the sample. This effect stands out when compared to blue light and control conditions, where the heart rate tends to be more stable and less affected by lighting. The fish's heart rate in control 1 was recorded to range from 32 to 62 beats per minute (bpm) with an average ±SE of 48.6±0.61 bpm. In the blue light treatment, the heart rate reached 40 to 54 bpm (46.6±0.90 bpm), and in the 2-heart rate control, it was recorded at 40 to 62 bpm (51.2±0.60 bpm). However, in the white light treatment, the heart rate increased significantly to 66 to 74 bpm (69.6±0.92 bpm), which showed a significant effect of white light on the increase in heart rate. The variation in eel heart rates under control 1, blue light, control 2, and white light across different length categories is shown in Figure 3.

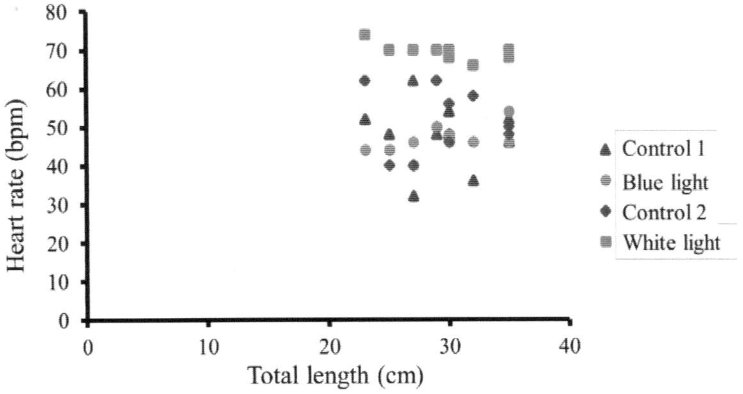

Fig. 3. Comparison of heart rate of eel at different length and light color.

Statistically, the white light treatment produced the highest mean heart rate and showed a significant difference ($P<0.05$) compared to the blue light treatment and control. Control 2 and control 1 showed intermediate values, while blue light produced the lowest heart rate (Table 1). These results indicate that blue light favors a stable physiological condition for eel fish than white light.

Table 1. Average heart rate of eels under different treatments.

Treatment	Heart rate (beats/minute) ± SE	Sample size (n)
Control 1	48.6 ± 0.61a	10
Blue Light	46.6 ± 0.90a	10
Control 2	51.2 ± 0.60a	10
White Light	69.6 ± 0.92b	10

The box plot graph shows the eel heart rate at four treatments: blue light (CB), white light (CP), initial control (K1), and control after blue treatment (K2). White light produces the highest median heart rate (~70 bpm) with low variation and one outlier. Blue light has the lowest median (~45 bpm) with a stable data distribution. Controls 1 and 2 showed a median of about 50 bpm, but K2 had the widest data distribution and few outliers, indicating high variability (Figure 4). Overall, white light was the most powerful in influencing the increase in heart rate, while blue light showed the most stable effect.

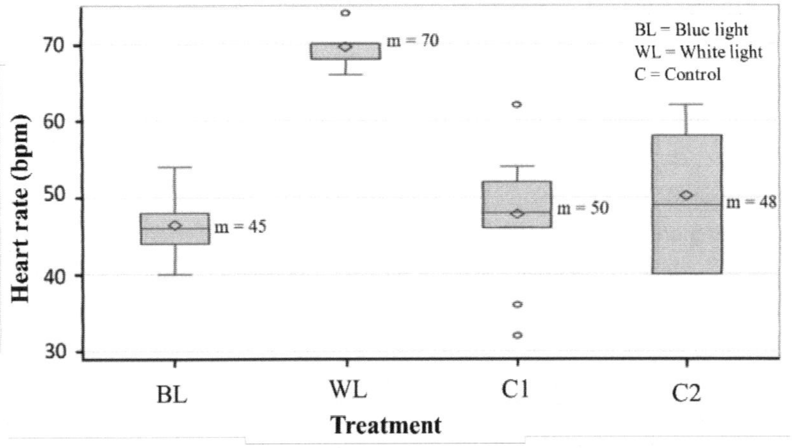

Fig. 4. Box plot the relationship between light treatment and heart rate.

Previous research has explored how fish heartbeats react to light, using different species and experimental conditions. The variations in heart rate observed under various light intensities are summarized in Table 2.

Table 2. Fish heartbeat in different light in some studies

Species	English Name	Fish size	Light	Heart rate (bits/minute)	Reference
Danio rerio	Zebra fish	Embryo to adult	14-hour light/10-hour dark (LD) cycle	150-180 bpm in light conditions, 120-140 bpm in dark conditions	Previous studies
		Embryo to adult	Glide LD (Light-Dark)	150-180 bpm in light conditions, down to 120-140 bpm in dark conditions	Previous studies
Oryzias latipes	Medaka	Embryo to adult	12-hour day/12-hour dark cycle (LD)	120-150 bpm in light conditions, 100-130 bpm in dark conditions	Previous studies
Cyprinus carpio	Common carp	Adult	LD (Light-Dark) and dark constant (DD) cycles	110-130 bpm in light conditions, 90-110 bpm in dark conditions	Previous studies
Anguilla bicolor	Indonesian shortfin eel	29.3±3.97 cm	Blue Light	46.2±3.52 bpm	This research
			White Light	69.6±1.92 bpm	

4 Discussion

Light characteristics in aquatic environments are particular and play a critical role in shaping fish behavior and physiology. Fish generally exhibits two types of phototactic responses: positive phototaxis, where individuals are attracted to light sources, and negative phototaxis, where they actively avoid them [14]. These responses vary across species and are influenced by phylogenetic lineage, ecological adaptations, and the physical properties of the light particularly its intensity and wavelength.

In *Anguilla* spp., light-induced responses are further categorized into primary responses, involving direct chromatophore reactions independent of visual input, and secondary responses, which are mediated by visual perception and regulated by the nervous system [15]. The present study demonstrated that white light elicited a higher frequency of heart rate peaks in elver-stage eels than blue light, indicating a stronger physiological reaction to white light. This is likely attributable to its broader spectral range and higher intensity, which may be perceived as more intrusive or stimulating.

The elevated heart rate observed under white light conditions suggests increased metabolic demand and a potential stress response. Full-spectrum lighting activates fish' behavioral and physiological stress pathways, including heightened cortisol secretion and accelerated metabolic activity. These effects are particularly pronounced in nocturnal species such as eels, which are evolutionarily adapted to low-light environments. Exposure to bright light may be interpreted as a threat or disturbance, thereby triggering cardiovascular activation.

Conversely, blue light was associated with a more stable and lower heart rate, consistent with findings by Riyanto et al. [10], which reported reduced activity in glass eels under blue illumination. This response likely reflects the spectral sensitivity of eel photoreceptors, which are adapted to deep or turbid waters where blue wavelengths dominate and are less disruptive. Interestingly, behavioral differences between developmental stages were noted: glass eels exhibited avoidance behavior, while elvers responded through physiological modulation, such as changes in heart rate.

From an ecological perspective, these findings reinforce the role of light as a key environmental cue in eel migration. Light influences spatial orientation, habitat selection, and diel vertical migration in anguillid species. Therefore, artificial manipulation of lighting in laboratory or aquaculture settings can significantly affect behavioral and physiological outcomes.

Regarding practical applications, the results offer valuable insights into eel fisheries management. White light may be effective as an attractant in fishing operations; however, it should be used judiciously to avoid inducing chronic stress. In contrast, blue light appears more suitable for conservation and restocking programs, as it promotes a calmer environment conducive to long-term survival and welfare.

5 Conclusions and recommendations

5.1 Conclusion

Exposure to white light significantly elevates heart rate in elver-stage *Anguilla* bicolor compared to blue light and dark control conditions, indicating a heightened physiological stress response. In contrast, blue light was associated with more stable cardiac activity, suggesting its potential to minimize stress and maintain physiological equilibrium.

These results highlight how adjusting light conditions can shape the physiological reactions of eels, especially when they are in vulnerable developmental phases. The differential effects of light spectra provide a foundation for further research to optimize environmental conditions in both experimental and applied settings. Specifically, blue light may offer advantages in conservation and restocking programs by promoting a less stressful environment. In contrast, white light may be strategically utilized in capture operations to enhance efficiency.

5.2 Recommendations

White light may be utilized as a short-term attractant in eel capture operations due to its ability to stimulate physiological responses. However, its prolonged application should be avoided to minimize stress and potential adverse effects on health and welfare. In contrast, blue light is recommended for use in conservation and restocking programs, as it promotes physiological stability and reduces stress, thereby supporting post-release survival and long-term well-being.

Future studies are encouraged to explore how different light intensities, exposure periods, and mixed wavelengths influence eel physiology and behavior at various stages of their development. Such studies will contribute to a more comprehensive understanding of light-based environmental manipulation and its implications for sustainable eel fisheries management and conservation strategies.

5.3 Acknowledgments

The author expressed his gratitude to IPB University and all parties who have supported this research.

References

1. L.J. McKinnon, *A review of eel biology: Knowledge and gaps*. Report to EPA Victoria (2006)
2. E.M. White, B. Knights, Environmental factors affecting migration of the European eel in the Rivers Severn and Avon, England. *J. Fish Biol.* **50**, 1104–1116 (1997)
3. T. Arai, S.R. Abdul Kadir, Opportunistic spawning of tropical anguillid eels *Anguilla bicolor bicolor* and *A. bengalensis bengalensis*. *Sci. Rep.* **7**, 41649 (2017)
4. B.A. Barton, Stress in fishes: a diversity of responses with particular reference to changes in circulating corticosteroids. *Integr. Comp. Biol.* **42**, 517–525 (2002)
5. B. Blanco-Vives, F.J. Sánchez-Vázquez, Blue light as a stimulus for fish behavior. *Fish Physiol. Biochem.* **35**, 469–472 (2009)
6. S.M. Noureldin, A.M. Diab, A.S. Salah, R.A. Mohamed, Effect of different monochromatic LED light colors on growth performance, behavior, and immune-physiological responses of goldfish *Carassius auratus*. *Aquaculture* **538**, 736532 (2021)
7. M. Matsuo, Y. Kamei, S. Fukamachi, Behavioural red-light sensitivity in fish according to the optomotor response. *R. Soc. Open Sci.* **8**, 210415 (2021)
8. G.L. Volpato, R.E. Barreto, Environmental blue light prevents stress in the fish Nile tilapia. *Braz. J. Med. Biol. Res.* **34**, 1041–1045 (2001)
9. I. Yasir, J.G. Qin, Effect of light intensity on color performance of false clownfish *Amphiprion ocellaris*. *J. World Aquac. Soc.* **40**, 337–350 (2009)
10. M. Riyanto, A.S. Aminah, W. Mawardi, R.I. Wahju, Response of *Anguilla bicolor* glass eel to different light colors. *Omni-Akuatika* **19**, 61–68 (2023)
11. W. Mawardi, R.I. Wahju, M. Riyanto, A.H. Uliyah, Biorhythms of eel (*Anguilla bicolor*) and its preferences for different baits under laboratory conditions. *ALBACORE* **8**, 319–329 (2024)
12. M. Riyanto, K. Yanase, T. Arimoto, Temperature and fatigue effect on the maximum swimming speed of jack mackerel *Trachurus japonicus*. *Fish. Sci.* **80**, 53–59 (2014)
13. Nofrizal, K. Yanase, T. Arimoto, Effect of temperature on the swimming endurance and post-exercise recovery of jack mackerel *Trachurus japonicus* as determined by ECG monitoring. *Fish. Sci.* **75**, 1369–1375 (2009)
14. G. Jékely, Evolution of phototaxis. *Philos. Trans. R. Soc. B* **364**, 2795–2808 (2009)
15. N. Oshima, Direct reception of light by chromatophores of lower vertebrates. *Pigment Cell Res.* **14**, 312–319 (2001)

Characteristics of candies made from Catfish (*Pangasius* sp.) skin collagen, gelatin and its hydrolysate

Mona Dianita Sari Putri[1], *Mala* Nurilmala[1*], and *Asadatun* Abdullah[2]

[1]Departement of Aquatic Product Technology, Faculty of Fisheries and Marine Sciences, IPB University, Jl Agatis Kampus IPB Dramaga, Bogor, West Jawa 16680, Indonesia

Abstract. Catfish skin is a by-product from the fishing industry that has potential as a source of collagen and gelatin. An enzymatic hydrolysis process was conducted using the commercial bromelain enzyme pineapple stem extract. The resulting collagen, gelatin and its hydrolysates were then used as basic ingredients in candy making. This study aims to determine the characteristics of candy made from collagen, gelatin and its hydrolysate; determine the effect of the type of raw material on the physicochemical properties of the candy; and determine the biological activity contained in it. The methods employed in this study consisted of the following stages: preparation of catfish skin, pre-treatment of catfish skin, extraction of catfish skin, hydrolysis of collagen and gelatin derived from catfish skin and candy formulation. The results showed that collagen hydrolysate produced through hydrolysis with pineapple stem extract is the best raw material for candy making. The hydrolysate had a pH value of 4.71±0.03; viscosity of 4.71±0.03 mPas; toxicity LD_{50} value of 1036.83 µg/mL; degree of hydrolysis of 57.42%; and molecular weight range of 9.8-12.7 kDa. The best characteristics were shown by collagen candy from hydrolyzate with pineapple stem extract. The candy produced a protein content of 64.03%; an ash content of 0.31%; a moisture content of 3.48%; and hedonic parameters appearance of 4.20, aroma of 4.73, flavor of 4.70, and texture of 3.00. Candy with the highest antioxidant activity was obtained from collagen hydrolyzed using commercial bromelain enzyme, which amounted to 98.62 µg/mL by the ABTS method and 382.33 µmol/g by the FRAP method.

1 Introduction

Catfish (*Pangasius* sp.) is one of the most economically important aquaculture commodities in Indonesia, with production increasing steadily each year. According to the Ministry of Marine Affairs and Fisheries, national catfish production reached 348,379 tons in 2024 [1]. Along with increased production, substantial amounts of by-products are generated, particularly fish skin from fillet processing industries. Fish skin has been widely recognized as a promising source of collagen due to its high protein content and well-developed connective tissue structure. Previous studies have reported that catfish skin contains approximately 30% collagen, highlighting its potential for conversion into high-value products such as collagen and gelatin [2]. Therefore, sustainable utilization of this by-product is essential to enhance its economic value and reduce processing waste.

Collagen is the most abundant structural protein in vertebrates and is characterized by a triple-helix structure rich in glycine, proline and hydroxyproline [3]. It can be extracted from

* Corresponding author: mnurilmala@apps.ipb.ac.id

various connective tissues, including skin, bones and scales and can be partially hydrolyzed to produce gelatin. Gelatin retains a similar amino acid composition to collagen and exhibits desirable functional properties, such as gelling, stabilizing, thickening and emulsifying capacities, which support its widespread application in food systems.

The use of collagen and gelatin derived from terrestrial animals has raised health and religious concerns, including risks associated with bovine spongiform encephalopathy and dietary restrictions. Consequently, fish skin has emerged as a safer and more widely accepted alternative source. Enzymatic hydrolysis further converts collagen and gelatin into low-molecular-weight peptides with improved digestibility and enhanced bioactivities, particularly antioxidant activity [4]. Proteolytic enzymes such as bromelain have been widely applied in collagen hydrolysis due to their effectiveness and stability under various processing conditions. Bromelain, a protease derived from pineapple, has been reported to exhibit superior hydrolytic performance and to enhance the antioxidant activity of collagen hydrolysates compared to other enzymes. Antioxidant peptides play an important role in scavenging free radicals and are increasingly valued in functional food with nutraceutical applications.

With the growing global demand for collagen- and gelatin-based functional products, the development of stable and economically feasible delivery systems is increasingly important. Hard candy represents a practical food matrix due to its long shelf life, ease of consumption, and low production cost. However, to the best of our knowledge, no previous studies have developed hard candy formulations incorporating collagen, gelatin and their hydrolysates derived from catfish (*Pangasius* sp.) skin. Therefore, this study aimed to evaluate the physicochemical characteristics and biological activities of hard candies formulated using different collagen-based raw materials from catfish skin.

2 Materials and methods

2.1 Equipments and materials

The equipment used in this study included a digital balance (Ohaus Corp., New Jersey, USA), cotton, calico cloth, pH indicator, electric stove, measuring cylinder (Pyrex, Asahi Glass, Thailand), Erlenmeyer flask (Pyrex, Asahi Glass, Thailand), water bath shaker (Depolab, Seoul, Korea), aluminum foil, thermometer, oven (Blue M, Jiangsu Zhengji Instruments, Jintan, China), universal pH indicator and blender (Miyako BL-151, Indonesia).

The main raw material used was catfish (*Pangasius* sp.) skin, obtained from PT. Adib Global Food (Indonesia). Chemicals used for collagen hydrolysate production included hydrochloric acid (HCl) (CV. Kimia Jaya Laboran, Cilacap Utara, Indonesia), sodium hydroxide (NaOH) (Merck KGaA, Germany), acetic acid (CH_3COOH) (Toko Central Kimia, Bogor, Indonesia), distilled water (Toko Central Kimia, Bogor, Indonesia), and bromelain enzyme $\geq$3 units (Sigma-Aldrich, USA). Additional ingredients used in the candy formulation were commercial sugar and pineapple stem obtained from a traditional market in Bogor, West Java, Indonesia. For hydrolysis degree analysis, trichloroacetic acid (TCA 20%) was used. Antioxidant activity was determined using the ABTS assay with 2,2-*azino-bis* (3-*ethylbenzothiazoline*-6-*sulphonic acid*) as the radical solution and $K_2S_2O_8$-ethanol as the blank.

2.2 Work procedure

The production of hard candy consisted of several sequential steps. The first stage was the preparation of catfish (*Pangasius* sp.) skin, followed by pre-treatment, extraction, enzymatic hydrolysis of collagen, gelatin and candy formulation.

2.2.1 Preparation of Catfish skin

The sample used was catfish (*Pangasius* sp.) skin that had been thoroughly cleaned. The skin was cut into square pieces and the purpose of this preparation was to remove residual fat. The skin was washed five times with water, followed by weighing according to the required sample amount.

2.2.2 Pre-treatment of Catfish skin

The pre-treatment process was carried out to remove non-collagenous components [5]. For collagen pre-treatment, catfish skin was soaked in 0.05 M NaOH solution at a ratio of 1:10 (w/v) for 1.5 hours. The solution was then neutralized to pH 6–7. Subsequently, the skin was immersed in 0.1 M acetic acid (CH_3COOH) solution at a ratio of 1:10 (w/v) for 2 hours at 4°C and again neutralized to pH 6–7. For gelatin pre-treatment, the skin was soaked in 0.05 M NaOH solution at a ratio of 1:10 (w/v) for 1.5 hours, followed by washing until neutral pH (6–7). The next step involved soaking the skin in HCl solution at a ratio of 1:5 (w/v) for 2 hours, after which it was washed until neutral pH (6–7) was achieved.

2.2.3 Extraction of Catfish skin

The extraction process was carried out in two stages, namely collagen and gelatin extraction from catfish (*Pangasius* sp.) skin [6]. Collagen extraction was performed using a water bath shaker at 40°C at a water–sample ratio of 1:1 (w/v) and maintained for 6 hours. Gelatin extraction was conducted at 70°C using the same proportion of 1:1 (w/v) and continued for 8 h. Following extraction, the solution was filtered using cotton and calico cloth. Collagen was dried in an oven at 40°C, while gelatin was dried at 50°C for 24 hours.

2.2.4 Hydrolysis of collagen and gelatin from Catfish skin

The hydrolysis process was conducted using an enzymatic method [7]. Collagen and gelatin (10 g) were hydrolyzed with pineapple core extract and commercial bromelain enzyme. Hydrolysis was conducted at 40°C and allowed to proceed for 3 hours, after which the enzyme was inactivated at 85°C for a duration of 15 minutes. The resulting liquid collagen and gelatin hydrolysates were subsequently oven-dried at 40°C for a period of 24 hours. The dried forms of collagen and gelatin hydrolysates were then ground into powder using a grinder.

2.2.5 Candy formulation

The candy formulation was adapted from the method of Wijaya [8] with modifications by incorporating 10 g of collagen, gelatin, and their respective hydrolysates. The candy preparation began with mixing the raw materials, including collagen powder, gelatin and their hydrolysates obtained from pineapple core extract and commercial bromelain enzyme for hydrolysis. Subsequently, 2 g of stevia sweetener and 88 mL of distilled water were added. The mixture was subjected to mild heating at 40°C for 90 minutes. The candies were dried in an oven at 40°C for 24 hours.

2.3 Analysis procedure

The analytical procedures in this study included sensory evaluation (hedonic test), determination of moisture, ash, protein levels, the extent of hydrolysis, viscosity, acidity, cytotoxicity using the Brine Shrimp Lethality Test (BSLT), antioxidant activity assays using ABTS and FRAP methods, as well as molecular weight analysis by SDS-PAGE.

2.3.1 Moisture percentage

Moisture percentage analysis [9] was conducted to quantify the water content present in the sample. Initially, a crucible was dried in an oven at 105°C for 1 hour, cooled in a desiccator for 15 minutes to reach room temperature and weighed. Approximately 3 g of the sample was placed in the crucible and dried in an oven at 105°C for 5 hours. The dried sample was then cooled in a desiccator to room temperature and weighed. Moisture content was calculated using the following equations:

$$\text{Weight loss (g)} = \text{Initial sample weight (g)} - \text{Final sample weight (g)} \quad (1)$$

$$\text{Moisture content (\%)} = \frac{Weight\ loss\ (g)}{Initial\ sample\ weight\ (g)} \times 100\% \quad (2)$$

2.3.2 Ash content

Ash content analysis [9] was performed to assess the mineral composition of the sample. A crucible was dried in an oven at 105°C for 1 hours, cooled in a desiccator for 15 minutes, and weighed. Approximately 1 g of the sample was placed in the crucible and incinerated over a Bunsen flame until no smoke was observed. The crucible containing the sample was then placed in a muffle furnace at 600°C for 6 hours. Following combustion, the crucible was allowed to cool to room temperature in a desiccator before being weighed. The ash content was then calculated using the following equations:

$$\text{Ash weight (g)} = \text{Final weight of crucible + sample (g)} - \text{Weight of empty crucible (g)} \quad (3)$$

$$\text{Ash content (\%)} = \frac{Ash\ weight\ (g)}{Sample\ weight\ (g)} \times 100\% \quad (4)$$

2.3.3 Protein content

Protein content was determined using the Kjeldahl method [9], which consists of three main stages: digestion, distillation and titration. Approximately 2 g of sample was placed into a 50 mL Kjeldahl flask, followed by the addition of 7 g K_2SO_4, 0.005 g HgO, 15 mL concentrated H_2SO_4, and 10 mL H_2O_2 added slowly, and left for 10 minutes in a fume hood. Digestion was performed at 410°C for approximately 2 hours until the solution turned yellowish green. The digested solution was diluted with 50 mL distilled water and 20 mL of 40% NaOH was added. The mixture was distilled at 100°C, and the distillate was collected in a 125 mL Erlenmeyer flask containing bromocresol green–methyl red indicator. Distillation was continued until 150 mL of green-colored distillate was obtained. Titration was performed as the final step using 0.1 N HCl until the endpoint (pink color) was reached. A blank solution was also analyzed. Protein content was calculated using the following formula:

$$\text{Protein content (\%)} = \frac{(VHCl - Vblank)\ x\ N\ x\ 14.007\ x\ F}{Sample\ weight\ (mg)}\ \text{x } 100\% \quad (5)$$

2.3.4 Degree of hydrolysis

The percentage of the degree of hydrolysis (DH) [9] was determined through protein fractionation using 20% trichloroacetic acid (TCA), which separates soluble and insoluble protein fractions at 10% each. A total of 500 μL of collagen hydrolysate sample was homogenized with 500 μL of 20% TCA and incubated at 4°C for 30 minutes. The mixture was then centrifuged at 3,000 g for 20 minutes. The supernatant was collected and its soluble protein content was analyzed using bovine serum albumin (BSA) as the protein standard. The degree of hydrolysis was calculated using the following formula:

$$\text{DH (\%)} = \frac{Nitrogen\ content\ in\ supernatant}{Total\ nitrogen\ content\ in\ sample}\ \text{x } 100\% \quad (6)$$

2.3.5 The degree of acidity

To determine the degree of acidity [9], a total of 1 g of sample was dissolved in 20 mL of distilled water, followed by the addition of 50 mL of distilled water, and homogenized thoroughly. The pH meter was turned on and allowed to stabilize before measurement. The electrode was immersed in the sample solution until a stable reading was displayed on the pH meter.

2.3.6 Antioxidant activity by ABTS method

To determine antioxidant activity, an ABTS stock solution [10] was prepared by reacting 7 mM ABTS solution with 2.45 mM potassium persulfate at a 1:1 ratio and incubating in the dark at room temperature for 16 hours. The solution was then diluted with ethanol until an absorbance of 0.700 ± 0.02 at 734 nm was obtained. Subsequently, 1 mL of sample solution (1 mg/mL) was added to 2 mL of ABTS solution and incubated at room temperature for 10 minutes. The absorbance of the mixture was measured at 734 nm. Antioxidant activity was expressed as IC_{50} values, calculated from the regression curve of ABTS radical scavenging inhibition across different sample concentrations.

2.3.7 Antioxidant activity by FRAP method

The FRAP reagent [11] was prepared using 300 mM acetate buffer (8 mL CH_3COONa and 92 mL CH_3COOH, pH 3.6), 10 mM TPTZ (2,4,6-*tripyridyl-s-triazine*) solution in 40 mM HCl, and 20 mM $FeCl_3 \cdot 6H_2O$. The working reagent was freshly prepared by mixing 35 mL acetate buffer, 3.5 mL TPTZ solution and 3.5 mL $FeCl_3 \cdot 6H_2O$ solution. For the assay, 500 μL of sample, 600 μL of distilled water and 3 mL of FRAP reagent were combined, vortexed and incubated in a water bath at 37°C for 30 minutes. The absorbance was measured at 593 nm. A standard calibration curve was constructed using ascorbic acid solutions at various concentrations.

2.3.8 Toxicity assay by Brine Shrimp Lethality Test (BSLT)

The BSLT method was used to evaluate the toxicity of the sulfated polysaccharide extracts [9]. Brine shrimp larvae were obtained by hatching 50 mg of shrimp eggs in 300 mL of seawater in a 500 mL Erlenmeyer flask. Hatching was carried out under continuous aeration and illumination with a 14 W white lamp for 48 hours at room temperature. Five serial dilutions of the test sample were prepared (1000 μg/mL, 500 μg/mL, 250 μg/mL, 125 μg/mL and 62.5 μg/mL) using seawater as the diluent. The assay was conducted in a 96-well plate, with each well containing 10 shrimp larvae and the respective test concentration. Each concentration was tested in triplicate. The plates were incubated under a 14 W white lamp at room temperature for 24 hours. After incubation, the number of surviving larvae was recorded. A negative control was prepared using the same procedure without the extract. Toxicity (IC_{50}) was calculated using probit analysis based on the following formula:

$$\text{Mortality (\%)} = \frac{NNumber\ of\ dead\ larvae}{Total\ larvae} \times 100\% \quad (7)$$

2.3.9 Sensory evaluation

Sensory evaluation was conducted using human senses to assess the quality attributes of the candy samples based on several indicators, including appearance, color, aroma, taste, and texture. A semi-trained panel of 30 participants was recruited. Panelists evaluated the samples using a 5-point hedonic scale: (1) dislike very much; (2) dislike; (3) neutral; (4) like; (5) like very much.

2.3.10 Viscosity

A sample of 3.33 g was dissolved in 50 mL of distilled water and homogenized. The viscosity of the solution [9] was measured using a Brookfield Synchro-Lectric Viscometer at room temperature with a shear rate of 60 rpm. Spindle No 1 was used in this assay and the viscosity values were expressed in mPas.

2.3.11 Molecular weight determination (SDS-PAGE)

Sodium Dodecyl Sulfate–Polyacrylamide Gel Electrophoresis (SDS-PAGE) was used to determine molecular weights [12]. Gelatin samples (2 mg) were dissolved in 1 mL of 5% SDS solution and heated at 85°C for 1 hour. The solution was centrifuged at 12,400 × g for 5 minutes and 20 μL of sample buffer was collected. The sample was reheated at 85°C before being loaded into the gel wells. For gel preparation, 5 mL of separating gel containing 12.5% acrylamide was prepared, followed by the addition of 1 mL distilled water. The gel was incubated at room temperature for 30 minutes until solidified and the excess water was removed. Subsequently, 1 mL of stacking gel was added and incubated at room temperature for 2 hours. Denatured protein samples and molecular weight marker 245 (BIO-RAD) were loaded into the gel wells (5 μL each). Electrophoresis was carried out at a constant voltage of 100 V for 3 hours. After electrophoresis, the gel was removed from the cast, stained with 25 mL *Coomassie Brilliant Blue* for 2 hours, rinsed with water and destained with 50 mL destaining solution until the protein bands became clearly visible.

2.4 Data analysis

The experimental design used in this study was a factorial Completely Randomized Design (CRD) consisting of two stages, namely:
Stage 1
Evaluation of raw material quality of hydrolysates consisting of six treatments, i.e.:

1. C = Collagen
2. G = Gelatin
3. CHP = Collagen hydrolysate (pineapple core extract hydrolysis)
4. GHP = Gelatin hydrolysate (pineapple core extract hydrolysis)
5. CHB = Collagen hydrolysate (commercial bromelain hydrolysis)
6. GHB = Gelatin hydrolysate (commercial bromelain hydrolysis)

Stage 2
Evaluation of candy product quality was carried out on six types of raw materials used in Stage 1, consisting of six treatments, i.e.:

1. P1 = Collagen candy
2. P2 = Gelatin candy
3. P3 = Collagen hydrolysate candy obtained from pineapple core extract hydrolysis
4. P4 = Gelatin hydrolysate candy obtained from pineapple core extract hydrolysis
5. P5 = Collagen hydrolysate candy obtained from commercial bromelain enzyme hydrolysis
6. P6 = Gelatin hydrolysate candy obtained from commercial bromelain enzyme hydrolysis

Each treatment in both stages was replicated three times. Data analysis was performed separately for each stage using Analysis of Variance (ANOVA). If a significant difference was observed ($P<0.05$), Duncan's Multiple Range Test (DMRT) was applied at a 95% confidence level. Sensory data were analyzed using a non-parametric test, namely the Kruskal–Wallis test. If significant differences were found, a Multiple Comparison Test was performed using the Mann–Whitney U test.

3 Results and discussion

3.1 Degree of acidity of collagen, gelatin and its hydrolysates

The measurement of acidity was conducted to determine the acid–base level of the samples. Acidity of collagen- and gelatin-based products is a key parameter in assessing their stability across various application fields. The acidity values of collagen, gelatin and its hydrolysates derived from catfish skin are presented in Fig. 1.

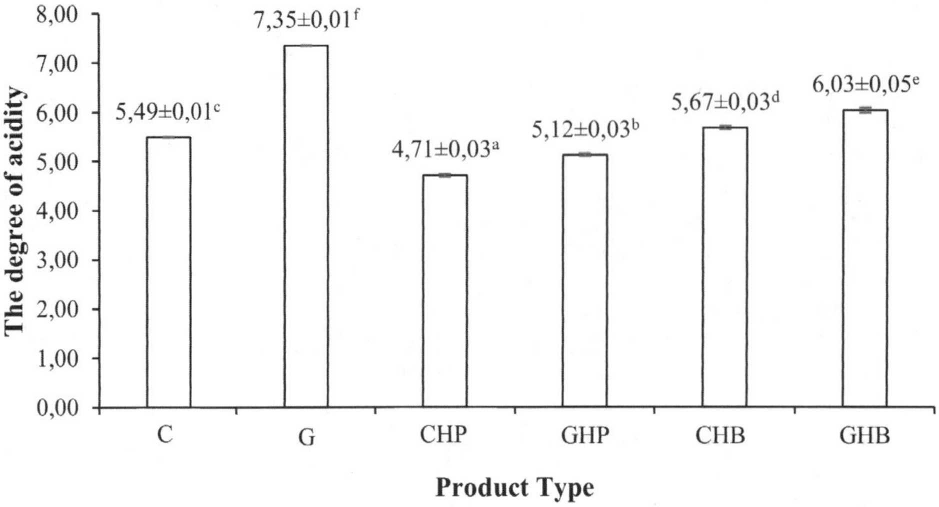

Fig.1. The degree of acidity of Collagen (C); Gelatin (G); Collagen Hydrolysate with pineapple extract (CHP); Gelatin Hydrolysate with pineapple extract (GHP); Collagen Hydrolysate with bromelain enzyme (CHB); Gelatin Hydrolysate with bromelain enzyme (GHB).

In the degree of acidity measurements shown in Figure 1, values ranged from 4.71 to 7.35. Significant differences were observed among the treatments (ANOVA, $P<0.05$). These results indicate that collagen and gelatin derived from catfish skin are suitable for use as ingredients in confectionery applications. The degree of acidity values of collagen obtained in this study tended to be slightly below neutral. This condition may be attributed to the soaking process using acetic acid solution (CH_3COOH) during the swelling stage of catfish skin. The hydrogen ions (H^+) from acetic acid are not fully absorbed into the collagen matrix. Repeated neutralization of the skin cannot completely eliminate residual acid retained within the collagen fibril network.

3.2 Viscosity of collagen, gelatin and its hydrolysates

Viscosity is one of the key parameters for assessing the physical properties of collagen-based materials, gelatin, collagen hydrolysate, as well as gelatin hydrolysates. The viscosity values for collagen, gelatin and their hydrolysates obtained from catfish skin tissue are presented in Fig. 2.

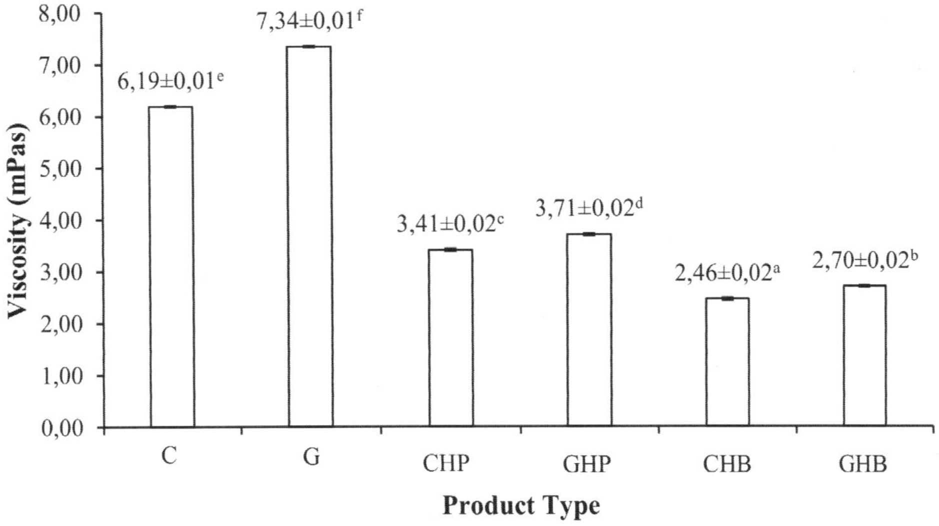

Fig.2. Viscosity of Collagen (C); Gelatin (G); Collagen Hydrolysate with pineapple extract (CHP); Gelatin Hydrolysate with pineapple extract (GHP); Collagen Hydrolysate with bromelain enzyme (CHB); and Gelatin Hydrolysate with bromelain enzyme (GHB).

Gelatin exhibited higher viscosity values compared to collagen extracted from catfish skin (Fig. 2). Different raw material types significantly influenced the viscosity values (ANOVA, $P<0.05$). Variations in viscosity between collagen, gelatin and their hydrolysates obtained from catfish skin tissue may be attributed in part to soaking duration. Longer soaking times tend to decrease the viscosity of the final product. This phenomenon is likely associated with the breakdown of polypeptide chains in gelatin into shorter peptide chains.

3.3 Toxicity of collagen, gelatin and its hydrolysates

Toxicity testing is essential to ensure the safety of a product. The Brine Shrimp Lethality Test (BSLT) is widely employed as an initial bioassay method because it is practical, cost-effective and capable of providing a preliminary indication of the presence of toxic compounds. The toxicity results of collagen, gelatin and its hydrolysates from catfish skin are summarized in Table 1.

Table 1. Toxicity of collagen, gelatin and its hydrolysates

Sample type	LC_{50} value (μg/mL)	Category
Collagen	1091.86	Non-toxic
Gelatin	1421.87	Non-toxic
Collagen hydrolysate (pineapple extract)	1036.83	Non-toxic
Gelatin hydrolysate (pineapple extract)	1138.94	Non-toxic
Collagen hydrolysate (bromelain enzyme)	1073.89	Non-toxic
Gelatin hydrolysate (bromelain enzyme)	1229.27	Non-toxic

The toxicity results revealed that all treatments were categorized as non-toxic and safe for consumption. A product is considered highly toxic when $LC_{50} < 30$ ppm, toxic when LC_{50} ranges between 31–200 ppm, mildly toxic when LC_{50} falls within 201–1000 ppm and non-toxic when $LC_{50}>1000$ ppm. An increase in the lethal dose (LC_{50}) value indicates a reduced toxic effect on larval mortality of *Artemia salina*. The BSLT method offers several advantages compared to other toxicity testing approaches. It requires a relatively short time to perform, is cost-effective, practical, does not necessitate aseptic techniques or special maintenance, requires only a small amount of sample and does not involve the use of animal serum. The findings indicate that collagen and gelatin possess potential to be developed into nutraceutical candy products. This is attributed to the non-toxic nature of catfish skin collagen and gelatin proteins within the human body.

3.4 Degree of hydrolysis of collagen and gelatin

The degree of hydrolysis (DH) is a crucial indicator used to assess the extent to which protein peptide bonds are cleaved during the hydrolysis process. DH has a major influence on determining the functional properties and biological activities associated with collagen and gelatin hydrolysates. The results of DH measurement are presented in Fig. 3.

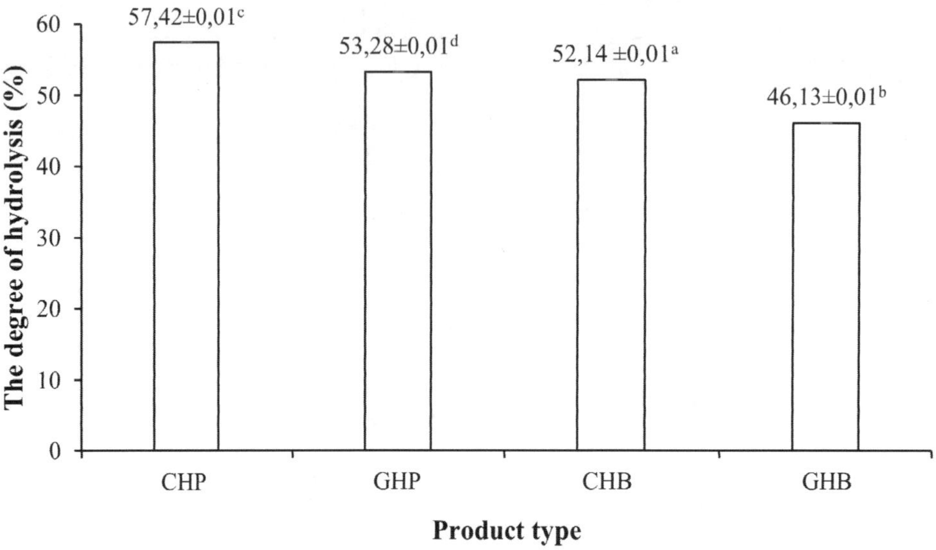

Fig.3. The degree of hydrolysis of Collagen Hydrolysate with pineapple extract (CHP); Gelatin Hydrolysate with pineapple extract (GHP); Collagen Hydrolysate with bromelain enzyme (CHB); and Gelatin Hydrolysate with bromelain enzyme (GHB).

The DH values displayed significant differences among treatments ($P<0.05$). The highest DH value was observed in CHP (collagen hydrolysate obtained from pineapple core extract hydrolysis), reaching 57.42%, while the lowest was found in GHB (gelatin hydrolysate obtained from commercial bromelain enzyme hydrolysis) at 46.13%. These results demonstrate that collagen substrates produced hydrolysates with higher DH values compared to gelatin. Variations in DH values may be affected by factors such as acidity, processing temperature and enzyme concentration, which directly influence enzymatic efficiency. Higher DH values indicate a more extensive protein hydrolysis process.

3.5 Molecular weight (SDS-PAGE)

The molecular weight profiles of collagen, gelatin and their hydrolysates were determined using sodium dodecyl sulfate–polyacrylamide gel electrophoresis (SDS-PAGE). This method separates protein biomolecules according to their molecular size and electrical charge, allowing visualization of protein band patterns as indicators of protein degradation resulting from enzymatic hydrolysis. The molecular weight profiles of the raw materials are illustrated in Fig. 4.

Catfish skin collagen (C) had molecular weights of 227.59 kDa (α1), 180.40 kDa (α2) and 93.81 kDa (β), while gelatin (G) had molecular weights of 229.62 kDa (α1), 192.67 kDa (α2) and 91.38 kDa (β). The molecular weights of pure collagen and gelatin exhibited different values compared to their hydrolysates, which had undergone the hydrolysis process. The molecular weights of the hydrolysates ranged between 9.8–34.2 kDa. Variations in molecular weight resulting from hydrolysis can affect the level of antioxidant activity. Compounds with lower molecular weights tend to have higher antioxidant activity. The SDS-PAGE profiles confirmed the existence of β, α1 and α2 chains in collagen as well as gelatin derived from catfish skin. The presence of two α chains observed in this study further confirmed that the collagen sample belonged to type I collagen, which is the dominant structural collagen commonly found in connective tissues. Type I collagen is characterized by its high molecular weight, consisting of α-chain monomers and β-chain dimers [13].

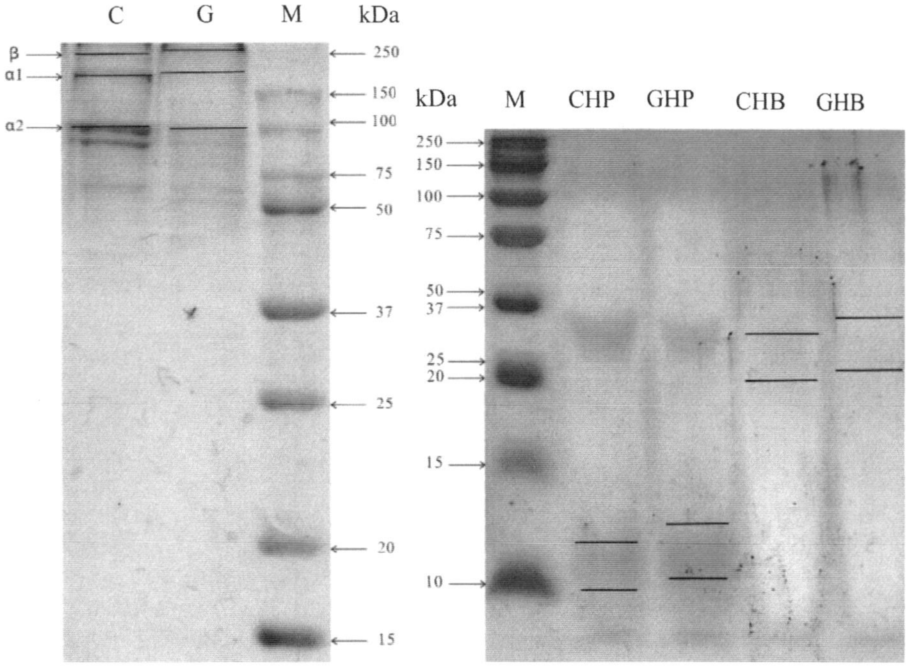

Fig.4. SDS-PAGE Profile of Collagen (C); Gelatin (G); Collagen Hydrolysate with pineapple extract (CHP); Gelatin Hydrolysate with pineapple extract (GHP); Collagen Hydrolysate with bromelain enzyme (CHB); and Gelatin Hydrolysate with bromelain enzyme (GHB). Standard molecular weight markers are shown in lane M.

3.6 Protein, ash, and moisture composition of candy

Proximate analysis aims to identify the chemical composition of a food product. The determination of protein, ash and moisture contents was performed to assess the nutritional value, quality and stability of the final product. The results of the candy chemical composition analysis are summarized in Table 2.

Table 2. Protein, ash and moisture composition of candy in accordance with SNI 3547.1:2008.

Parameter (%)	P1	P2	P3	P4	P5	P6	SNI (%)
Protein content	67.47 ± 0.06	67.33 ± 0,11	64.03 ± 0.16	65.70 ± 0.16	62.29 ± 0.01	60.75 ± 0.01	-
Ash content	0.48±0.03	0.40±0.06	0.31±0.12	0.44±0.09	0.34±0.01	0.47±0.04	Max. 2.0
Moisture content	3.17±0.03	3.23±0.04	3.48±0.04	3.41±0.04	3.37±0.01	3.30±0.02	Max. 3.5

Note: P1: Collagen candy; P2: Gelatin candy; P3: Collagen hydrolysate candy obtained from pineapple extract hydrolysis; P4: Gelatin hydrolysate candy obtained from pineapple extract hydrolysis; P5: Collagen hydrolysate candy obtained from bromelain enzyme hydrolysis; P6: Gelatin hydrolysate candy obtained from bromelain enzyme hydrolysis

Appearance is among the parameters used to assess the visual quality of candy products. The highest hedonic score for appearance was obtained in treatment P3 (candy formulated with collagen hydrolysate obtained using pineapple extract). This finding indicates that the hydrolysis of collagen using pineapple extract was able to enhance the visual appearance of the product. The next parameter is aroma. Treatment P3 received the highest hedonic score for aroma, with a value of 4.73. These findings indicate that both the raw material type and the hydrolysis method have a significant role in influencing the aroma quality of candy, where hydrolysis using pineapple extract contributes positively to aroma enhancement, whereas hydrolysis with bromelain enzyme tends to produce a less favorable aroma.

Flavor is another key factor in determining consumer acceptance of food products. The quality of flavor is not solely influenced by the composition of the raw materials, but is also affected by the processing method. The most preferred candy flavor was observed in treatment P3, which obtained a score of 4.70. The hedonic test results for candy color showed significant differences among treatments, with scores ranging from 3.00 to 4.30. The highest acceptance level for color was also found in treatment P3. Furthermore, texture is another crucial parameter that determines the overall quality of a product. Table 2 shows that the texture of each treatment varied significantly. Treatment P5 (candy formulated with collagen hydrolysate obtained using bromelain enzyme) received the highest preference score for texture, with a value of 4.57. The texture characteristics of candy are affected by the composition of the raw materials and the candy-making process.

3.7 Antioxidant activity of candy

The antioxidant activity of the hard candy samples was evaluated using two analytical methods, namely the ABTS and FRAP assays. The principle of ABTS is to evaluate the change in color of radical compounds after reacting with antioxidants, as determined by absorbance values using spectrophotometry. The antioxidant activity of the candies as measured by the ABTS method is presented in Fig. 5.

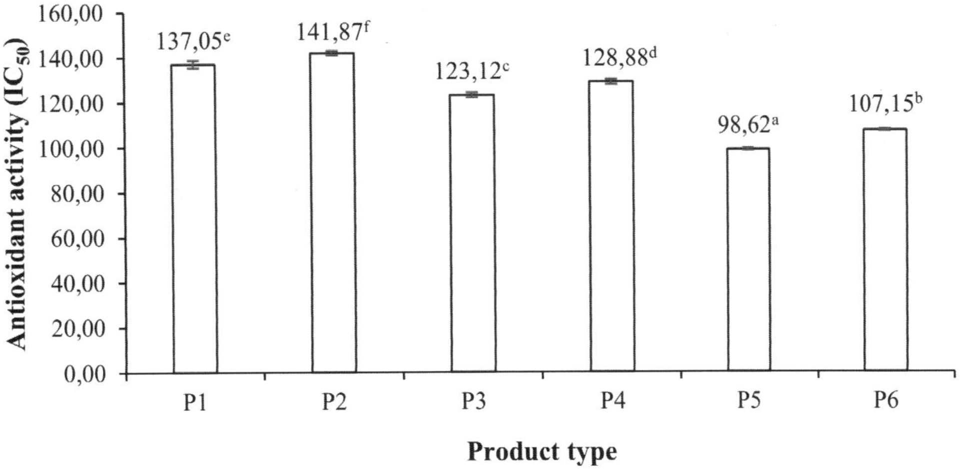

Fig.5. Antioxidant activity of candies; Collagen candy (P1), Gelatin candy (P2), Candy made from collagen hydrolysate obtained from pineapple extract hydrolysis (P3), Candy made from gelatin hydrolysate obtained from pineapple extract hydrolysis (P4), Candy made from collagen hydrolysate obtained from bromelain enzyme hydrolysis (P5), Candy made from gelatin hydrolysate obtained from bromelain enzyme hydrolysis (P6).

The IC_{50} values of antioxidant activity in the candies ranged from 98.62 to 141.87 µg/mL. The candy with the strongest antioxidant activity (lowest IC_{50} value) was the collagen hydrolysate candy obtained through enzymatic hydrolysis using bromelain, which exhibited an IC_{50} of 98.62 µg/mL. A higher IC_{50} value indicates weaker antioxidant activity in a product. There were significant differences (ANOVA, $P<0.05$) among treatments with respect to IC_{50} values of antioxidant activity. Hydrolysis of collagen and gelatin, whether by pineapple extract or bromelain enzyme, was found to enhance the antioxidant activity of the candy products. Variations in antioxidant activity can be affected by various factors, including the source of raw materials, amino acid composition, type of enzyme applied and extraction solvent used [14].

The Ferric Reducing Antioxidant Power (FRAP) assay operates based on the principle of reducing Fe^{3+} within the Fe^{3+}–TPTZ complex to the Fe^{2+}–TPTZ form through electron donation. The intensity of the resulting complex color is measured using spectrophotometric analysis and serves as an indicator of the reducing capacity of antioxidant compounds. The results of antioxidant activity measurements of both raw materials and candy products are presented in Fig. 6. The antioxidant capacity of raw materials prior to candy production ranged from 393.33 to 450.33 µmol/g. The data indicates that hydrolysis treatments, particularly using pineapple extract, enhanced antioxidant capacity compared to pure collagen or gelatin. Significant differences existed among the treatments with respect to the antioxidant activity of the raw materials (ANOVA, $P<0.05$). Likewise, antioxidant capacity measurements of the candy samples demonstrated statistically significant variations among treatments (ANOVA, $P<0.05$). The measured antioxidant capacity of candies measured using the FRAP method ranged from 335.33 to 382.33 µmol/g. All treatments exhibited strong antioxidant capacity, falling within the standard range of 100–400 µmol/g. The highest antioxidant capacity was observed in treatment P5, reinforcing the notion that hydrolysis plays a crucial role in breaking peptide bonds into smaller, bioactive fragments, thereby generating strong antioxidant activity [15].

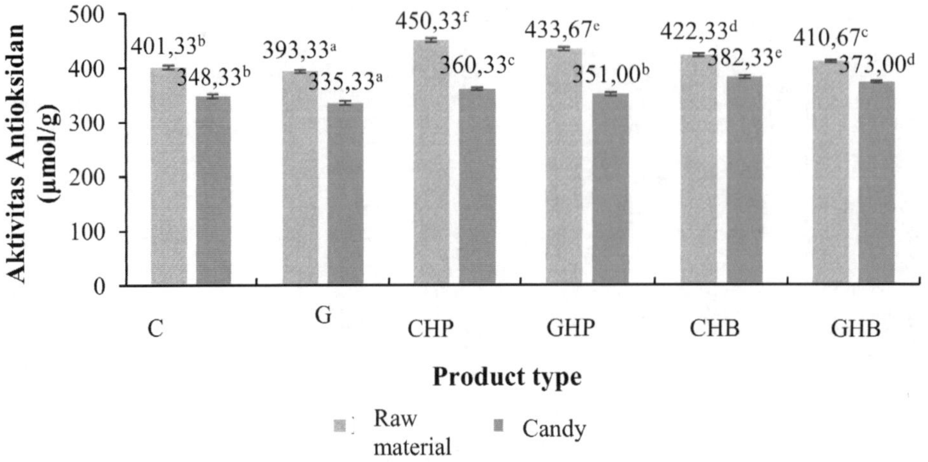

Fig.6. Antioxidant activity (FRAP) of raw materials and candy products Collagen (C); Gelatin (G); Collagen Hydrolysate with pineapple extract (CHP); Gelatin Hydrolysate with pineapple extract (GHP); Collagen Hydrolysate with bromelain enzyme (CHB); and Gelatin Hydrolysate with bromelain enzyme (GHB).

4 Conclusion

The differences in raw materials, namely collagen, gelatin, collagen hydrolysate and gelatin hydrolysate, had a significant influence on the characteristics as well as the antioxidant properties of the candy. Collagen hydrolysate produced through enzymatic hydrolysis with pineapple core was identified as the most appropriate raw material for candy formulation, as it demonstrated physicochemical properties that support overall product quality and safety in the finished product. The most favorable candy characteristics were observed in collagen based candy derived from this hydrolysate. Meanwhile, the strongest antioxidant activity was observed in collagen hydrolysate produced using commercial bromelain enzyme.

References

1. [KKP] Kementerian Kelautan dan Perikanan, Angka Konsumsi Ikan Nasional. Jakarta: Kementerian dan Kelautan dan Perikanan (2023).
2. M. Nurilmala, H. Suryamarevita, H.H. Hizbullah, AM. Jacoeb, Y. Ochiai, Fish skin as a biomaterial for halal collagen and gelatin. Saudi J. Biol. Sci. **29**, 1100–1110 (2022). https://doi.org/10.1016/j.sjbs.2021.09.056
3. V. Ferarro, M. Anton, V.S. Lhoutellier, The "sisters" α-helices of collagen, elastin and keratin recovered from animal by-products. Trends Food Sci. Technol. **51**, 65–75 (2016). https://doi.org/10.1016/j.tifs.2016.03.006
4. L.M.S. Lopera, J.E.Z. Montoya, Optimization of enzymatic hydrolysis of red tilapia scales (*Oreochromis* sp.) to obtain bioactive peptides. Biotechn. Rep. **30**, 244–262 (2021). https://doi.org/10.1016/j.btre.2021.e00611
5. U.M.R. Hartina, H. Qhairul, Q.I.N. Nor, M. Hasmadi, Properties of hydrolysed collagen from the skin of milkfish (*Chanos chanos*) as affected by different enzymatic treatments. Int. J. Res. Sci. Manag. **6**, 34–41 (2019). http://ijrsm.com/index.php/journal-ijrsm/article/view/186

6. M. Nurilmala, A.M. Jacoeb, R.A. Dzaky, Karakteristik gelatin kulit ikan tuna sirip kuning. Jurnal Pengolahan Hasil Perikanan Indonesia **20**, 2 (2017). https://doi.org/10.17844/jphpi.v20i2.18049
7. M. Nurilmala, H.H. Hizbullah, E. Karnia, E. Kusumaningtyas, Y. Ochiai, Characterization and antioxidant activity of collagen, gelatin, and the derived peptides from yellowfin tuna (*Thunnus albacares*) skin. Multidisciplinary Digital Publishing Institute **18**, 98 (2020). https://doi.org/10.3390/md18020098
8. C.H. Wijaya, Komposisi permen cajuput yang dapat menghambat pertumbuhan mikroba penyebab karies gigi dan proses pembuatannya. Master Thesis, IPB University, Bogor (2007).
9. [AOAC] Association of Official Analytical Chemist. *Official Method of Analysis of The Association of Official Analytical Chemist*. Virginia: Association of Official Analytical Chemist, Inc (2005).
10. G. Luisi, A. Stefanucci, G. Zengin, M.P. Dimamito, A. Mollica, Antioxidant and tyrosinase inhibitory in vitro activity of amino acids and small peptides: new hints for the multifaceted treatment of neurologic and metabolic disfunctions. Multidisciplinary Digital Publishing Institute **8**, 1–14 (2018). https://doi.org/10.3390/antiox8010007
11. Y. Kumar, A.U. Muzaddadi, V. Kumar, S. Singhal, D. Kumar, A.K. Biswas, Extraction and characterization of collagen hydrolysate from buffalo (*Bubalus Bubalis*) skin including its antioxidant properties and antiarthritic effect. Waste Biom. Valor. 15, 4159–4173 (2024). https://doi.org/10.1007/s12649-024-02462-y
12. U.K. Laemmli, Cleavage of structural proteins during assembly of head of bacteriophage T4. Nature **277**, 680–685 (1970). https://doi.org/10.1038/227680a0
13. S. Sharma, S. Dwivedi, S. Chandran, A. Srivastava, P. Vijay, Collagen brief analysis. Oral Maxillofac. Pathol. J. **10**, 210-222 (2019). https://doi.org/10.5005/jpjournals-10037-1143
14. R. Wu, C. Wu, D. Liu, X. Yang, J. Huang, J. Zhang, H. Li, Overview of antioxidant peptides derived from marine resources: The sources, characteristic, purification, and evaluation methods. Appl. Biochem. Biotechn. **176**, 1815–1833 (2019). https://doi.org/10.1007/s12010-015-1689-9
15. Y.J. Lin, G.W. Le, J.Y. Wang, Y.X. Li, Y.H. Shi, J. Sun, Antioxidative peptides derived from enzyme hydrolysis of bone collagen after microwave assisted acid pre-treatment and nitrogen protection. Intern. J. Mol. Sci. **11**, 4297–4308 (2010). https://doi.org/10.3390/ijms11114297

Morphometric diversity of Hawksbill Turtle (*Eretmochelys imbricata*) at the Kelapa Island and Harapan Island Turtle Conservation Centers

Hanita Sofya Yusrianti[1*], *Adriani*[1], *and Dea Fauzia* Lestari[1]

[1]Departement of Marine Science and Technology, Faculty of Fisheries and Marine Science, IPB University, Jl. Raya Dramaga, Bogor, West Java, 16680, Indonesia

Abstract. Kepulauan Seribu National Park (TNKpS) is one of the marine nature conservation areas located in Pulau Seribu, DKI Jakarta Province. Kelapa Island and Harapan Island are centers for hawksbill turtle conservation, as several islands in the TNKpS are their landing and nesting areas. The study aimed to analyze the morphometric characteristics of hawksbill turtles based on their life stages. Morphometric measurements were carried out on 54 samples of hawksbill turtles of different ages with 13 parameters to differentiate cohorts, then a correlation analysis was conducted between carapace length and hawksbill turtle weight and nMDS analysis. Hawksbill turtles (*Eretmochelys imbricata*) in SPTN Region I Kelapa Island, and Region II Harapan Island showed a negative allometric growth pattern, with a correlation level of 0.001 for hatchlings (very low) to 0.938 for juveniles (very strong). The nMDS analysis showed similarities in characteristics from the early hatchling to early juvenile phases and there were 3 cohorts of hawksbill turtles conserved in SPTN Regions I and II TNKpS, based on carapace length.

1 Introduction

1.1 Background

Kepulauan Seribu National Park (TNKpS) is a marine nature conservation area located on the Thousand Islands, DKI Jakarta Province, which is geographically located between 5°23'-5°40'S and 106°37'E north of Jakarta [1]. The management of the Thousand Islands Marine National Park is carried out by the Thousand Islands National Park Office (BTNKpS). The management of the TNKpS area is divided into three National Park Management Section (SPTN) areas, namely SPTN Region I Kelapa Island, SPTN Region II Harapan Island and SPTN Region III Pramuka Island (BTNKpS). There are 2 types of turtles in the Thousand Islands National Park, namely the hawksbill turtle (*Eretmochelys imbricata*) and the green turtle (*Chelonia mydas*) [1].

Sea turtles are marine biota that utilize land areas for egg-laying every season. Six of the seven sea turtles are found in Indonesian waters: the olive ridley turtle (*Lepidochelys olivacea*), the green turtle (*C. mydas*), the flatback turtle (*Natator depresus*), the hawksbill turtle (*E. imbricata*), the leatherback turtle (*Dermochelys coriacea*), and the loggerhead turtle (*Caretta caretta*) [2].

* Corresponding author: hanitasfyyusrianti@apps.ipb.ac.id

SPTN Region I Kelapa Island and SPTN Region II Harapan Island are conservation centers for hawksbill turtles that successfully lay eggs in the TNKpS area with a higher number of hawksbill turtles than other TNKpS areas. TNKpS is a landing area and habitat for hawksbill turtles because of the relatively few predators, good beach morphology, and tidal patterns that approach vegetation areas, making it easier for hawksbill turtles to reach the nesting area.

The hawksbill turtle (*E. imbricata*) is a sea turtle characterized by its beak-shaped snout, a downward-curving upper jaw, and a relatively sharp, eagle-like shape, often referred to as the "Hawksbill turtle". Hawksbill turtles are listed in the IUCN Red Data Book as critically endangered [3]. One of the most visible characteristics of the hawksbill turtle is the arrangement of the segments that adorn its carapace. The average adult hawksbill turtle can grow to one meter and weigh around 80 kg. The largest hawksbill turtle ever caught weighed 127 kg [1]. To understand and support the conservation of hawksbill turtles (*E. imbricata*), research on the morphometric diversity of hawksbill turtles in the Seribu Islands National Park in SPTN Region I, Kelapa Island, and SPTN Region II, Harapan Island, is crucial. By understanding and knowing the variations in the morphometric characteristics of the hawksbill turtle, we can take more appropriate actions to protect this endangered species.

1.2 Purpose

This study aims to analyze the morphometric characteristics and correlation between carapace length and weight of hawksbill turtles in SPTN Region I Kelapa Island and SPTN Region II Harapan Island, Kepulauan Seribu National Park.

2 Methods

2.1 Time and place

In-situ data collection was conducted from March 17 to 24, 2025, at SPTN Region I Kelapa Island and SPTN Region II Harapan Island, Kepulauan Seribu National Park. There were 54 hawksbill turtle samples, including 15 samples of 2-day-old hatchlings, 10 samples of 10-day-old juveniles, 10 samples of 15-day-old juveniles, 10 samples of 90-day-old juveniles, and 9 samples of young turtles. The data collection locations can be seen in Figure 1. Data processing was carried out at the Department of Marine Science and Technology, Faculty of Fisheries and Marine Science, IPB University.

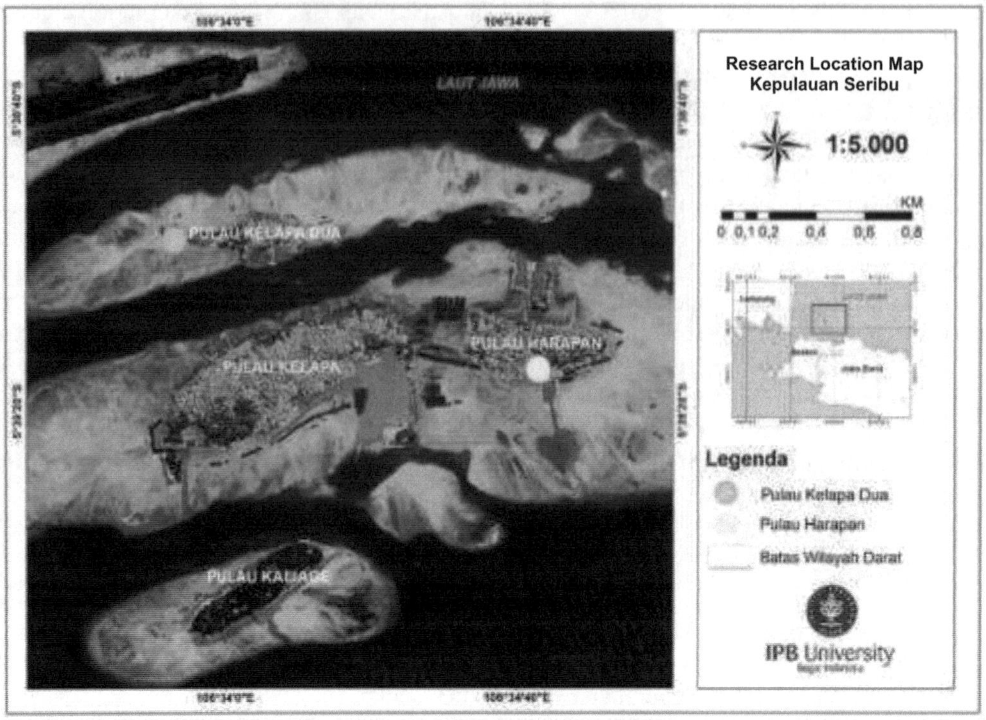

Fig. 1. Map of hawksbill turtle morphometric research sites in TNKpS.

2.2 Tools and materials

The tools and materials used in the morphometric analysis of hawksbill sea turtles are presented in Table 1.

Table 1. Tools and materials for morphometric research on hawksbill sea turtles.

No	Tools and Material	Purpose
1	Measuring tape	Measuring carapace length, carapace width, and other measurements
2	Scales	Measuring the weight of the turtle
3	Vernier	Measuring hard-to-reach parts of the body
4	Camera	Documenting the measurement process and the physical condition of the turtle
5	Data sheet	Recording the measurement data obtained
6	Stationery	Writing the data on the data sheet
7	*Microsoft Excel*	Data Processing and Analysis
8	*Microsoft Word*	Data Processing and Analysis
9	*Software PAST*	Data Processing and Analysis
10	Laptop/PC	Data Processing and Analysis

2.3 Work procedures

This research was conducted in several stages, data collection, additional data collection, and data analysis. Beginning with morphometric measurements, the obtained data were explained for their morphometric variations. The results of the morphometric analysis were then used to analyze the relationship between straight carapace length and weight and to analyze the nMDS. The results produced an analysis diagram that was then interpreted.

2.4 Data analysis

2.4.1 Morphometric variation analysis of hawksbill turtles

Morphometric variation analysis of sea turtles was conducted by measuring their bodies. Measurements were taken based on the turtle life stages. The measurement data obtained was collected based on stage and averaged to determine the morphometric mean value for each stage of the turtle life. Hawksbill sea turtles go through several stages in their lives, namely hatchling, juvenile, young turtle, and adult turtle, with the following descriptions:

1. Hatchling = The youngest stage, where the umbilical cord is still clearly visible, with a carapace size of < 10 cm.
2. Juvenile = Umbilical cord absent, carapace length 10 – 40 cm.
3. Young turtle = After juvenile but not yet mature enough to lay eggs, carapace length between 40 – 80 cm.
4. Adult turtle = Carapace length > 80 cm.

Morphometric measurements of sea turtles generally include 12 measurement parameters, namely straight carapace length (SCL), straight carapace width (SCW), curved carapace length (CCL), curved carapace width (CCW), front limb length (FLL), hind limb length (HLL), head length (HL), head width (HW), head circumference (HC), total tail length (TTL), length between cloaca and tail tip (CETL), length between plastron tip and cloaca (PCL) [4].

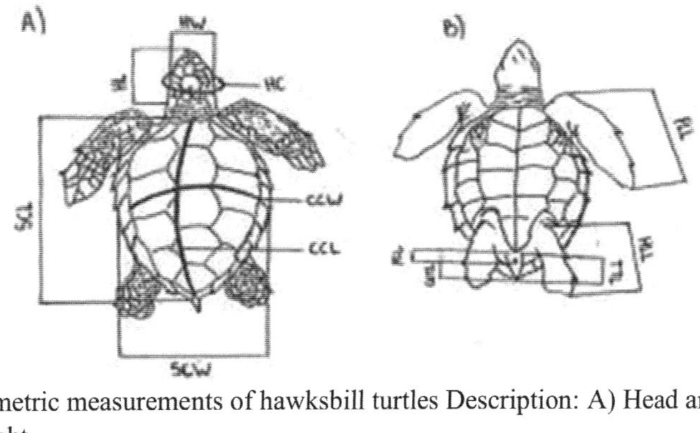

Fig. 2. Morphometric measurements of hawksbill turtles Description: A) Head and body, B) Limb, tail, weight.

The measurement data were processed using descriptive statistics. Mathematically, the average (X) is formulated as follows:

$$X = \frac{\Sigma xi}{N} \quad (1)$$

Description:
Σxi = Sum of all data values
N = Number of data points

2.4.2 Correlation analysis of straight carapace length and weight of hawksbill turtles

Carapace length and weight data can be used to observe growth patterns or relative growth. The data analysis used is regression analysis by comparing tabulations from data processing using Microsoft Excel to find the relationship between carapace length and turtle weight. The regression analysis formula:

$$W = aL^b \quad (2)$$

Description:
W = Body weight (gram)
L = Body length (cm)
a and b = Constant

This equation can be transformed into a linear form, as follows:

$$Log\ y = a\ log\ log\ x\ + log\ log\ b \quad (3)$$

$$Log\ b = \frac{\Sigma loglog\ y\ .\ \Sigma(loglog\ x).\Sigma(loglog\ x\ .loglog\ y)}{\Sigma(loglog\ x)^2 - \Sigma(loglog\ x)^2} \quad (4)$$

Description:

$$Log\ b = \frac{\Sigma loglog\ x - Nloglog\ b}{\Sigma loglog\ X} \quad (5)$$

Applicable:

$b = 3,$ depicts isometric growth
$b > 3,$ describes positive allometric growth or weight increases faster than length.
$b < 3,$ describes negative allometric growth or weight increases more slowly than length.

Correlation analysis is used to determine the relationship between two or more variables. Guidelines for interpreting the correlation coefficient (r) are presented in Table 2.

Table 2. Interpretation of correlation coefficient results.

Coefficient Interval	Relationship Level
0,00 – 0,199	Very Low
0,20 – 0,399	Low
0,40 – 0,599	Average

Coefficient Interval	Relationship Level
0,60 – 0,799	Strong
0,80 – 1,000	Very Strong

2.4.3 nMDS analysis

nMDS analysis (non-metric Multidimensional Scaling) is one of the multivariate statistical analyses used to determine the position of data based on similarity assessment, and is used to understand interdependence or mutual dependence among data. nMDS analysis is conducted using the Bray-Curtis distance matrix to reduce multivariate data into a two-dimensional space. Points that are close indicate similar composition or characteristics, while points that are far apart indicate differences in straight carapace length (SCL) and weight of the scaly turtle at each life stage. The analysis is conducted using PAST software.

3 Result and discussion

3.1 Result of morphometric variation analysis of hawksbill turtles

The results obtained from the research at the Turtle Conservation Center SPTN Region I Pulau Kelapa and SPTN Region II Pulau Harapan consist of combined data from morphometric measurements of hawksbill turtles in the hatchling, juvenile, and young turtle phases from both locations, which can be seen in Appendix 1 to Appendix 3. The measurement data is combined because the number of hawksbill turtles in SPTN Region I Pulau Kelapa at each phase is limited compared to SPTN Region II Pulau Harapan.

The measurement parameters conducted include straight carapace length (SCL), straight carapace width (SCW), curved carapace length (CCL), curved carapace width (CCW), front flipper length (FLL), hind flipper length (HLL), head length (HL), head width (HW), head circumference (HC), total tail length (TTL), length between the cloaca and the tail tip (CETL), length between the plastron tip and the cloaca (PCL), and weight measurements were also taken. The results of the morphometric measurements are presented in the following table.

3.1.1 Morphometric variation analysis of hawksbill turtles part A (head and body)

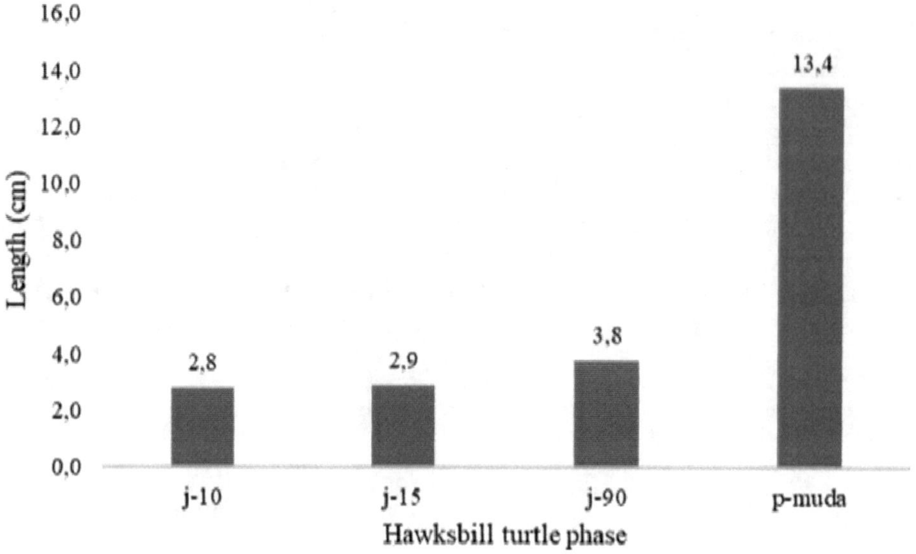

Fig. 3. Comparison of morphometric variations in head length (HL) by life stage in TNKpS.

Morphometric measurements of head length (HL) were conducted on 30 juvenile hawksbill turtle samples and 9 juvenile turtle samples. Figure 4 shows the average HL of 10- day-old juveniles at 2.8 cm, 15-day-old juveniles at 2.9 cm, 90-day-old juveniles at 3.8 cm, and 13.4 cm for young turtles.

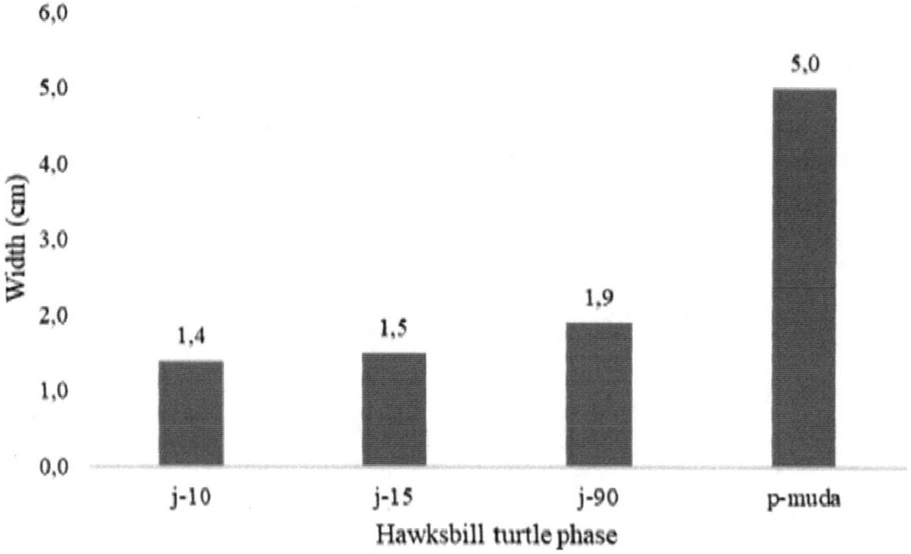

Fig. 4. Comparison of morphometric variations in head width (HW) by life stage in TNKpS.

Morphometric measurements of head width (HW) were conducted on 30 juvenile hawksbill turtle samples and 9 juvenile turtle samples. Figure 5 shows the average head circumference (HW) of 10-day-old juveniles at 1.4 cm, 15-day-old juveniles at 1.5 cm, 90-day-old juveniles at 1.9 cm, and young turtles at 5.0 cm.

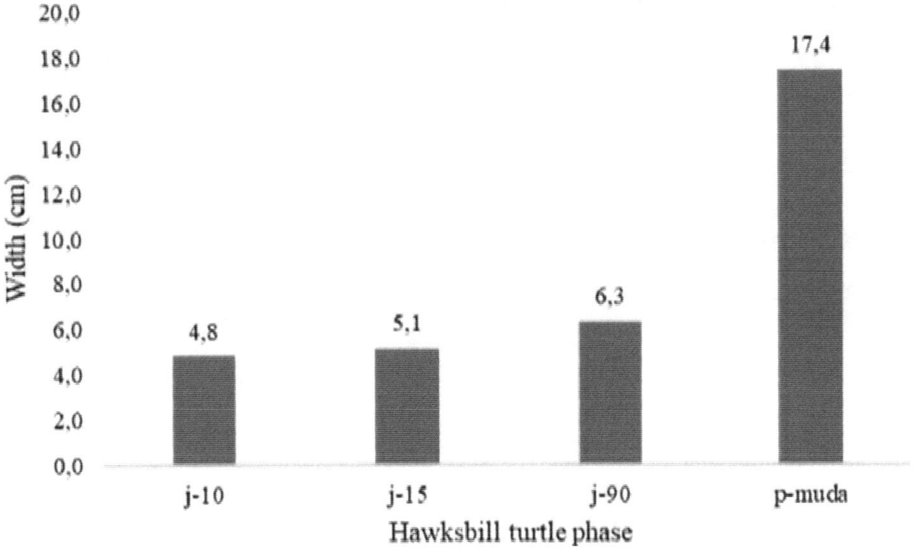

Fig. 5. Comparison of morphometric variations in head circumference (HC) by life stage in TNKpS.

Morphometric measurements of head circumference (HC) were conducted on 30 juvenile hawksbill turtle samples and 9 juvenile turtle samples. Figure 6 shows the average HW of 10-day-old juveniles at 4.8 cm, 15-day-old juveniles at 5.1 cm, 90-day-old juveniles at 6.3 cm, and young turtles at 17.4 cm.

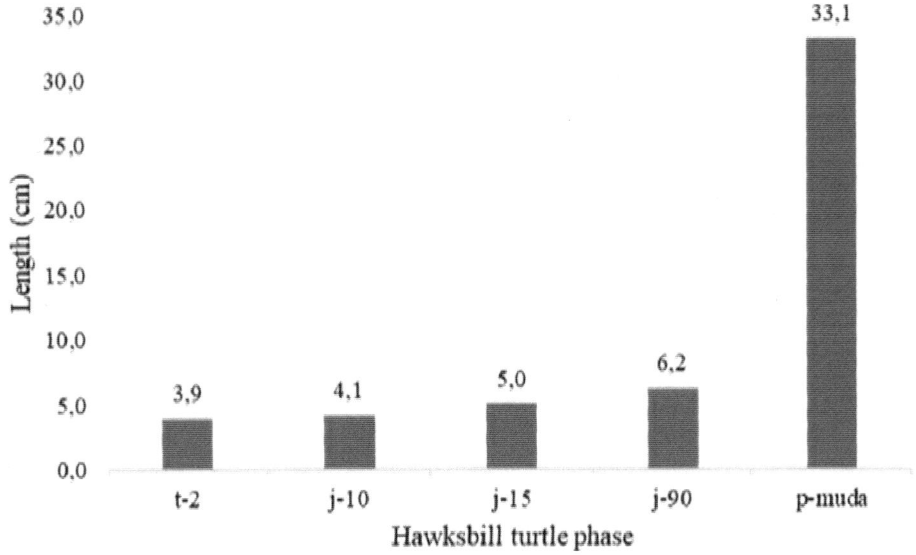

Fig. 6. Comparison of morphometric variations in straight carapace length (SCL) based on life stage in TNKpS.

Morphometric measurements of straight carapace length (SCL) were conducted on 15 hatchling samples, 30 juvenile hawksbill turtle samples, and 9 juvenile turtle samples. Figure 7 shows the average SCL of 2-day-old hatchlings at 3.9 cm, 10-day-old juveniles at 4.1 cm, 15-day-old juveniles at 5.0 cm, 90-day-old juveniles at 6.2 cm, and young turtles at 33.1 cm.

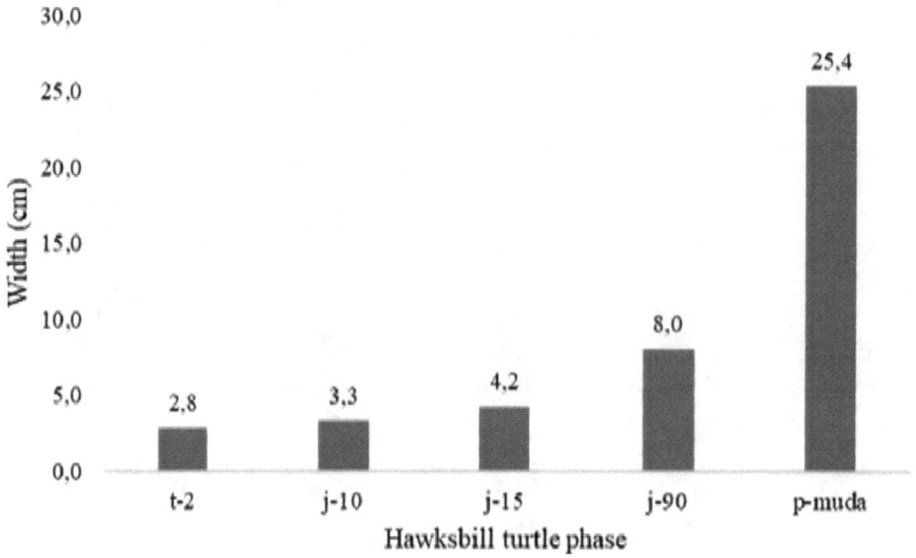

Fig. 7. Comparison of morphometric variations in straight carapace width (SCW) based on life stage in TNKpS.

Morphometric measurements of straight carapace width (SCW) were conducted on 15 hatchling samples, 30 juvenile hawksbill turtle samples, and 9 juvenile turtle samples. Figure 8 shows the average SCW size of 2-day-old hatchlings at 2.8 cm, 10-day-old juveniles at 3.3 cm, 15-day-old juveniles at 4.2 cm, 90-day-old juveniles at 8.0 cm, and young turtles at 25.4 cm.

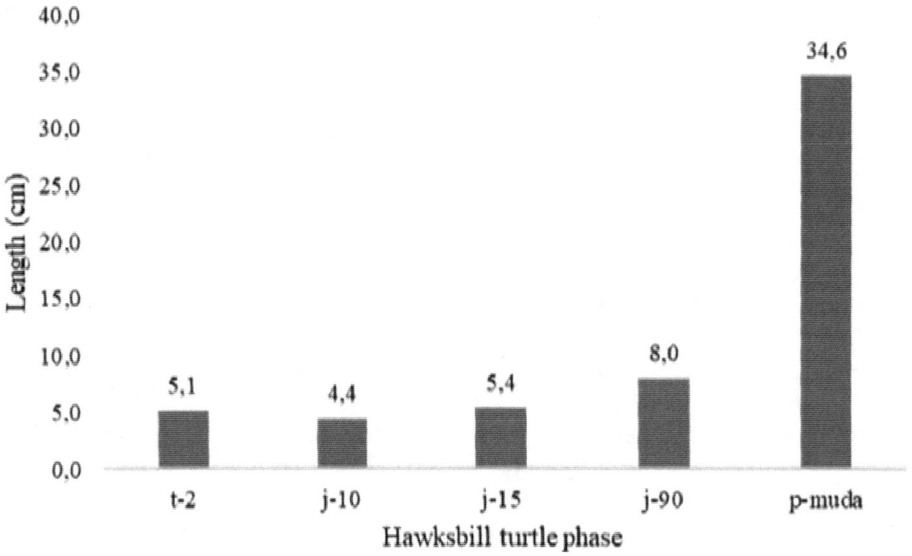

Fig. 8. Comparison of morphometric variations in curved carapace length (CCL) by life stage in TNKpS.

Morphometric measurements of curved carapace length (CCL) were conducted on 15 hatchling samples, 30 juvenile hawksbill turtle samples, and 9 juvenile turtle samples. Figure 9 shows the average CCL size of 2-day-old hatchlings at 5.1 cm, 10-day-old juveniles at 4.4 cm, 15-day-old juveniles at 5.4 cm, 90-day-old juveniles at 8.0 cm, and young turtles at 34.6 cm.

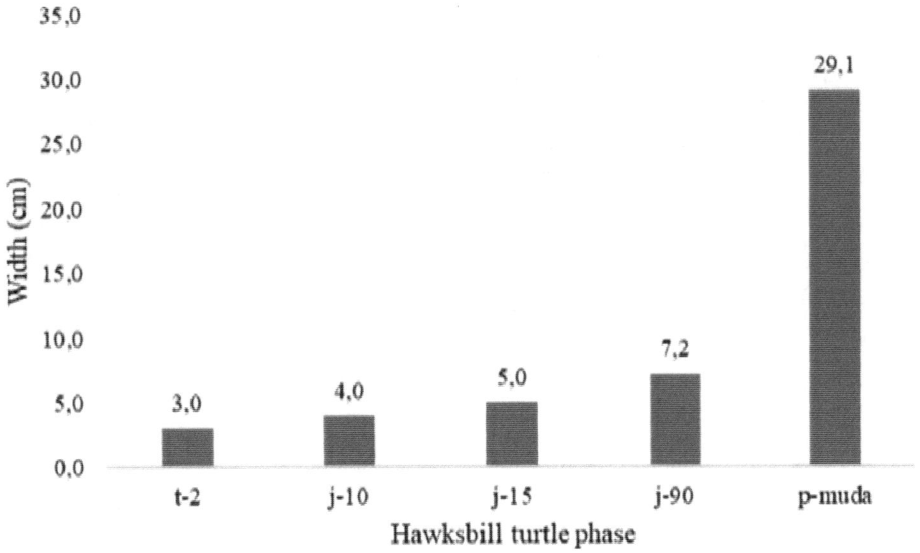

Fig. 9. Comparison of morphometric variations in curved carapace width (CCW) by life stage in TNKpS.

Morphometric measurements of curved carapace length (CCW) were conducted on 15 hatchling samples, 30 juvenile hawksbill turtle samples, and 9 juvenile turtle samples. Figure 10 shows the average CCW of 2-day-old hatchlings at 3.0 cm, 10-day-old juveniles at 4.0 cm, 15-day-old juveniles at 5.0 cm, 90-day-old juveniles at 7.2 cm, and young turtles at 29.1 cm.

3.1.2 Morphometric variations in hawksbill turtles Part B (limb, tail, and weight)

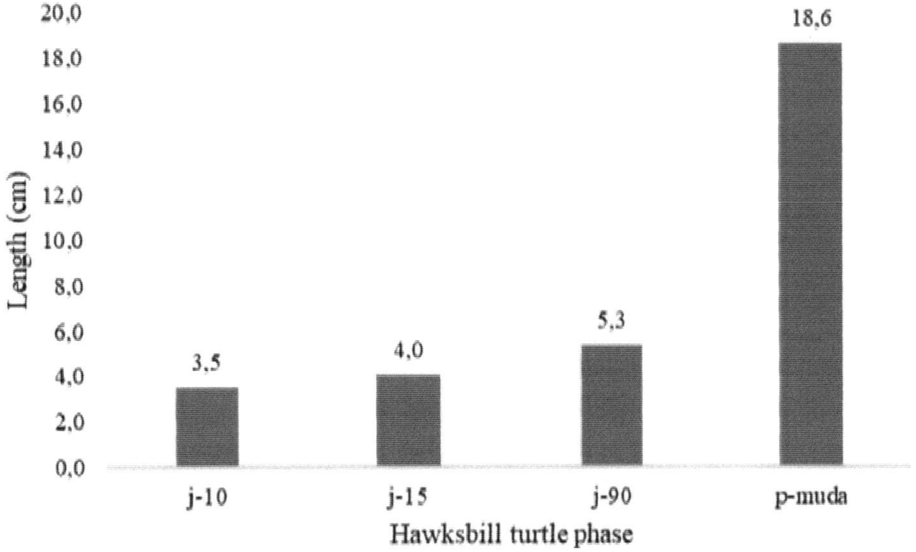

Fig. 10. Comparison of morphometric variations in front limb length (FLL) by life stage in TNKpS

Morphometric measurements of front limb length (FLL) were conducted on 30 juvenile hawksbill turtle samples and 9 juvenile turtle samples. Figure 11 shows the average length of the front limb length (FLL) for 10-day-old juveniles at 3.5 cm, 15-day-old juveniles at 4.0 cm, 90-day-old juveniles at 5.3 cm, and young turtles at 18.6 cm.

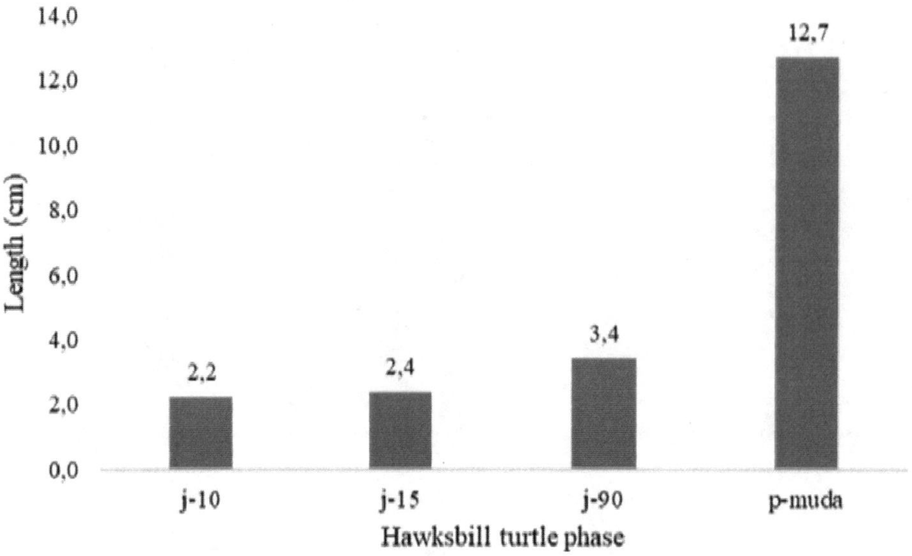

Fig. 11. Comparison of morphometric variations in hind limb length (HLL) by life stage in TNKpS.

Morphometric measurements of hind limb length (HLL) were conducted on 30 juvenile hawksbill turtle samples and 9 juvenile turtle samples. Figure 12 shows the average length of the HLL for 10-day-old juveniles at 2.2 cm, 15-day-old juveniles at 2.4 cm, 90-day-old juveniles at 3.4 cm, and young turtles at 12.7 cm.

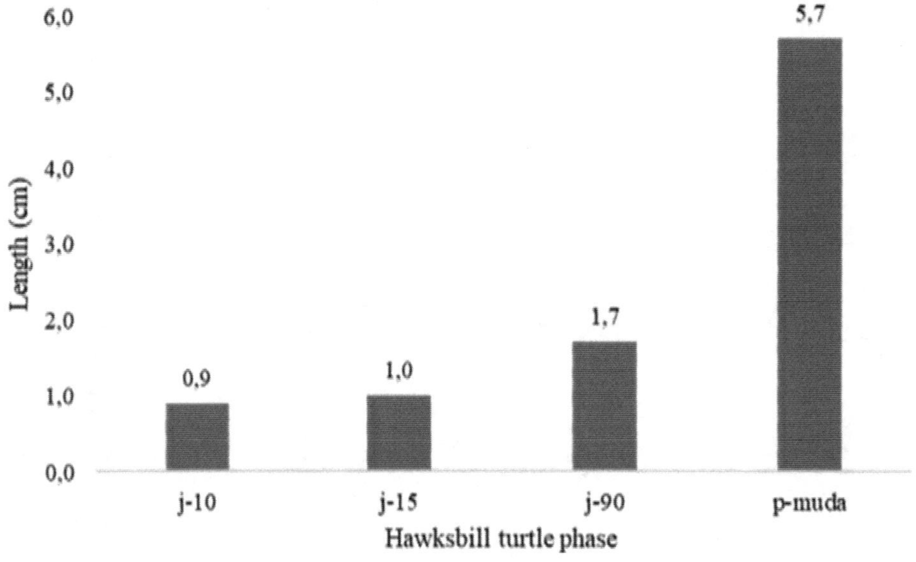

Fig. 12. Comparison of morphometric variations in total tail length (TTL) based on life stage in TNKpS.

Morphometric measurements of total tail length (TTL) were conducted on 30 juvenile hawksbill turtle samples and 9 juvenile turtle samples. Figure 13 shows the average TTL of 10-day-old juveniles at 0.9 cm, 15-day-old juveniles at 1.0 cm, 90-day-old juveniles at 1.7 cm, and 5.7 cm for young turtles.

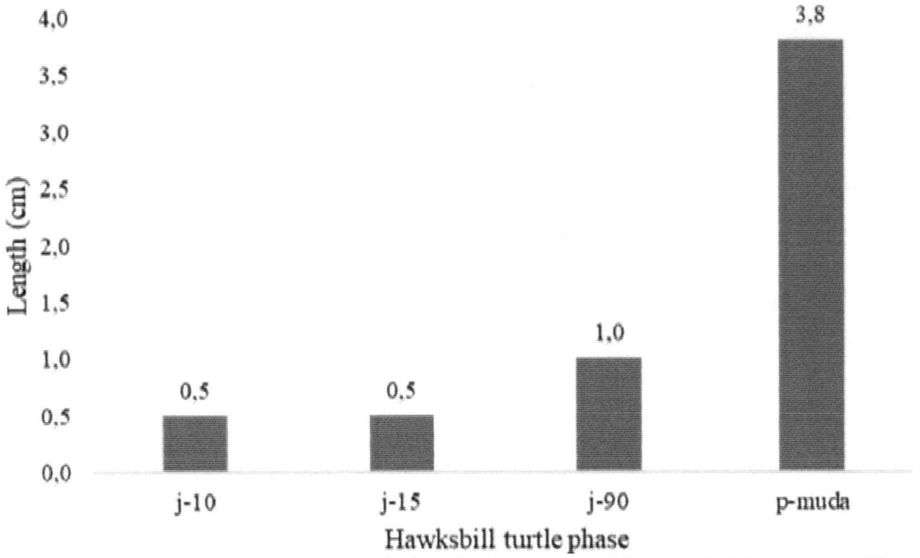

Fig. 13. Comparison of morphometric variations in cloaca-tip-tail length (CETL) based on life stage in TNKpS.

Morphometric measurements of cloaca tip tail length (CETL) were conducted on 30 juvenile hawksbill turtle samples and 9 juvenile turtle samples. Figure 14 shows the average CETL size of 10-day-old juveniles at 0.5 cm, 15-day-old juveniles at 0.5 cm, 90-day-old juveniles at 1.0 cm, and young turtles at 3.8 cm.

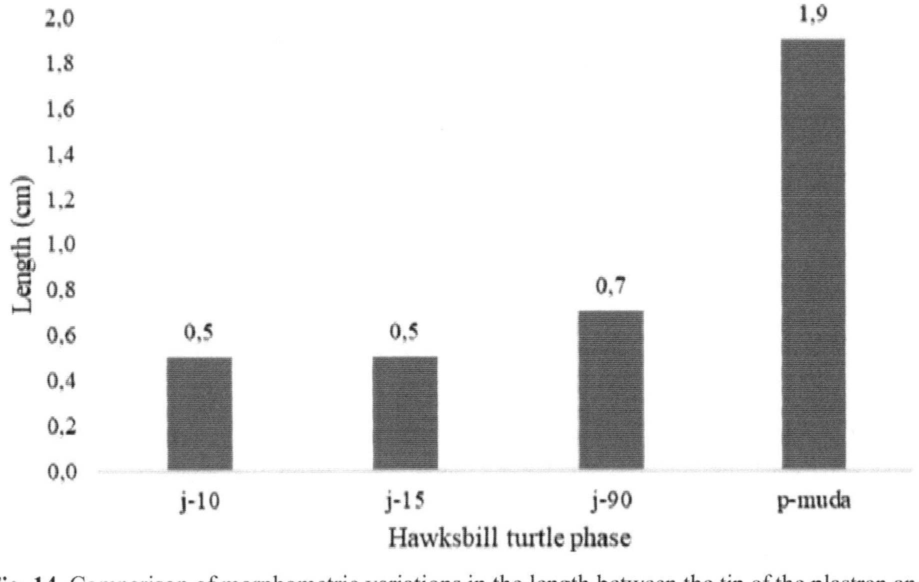

Fig. 14. Comparison of morphometric variations in the length between the tip of the plastron and the cloaca (PCL) based on life stage in TNKpS.

Morphometric measurements of the length between the tip of the plastron and the cloaca (PCL) were conducted on 30 juvenile hawksbill turtles and 9 juvenile turtles. Figure 15 shows the average PCL size of 10-day-old juveniles at 0.5 cm, 15-day-old juveniles at 0.5 cm, 90-day-old juveniles at 0.7 cm, and 1.9 cm for young turtles.

Table 3 Comparison of body weight variations by life stage in TNKPs

Sample	Weight-h2 (gr)	Weight-j10 (gr)	Weight-j15 (gr)	Weight-j90 (gr)	Weight-y.turtle (gr)
1	12,00	13,26	22,76	62,30	1150
2	13,17	15,16	22,16	70,07	3750
3	11,76	16,14	23,57	60,18	3450
4	13,21	16,90	25,78	76,23	3450
5	11,77	16,25	24,44	84,08	4250
6	12,90	16,24	21,88	59,75	3100
7	13,09	15,98	24,23	79,31	2700
8	12,87	14,49	20,36	57,26	5000
9	13,76	12,26	19,01	41,52	3750
10	12,72	14,13	23,08	50,00	
11	11,82				
12	13,15				
13	12,23				
14	13,96				
15	13,06				
Average	12,76	15,08	22,73	64,07	3400

Body weight measurement was conducted on 15 samples of hatchlings aged 2 days (t 2), 30 samples of juvenile hawksbill turtles aged 10 days (j 10), juvenile aged 15 days (j 15), juvenile aged 90 days (j 90), and 9 samples of young turtle phase (p young). Table 4 indicates that body weight increases with age. The average weight indicates for hatchlings aged 2 at 12.76 grams, juvenile aged 10 days at 15.08 grams, juvenile aged 15 days at 22.73 grams, while juvenile aged 90 days at 64.07 grams, and young turtle at 340.0 grams.

Table 4 Result of the calculation of the relationship between straight carapace length and weight of the hawksbill turtle in TNKpS.

Phase	n	a	b	r	R^2	Growth Pattern
t-2	15	11,726	0,0617	0,035	0,001	Negative allometric
j-10	10	0,678	2,1919	0,617	0,381	Negative allometric
j-15	10	0,782	2,0870	0,844	0,712	Negative allometric
j-90	10	0,145	2,9898	0,969	0,940	Allometrik negative
p-muda	9	1,171	2,2692	0,969	0,938	Allometrik negative

Description: Hatchling aged 2 days (t-2), juvenile aged 10 days (j-10), juvenile aged 15 days (j-15), juvenile aged 90 days (j-90), young turtle (p-muda).

3.2 Result of correlation analysis of straight carapace length and weight of hawksbill turtles

Growth observation on hawksbill turtles can be seen from the correlation of straight carapace length and weight. Hawksbill turtles in the Turtle Conservation Center SPTN Region I Pulau Kelapa and SPTN Region II Pulau Harapan consist of 3 life phases. The life phases of hawksbill turtles include hatchlings, juveniles, and young turtles. The results of the correlation of straight carapace length (SCL) and weight of hawksbill turtles can be seen based on the following figure. Observations show a varying pattern according to the growth phase of the turtles. The results of linear regression tests and the coefficient of determination

(R^2) illustrate the strength of the relationship between length and weight that differs in each life phase.

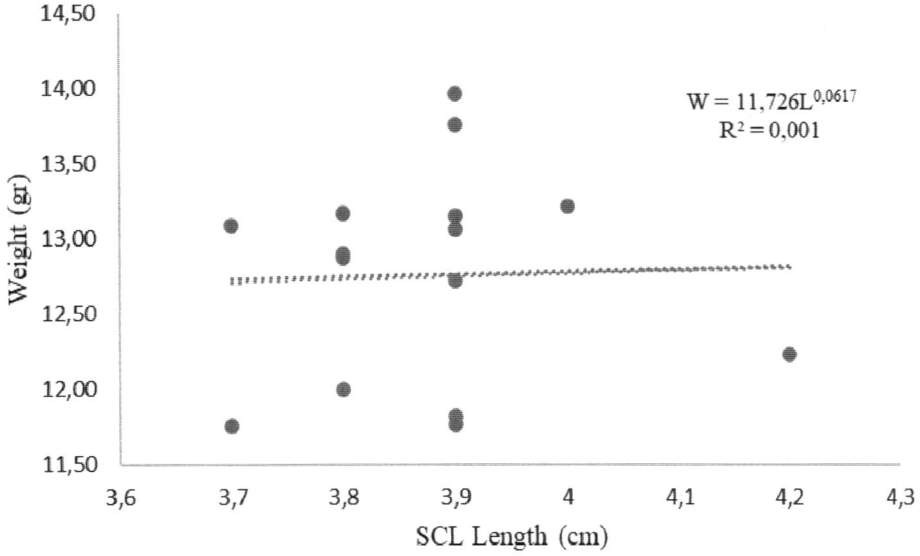

Fig. 15. Correlation between straight carapace length (SCL) and weight of 2-day-old hatchlings in TNKpS.

The results of the calculations on 15 samples of 2-day-old hatchlings can be seen in Appendix 9, with straight carapace lengths ranging from 3.7 to 4.2 cm and weights ranging from 11.76 to 13.96 grams, resulting in a correlation value as shown in Figure 16. The results indicate the equation $W = 11.726L^{o}.0617$ with (R^2) 0.001, which means the relationship between straight carapace length and weight is very low.

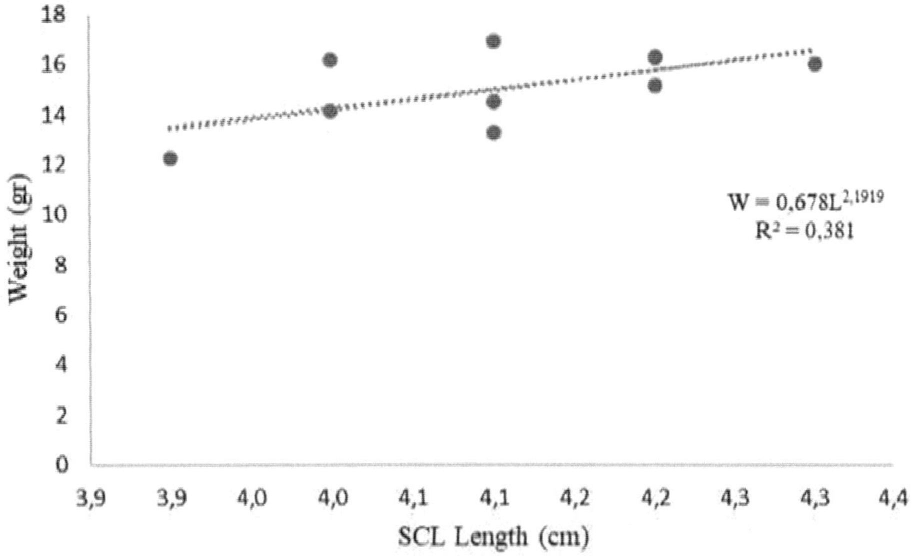

Fig. 16. Correlation between straight carapace length (SCL) and weight of 10-day old juveniles in TNKpS

The results of morphometric measurements on 10 samples of 10-day-old juveniles can be seen in Appendix 10 with a straight carapace length ranging between 3.9 - 4.3 cm and a weight ranging between 12.26 - 16.9 grams producing a correlation value as in Figure 17. The results show the equation $W = 0{,}678L2{,}1919$ with $(R2)$ 0.381 Figure 17 Correlation between straight carapace length (SCL) and juvenile weight 17 10 days in TNKpS which means that the length of the carapace is straight and the weight has a low relationship.

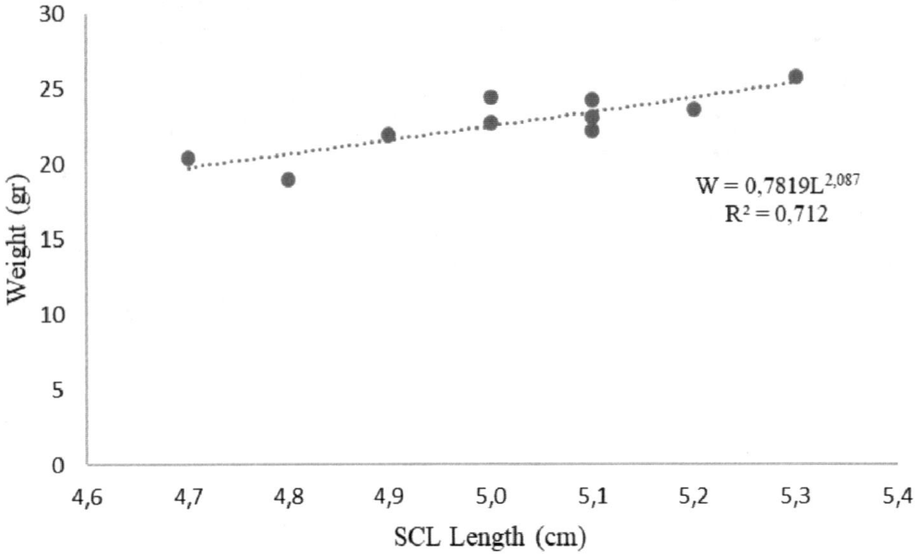

Fig. 17. Correlation between straight carapace length (SCL) and weight of 15-day-old juveniles in TNKpS.

The results of morphometric measurements on 10 juvenile samples aged 15 days can be seen in Appendix 11 with a straight carapace length ranging between 4.7 – 5.3 cm and a weight ranging between 19.01 – 25.78 grams producing a correlation value as in Figure 18. The results show an equation $W = 0{,}7819L2{,}087$ with $(R2)$ 0.712 which means that the straight carapace length and weight have a strong relationship.

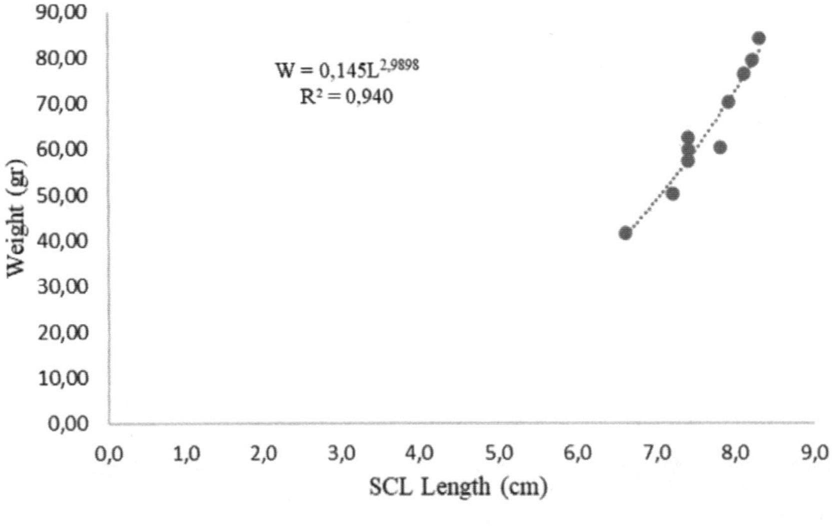

Fig. 18. Correlation between straight carapace length (SCL) and weight of 90-day-old juveniles in TNKpS.

The results of morphometric measurements on 10 juvenile samples aged 90 days can be seen in Appendix 12 with a straight carapace length ranging from 6.6 – 8.3 cm and a weight ranging from 41.52 – 84.08 grams producing a correlation value as in Figure 19. The results show an equation $W = 0{,}145L2{,}9898$ with ($R2$) 0.940 which means that the straight carapace length and weight have a very strong relationship.

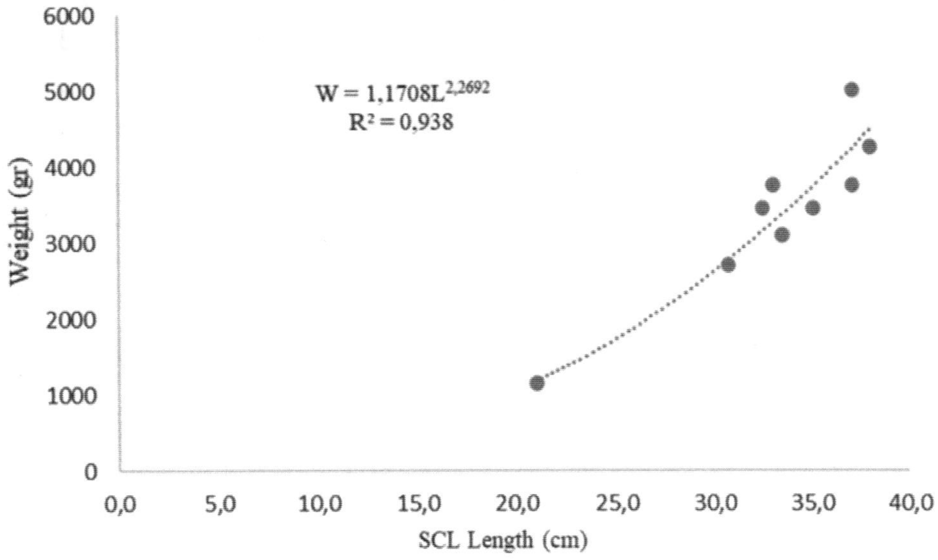

Fig. 19. Correlation between straight carapace length (SCL) and weight of young turtles in TNKpS.

The results of morphometric measurements on 9 samples of young turtles aged 2.5 years to 3 years can be seen in Appendix 13 with a straight carapace length. Ranges from 21.0 – 38.0 cm and weighs between 1150 – 5000 grams produces a correlation value as in Figure 20. The results show the equation $W = 1{,}1708L2{,}2692$ with ($R2$) 0.938 which means that the straight carapace length and weight have a very strong relationship.

3.3 Discussion of morphometric variation of hawksbill turtles

Morphometric measurements of head length (HL) in the juvenile phase start at around 2–3 cm at 10 days of age, increase to 3 cm at 15 days of age, then 3.5–4.3 cm at 90 days of age, and finally 9–15.7 cm in the young turtle phase. Research by [5] analyzed the skulls of green sea turtles (C. mydas), comparing the size of juvenile and adult individuals. They found that changes in the skull, particularly relative head length and width, reflect adaptations to shifting dietary patterns during development. This suggests that the progressive increase in head length (HL) from juvenile to young turtle stages is not only due to general growth but also ecological adaptations leading to improved foraging efficiency.

Head circumference (HC) measurements showed a significant jump from 4.5 – 4.9 cm at 10 days of age, 4.9 – 5.3 cm at 15 days of age, to 6 – 6.5 cm at 90 days, and finally 13.5 – 20.1 cm in the juvenile turtle phase. Head width (HW) also showed a consistent increase from 1.3 cm at 10 days of age, to 6.0 cm in the juvenile turtle phase. In line with the study of [6] on adult hawksbill turtles in Bulukumba waters, which included head measurements and reported significant differences between age classes and described exponential growth throughout the juvenile and juvenile turtle phases. The proportional increase in head size with growth demonstrates the role of head width in ecology and jaw strength in line with the pattern of increasing head width (HW) from the juvenile to the juvenile turtle phase.

Morphometric measurements of straight carapace length (SCL) and straight carapace width (SCW) increase with age. Straight carapace length (SCL) increases from a range of 3.8 – 4.2 cm in 2-day-old hatchlings to 21.0–38.0 cm in the juvenile turtle phase, while straight carapace width (SCW) increases from 2.5–2.9 cm to 14.5–29.5 cm. Growth reflects healthy development and structural adaptations to support swimming activities and body protection. The morphometric growth of hawksbill turtles can be greatly influenced by dietary factors and protective habitat. Research by [7] examined the effect of feed type on the growth of hawksbill turtle hatchlings at Tikus Emas Beach, Sungailiat, Bangka. Results showed that different feed compositions significantly varied the increase in parameters such as straight carapace length (SCL) and straight carapace width (SCW).

Measurements of curved carapace length (CCL) and curved carapace width (CCW) showed an increase with age, reflecting measurable and consistent growth in hawksbill turtles. Research by [8] showed that variations in feed composition between lemuru and seaweed significantly affected the curved carapace length (CCL) and curved carapace width (CCW) of hawksbill turtle hatchlings. Research shows that carapace width growth continues to follow the increase in length proportionally.

The length of the forelimbs (FLL) and the length of the hindlimbs (HLL) in hawksbill turtles show negative allometric growth, where the growth in the length of the forelimbs and hindlimbs is slow compared to the body growth. Morphometric data from the 10-day juvenile phase showed that the FLL ranged from 3.0 to 4.0 cm, increasing to 3.7 to 4.1 cm at 15 days of age, and reaching 4.5 to 5.9 cm at 90 days of age, while the juvenile turtle phase had an FLL between 12.0 and 23.0 cm. HLL measurements for the 10-day juvenile phase ranged from 2.0 to 25 cm, increasing slightly to 2.0 to 2.8 cm at 15 days of age, then 3.0 to 4.0 cm at 90 days of age, and reaching 9.0 to 17.5 cm in the juvenile turtle phase. This is in line with the results of research by [9] which documented similar FLL and HLL ranges in hawksbill turtles caught by catch in Sambas Waters, West Kalimantan, and showed that limb growth accelerated with age.

Tail morphometrics in hawksbill turtles show clear growth with age, with total tail length (TTL) increasing from 0.8–1.0 cm in 10-day-old juveniles to 4.5–7.0 cm in the juvenile phase. Meanwhile, the length from the cloaca to the tip of the tail (CETL) increases from 0.4–0.5 cm in 10-day-old juveniles to 2.5–5.0 cm in young turtles. Similarly, the length between the plastron and the cloaca (PCL) grows from 0.4–0.5 cm to 1.0–2.5 cm. [10] study in northwestern Mexico measured total tail length (TTL) along with curved carapace length (CCL) and found juveniles had tails longer than 5 cm. This consistently indicates sexual dimorphism and a transition to maturity.

Analysis of the body weight of hawksbill turtles (E. imbricata) shows an increase with age, from hatchling to juvenile. The weight of 2-day-old hatchlings ranges from 11.76 to 13.96 grams, then increases to 12.26 to 16.90 grams in 10-day-old juveniles, and reaches 19.01 to 25.78 grams at 15 days. At the age of 90 days, the juvenile body weight increases drastically to reach 41.52 – 84.08 grams. In the young turtle phase, body weight jumps to 1150 – 5000 grams. This increasing pattern indicates a healthy and progressive growth process. Research by [7] treated hawksbill turtle hatchlings with different feed treatments for 12 weeks. The group fed lemuru fish reached a weight of 211.8 grams, while the group fed pellets reached 95.3 grams. This confirms that adequate nutrition and age are important factors in supporting the rate of body weight growth of hawksbill turtles in conservation settings.

3.4 Discussion of the correlation between of straight carapace length and weight of hawksbill turtles

The correlation between hatchlings at 2 days of age was very low, at 0.001. This indicates that in this early stage of life, carapace length has little correlation with weight. Research by

[9] showed that in the early stages of life, carapace length growth is faster than weight growth. This is quite normal because hatchling development is still influenced by energy reserves from the egg yolk, which have not yet been fully absorbed, and there has not yet been any accumulation of body mass from independent feeding activities. In the juvenile phase at 10 days of age, the correlation increased, although still in the low category, with a value of 0.381. This indicates the beginning of a slightly more consistent growth pattern between length and weight, possibly because the hatchlings began to actively feed and adapt to the conservation environment. Research by [13] showed that although there was an increase in specific growth rate, there was no significant difference between feeding treatments on hatchling growth. This indicates that other factors such as feed quality and environmental conditions also plays an important role in the growth of hawksbill turtles.

The juvenile phase at 15 days and 90 days old had values of 0.712 and 0.940, respectively, indicating a stronger correlation between straight carapace length and weight. This suggests that with increasing age and size, hawksbill turtle growth becomes more proportional between carapace length and weight. Research by [13] confirmed that a combination of feed with optimal protein and calcium content can increase the efficiency of hawksbill turtle morphometric growth during this phase. The correlation between the juvenile turtle phase and the age range of 2.5 to 3 years showed a very strong relationship with a value of 0.938. This indicates that the turtle's growth during this juvenile turtle phase is proceeding well, as can be seen in Appendix 14. Environmental factors and the availability of suitable habitat greatly influence the successful growth of hawksbill turtles. According to [14] the importance of habitat quality and bio-physical factors supporting the growth and survival of hawksbill turtles.

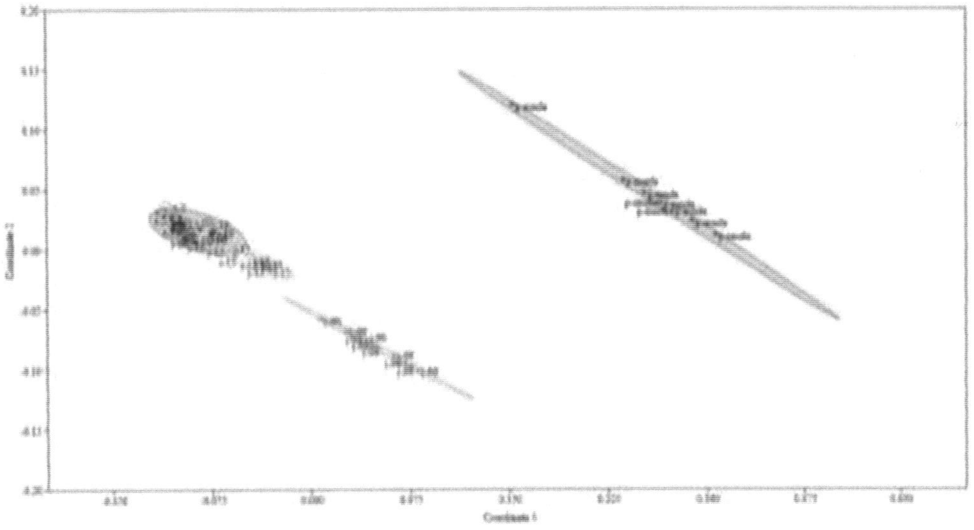

Fig. 20. Nmds analysis of straight carapace length (SCL) and weight based on age of hawksbill turtles in TNKpS

The results of the nMDS analysis in Figure 21 using the Bray-Curtis index indicate the presence of morphometric clustering of hawksbill turtles. 2-day-old hatchlings (t-2), 10-day-old juveniles (j-10), and 15-day-old juveniles (j-15) cluster tightly in the lower left corner. 90-day-old juveniles (j-90) begin to shift away from the hatchling and young juvenile clusters. Young turtles (p-young) form their own cluster far to the right. The nMDS plot with the Bray-Curtis index shows a clear growth pattern in hawksbill turtles based on two main morphometric variables, namely straight carapace length (SCL) and body weight. The hatchling phase (t-2), 10-day-old juveniles (j-10), and 15-day-old juveniles (j-15) cluster very tightly, indicating that in the early stages of growth, variation between individuals is still low, and changes in carapace size and weight are not very different between individuals. At around 90 days (j-10), the clusters began to move away from the early phase clusters, indicating that growth between samples began to differ significantly, possibly due to variations in individual growth rates or the influence of the environment and diet that began to appear. Indonesian research by [7] stated that the type of food can affect the growth rate of hawksbill turtle hatchlings at Tikus Emas Beach, Sungailiat.

The group of young turtles (p-young) is separated significantly from the other clusters, indicating that this phase of size and weight increase has reached a level where individual morphometrics are very different from the initial phase as a whole. These findings align with research by [3] who reported that the growth rate of juvenile hawksbill turtles varies with size and population density in the Chagos Archipelago. The early juvenile period, at 10 to 15 days of age, is homogeneous in size and weight, but after passing a certain threshold of around 90 days, the growth process becomes more diverse, possibly due to differences in food intake and other factors. Research by [15] who analyzed the morphometrics of olive ridley turtle hatchlings in the Mentawai Islands, showed a homogeneous pattern in the early phase but developed into greater size variation with age.

3.5 Conclusion and suggestion

3.5.1 Conclusion

Hawksbill turtles (*E. imbricata*) in SPTN Region I, Kelapa Island, and Region II, Harapan Island exhibit negative allometric growth from hatchling to juvenile turtle stage. Measurement results show that carapace length and width increase with age, but body weight increases much faster. The correlation between straight carapace length and body weight also increases with age, starting from a very low correlation in 2-day-old hatchlings to a very strong correlation in juvenile turtles aged 2.5-3.5 years. There are three cohorts of hawksbill turtle growth in the turtle conservation center of SPTN Region I, Kelapa Island, and SPTN Region II, Harapan Island, TNKpS.

3.5.2 Suggestion

Long-term monitoring is necessary and consistency in the use of measuring instruments is required during morphometric measurements get a more complete picture of growth. Further research in-depth analysis can be done by observing the influence of feed nutrition and environmental quality parameters to optimize data and conservation strategies.

References

1 Afifah AN, Sabila F, Hardi OS. Analisis karakteristik habitat penyu sisik Taman Nasional Kepulauan Seribu, Pulau Pramuka, Kabupaten Kepulauan Seribu, Provinsi DKI Jakarta. J. Siliwangi. 5,1: 23-27 (2019).

2 Pratama AA, Romadhon A. Karakteristik habitat peneluran penyu di Pantai Taman Kili-Kili Kabupaten Trenggalek dan Pantai Taman Hadiwarno Kabupaten Pacitan. JUVENIL. 1,2: 198-209 (2020).

3 Mortimer JA, Esteban N, Laloe JO, Stokes HJ, Tromp JJ, Hays GC. Low growth rates at high population densities in sea turtles. Marine Biology. 172,79: 1-14 (2025).

4 Sönmez B, Turan C, Ozdilek YS, Turan F. Sex determination of green sea turtle (Chelonia mydas) hatchlings on the bases of morphological characters. J. Black Sea/Mediterranean Environment. 22:93-102 (2016).

5 Nishizawa H, Asahara M, Kamezaki N, Arai N. Differences in the skull morphology between juvenille and adult green turtles: implications for the ontogenetic diet shift. Current Herpetology. 29,2: 97-101 (2010).

6 Nurdin MA, Anshar AR, Basrul Z, Sulaiman SY, Jamil NA, Rozalinda, Azisya N. Study of morphometric Measurement and physiological status in hawksbill turtles (Eretmochelys imbricata). J. of the Indonesian Veterinary Research. 6,2: 67-72 (2022).

7 Rinanda D, Afriyansyah B, Syarif AF. Pemberian pakan berbeda terhadap pertumbuhan tukik penyu sisik (Eretmochelys imbricata) di Pantai Tikus Emas, Sungailiat. J. Sumberdaya Akuatik Indopasifik. 6,1: 47-58. (2022).

8 Puteri FR, Afiati N, Widyorini N. Pengaruh komposisi jenis pakan yang berbeda terhadap pertumbuhan tukik penyu sisik (Eretmochelys imbricata) di penetasan semi-alami penyu Taman Nasional Karimunjawa. Indonesian J. of Fisheries Science and Technology. 14,2: 110-114 (2019).

9 Fajar S, Kushartono ED, Redjeki S. Morfometri penyu yang tertangkap secara Bycatch di Perairan Sambas, Kalimantan Barat. J. of Marine Research. 7,2: 125- 132 (2018).

10 Estevez LM, Angulo AA, Astorga ME, Becerra CD, Leyva NC, Amador FC, Amador JBC, Carvajal TVC, Robledo AF, Gaos AR, Hart CE, Weaver AH, Lopez JL, Lucero J, Llamas I, Mancini A, Oceguera K, Seminoff JA, Tershy BR, Yanez IL, Norzagaray AZ, Croll DA. Exploring the demography and conservation needs of hawksbill sea turtles Eretmochelys imbricata in north-west Mexico. Oryx. 57,3: 392-400 (2022).

11 Gatto CR, Jones TT, Imlach B, Reina RD. Ontogeny and ecological significance of metabolic rates in sea turtle hatchlings. Frontiers in Zoology. 19,6: 1-22 (2022).

12 Tezak B, Bentley B, Arena M, Mueller S, Snyder T, Romero IS. Incubation environment and parental identity affect sea turtle development and hatchling phenotype. Oecologia. 192,1: 1-13 (2020).

13 Lisandari G, Adibrata S, Supratman O. Pengaruh pemberian pakan yang berbeda terhadap pertumbuhan tukik penyu sisik (Eretmochelys imbricta) di Konservasi Penyu Aik Batu Banyak Desa Keciput Kabupaten Belitung. J. Ilmu Perairan. 2(2): 7-15 (2020).

14 Ubaydillah YZ, Yona D, Kasitowati RD. Analisis kesesuaian habitat peneluran penyu sisik (Eretmochelys imbricata) di Pantai Batu Hitam dan Pantai Bima, Taman Nasional Baluran, Situbondo. J. Kelautan Tropis. 26,2: 203-214 (2023).

15 Sababalat SR, Damanhuri H, Suparno. Analisis morfometrik tukik penyu lekang (Lepidochelys olivacea) dan kesesuaian habitat peneluran di Pantai Buggei Siata Desa Betumonga, Kabupaten Kepulauan Mentawai. E-Jurnal Universitas Bung Hatta. 25,3 (2024).

BIO Web of Conferences **221**, 03002 (2026) https://doi.org/10.1051/bioconf/202622103002
7th EMBRIO International Symposium (EIS 2025)

The relationship between Ascidians and plastic pollution: a systematic review of interaction mechanisms, biological impacts, and ecological roles

Gabriel Suripatty[1*], *Angelina* Br. Hombing[1], *Meutia* S. Ismet[2], *Beginer* Subhan[2], *Muhammad* R. Cordova[3,4], and *Neviaty* P. Zamani[2,5,1],

[1]Marine Science Study Program, Postgraduate Program, Departement of Marine Science and Technology, Faculty of Fisheries and Marine Science, IPB University (Bogor Agricultural University), Bogor, West Java, Indonesia
[2]Departement of Marine Science and Technology, Faculty of Fisheries and Marine Science, IPB University (Bogor Agricultural University), Bogor, West Java, Indonesia
[3]Research Center for Oceanography, the Indonesian National Research and Innovation Agency, Ancol, Jakarta, Indonesia
[4]Research Center for Oceanology, the Indonesian National Research and Innovation Agency, Jakarta, Indonesia
[5]Center for Transdisciplinary and Sustainability Sciences, IPB University (Bogor Agricultural University), Bogor, West Java, Indonesia

Abstract. Microplastic (MP) and nanoplastic (NP) pollution are global threats that permeate marine ecosystems. As sessile filter-feeders, ascidians interact heavily with these pollutants. This systematic review synthesizes the literature regarding the interaction between ascidians and plastics, focusing on their role as bioindicators, physiological impacts, status as model organisms, and potential in vertical pollutant transport. Sixteen primary studies indicate that various ascidian species consistently accumulate MP, highlighting their strong potential as bioindicators. Significant sublethal impacts, including disturbances in reproduction, development, immune response, and oxidative stress, have been widely reported. Laboratory studies confirm that ascidians, particularly from the genera Ciona and Phallusia, are valid model organisms for tracking particle pathways, toxicity, and underlying molecular mechanisms. Moreover, evidence suggests that ascidians play a role in the vertical transport of MP from the water column to sediments through fecal pellets, underscoring their dual ecological function. In conclusion, ascidians are not merely passive victims but serve as a unique and multifaceted model system for plastic research. Further research is needed to standardize bio-monitoring protocols and validate their bioremediation potential in the field.

1 Introduction

Marine debris has become an escalating global crisis, threatening marine life, ecosystems, and human health. Marine debris refers to solid, persistent materials created by humans, either directly or indirectly, intentionally or unintentionally discarded or left in the marine environment. About 80% of marine debris is plastic. Since its production began in the 1950s,

[*] Corresponding author: neviaty@apps.ipb.ac.id

the amount of plastic waste has reached 368 million tons, with approximately 11 million tons ending up in the oceans. 90% of this waste originates from land-based activities, such as household, industrial, and agricultural waste, while the remaining 10% comes from coastal and marine activities, including fisheries, shipping, and beach recreation. This waste enters the oceans through rivers, surface runoff, ocean currents, and wind. Approximately 30 million tons of plastic waste are projected to end up in the oceans by 2040.

The durability of plastic, which takes a long time to degrade, causes its accumulation in marine environments. This substance can be found in almost all marine environments, from the coastline to the surface, water column, and seafloor. Its presence poses a serious threat to aquatic organisms. Some of the impacts caused by plastic waste include entanglement, ingestion, serving as a carrier for invasive species, and accumulating toxic compounds. Small plastic particles (microplastics 1 - 5 mm and nanoplastics < 1 μm), when ingested, can cause oxidative stress, digestive, reproductive, growth, immune system disorders, and even genetic damage. This makes it a real threat to organisms living in the oceans.

Ascidians, also known as "sea squirts," are marine organisms belonging to the phylum Chordata. They are commonly found attached to hard substrates such as rocks and coral reefs in shallow to deep waters. These organisms exhibit a wide range of sizes and forms, from solitary species to colonial groups that can form large, dense aggregations. Ascidians play a vital role in marine ecosystems, functioning as filter feeders that help process organic matter in the water, and serving as a food source for many other organisms. Their filtration activities contribute to improving water clarity and nutrient cycling, which is essential for the overall health of marine environment. Furthermore, ascidians can act as habitats for other smaller organisms, thus enhancing biodiversity within their ecosystems.

However, the increasing presence of plastic pollution in the oceans has raised concerns about how such contaminants interact with marine organisms, particularly filter feeders like ascidians. In recent years, the relationship between ascidians and plastic pollution has begun to attract scientific attention, given the high potential for interactions between these organisms and various forms of plastic waste entering the oceans. Research has shown that plastics can impact ascidians in several ways, both directly and indirectly. Floating plastics or those that sink to the seafloor may attach to the bodies of ascidians, disrupting their filtration processes and worsening the water quality they manage. Additionally, exposure to micro/nano-plastics on the bodies of ascidians may potentially lead to physiological disturbances and toxicity in these organisms.

Although the impact of plastic pollution on various marisne organisms has been widely studied, there is limited research specifically examining the interactions between ascidians and plastic pollution. Understanding the mechanisms of interaction, biological impacts, and ecological roles of ascidians in plastic-contaminated ecosystems remains underexplored. Therefore, a systematic review is needed to further investigate this relationship by reviewing existing literature. This study aims to conduct a systematic literature review and bibliometric analysis of research examining the relationship between ascidians and plastic pollution, exploring interaction mechanisms, biological impacts, and ecological roles. Through this approach, it is expected that a clearer understanding of the effects of plastic pollution on ascidians will be achieved, as well as their contributions to maintaining the balance of marine ecosystems increasingly threatened by plastic pollution.

2 Methodology

This study follows the PRISMA (Preferred Reporting Items for Systematic Reviews and Meta- Analyses) framework, in conjunction with bibliometric analysis, to identify trends in research within a particular sector (Figure 1). This approach allows for the identification of frequently explored topics and areas that have received less attention. The combination of systematic literature review and bibliometric analysis is widely utilized because it helps researchers uncover research gaps and discover potential topics for future study. For the analysis, several software tools were utilized, including Excel and Biblioshiny through R. These tools are particularly useful for summarizing prior research both qualitatively and quantitatively. Additionally, they assist in the data visualization process, making the collected data easier to interpret and understand. The research articles included in this study were sourced from the Scopus database (https://www.scopus.com) and the Web of Science Core Collection database (https://www.webofscience.com). The search focused on the following keywords: "Microplastics" OR "micro litter" OR "microplastic*" OR "nanoplastic*" AND "Ascidia" OR "Ascidian" OR "Ascidiacea" OR "Tunicata" OR "Tunicate." All articles, reviews, and conference proceedings in English from these databases, published from the inception of the database until 2024, were eligible for inclusion. The data were first extracted as Bibliography TeX (BibTeX) files, then processed using the RStudio application with the Biblioshiny (bibliometrix) package. Afterward, the datasets were merged, and duplicate entries were removed to ensure data integrity.

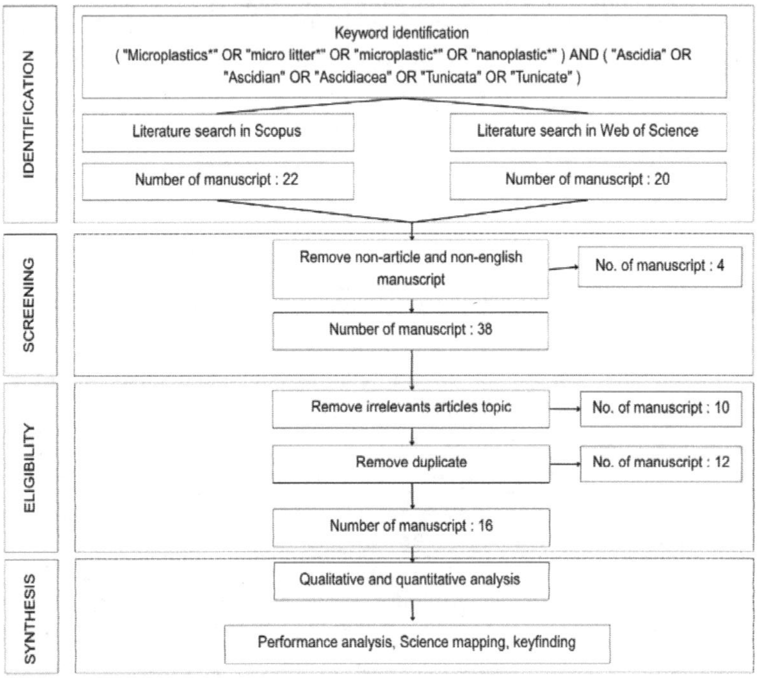

Fig. 1. PRISMA research framework.

3 Result and discussion

3.1 Performance analysis

The first step in this study is to conduct a research performance analysis to identify developments and provide an overview of research on the interaction between ascidians and plastic pollution. This topic has not been extensively studied yet. From 2018 to 2024, there were 16 research publications produced by 64 researchers, sourced from 12 different publishers. Research on the interaction between ascidians and plastic pollution has demonstrated a positive impact, with an average citation rate of 33.19 per document and an annual growth rate of 6.99%. However, despite the high citation impact, the overall number of publications remains limited, indicating that while the topic attracts significant attention in terms of citations, it has not yet generated a substantial volume of research. However, further investigation is needed to explore current trends and uncover potential research gaps for future studies. A summary of the research performance in the area of ascidian-plastic pollution interaction is provided in Table 1.

Table 1. Descriptive analysis of research performance.

Component	Description	Result
Publication	Total number of research publications	16
Productivity	Annual Growth Rate %	6.99
Source	Total number of journal sources in related fields	12
Total citation	Total citation in all publication	531
Average Citations per Document	Total citation/publication	33.19
Total Author	Total research authors contributing to the field	64
Author's Keywords	Total publication keywords	71

3.2 Annual scientific production

Figure 2 depicts the growth in yearly publications related to the interaction between plastic pollution and ascidians from 2018 to 2024. The dashed line represents the linear regression analysis conducted on the total number of publications. The R^2 (coefficient of determination) is calculated to be 0.54, indicating that the linear trend line accounts for approximately 54% of the increase in publications. The trend line shows a decline in 2019-2020, likely due to the COVID-19 pandemic and lockdowns. However, the number of publications increased again in 2021 and continued to rise, reaching 3 publications in 2022-2024. Despite this increase, the overall volume of research remains relatively low, indicating that the academic and scientific community has not extensively focused on this topic yet.

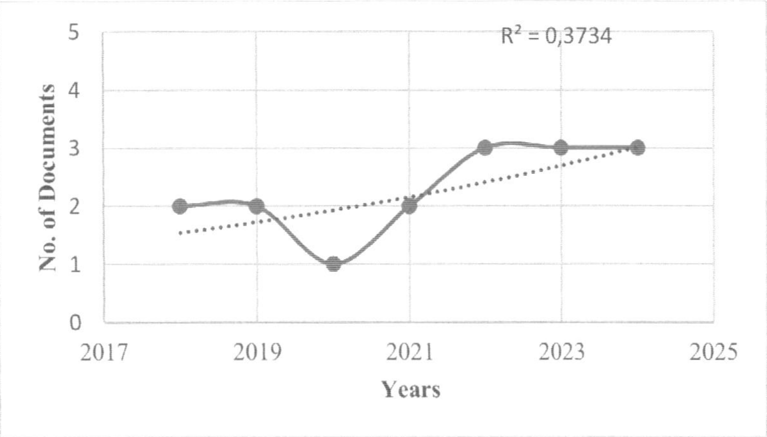

Fig. 2. Annual publication trend from 2018 to 2024.

3.3 Analysis of journals

The bibliometric data obtained highlights the contributions of various sources related to research on the interaction between ascidians and plastic pollution (Table 2 & Figure 3).

Among the listed journals, Environmental Pollution has the highest number of publications (4 articles) and a significant total citation count of 239, with an H-index of 4, indicating a strong influence in this field. The next source, Marine Pollution Bulletin, published 2 articles with a total citation count of 123 and an H-index of 2, showing relevant contributions, albeit fewer than Environmental Pollution. Other sources, such as Biomolecules and Chemosphere, each published 1 article with lower citation counts of 8 and 3, respectively, and an H-index of 1, reflecting a more limited influence. Additionally, European Zoological Journal with 1 article has a total citation count of 28 and an H-index of 1, indicating a moderate contribution in the related field. Other journals with a single publication include Frontiers in Immunology (6 citations), Journal of Aquatic Food Product Technology (51 citations), Journal of Experimental Zoology Part A-Ecological and Integrative Physiology (4 citations), Journal of Xenobiotics (2 citations), and Microplastics and Nanoplastics (28 citations), each with an H-index of 1. Overall, although most sources have published only one article, their citation counts and h-indices demonstrate their influence on the development of research in plastic pollution and its interaction with ascidians.

Table 2. Top ten most productive Journals and their Citations and H-Index.

No	Sources	Articles	Total Citation	H-Index
1	Environmental Pollution	4	239	4
2	Marine Pollution Bulletin	2	123	2
3	Biomolecules	1	8	1
4	Chemosphere	1	3	1
5	European Zoological Journal	1	28	1
6	Frontiers in Immunology	1	6	1
7	Journal of Aquatic Food Product Technology	1	51	1
8	Journal of Experimental Zoology Part A-Ecological and Integrative Physiology	1	4	1
9	Journal of Xenobiotics	1	2	1
10	Microplastics and Nanoplastics	1	28	1

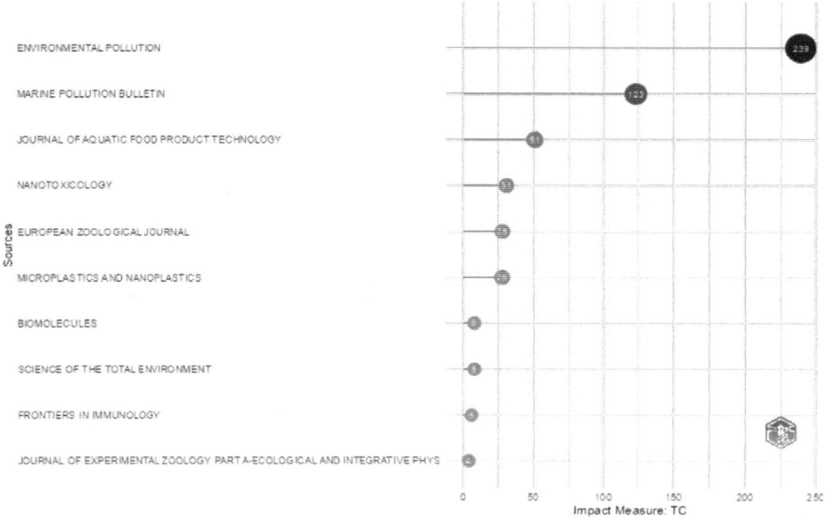

Fig. 3. Top ten most cited journals.

3.4 Most relevant countries

The bibliometric data reveals significant variations in research contributions and collaboration among countries (Table 3). Italy, with 11 articles and 318 citations, leads in both publication volume and citation impact, with notable international collaboration reflected in 3 Multi-Country Publications (MCP). Israel, contributing 4 articles with 205 citations, focuses primarily on national research, as evidenced by the absence of MCP. China, with just 1 article and 8 citations, also has no MCP, indicating a more isolated approach to its research. Overall, the data highlights Italy's strong international engagement, while Israel and China show more self-contained research activities. Research into the interaction between ascidians and plastic particles is a geographically concentrated field, largely driven by the unique environmental conditions of the Mediterranean Sea. This region is a recognized hotspot for both ascidian biodiversity, with approximately 229 species recorded, and for plastic pollution, containing an estimated 7% of the world's MPs. This unfortunate confluence makes it a critical natural laboratory, spurring focused research from bordering nations. Overall, the data illustrates how the severe MP pollution in the Mediterranean and its rich ascidian fauna have positioned regional actors like Italy as highly collaborative international leaders in this field, while the contributions from other nations like Israel and China remain more insular.

Table 3. Highest producing and cited nation.

Country	Articles	Total Citation	SCP	MCP
Italy	11	318	8	3
Israel	4	205	4	0
China	1	8	1	0

3.5 Most frequent and trending keywords

This section investigates the primary author's keywords in the study of asidians and plastic pollution, with a notable trend in the prevalence of MP (Figure 4). Although the broader topic is plastic pollution, the term "microplastic" appears most frequently (6 times), highlighting the increasing focus on MP within the field. This suggests that, while plastic pollution encompasses a wide range of pollutants, MP have garnered more attention in recent research due to their widespread distribution and potential ecological impacts. The term "ascidian," with 5 occurrences, emphasizes the significance of marine organisms in studies of plastic pollution, particularly in terms of their interactions with pollutants like MP. Other terms such as "polystyrene nanoparticles," "bioaccumulation," and "*Ciona robusta*" reflect common areas of research focused on the bioaccumulation of plastic pollutants in marine organisms. This concentration on MP highlights a critical area of concern within plastic pollution research, emphasizing the need for further studies to understand their effects on marine biodiversity and ecosystems.

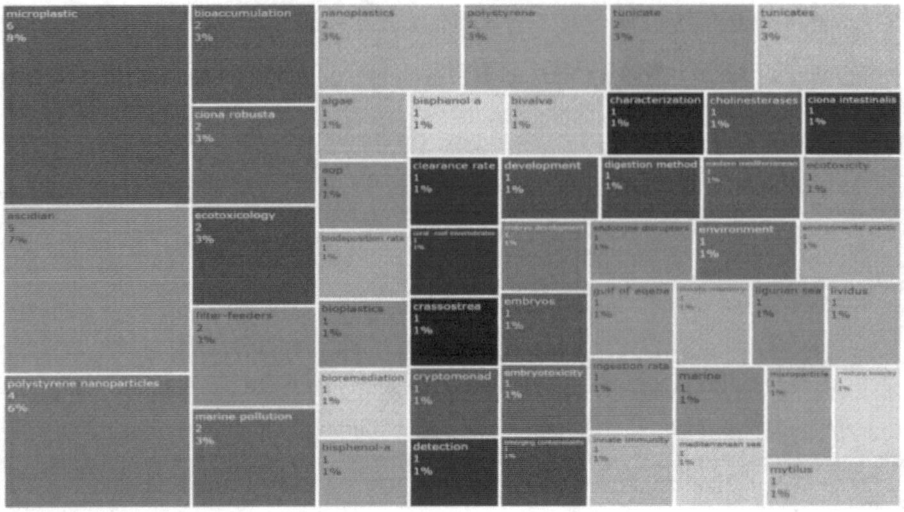

Fig. 4. Most frequent keywords.

3.6 Report finding on research of ascidian and plastic pollution interaction

1.6.1 Mechanisms of interaction

As non-selective suspension feeders, ascidians are highly vulnerable to the unintentional ingestion of plastic particles. They pump large volumes of water through their oral siphon, capturing food particles such as phytoplankton and organic detritus in mucus webs that line their branchial sac [1]. Since MP and NP often overlap in size with their natural food sources, these particles can easily enter the digestive system of ascidians. The efficiency of plastic capture and retention is not uniform across all ascidian species and is strongly influenced by the morphology of their filtering apparatus. The filtration rates in ascidians are primarily determined by the area of the branchial basket and the length of the ciliary band lining the stigmatal openings, as observed in a study of seven ascidian species. Filtration rates were measured by the exponential reduction in algal cell concentration, with no impact from the form of the stigmatal opening itself on the filtration efficiency [1]. These findings suggest that simpler filtering structures may be less efficient at sorting and rejecting non-food particles, leading to higher ingestion rates of plastics. This highlights the importance of considering species-specific anatomical traits when assessing susceptibility to MP pollution.

The fate of ingested plastics in ascidians varies depending on their size. Larger MP (e.g., >10 μm) largely remain in the digestive tract and are eventually excreted as fecal pellets [2]. However, smaller particles can cross biological barriers, leading to systemic contamination. Crucial experimental evidence from *Ciona intestinalis* demonstrated that 1 μm polystyrene particles could migrate from the gut cavity into the hemocoel (the circulatory cavity) within just eight days of exposure [2]. This finding fundamentally alters the understanding of plastics from mere gut content to persistent internal contaminants. Once in the circulatory system, these particles may be phagocytosed by immune cells (amoebocytes), which acts as a defense mechanism but also serves as a pathway for further distribution throughout the body tissues [2]. The ability of NP to cross the gut barrier represents a "Trojan Horse" mechanism for cellular toxicity. NP have a high surface-area-to-volume ratio, making them highly effective sorbents for chemical pollutants, including additives released from plastic matrices (e.g., phthalates, BPA) and environmental toxins [3]. When these contaminant-laden nanoparticles are translocated into the hemocoel and taken up by cells, they deliver a concentrated dose of toxins directly to internal tissues. This delivery mechanism can bypass key detoxification pathways located in the gut, potentially resulting in much higher cellular toxicity compared to exposure to the same chemicals dissolved in water.

Beyond direct ingestion, plastic debris serves as a potent ecological vector. Its high durability and buoyancy make it a new, long-lasting substrate for the transportation of marine organisms over long distances, a process known as "rafting" [4]. This process dramatically increases the dispersal opportunities for many species, including ascidians, allowing them to cross previously insurmountable biogeographical boundaries [4]. Many ascidian species are notorious invasive species that often spread via ship hull fouling and thrive in human-modified environments [4]. Rafting plastics provide an additional dispersal pathway, potentially accelerating the rate and range of biological invasions. This creates a feedback loop between two major global change drivers. First, plastic pollution facilitates the spread of invasive ascidians into new habitats. Second, many of these invasive species are highly tolerant of polluted conditions and often form dense populations in harbors and marinas, the same areas where plastic concentrations tend to be high [4]. Third, these invasive species can be highly efficient accumulators of MP. As such, successful invasive ascidians can become new hubs for the accumulation and processing of MP in the ecosystems they colonize, potentially altering the fate and bioavailability of local plastics for other organisms. This synergy likely exacerbates the ecological impacts of both stressors.

The surface of plastic debris in the ocean is rapidly colonized by unique microbial communities known as the "plastisphere" [5]. This biofilm community, consisting of bacteria, diatoms, and other protists, differs significantly from the microbial communities in the surrounding water column. The plastisphere can harbor pathogenic species (e.g., *Vibrio* spp.), harmful algae, and bacteria carrying antibiotic resistance genes, turning plastic debris into vectors for disease and genetic material [5]. Additionally, this biofilm alters the chemical properties of the plastic surface, potentially increasing its attractiveness to organisms by mimicking the chemical cues of natural food sources.

1.6.2 Biological impacts

Ingestion of MP can significantly disrupt basic physiological functions in tunicates. A study on the pelagic tunicate Dolioletta gegenbauri showed that exposure to environmentally relevant concentrations of microbeads caused a sharp decline in feeding rate (58%), growth rate (85%), and oxygen consumption (33%) [6]. These effects are primarily due to false satiety, where the digestive tract fills with indigestible plastic, reducing energy intake and causing a negative energy balance. The diversion of energy from vital functions such as growth and reproduction is a key mechanism of sublethal damage in filter-feeding organisms [6]. Although comparable quantitative data for benthic ascidians is still limited, these physiological principles are likely applicable, suggesting that chronic plastic exposure may harm energy allocation and overall fitness in ascidian populations.

Plastic pollution presents a multifaceted threat to the life cycle of ascidians, affecting development and reproduction through both physical particle effects and chemical toxicity. Physically, exposure to 10 μm polystyrene MP in *C. intestinalis* has been shown to cause significant delays in juvenile growth and metamorphosis [2]. The post-metamorphic juvenile stage appears to be particularly vulnerable, likely due to its high energy requirements for growth and organ reorganization. In contrast, pre-metamorphic larval development appears unaffected, indicating a stage-specific vulnerability to particle-induced stress [2]. Chemically, plastic additives are a significant concern. Phthalates, common plasticizers, have been detected in high concentrations in ascidian tissues from the Mediterranean Sea and the Red Sea, confirming the uptake of these chemicals from the environment [7]. Phthalates are endocrine-disrupting chemicals (EDCs) known to interfere with the hormonal systems regulating reproduction in various marine organism [8]. While direct studies on endocrine disruption by phthalates in ascidians are lacking, extensive evidence from other marine organisms including disrupted gametogenesis, decreased fecundity, and developmental abnormalities provides a strong framework to hypothesize similar long-term risks for ascidian populations [9]. Studies investigating the combined effects of NP and other additives, such as Bisphenol A (BPA), on *C. robusta* embryos found that while BPA itself caused developmental defects (abnormal pigmentation of sensory organs), NP did not show synergistic effects in seawater [10]. This suggests that the high ionic strength of seawater may alter the surface properties of nanoparticles, limiting their role as vectors for BPA in this context and highlighting the complexity of pollutant interactions in the marine environment [10].

Exposure to plastics triggers complex responses at the molecular and cellular levels in ascidians, beyond observable physiological effects, providing insights into the mechanisms of toxicity. Transcriptomic analysis of *C.robusta* embryos exposed to amino-modified polystyrene NP has revealed widespread changes in gene [3]. Major molecular pathways affected include glutathione metabolism (involved in antioxidant defense), immune defense, nervous system development, and energy metabolism [11]. These findings offer clear molecular evidence that NP induce multi-system stress responses, even during early developmental stages. Additionally, proteomics, the study of the entire set of proteins, has been employed to identify stress biomarkers in ascidians collected from the field. Studies on species such as M. exasperatus and C. robusta have highlighted proteins that are differentially expressed in response to various environmental stressors [12]. These proteins are often linked to cytoskeletal structure, energy metabolism, and general stress responses (such as heat shock proteins). Changes at the protein level serve as highly sensitive early indicators of stress, occurring well before observable impacts like growth reduction or mortality, thus underscoring the value of "-omic" approaches in biomonitoring. Some of the effects that may result from plastic particles exposure in ascidian organisms are summarized in Table 4.

Table 4. Plastic particles effect in ascidians or related organsims.

Biological Organization Level	Organisms Studied	Plastic Type	Key Findings	References
Molecular/ Cellular	*C. robusta*, *C. intestinalis*	PS NPs and MPs	Gene transcription related to glutathione metabolism and immune defense altered; translocation from the gut to the hemocoel within 8 days.	[2]
Physiological	*Dolioletta gegenbauri* (pelagic tunicate)	Microbeads	Feeding rate reduced by up to 58% and oxygen consumption by up to 33%.	[6]
Developmental	*C. robusta*, *C. intestinalis*	PS MPs, BPA	Growth affected at high concentrations; larval development unaffected; BPA causes abnormal pigmentation of sensory organs.	[2]
Behavioral	*C. robusta*	PS NPs	Disrupted swimming behavior identified as a key event before adverse outcomes at the organism level.	[11]

1.6.3 Ecological roles

Ascidians have many characteristics that make them ideal bioindicator or sentinel organisms for plastic pollution. Their sessile nature ensures that the contamination they accumulate reflects the local environmental conditions. Their high filtration rate effectively concentrates particles from the water column, enabling the detection of pollutants even at low environmental concentrations [1]. Additionally, their global distribution, including the ability of many invasive species to thrive in heavily polluted coastal areas such as ports and marinas, makes them accessible for monitoring programs worldwide [1]. Field studies have successfully used solitary ascidian species such as *H. momus, M. exasperatus*, and *S. plicata* to measure MP levels and chemical additives such as phthalates in the Eastern Mediterranean and Red Sea, demonstrating the feasibility of this approach [7]

While the potential is vast, the utilization of ascidians within the framework of formal Environmental Risk Assessment (ERA) remains minimal, with ascidians appearing in less than 0.13% of ERA studies [1]. The major barrier to broader adoption is the lack of standardized methodologies [1]. Currently, there is no universally aligned protocol for ascidian sampling, plastic particle extraction from their tissues, chemical analysis, or data reporting. This methodological inconsistency makes it difficult to compare results between studies and hinders the development of a robust global dataset [13]. Therefore, the lack of standardization is not only a scientific challenge but also a critical policy barrier. Without comparable and reliable data, ascidians cannot be effectively integrated into large-scale monitoring programs, such as the Global Ocean Observing System (GOOS), or used to inform and assess the effectiveness of international agreements aimed at reducing plastic pollution [13]

Ascidians could also act as an important link in the movement of plastics through marine ecosystems. They can facilitate the trophic transfer of MP from lower trophic levels to higher ones. A significant experimental study showed that juvenile *C. intestinalis* accumulated significantly more MP when they consumed mixotrophic flagellates that had previously ingested plastic beads [14]. This provides direct evidence of trophic transfer at the base of the food web, where ascidians act as secondary consumers of plastic by eating contaminated protists. Additionally, ascidians are active "ecological transformers" of plastic pollution, significantly altering its physical properties and fate. Through the digestive process, they repackage small suspended MP into larger, denser fecal pellets [15]. Studies on *S. plicata* show that these fecal pellets containing MP sink much faster than normal feces. This process creates a rapid pathway for transporting MP from the surface water column to the seafloor, a process known as benthic-pelagic coupling [15]. By accelerating this vertical flux, ascidians effectively move plastics from the pelagic environment to the benthic environment, where they become available to deposit-feeding organisms and can be buried in sediments. Furthermore, the digestive process coats the plastic particles with organic material, which may obscure their synthetic nature and potentially make them more attractive to other organism [15].

4 Conclusion

Research on the interaction between ascidians and plastic pollution remains limited, with only 16 publications by 64 researchers between 2018 and 2024. Despite the small number of publications, the research demonstrates significant impact, with an average of 33.19 citations per publication and an annual growth rate of 6.99%. Ascidians, acting as filter feeders, interact with plastic particles by ingesting them, which disrupts their filtration mechanisms. This leads to physiological effects, reduction in feeding rate and decrease in oxygen consumption. Moreover, ascidians accumulate MP, causing biological impacts such as oxidative stress and developmental delays. Ecologically, they serve as bioindicators of plastic pollution and play a crucial role in the vertical transport of plastics through the production of fecal pellets that move MP from the water column to the seafloor. Looking ahead, standardization of methods for global monitoring, development of molecular biomarkers for early detection of MP exposure, and further exploration of ascidians bioremediation potential are necessary. With these advancements, ascidians could play a broader role as agents of bioremediation and monitoring, helping to mitigate the growing plastic pollution in marine ecosystems.

This activity is supported by the Directorate General of Research and Development, Ministry of Education, Culture, Research, and Technology of the Republic of Indonesia, under the Master's to Doctoral Education Program for Excellent Bachelor's Degree Holders (PMDSU) Scholarship, in accordance with the Research Program Implementation Contract for the 2025 Fiscal Year, Number: 006/C3/DT.05.00/PL/2025.

References

1 A. Rosner and B. Rinkevich, Environments - MDPI **11**, 232 (2024)

2 S. Messinetti, S. Mercurio, G. Scarì, A. Pennati, and R. Pennati, Eur Zool J **86**, 189 (2019)

3 M. C. Eliso, I. Corsi, A. Spagnuolo, and R. Dumollard, J Xenobiot **15**, 10 (2025)

4 O. T. Streit, G. Lambert, P. M. Erwin, and S. López-Legentil, Mar Pollut Bull **167**, 112262 (2021)

5 R. K. Naik, M. M. Naik, P. M. D'Costa, and F. Shaikh, Mar Pollut Bull **149**, 110525 (2019)

6 G. A. Paffenhöfer and M. Köster, Arch Environ Contam Toxicol **78**, 94 (2020)

7 G. Vered, A. Kaplan, D. Avisar, and N. Shenkar, Mar Pollut Bull **138**, 618 (2019)

8 S. Ullah, S. Ahmad, X. Guo, S. Ullah, S. Ullah, G. Nabi, and K. Wanghe, Front Endocrinol (Lausanne) **13**, 1084236 (2023)

9 Y. Geng, Z. Liu, R. Hu, Y. Huang, F. Li, W. Ma, X. Wu, H. Dong, K. Song, X. Xu, Z. Zhang, and Y. Song, Front Physiol **14**, 1254886 (2023)

10 E. Ferrari, M. C. Eliso, A. Bellingeri, I. Corsi, and A. Spagnuolo, Biomolecules 2022, Vol. 12, Page 1661 **12**, 1661 (2022)

11 M. C. Eliso, E. Bergami, L. Bonciani, R. Riccio, G. Belli, M. Belli, I. Corsi, and A. Spagnuolo, Environmental Pollution **318**, (2023)

12 X. Li, S. Li, J. Cheng, R. Fu, and A. Zhan, Front Mar Sci **8**, 761628 (2021)

13 M. S. Savoca, N. A. Abreo, A. H. Arias, L. Baes, M. Baini, E. Bergami, S. Brander, M. Canals, C. A. Choy, I. Corsi, B. De Witte, C. Domit, S. Dudas, E. M. Duncan, C. E. Fernández, M. C. Fossi, O. Garcés-Ordóñez, B. J. Godley, D. González-Paredes, V. G. Carman, B. M. Hamilton, B. D. Hardesty, S. H. Hong, S. Kahane-Rapport, L. M. Kashiwabara, M. B. Lacerda, G. Luna-Jorquera, C. Manno, S. E. Nelms, C. Panti, D. J. Pérez-Venegas, C. K. Pham, J. F. Provencher, S. Purca, H. Rashid, Y. Rodríguez, C. Sparks, C. J. Sun, M. Thiel, C. Tsangaris, and R. G. Santos, Environmental Science: Advances **4**, 10 (2025)

14 R. Pennati, C. Castelletti, M. Parolini, G. Scarì, and S. Mercurio, J Exp Zool A Ecol Integr Physiol **337**, 639 (2022)

15 E. Harel, I. Zucker, and N. Shenkar, Chemosphere **367**, (2024)

Colony size variation of Acroporidae Family in Spermonde Island, Makassar, South Sulawesi

Muhammad Belva Mahardikaputra[1*], *Suryo* Kesumo[2], *Meutia* Samira Ismet[1], and *Dea* Fauzia Lestari[1]

[1]Departement of Marine Science and Technology, Faculty of Fisheries and Marine Science, IPB University, Jl. Raya Dramaga, Bogor, West Java, 16680, Indonesia

[2]Yayasan Karang Lestari Indonesia (YKLI)

Abstract. The Acroporidae family plays a key role in maintaining the balance of marine ecosystems, particularly in coral reef environments. One important parameter is colony diameter, which reflects growth, colony age, and the overall health of the marine environment. This study aims to assess the potential stock of Acroporid coral populations across 11 stations in the Spermonde Archipelago, Makassar, South Sulawesi. The research employed the underwater photogrammetry method, producing data. Data were analyzed using mean colony diameter, standard deviation, heatmap analysis, and significance testing. The results showed that *Acropora* was the dominant genus at most stations, with the highest dominance at Station 3 (79%). *Anacropora* had the largest colony diameter, while *Montipora* and *Alveopora* tended to be smaller. Station 3 was identified as the site with the best population stock. These findings indicate that Acroporidae coral populations are heterogeneous and influenced by local environmental conditions.

1 Introduction

Coral reefs are among the most complex and productive tropical marine ecosystems in the world. In addition to providing habitat for a wide variety of marine organisms, coral reefs play an essential role in maintaining coastal stability, sequestering carbon, and serving as an economic resource through fisheries, tourism, and the ornamental coral trade [1]. One of the dominant coral families in reef ecosystems is Acroporidae, characterized by its complex branching morphology and rapid growth rate [2]. The ecological role of Acroporidae is highly significant in maintaining the balance of aquatic ecosystems, particularly in coral reef environments. This family is recognized as a primary builder of reef structure, offering habitat and protection for a diverse array of marine species.

The Spermonde Archipelago, located in the southwest of Makassar Strait, is a cluster of islands with exceptionally rich coral reef ecosystems but faces substantial pressures. This area serves as a hub for fishing, maritime transportation, and the collection of ornamental corals. As a megadiverse country, Indonesia hosts more than 500 coral species, including members of the Acroporidae family such as *Acropora*, *Montipora*, *Isopora*, and *Anacropora*. These coral populations continue to experience stress due to climate change, coral bleaching,

* Corresponding author: muhammadbelva@apps.ipb.ac.id

illegal harvesting, and human activities such as reclamation and sedimentation [3]. Effective management based on robust data—particularly through coral stock assessments—is essential for understanding community structure, regeneration potential, and the natural reproductive capacity of coral reefs. The Spermonde Archipelago, located in the southwestern waters of Makassar, is a cluster of islands with exceptionally rich coral reef ecosystems but faces substantial pressures. This area serves as a hub for fishing, maritime transportation, and the collection of ornamental corals.

Assessing coral stock populations, particularly those of the Acroporidae family, is crucial in this region to understand the current community structure and to design long-term management and conservation strategies. In the context of resource management, stock information is highly valuable for setting exploitation quotas, identifying broodstock, and supporting restoration programs focused on priority species. One of the approaches to evaluate stock populations is by measuring variations in coral colony size, particularly colony diameter, which serves as an indicator of growth phase and reproductive maturity. Larger colonies generally have greater reproductive capacity and provide more larvae for ecosystem regeneration [2].

A recent data collection method for obtaining colony size in a non-invasive manner is Underwater Photogrammetry. This technique has proven effective for mapping coral communities with high precision, as it can capture comprehensive information on coral colonies without damaging the natural substrate [5]. It also facilitates documentation, diameter measurements, and coral morphology classification from multiple perspectives. Through dedicated software processing 2D morphometric and community structure analyses can be produced for coral stock data.

2 Methods

2.1 Time and place

This study was conducted in July 2025 in the Spermonde Archipelago, Makassar, South Sulawesi, as shown in Figure 1.

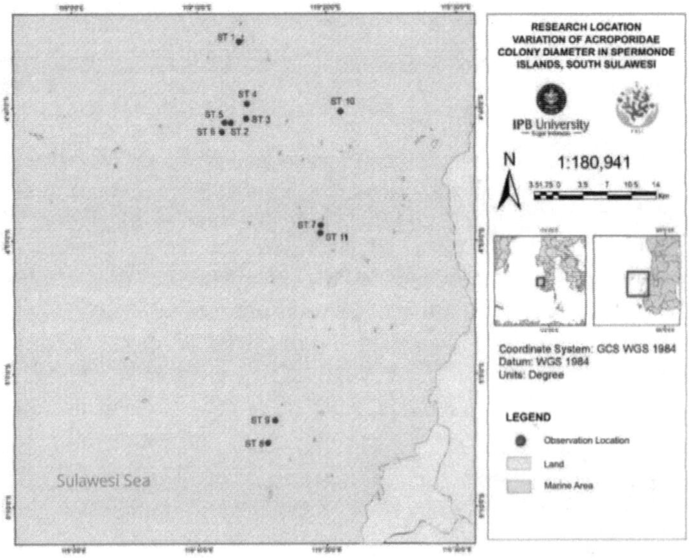

Fig. 1. Map of the study location in the Spermonde Archipelago, Makassar.

The method employed in this study was a quantitative approach using photogrammetry, with one replication conducted at each sampling point. Data collection was carried out at 11 stations located in different areas. The data were collected by a team from the Karang Lestari Indonesia Foundation, which was divided into four groups, each assigned to different sampling locations.

2.2 Tools and materials

Tools and materials used in this study were employed to collect data on colonies of the family *Acroporidae* using the underwater photogrammetry method, as presented in Table 1 below.

Table 1. Tools and materials for stock population Acroporidae.

No	Tools and Material	Purpose
1	Scuba SeT	One set
2	*Underwater Camera*	Olympus TG-5 with waterproof housing
3	*Belt Transect*	Measurement comparison
4	*Roll* meter	Measurement comparison
5	*Garmin dive com*	Measuring depth

2.3 Work procedures

This study consisted of four stages, beginning with in-situ sampling conducted by a team from the Karang Lestari Indonesia Foundation. The next stage involved the analysis of coral data through processing using the PhotoQuad software. The resulting data were then analyzed using the mean diameter variation formula to determine the average size of the coral colonies.

2.4 Underwater photogrammetry

The data collection method used in this study was underwater photogrammetry. This technique involves capturing high-resolution images to create a digital surface model, allowing for detailed analyses of coral structure and growth variation over time (5). Field observations were conducted by a team from the Karang Lestari Indonesia Foundation in the waters of Makassar, South Sulawesi [6]. Data were collected using an Olympus TG-5 underwater camera to record images beneath the surface, along a 50-meter belt transect positioned horizontally by divers, with the camera directed downward. The diver's field of view extended approximately one meter to the right and left of the belt transect to ensure even coverage of the observed area. An illustration of the data collection process is presented in Figure 2 below.

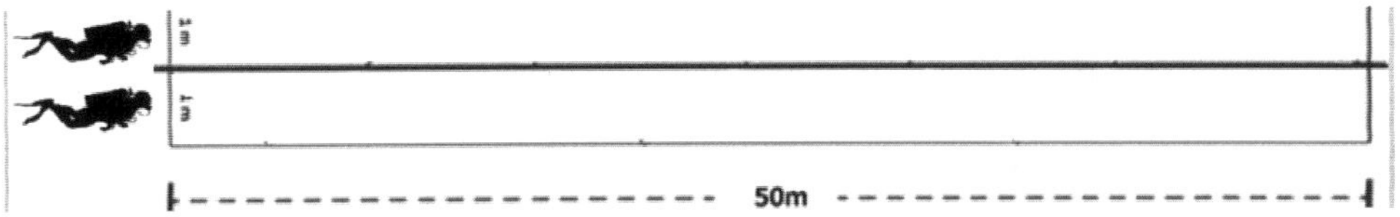

Fig. 2. Illustration of photogrammetry data collection.

2.5 Average variation in coral colony diameter

The data analysis in this study aimed to determine the mean variation in the diameter of the family *Acroporidae*. The resulting data were then subjected to statistical testing to assess differences in diameter among stations.

2.5.1 Combined Mean Size of Coral Diameter Variation

The mean size of corals in this study was calculated using the formula proposed by [7] as follows:

$$X = \frac{\sum x}{n} \quad (1)$$

Description:
n = Number of stations
$\sum x$ = Number of colony diameter measurements
$\bar{X}$ = Mean Variation in Coral Diameter

2.5.2 Standar Deviation

$$\int = v \frac{\sum(xi-\underline{x})^2}{n-1} \quad (2)$$

Description :
$\int$ = Standar deviation
xi = The value of α for the i-th substrate at the n-th station
x̄ = Average

3 Result and Discussion

3.1 Result of colony occurrence coral

Table 2. shows the dominance of coral colonies by genus at each station. *Acropora* was most abundant at Station 5, with 78 colonies, while *Anacropora* dominated Station 1 with eight colonies. *Alveopora* reached its highest abundance at Station 7 with 56 colonies, and *Isopora* was most numerous at Stations 6 and 9, each with 13 colonies. *Montipora* recorded its greatest number at Station 5, with 72 colonies.

Table 2. The presence of coral genus colonies in the research stations.

Lokasi	*Acropora*	*Anacropora*	*Alveopora*	*Isopora*	*Montipora*
Station 1	40	8	-	-	10
Station 2	4	-	1	-	2
Station 3	8	-	-	1	20
Station 4	10	-	-	-	11
Station 5	78	-	-	-	72
Station 6	29	2	-	13	32
Station 7	1	-	56	-	-

Lokasi	*Acropora*	*Anacropora*	*Alveopora*	*Isopora*	*Montipora*
Station 8	21	-	-	-	9
Station 9	23	1	5	13	38
Station 10	19	4	-	1	6
Station 11	9	-	-	-	3
Total	242	15	62	28	203

Acropora occurred at every station due to its faster growth compared with other genera. Haerul [6] reported that Indonesia is the center of *Acropora* diversity, with more than 90 species, many of which remain unnamed. *Montipora* was found at 10 stations; like *Acropora*, it is a key reef-building genus in tropical regions such as Indonesia and has high regenerative capacity, allowing it to occupy available substrate more rapidly than other genera.

The remaining genera (*Anacropora*, *Alveopora*, and *Isopora*) occurred at fewer stations. (7), *Anacropora* and *Alveopora* exhibit slower growth, are more sensitive to environmental change, and have narrower geographic and ecological distributions.

3.2 Percentage colony occurrence acroporidae

Based on the distribution of *Acroporidae* coral genera by station, it is evident that Acropora is the most dominant genus at most sites. *Acropora* plays an important role in stock population structure due to its rapid growth and high regenerative capacity [2]. The dominance of *Acropora* at many locations indicates a high potential stock population and strong ecosystem recovery capacity.

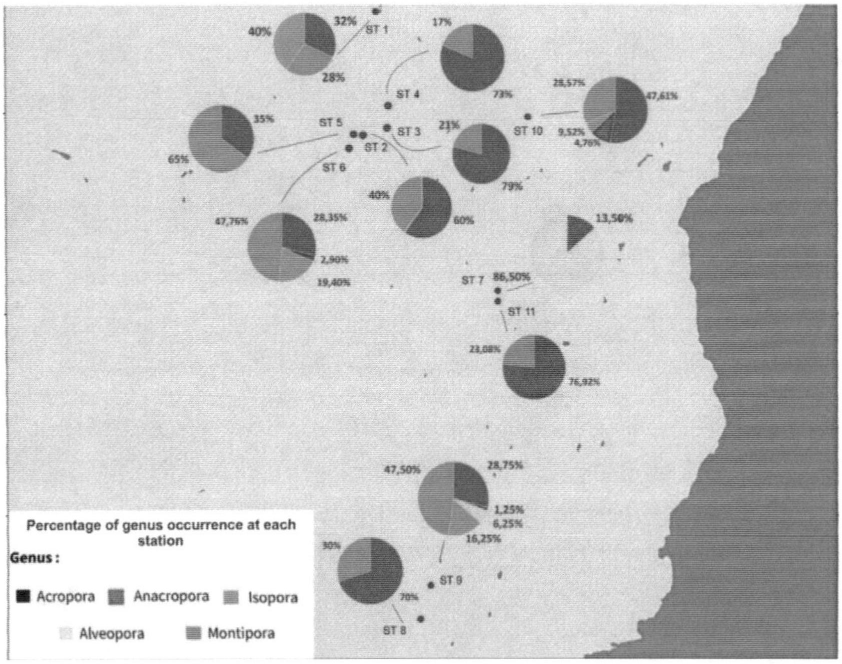

Fig. 3. Distribution of coral genera in the research stations.

Montipora shows a relatively even percentage of occurrence but is not dominant. This genus has a slower growth rate and a stable survival strategy, thus contributing more to maintaining community continuity rather than serving as a major source of broodstock [8]. *Anacropora* and *Isopora* appear only in limited numbers and at a few sites, indicating that the stock populations of these genera are small and their habitats highly specific. Such genera tend to be vulnerable to disturbance and are not suitable for utilization without a conservative approach.

Overall, this distribution pattern suggests that coral stock populations in the Spermonde Archipelago are heterogeneous: some locations have high potential as seed sources (*Acropora*), while others require protection because they harbor rare genera.

3.3 Variation in the diameter size of acroporidae colonies

Figure 4 shows the diameter distribution of five coral genera in the Spermonde Archipelago, reflecting environmental and human impacts. Large *Acropora* colonies dominate Stations 1, 3, 5, and 10, indicating stable habitats and good regeneration potential. *Montipora* occurs widely but mostly as small colonies, suggesting high recruitment yet slow growth. Station 10, within a conservation zone, hosts sizable *Anacropora* and *Isopora* colonies, highlighting well-protected stocks. In contrast, Station 9, along a shipping route, has small colonies across genera, indicating stress and low recovery potential.

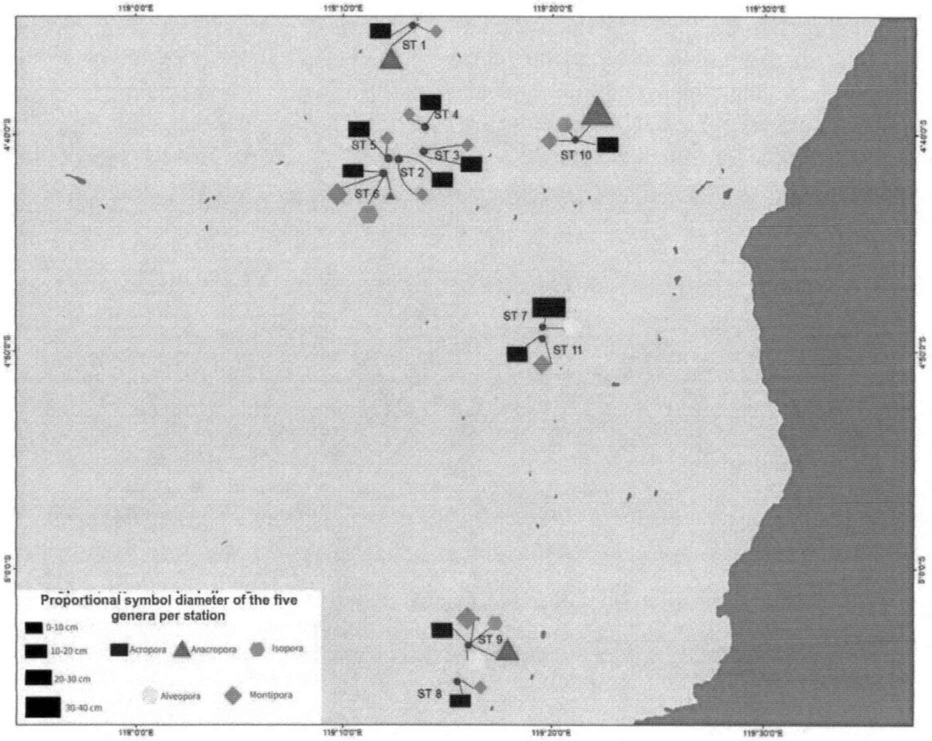

Fig. 4. Map of the percentage distribution of *Acroporidae* colony diameters in the Spermonde Archipelago.

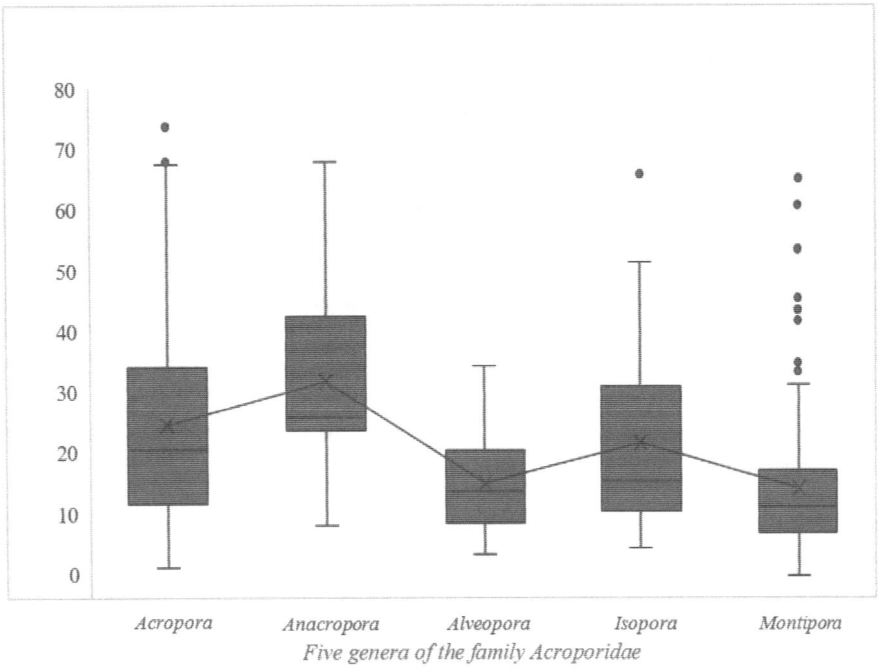

Fig. 5. Boxplot of colony diameter for five genera of the Family *Acroporidae* across 11 stations.

Figure 5 shows variation in the colony diameter of the family *Acroporidae*. Large colonies of *Anacropora* and *Acropora* indicate advanced growth stages and high broodstock potential. Larger colonies generally have greater reproductive capacity, both asexual (fragmentation) and sexual (gamete release), and are therefore essential for maintaining natural community regeneration [9]. The wide size distribution of *Acropora* suggests a healthy population spanning juvenile to adult stages.

Montipora and *Alveopora* tend to have small colonies with low median values, indicating populations dominated by young or suboptimally developed individuals. This may reflect their inherently slow growth strategies or environmental pressures such as sedimentation and anthropogenic disturbance, which limit colony development [10]. Despite their size, smaller genera such as *Montipora* play an important role in maintaining community structure and genetic diversity, particularly under fluctuating environmental conditions.

Isopora exhibits a broad but uneven size distribution with many outliers, suggesting that although some colonies grow large, most remain medium-sized, possibly due to irregular regeneration or growth disturbances across sites.

Mean colony diameter was analyzed to determine the average size of each genus at each station. Standard deviation described the degree of variation in colony diameters from their mean, reflecting the heterogeneity of coral growth within stations. The average diameters of *Acroporidae* colonies are presented in Table 3 below.

Table 3. Mean colony diameter (cm) of corals in the Family *Acroporidae*.

Station	*Acropora*	*Anacropora*	*Alveopora*	*Isopora*	*Montipora*
1	18,74 ±3,83	35,16 ± 15,27		-	20,7 ± 6,41
2	24,15 ± 2,71	-	22,16	-	17,63 ± 5,28
3	25,61 ±3,97	-	-	8,5± 8,3	11,5 ± 4,33
4	31,87 ± 3,15	-	-	-	21,39 ± 12,1
5	29,87 ± 5,04	-	-	-	13,5 ± 10,33

Station	*Acropora*	*Anacropora*	*Alveopora*	*Isopora*	*Montipora*
6	19,72 ± 10,8	13,5 ± 10,33	-	8,5± 8,3	8,5± 8,3
7	45,82 ± 13,2	-	17,63 ± 5,28	-	25,61 ±3,97
8	21,39 ± 12,1	-	-	-	19,72 ± 10,8
9	31,43 ± 9,39	28,1 ± 8,72	21,84 ± 2,97	41,52	24,15 ± 2,71
10	20,7 ± 6,41	42,82 ± 10,3	-	13,7 ± 4,52	20,7 ± 6,41
11	19,75 ± 0,67	-	-	-	24,15 ± 2,71
Average	18,74 - 45,82	13,5 - 42,82	11,5 – 21,84	8,5 – 31,77	12,16-23,55

The analysis above shows that the genus *Acropora* has the largest mean colony diameter, indicating its dominance at each station due to the wide range of colony sizes. *Anacropora* generally has a high standard deviation at all sites, reflecting broad size variation; although less common, this genus can grow large under favorable conditions. *Alveopora* shows the smallest diameters with limited variation, suggesting small colony sizes and restricted distribution due to its specific habitat preferences [11]. *Isopora* presents moderate mean diameters but high standard deviation, indicating significant differences between small and large colonies at some stations. *Montipora* has relatively even mean diameters and lower standard deviation compared to other genera, suggesting more uniform and stable colony sizes and good adaptability to different environmental conditions [1].

To support the heatmap analysis, statistical tests (ANOVA and Kruskal–Wallis) were applied to assess differences in colony diameter among stations. Results were interpreted based on the significance value (P-value): values < 0.05 indicate significant differences, while values > 0.05 suggest no significant diameter differences among genera across stations. The results are presented in the following table.

Table 4. Results of the Statistical Test on Colony Diameter among Genera in the Spermonde Archipelago.

Genus	df	Sig. (P Value)
Acropora	10	0.001
Anacropora	3	0.182
Alveopora	2	0.336
Isopora	3	0.058
Montipora	9	0.002

Table 4 shows differences in colony diameter among genera at each station. The statistical test for *Acropora* yielded an Asymp. Sig. (P-value) of 0.001 (< 0.05), indicating significant differences in colony diameter among stations. For *Anacropora*, the P-value was 0.182 (> 0.05), suggesting no significant differences among stations. *Alveopora* also showed no significant differences (P = 0.366 > 0.05). *Isopora* had a P-value of 0.058 (> 0.05), likewise indicating no significant differences.

By contrast, *Montipora* recorded a P-value of 0.002 (< 0.05), revealing significant diameter differences among stations. The significant variation in *Acropora* and *Montipora* is attributed to their high frequency of occurrence across most stations, resulting in wide diameter ranges between sites.

4 Conclusion

The *Acroporidae* coral community in the Spermonde Archipelago shows population structures that vary among stations. *Acropora* dominates most sites, particularly Station 10, indicating the highest stock potential, supported by large colony sizes and an even distribution of both *Acropora* and *Anacropora*. *Anacropora* also exhibits the largest colonies, highlighting its role as an important broodstock. *Montipora* and *Alveopora* display smaller colonies with limited distribution. Overall, the variation in size and genus composition reflects a heterogeneous stock, with certain locations serving as key sources for natural recruitment. Station 10, which shows the dominance of *Acropora* with large colony sizes, can be designated as a priority conservation zone or a source of natural recruits, given its potential as a coral stock. Similar analyses are needed for other coral families present in the Spermonde Archipelago, Makassar, South Sulawesi..

References

1. T.P. Hughes, A.H. Baird, D.R. Bellwood, M. Card, S.R. Connolly, C. Folke, J. Roughgarden, Climate change, human impacts, and the resilience of coral reefs. *Science*. **301**, 929–933 (2003).
2. E.S. Darling, L. Alvarez-Filip, T.A. Oliver, T.R. McClanahan, I.M. Côté, Evaluating life-history strategies of reef corals from species traits. *Ecol. Lett.* **15**, 1378–1386 (2012).
3. T.R. McClanahan, E.S. Darling, J.M. Maina, *et al.*, Highly variable taxonomic and functional coral reef recoveries in the Anthropocene. *Sci. Adv.* 7, eabe4419 (2021).
4. D. Dima, R.E. Hamid, D. Meye, V.M. Ati, F.M. Ike, A.N. Momo, Tingkah laku bertelur dan morfometrik penyu lekang (*Lepidochelys olivacea*) di Taman Wisata Alam Menipo, Kecamatan Amarasi Timur, Kabupaten Kupang. *J. Kalwedo Sains*. **3** (2) (2022).
5. P. Hemholz, D. Adams, D. McMeekin, A. Woods, Improving underwater photogrammetric 3D reconstruction processing of shipwreck sites. *ISPRS Ann. Photogramm. Remote Sens. Spatial Inf. Sci.* **X-4**, 7–14 (2024). https://doi.org/10.5194/isprs-annals-X-4-2024-7
6. Suharsono, Giyanto, A formulation approach to quantify the abundance of coral genera. *Mar. Res. Indonesia* **31**, 1–11 (2006). https://doi.org/10.14203/mri.v31i0.424
7. J. Fowler, L. Cohen, P. Jarvis, *Practical statistics for field biology*, 2nd ed., (John Wiley & Sons Ltd., Chichester, England, 1998).
8. Z.T. Richards, R.A. Garcia, C.C. Wallace, N.L. Rosser, P.R. Muir, A diverse assemblage of reef corals thriving in a dynamic intertidal reef setting (Bonaparte Archipelago, Kimberley, Australia). *PLoS ONE*. **13**, e0195161 (2018).
9. P.T. Pellitier, Variation in the size–structure of dominant branching coral taxa (*Acroporidae: Acropora* and *Pocilloporidae: Pocillopora*) in New Ireland Province, Papua New Guinea. *Pac. Sci.* **74**, 283–299 (2021). https://doi.org/10.2984/74.3.6
10. F. Moberg, C. Folke, Ecological goods and services of coral reef ecosystems. *Ecol. Econ.* **29**, 215–233 (1999).
11. A.E. Magurran, *Measuring Biological Diversity*, (Blackwell Publishing, United States, 2004).
12. M.L. Trapon, M.S. Pratchett, A.S. Hoey, Spatial variation in abundance, size and orientation of juvenile corals related to the biomass of parrotfishes on the Great Barrier

Reef, Australia. *PLoS ONE*. **8**, e57788 (2013). https://doi.org/10.1371/journal.pone.0057788

Bacterial associations in seagrass ecosystems: a systematic review of microbial diversity and ecological roles

Anjela K. Amalia[1], *Diva S.* Indira[1], *M. Z.* Ramadhian[1], *Gabriel* Suripatty[1], *Rastina*[2], *Yeti* Darmayati[3] and *Meutia S.* Ismet[2*]

[1] Marine Science Study Program, Postgraduate Program, Department of Marine Science and Technology, Faculty of Fisheries and Marine Science, IPB University (Bogor Agricultural University), Bogor, West Java, Indonesia
[2] Department of Marine Science and Technology, Faculty of Fisheries and Marine Science, IPB University (Bogor Agricultural University), Bogor, West Java, Indonesia
[3] Research Center for Oceanography, National Research and Innovation Agency (BRIN), Ancol, Jakarta, Indonesia

Abstract. Seagrasses are important marine plants that support coastal ecosystems, yet their bacterial associations remain insufficiently explored. This bibliometric study and meta-analysis examine bacterial communities associated with seagrass species, focusing on their roles in plant health, nutrient cycling, and ecosystem functioning. A comprehensive search was conducted using Scopus and Web of Science databases, applying the Boolean keywords: ("bacteria*" OR "bacterial*" OR "microbe*" OR "microbial*") AND ("seagrass*" OR "*Phyllospadix**" OR "*Zostera**" OR "*Enhalus**" OR "*Halophila**" OR "*Thalassia**" OR "*Posidonia**" OR "*Amphibolis**" OR "*Cymodocea**" OR "*Halodule**" OR "*Syringodium**" OR "*Thalassodendron**"). The review reveals a wide diversity of bacterial groups, including endophytic, epiphytic, and rhizosphere bacteria, with significant functions in nutrient cycling, disease resistance, and plant-microbe interactions. However, research gaps persist,particularly in understanding the functional roles of these microbial communities under varying environmental conditions. This study provide a comprehensive systhesis that highlights the importance of bacterial diversity associations in seagrass ecosystem. Crucially, this review presents a profound new perspective oe critical new insight into their ecological roles, solidifying the foundation for future microbe-based conservation and management strategies.

1 Introduction

Seagrass is aflowering plant that thrives entirely underwater forming extensive meadows in shallow coastal waters across the globe [1]. As one of the most productive coastal ecosystems, seagrass meadows play a vital role in sustaining the ecological balance of marine environments by providing a range of critical ecosystem services. These include substantial blue carbon storage, which contributes to climate change mitigation; nutrient regulation and recycling; shoreline stabilization and protection against erosion; as well as offering structurally complex and biodiverse habitats for numerous marine species [2, 3]. Furthermore, seagrass serves as a crucial nursery ground for a range of economically valuable

*Corresponding Author : meutiais@apps.ipb.ac.id

fish species, while also contributing to the reduction of greenhouse gas emissions through its remarkable ability to absorb and sequester substantial amounts of carbon [3]. Despite their exceptionally high ecological and socioeconomic value, seagrass meadows worldwide are currently under severe threat. The global rate of decline is estimated at about 7% annually, driven predominantly by anthropogenic pressures, including pollution from land-based activities, coastal infrastructure development, and the impacts of global warming, which contribute to rising sea surface temperatures [4, 5]. This degradation not only poses a significant risk to the long-term sustainability of marine biodiversity but also undermines the capacity of these ecosystems to deliver essential environmental services.

Within seagrass ecosystems, bacteria play a crucial role in maintaining ecological stability and functionality. These microorganisms participate in a range of critical biogeochemical processes, including nitrogen fixation to enhance ecosystem fertility, the decomposition of organic matter that contributes to nutrient cycling, and sediment stabilization, which supports the integrity of habitat structure. Additionally, specific bacterial communities serve as bioprotective agents by suppressing the proliferation of harmful pathogens [6, 7]. Symbiotic associations between bacteria and seagrasses have been shown to enhance host resilience against various environmental stressors, including eutrophication and elevated sea surface temperatures [8]. A prominent example is the occurrence of symbiotic bacteria in *Zostera marina*, which have been shown to inhibit the onset of vibriosis in nearby marine organisms, thereby playing a vital role in maintaining the health and stability of coastal ecosystem communities [6].

While the ecological significance of bacteria-seagrass interactions is well-recognized, the body of literature exploring them is still scarce. This gap is especially evident in the lack of bibliometric analyses, which are needed to map the evolution of the research field and highlight future directions[3]. Bibliometric approaches offer a systematic framework for this purpose by enabling researchers to evaluate literature, detect trends, and mapcollaboration networks[9]. Applying such an analysis to bacterial-seagrasss associations can provide valuable insights into the fields progression and uncover new opportunities for investigation. Accordingly, this study employs a bibliometric approach to identify patterns and trends in publications on this topic. The result are expected to deepen our understanding of microbe–seagrass interactions and guide future research toward the effective management and conservation of seagrass ecosystems.

2 Material and methods

This study utilizes a meta-analysis study that aims to analyze bacterial communities associated with seagrass species. The literature search was conducted using two major academic databases, Scopus and Web of Science, covering articles published between 1982 and 2024. The search was designed to identify peer-reviewed English-language articles focusing on the bacterial communities associated with seagrass species The following keywords were used to guide the search: ("bacteria*" OR "bacterial*" OR "microbe*" OR "microbial*") AND ("seagrass*" OR "*Phyllospadix**" OR "*Zostera**" OR "*Enhalus**" OR "*Halophila**" OR "*Thalassia**" OR "*Posidonia**" OR "*Amphibolis**" OR "*Cymodocea**" OR "*Halodule**" OR "*Syringodium**" OR "*Thalassodendron**")

From the initial search, the dataset was filtered to remove inaccessible literature and studies that did not meet the inclusion criteria, resulting in 201 articles on seagrass-associated bacteria ready for analysis. In this study, a bibliometric analysis was performed using Microsoft Excel and the Biblioshiny package in R, providing a comprehensive overview of research areas, author distribution, and overall productivity within the field. Results are presented both quantitatively and qualitatively, as well as visually, to identify emerging trends, dominant keywords, and critical gaps in the literature. These findings not only pinpoint the key bacterial species associated with seagrasses but also highlight thematic areas such as bioprospecting, nutrient cycling, disease resistance, and plant–microbe interactions, as well as the most prolific research collaborations. Together, they provide a solid foundation for future research recommendations. The complete research framework is depicted in Figure 1.

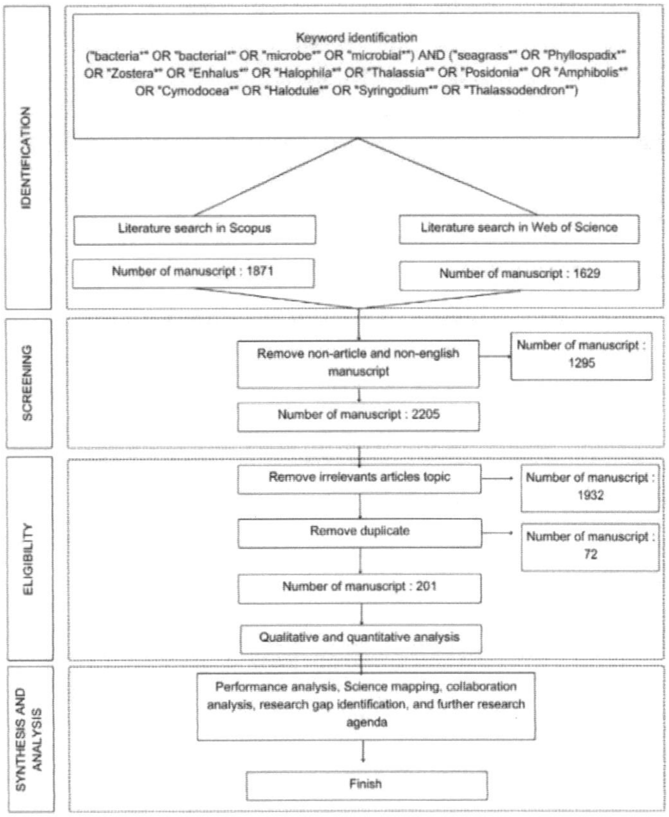

Fig. 1. Research framework shows literature search criteria.

3 Results and discussion

3.1 Performance analysis

Based on the bibliometric dataset collected, this section presents a performance analysis to evaluate research productivity, publication trends, and collaboration patterns in studies on seagrass-associated bacteria. The analysis covers the period from 1982 to 2024, representing more than four decades of scientific activity. A total of 201 documents were identified, reflecting a specialized thematic focus within the broader field of marine microbiology. The annual publication growth rate of 5.64% indicates a steady increase in research interest over time. The average document age of 9.84 years suggests that the dataset includes both foundational studies and recent advancements, providing a balanced overview of the field's evolution.In terms of research productivity and publication sources, studies on bacterial communities associated with seagrass species were distributed across 81 journals. This indicates that the topic spans multiple disciplines within marine and microbial ecology rather than being confined to a single specialized outlet. The literature is dominated by journal articles, accounting for 200 papers, with only one proceedings paper recorded. This suggests that most research outputs are disseminated through peer-reviewed journals rather than conference publications. The average citation rate of 31.01 citations per document reflects a moderate level of scholarly impact, showing that while the field attracts consistent academic attention, it has not yet reached the high citation influence observed in more established research areas. The authorship and collaboration aspects reveal a strong collective character, with a total of 922 authors contributing to the body of work. The high number of authors, relative to the modest number of documents, indicates a prevalent practice of collaborative research. Only three documents were single-authored, underscoring the inherently collaborative nature of this literature. The average number of co-authors per document is 5.78, a figure that reflects the involvement of large research teams and suggests the likelihood of multi-institutional or cross-disciplinary collaboration. The proportion of international collaboration stands at 27.86%, indicating that a substantial portion of this research involves cross-country partnerships, which can enhance both the visibility and the impact of the work. A summary of research performance in these areas is presented in Table 1 and Figure 2.

Table 1. The descriptive summary.

Component	Description	Result
Publication	Total number of research publications	201
Publication Period	Period Active period of research publications	1982:2024
Productivity	Annual Growth Rate %	5.64
Source	Total number of journal sources in related fields	81
Total citation	Total citation in all publication	6233
Average Citations per Document	Total citation/publication	31.01
Total Author	Total research authors contributing to the field	922
Single Author Publication	Individual research publications	3
Author's Keywords	Total publication keywords	610

3.2 Annual scientific production

Figure 2 depicts the growth in yearly publications related to bacteria associated with seagrass (leaf, sediment, or rhizomes) from 1982 to 2024. The dashed line represents the linear regression analysis conducted on the total number of publications. The R-square (coefficient of determination) is calculated to be 0.47, indicating that the linear trend line may account for around 47% of the increase in publications. The trend line also demonstrates a significant increase in publications since the early 2000s, with a notable surge after 2017. This indicates that the academic and scientific community has increasingly focused on research related to bacteria associated with seagrass in recent years.

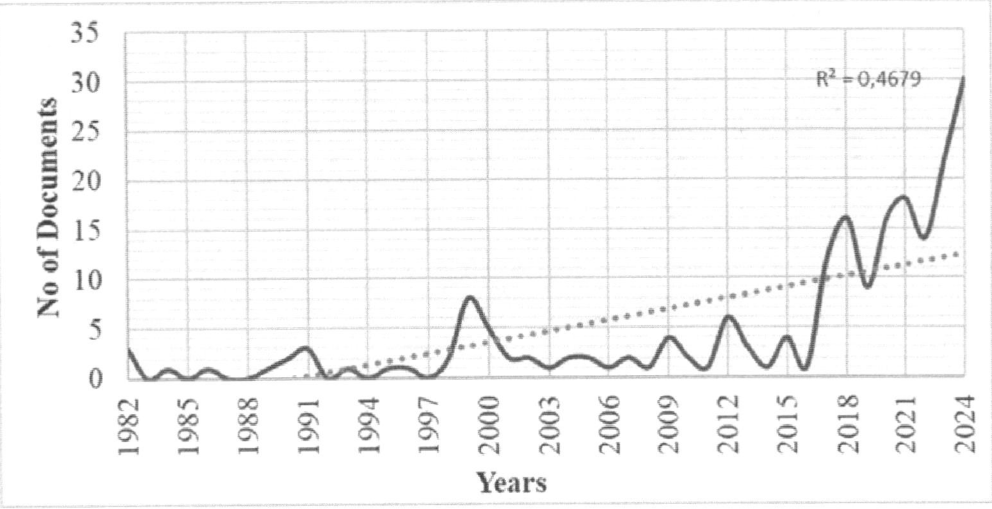

Fig. 2. Annual publication trend from 1982 to 2024.

3.3 Analysis of journals

Bibliometric data reveal that a total of 81 journals contain at least one article on bacteria associated with seagrass. Among these journals, a significant majority, 56.79% (46 journals), have published only a single article on the subject. This suggests that these journals do not have a specialized focus on bacteria associated with seagrass, possibly indicating that the topic is a cross-cutting theme relevant across various fields of knowledge. Furthermore, 20.98% of journals have published two articles, 11.11% have published three articles, and 3.7% have published four articles. Additionally, 7.4% of journals have published five or more articles, reflecting a comparatively higher level of engagement with the topic.

Table 2 presents the rankings, citations, and H-index of the ten most productive journals. These top-tier journals collectively account for 97 articles, constituting 49.25% of the total articles included in the study. Among them, the journal "Marine Ecology Progress Series" emerges as the most prolific contributor, with an impressive 28 articles. Notably, this journal also boasts the highest citation count, with a total of 1,846 citations, and the highest H-index, at 25. Following closely behind, the journal "Frontiers in Microbiology" ranks second with 18 articles, while the journal "Aquatic Microbial Ecology" holds the third position with 11 articles. The journal "Frontiers in Marine Science" ranks fourth, having published 13 articles. Notably, this journal also boasts an impressively high citation count of 144 and an H-index of 8.

Table 2. Top ten most productive journals and their citations and H-Index.

No	Source	Articles	Total Citation	H-Index
1	Marine Ecology Progress Series	28	1846	25
2	Frontiers in Microbiology	18	545	12
3	Aquatic Microbial Ecology	11	490	8
4	Frontiers in Marine Science	13	144	8
5	Applied and Environmental Microbiology	7	647	7
6	International Journal of Systematic and Evolutionary Microbiology	5	156	5

7	FEMS Microbiology Ecology	4	223	4
8	Scientific Reports	4	99	4
9	ISME Journal	3	175	3
10	Microbes and Environments	4	56	3

Figure 3 illustrates the prominent journals renowned for their high citation counts in the field of bacteria associated with seagrass. Notably, the journal "Marine Ecology Progress Series" emerges as the most cited journal, accumulating a substantial total of 1,846 citations. This exceptional citation count is primarily attributed to a single article by [10], which investigates the influence of marine plants on carbon decomposition, nutrient fluxes, and denitrification in Australian lagoons. Following closely, the journal "Applied and Environmental Microbiology" secures the second position, amassing a noteworthy 647 citations. Additionally, the journal "Frontiers in Microbiology" ranks third in terms of citations, with a respectable total of 545 citations.

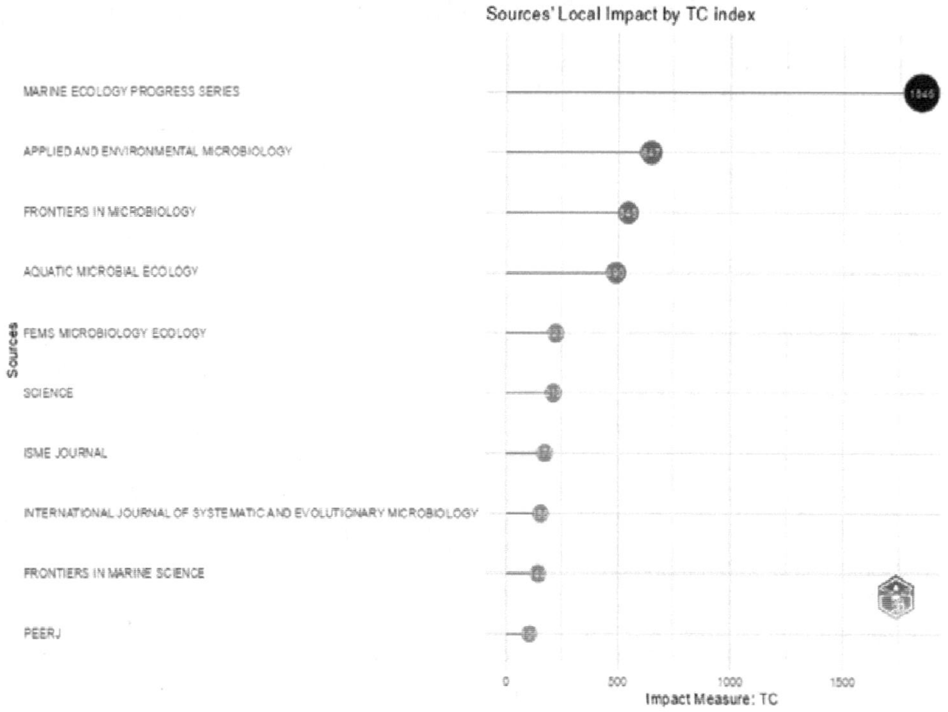

Fig. 3. Top 10 most cited journals.

3.4 Most relevant countries

Following the analysis of journal distribution and publication trends, it is equally important to investigate geographical distribution of research productivity to better understand global contribution disparities. Therefore, a country-level analysis was performed, focusing on author affiliations to determine which nations have contributed most to the study Following the analysis of journal distribution and publication trends, it is equally important to investigate geographical distribution of research productivity to better understand global contribution disparities. Therefore, a country-level analysis was performed, focusing on author affiliations to determine which nations have contributed most to the study. This analysis indicates that the global research landscape on seagrass-associated bacteria is still in its early stages of development, exhibiting an uneven distribution of contributions. Although there is cumulative participation from 32 countries, the data reveal a high concentration of research productivity within a limited number of nations. Specifically, publications are overwhelmingly dominated by a small group, with only four countries having published more than ten articles, and only three surpassing the twenty-five article mark. This limited international engagement clearly affirms that the research niche of bacteria-seagrass interactions has not yet become a primary focus, leaving the field wide open for further exploration and cross-border investigation across the globe.

Table 3 lists the 10 nations with the highest academic production and citations in the field of research on bacteria associated with seagrass. These countries collectively account for 79.6% of the total articles and 92.22% of the total citations. Notably, the United States stands out as the most productive nation, having published 39 articles and received the most citations (1,760). Australia ranks second in both productivity and citations, with 26 articles and 996 citations. China ranks third in productivity, with 26 articles, and fifth in terms of citations (314). This discrepancy highlights the significant impact and influence of India's limited contributions, suggesting that the articles it has published may have attracted substantial attention or addressed highly relevant research questions in the field of bacteria associated with seagrass.

Table 3. Highest producing and cited nation.

Highest Producing Nations			Highest Cited Nations		
Country	**Articles**	**Citations**	**Country**	**Citations**	**Articles**
USA	39	1760	USA	1760	39
Australia	26	996	Australia	996	26
China	26	314	Denmark	722	14
Denmark	14	722	Netherlands	409	6
Japan	13	249	Italy	316	12
Italy	12	316	China	314	26
Spain	11	309	Spain	309	11
Germany	7	208	Japan	249	13
Netherlands	6	409	India	239	2
Saudi Arabia	6	130	Germany	208	7

Fig. 4 illustrates the involvement of multiple-country (MCP) and single-country (SCP) collaborations among corresponding authors in publications related to research on bacteria associated with seagrass. The analysis reveals that the United States exhibits the highest productivity in SCP with 29 articles, while ranking second in MCP with 10 articles, indicating a significant focus on this research area. Australia appears as the nation that collaborates the most, providing the most MCP (11 articles) and ranking third in SCP with 15 articles. China follows closely behind, ranking second in SCP with 24 articles and ninth in MCP with only 2 articles. Notably, among the top ten countries, the majority are more engaged in single-country publications (SCP) compared to multi-country collaborations (MCP). This trend suggests that while there is substantial individual research output from these nations, there is still room for increased international cooperation in this field. The United States, Australia, and China, despite their leadership roles, exhibit varying levels of involvement in MCP, suggesting differences in research strategies or international partnerships.

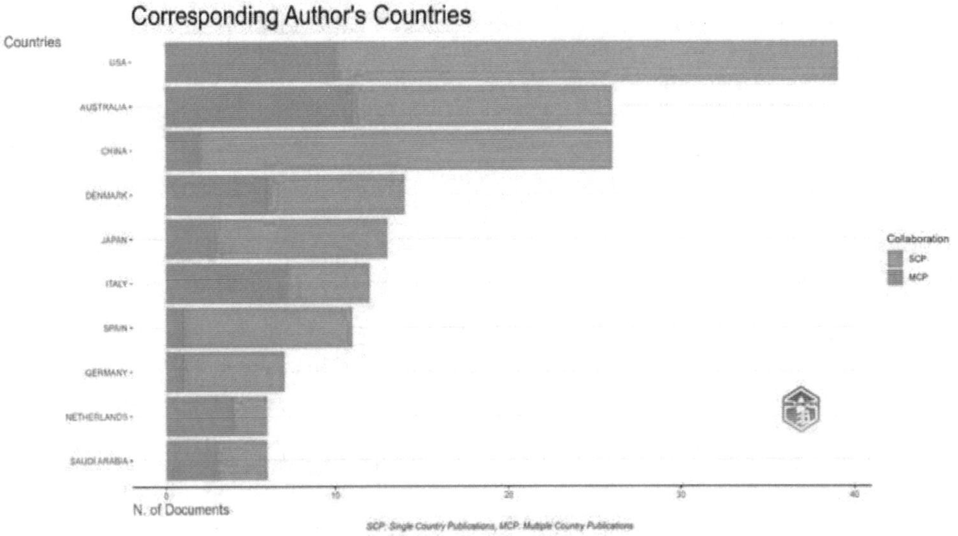

Fig. 4. Multiple-country (MCP) and single-country (SCP) collaboration.

3.5 Topmost relevant documents

The bibliometric analysis reveals a markedly skewed citation landscape in the literature on bacterial communities associated with seagrass species: of the 201 documents included, only 2 (≈0.99%) have surpassed 200 citations and 13 (6.47%) have more than 100 citations, showing that a tiny subset of publications accounts for the field's most significant citation influence. This concentration suggests the presence of a few obvious, potentially foundational studies that shape subsequent work. At the same time, the majority of papers receive relatively modest attention, an outcome that can reflect the field's niche size, varied methodologies, uneven journal visibility, or the long-time lags often required for ecological papers to accumulate citations. Practically, such a distribution has two important implications: first, reliance on a handful of high-impact studies may skew theoretical framing and methodological norms if those few papers are not broadly representative; second, many valuable but less-cited contributions may be under-recognized despite their potential to refine understanding. As research activity in this area continues to expand and open access, data sharing, and interdisciplinary collaborations increase, aggregate citation counts and the diversity of influential works are likely to grow. However, addressing current imbalances will also require deliberate efforts to improve visibility and reproducibility across the corpus. For a focused view of the most influential contributions, the ten most-frequently cited documents and brief summaries of their principal contributions are compiled in Table 4.

Table 4. Top ten most relevant documents on

No	Reference	Key Highlight	Total Citations
1	Van D H T et al. (2012), Science	This study reveals a three-way symbiotic relationship among seagrasses, lucinid bivalves, and sulfide-oxidizing bacteria, which functions as a key mechanism for mitigating sulfide stress in seagrass ecosystems. Through the release of oxygen from seagrass roots, the symbiosis not only promotes seagrass growth but also supports the metabolic needs of the bivalves and their bacterial partners. This interaction regulates sulfide levels in sediments, thereby preventing toxicity, maintaining sediment health, and facilitating favorable conditions for benthic communities. Such microbial faunal–plant partnerships demonstrate how biological interactions underpin critical biogeochemical processes in coastal environments.	213
2	Eyre and Ferguson (2002), Marine Ecology Progress Series	Denitrification efficiency in shallow coastal lagoons is strongly linked to carbon decomposition rates, with higher decomposition rates reducing nitrogen removal and increasing nitrogen recycling, which may drive eutrophication. Dissolved organic nitrogen is a significant component of nitrogen fluxes, while seagrass habitats, especially *Zostera capricorni* and seagrass epiphyte assemblages, maintain high nitrogen removal efficiency by limiting ammonium release. The productivity-to-respiration ratio predicts whether sediments act as nutrient sinks or sources, and benthic microalgae contribute to sediment oxygenation, thereby enhancing coupled nitrification and denitrification.	205
3	Weidner et al. (1996), Applied and Environmental Microbiology	A culture-independent approach using PCR–ARDRA revealed a wide range of uncultured bacterial phylotypes inhabiting the leaf surfaces of *Halophila stipulacea*. The community included biofilm-forming bacteria such as Hyphomonas and plastid-like sequences likely originating from epiphytic algae. These findings highlight the complex and multi-origin nature of microbial assemblages on seagrass leaves.	170
4	Kristensen et al. (2000), Aquatic Microbial Ecology	Microbial processes along the mangrove and seagrass gradient exhibited a clear shift from iron-reducing dominance in mangrove sediments to increased aerobic respiration in bioturbated mudflats and oxygen-based pathways in seagrass meadows. Nitrogen was largely retained through rapid bacterial assimilation, influenced by the high C:N ratio of organic matter from plant litter. These patterns highlight the roles of sediment type, iron availability, plant roots, and infaunal activity in regulating carbon and nitrogen cycling, aligning with the literature on microbial diversity and ecological functions in seagrass-linked ecosystems.	130

5	Welsh et al. (2000), Marine Ecology Progress Series	The intertidal *Zostera noltii* meadow functions as a highly autotrophic system where nitrogen uptake is driven mainly by seagrass and epiphyte assimilation, independent of ambient nutrient levels. Rhizosphere-associated bacteria play a central role through active nitrogen fixation that surpasses denitrification, resulting in a net gain of nitrogen and supporting sustained primary productivity. Biomass export and sediment burial, rather than microbial nitrogen loss, dominate nutrient removal, highlighting how plant–microbe interactions shape nutrient cycling and contribute to ecosystem stability in seagrass habitats.	122
6	Fahimipour et al. (2017), Applied and Environmental Microbiology	Leaves host microbiomes resembling the surrounding seawater with high variability, while roots maintain stable, distinct communities enriched in sulfur-metabolizing bacteria, indicating roles in sulfide detoxification. Aboveground communities are environmentally driven, whereas belowground microbiomes are shaped by host selection for functional traits, highlighting compartment-specific bacterial diversity and ecological roles in seagrass resilience.	119
7	Caffrey and Kemp (1990), Marine Ecology Progress Series	The presence of Potamogeton perfoliatus and Zostera marina enhances sediment microbial activity, particularly in the processes of ammonification, nitrification, and denitrification. Root oxygen release supports nitrification, while organic inputs from plant material stimulate ammonification. Nitrate is often reduced back to ammonium, conserving nitrogen within the system. Seasonal growth cycles influence the magnitude and coupling of these processes.	119
8	Ziegler and Benner (1999), Marine Ecology Progress Series	Nutrient dynamics in subtropical seagrass ecosystems are strongly mediated by interactions between the water column and benthos, with dissolved organic matter (DOM) released from vegetated sediments serving as a major driver of bacterioplankton activity and nitrogen regeneration. Bacterioplankton rely heavily on seasonally variable dissolved organic nitrogen (DON), which influences their growth efficiency and nutrient cycling. These results highlight the critical ecological role of bacteria–seagrass associations in maintaining nutrient availability, ecosystem productivity, and resilience, providing empirical evidence directly relevant to understanding microbial diversity and functions in seagrass ecosystems.	115
9	Holmer et al. (2004), Aquatic Microbial Ecology	Bacterial carbon sources in seagrass sediments shift from seagrass detritus in pristine meadows to seston and macroalgae in nutrient-enriched environments, reflecting high metabolic flexibility and influencing primary productivity and sediment biogeochemistry.	112
10	Küsei et al. (1999), Applied and Environmental Microbiology	The rhizosphere of *Halodule wrightii* hosts a diverse array of anaerobic bacteria, particularly sulfate-reducing bacteria and acetogens, which colonize both the root surface and the deep cortex layers, indicating adaptation to fluctuating oxygen conditions and involvement in key biogeochemical processes,	105

		such as sulfur and carbon cycling, as well as potential nitrogen fixation.	

3.6 Frequent trending keyword

This section examines the main keywords used by authors, their interrelationships, and the keywords currently trending within the field of Marine Biodiversity. The analysis is based on a comprehensive dataset comprising a total of 201 published documents, from which 1,732 keywords were identified and retrieved in total. Among all identified keywords, only a small proportion appeared with high frequency: one keyword (0.1%) occurred more than 50 times, 36 keywords (2.1%) occurred more than ten times, 82 keywords (4.7%) occurred more than five times, while the vast majority, 1,613 keywords (93.1%) appeared only once. Figure 5 presents a visual representation of the fifty most frequently used keywords. The analysis reveals that "seagrass" is the most dominant keyword, occurring 94 times, followed by "bacteria" (49 occurrences), "diversity" (30 occurrences), and "microbial community," "nitrogen-fixation," and "*zostera marina*," each with 27 occurrences.

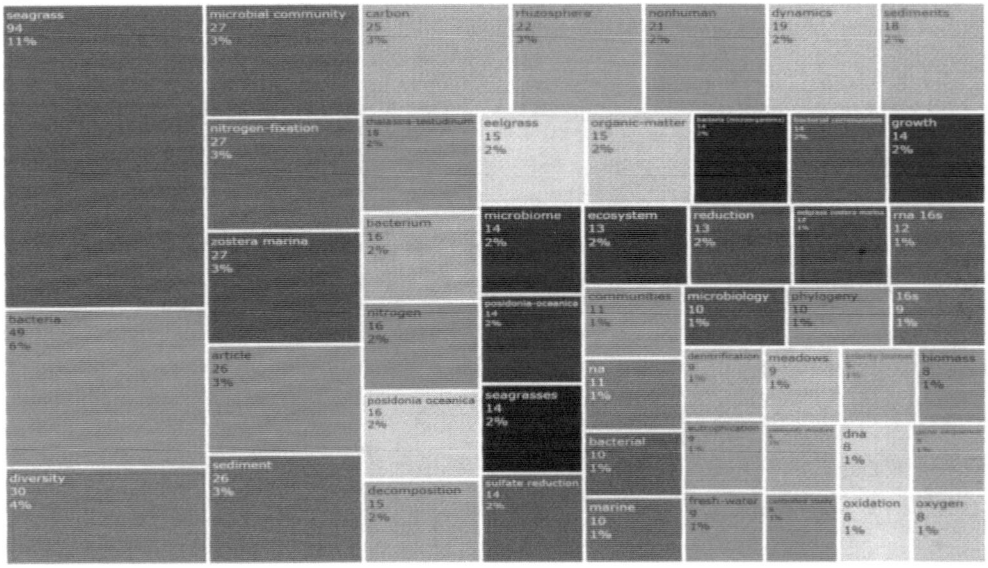

Fig. 5. Most frequent keywords.

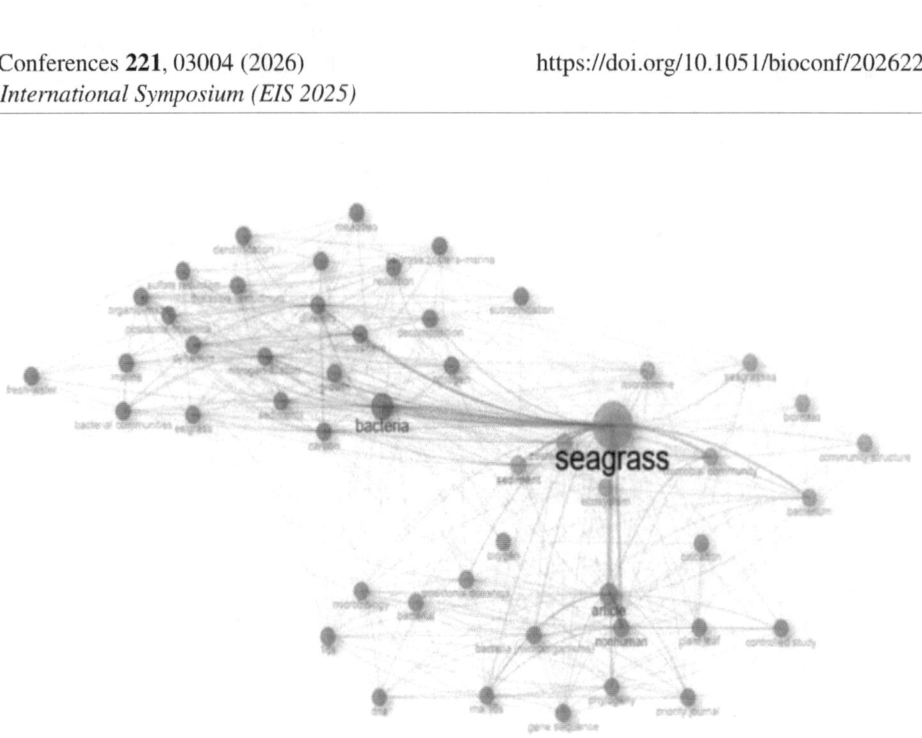

Fig. 6. Co-occurrence network of author's keywords.

In Figure 6, the visual network mapping, each circle represents a specific author keyword, with its size corresponding to the frequency of occurrence. The larger the circle, the more frequently the keyword appears in the literature. Connecting lines between circles indicate keywords that frequently co-occur, with thicker lines representing stronger associations. Circles sharing the same color denote keywords belonging to the same cluster of closely related topics. The analysis identified three main clusters, each assigned a distinct color.

The first cluster, represented in red, is characterized by the combination of "bacteria" and related terms such as "diversity," "nitrogen," and "organic-matter," which show more frequent co-occurrences. The second cluster, shown in green, is predominantly defined by the recurring keyword "seagrass," closely associated with topics such as "microbial communities," "sediment," and "biomass." The third cluster, represented in blue, highlights significant keywords including "article," "microbiology," and "controlled study." This clustering analysis reveals the thematic structure and co-occurrence patterns of the most dominant author keywords in the literature.

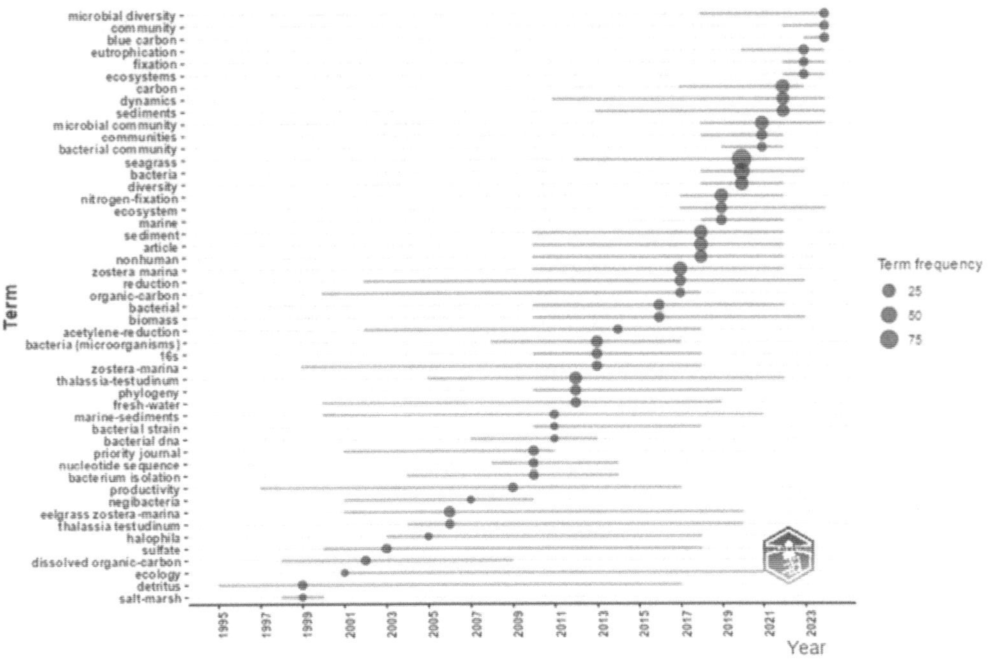

Fig. 7. Trending keywords of last ten year (1995–2024).

Figure 7 illustrates the twenty leading research topics, along with the most dominant author keywords used during the study period (1995–2024). The results show that "seagrass" (94 occurrences) is the most frequently used keyword, first appearing in 2012 and continuing through 2024. It is followed by "bacteria" (49 occurrences) and "diversity" (30 occurrences), both of which first appeared in 2018. In the most recent period, specifically in 2024, emerging trending terms include "microbial diversity," "community," and "blue carbon," reflecting the growing research interest in microbial associations and their ecological roles within seagrass ecosystems.

4 Discussion

4.1.1 Microbial diversity and ecological roles

Seagrass ecosystems support a broad taxonomic diversity of bacteria, including *Bacillus* spp., Pseudomonadaceae, Shewanellaceae, Rhodobacteraceae [11], Firmicutes, Actinobacteria, and various classes of Proteobacteria [12], as well as specialized groups such as sulfate-reducing bacteria (SRB) and iron-reducing bacteria [13]. These bacteria play crucial roles in maintaining seagrass health by mobilizing phosphorus and iron from nutrient-poor sediments, protecting roots from sulfide toxicity, and supporting root metabolism through symbiotic interactions [13]. Specific contributions include the production of hydrolytic enzymes (amylase, cellulase, protease) and antagonistic compounds that inhibit pathogen infection.

Seagrass-associated bacteria play a crucial role in driving nutrient cycling and maintaining overall ecosystem functioning. Groups such as Desulfobacterales, Thiotrichales [2], and Chromatiales [14] regulate sulfur and nitrogen cycles, maintaining sediment fertility and biogeochemical balance. Another critical role involves pollutant detoxification: sulfur-oxidizing bacteria (SOB) neutralize toxic H_2S that threatens seagrass health, while symbiotic bacteria detoxify heavy metals and enhance salinity adaptation [11]. Collectively, these microbial activities stabilize sediments and strengthen the resilience of tropical marine ecosystems.

Bioprospecting represents a critical aspect of seagrass-associated bacteria, particularly for pharmaceutical applications. *Streptomyces* sp. produces broad-spectrum anthraquinones effective against antibiotic-resistant bacteria [1], while *Pseudoalteromonas*, *Teredinibacter*, and members of the *Myxococcota* synthesize antibacterial, antiviral, and anticancer compounds with strong potential for drug development [2]. In aquaculture, symbiotic bacteria function as natural biocontrol agents, capable of reducing pathogenic *Vibrio* populations in less than one hour [9]. Furthermore, the production of proteolytic, amylolytic, and cellulolytic enzymes by root-associated bacteria presents promising opportunities for industrial applications and bioremediation.

4.1.2 Current Trend

Based on bibliometric data on research concerning bacterial associations in seagrass ecosystems, there has been a marked upward trend in publications from the early 1980s through 2024. Relatively few and fluctuating publications characterized the initial period (1982–1998), indicating that this research area remained sporadic and had yet to become a primary focus. From the late 1990s onward, the number of publications began to increase gradually, driven by a growing recognition of seagrasses as critical ecosystems for carbon storage, habitat provision, and biodiversity support, including the support of microbial communities. A more consistent surge has been evident since 2010, with a sharp peak occurring between 2017 and 2024. This growth is likely linked to heightened global attention to climate change, coastal habitat degradation, and advances in molecular technologies that have facilitated the analysis of bacterial communities.

In terms of journal contributions, *Marine Ecology Progress Series* ranks highest with an H-index of 25, a total of 1,846 citations, and 28 articles, reflecting its substantial influence in the field of marine ecology. Other journals such as *Frontiers in Microbiology* and *Aquatic Microbial Ecology* also make significant contributions, with total citations of 545 and 490, and article counts of 18 and 11, respectively. The presence of these journals indicates that research on seagrass-associated bacteria is inherently multidisciplinary, encompassing marine ecology, environmental microbiology, and biotechnology. The dominance of reputable international journals further underscores the global reach and high relevance of this research to marine ecosystem issues.

Country-level analysis shows that the United States and Australia lead in both publication output and citation counts, followed by China, Denmark, Italy, Spain, Japan, the Netherlands, Germany, India, and Saudi Arabia. The high contributions of these countries align with the presence of leading marine and ecological research centers in these regions, as well as adequate access to research funding. Denmark and the Netherlands, despite having relatively few publications, exhibit a high citation-per-article ratio, indicating strong research quality and significant scientific impact. This pattern is consistent with the concept of geographic concentration of innovation described by Hemminga and Duarte (2000) in *Marine Ecology Progress Series* (vol. 311, p. 233), whereby research on seagrasses and their associated organisms often clusters in locations with robust marine research infrastructure, good field access, and strong international collaboration networks.

The article by [3] in the *Journal of Experimental Marine Biology and Ecology* highlights the global distribution of seagrasses across six major bioregions, which is relevant to the distribution of countries producing publications in this field. Australia, located within the Tropical Indo-Pacific and Temperate Southern Oceans, harbors exceptionally high seagrass diversity and serves as a base for large-scale seagrass ecology research. The United States spans multiple bioregions, including the Temperate North Atlantic, Tropical Atlantic, and

Temperate North Pacific, enabling the study of seagrass-associated bacteria across a variety of ecosystem types. European countries, such as Denmark, the Netherlands, Italy, and Spain, focus on the temperate North Atlantic and Mediterranean regions, with research often utilizing *Zostera marina* as a model ecosystem for environmental microbiology studies.

Overall, the bibliometric data illustrate that research on bacterial associations within seagrass ecosystems has evolved from a limited area of study into a dynamic and strategically significant field. The notable growth in publications, contributions from reputable journals, and the dominance of countries with strong research infrastructure demonstrate that this topic is increasingly recognized as an integral component of marine ecology and conservation studies. Drawing on insights from the two reference articles, the sustainability of this research will depend heavily on international collaboration, the advancement of more sophisticated molecular methodologies, and the mapping of relationships between global seagrass distribution and their associated microbial communities.

The bibliometric data presented indicate that the United States (USA) ranks highest in both the number of publications (39 articles) and citations (1,760), followed by Australia with 26 articles and 996 citations. Other countries, including China, Denmark, Italy, Spain, Japan, the Netherlands, Germany, India, and Saudi Arabia, contribute to varying extents. However, some nations, such as Denmark and the Netherlands, display a high citation-to-article ratio, signifying substantial influence despite a relatively small publication volume. This pattern aligns with the concept of geographic concentration of innovation, as described by Feldman and Florida (1994), which posits that scientific productivity tends to cluster in regions with strong research infrastructure, extensive collaboration networks, and robust institutional support. Countries dominating in publication output generally host leading marine research centers, benefit from substantial funding, and maintain international connections that enhance the global visibility of their research.

These findings are also consistent with the study by [3] on seagrass distribution and diversity, which categorizes the world's seagrass ecosystems into six bioregions. Australia, for example, lies within the Tropical Indo-Pacific and Temperate Southern Oceans—two bioregions with the highest seagrass diversity globally—. In contrast, the USA spans several bioregions, including the Temperate North Atlantic, Tropical Atlantic, and Temperate North Pacific. European countries such as Denmark, the Netherlands, Italy, and Spain are situated mainly in the temperate North Atlantic and Mediterranean regions, with moderate diversity but a strong tradition of marine ecological research. Understanding the distribution of these bioregions is essential for interpreting variations in seagrass-associated bacterial communities, as geographic factors, the physico-chemical characteristics of the waters, and the identity of host species all influence microbial composition.

5 Conclusion

Research on bacterial associations in seagrass ecosystems has expanded rapidly since the early 2000s, with a marked increase in publications after 2017. Bacteria associated with seagrasses play vital roles in maintaining plant health, regulating nutrient cycles such as nitrogen, phosphorus, and sulfur, detoxifying pollutants, and enhancing ecosystem resilience to environmental disturbances. Several bacterial groups, including *Bacillus*, *Streptomyces*, *Pseudoalteromonas*, as well as sulfate- and iron-reducing bacteria, contribute not only to ecological functions but also hold significant potential for applications in biotechnology, bioremediation, and pharmaceuticals. Nevertheless, publication output remains concentrated in countries with advanced marine research centers, such as the United States, Australia, and Denmark, while contributions from many other regions remain limited. This highlights both geographic and thematic gaps that present opportunities for future research.

Future research should prioritize strengthening multidisciplinary approaches that integrate metagenomics, meta-transcriptomics, and metabolomics to gain a comprehensive understanding of bacterial community functions and link them to ecological parameters such as sediment quality, temperature, salinity, and pollutants. Studies should also address long-term monitoring seasonal and annual variations to assess bacterial responses to climate change and habitat degradation. Furthermore, it is essential to explore seagrass-associated bacteria for potential applications in biotechnology, bioremediation, and pharmaceuticals. Therefore, further microbial research in seagrass ecosystems is absolutely essential. This research is necessary not only to fully elucidate the bacteria's ecological functions and support ecosystem resilience, but also to develop a comprehensive global database that documents their distribution, functions, and potential applications, thereby guiding future efforts in marine conservation and management.

Reference

1 S. Ravikumar, M. Gnanadesigan, A. Saravanan, N. Monisha, V. Brindha, and S. Muthumari, "Antagonistic properties of seagrass associated Streptomyces sp. RAUACT-1: A source for anthraquinone rich compound," Asian Pac. J. Trop. Med., vol. 5, no. 11, pp. 887–890 (2012).

2 E. Rubio-Portillo, A.-B. Martin-Cuadrado, A. Á. Ramos-Esplá, and J. Antón, " Metagenomics Unveils Posidonia oceanica 'Banquettes' as a Potential Source of Novel Bioactive Compounds and Carbohydrate Active Enzymes (CAZymes) ," mSystems, vol. 6, no. 5 (2021).

3 F. Short, T. Carruthers, W. Dennison, and M. Waycott, "Global seagrass distribution and diversity: A bioregional model," J. Exp. Mar. Bio. Ecol., vol. 350, no. 1–2, pp. 3–20 (2007).10.1016/j.jembe.2007.06.012

4 M. Waycott *et al.*, "Accelerating loss of seagrasses across the globe threatens coastal ecosystems," *Proc. Natl. Acad. Sci.*, vol. 106, no. 30, pp. 12377–12381 (2009) https://doi.org/10.1073/pnas.0905620106

5 R. K. F. Unsworth *et al.*, "Global challenges for seagrass conservation," *Ambio*, vol. 48, no. 8, pp. 801–815 (2019). https://doi.org/10.1007/s13280-018-1115-y

6 J. B. Lamb *et al.*, "Seagrass ecosystems reduce exposure to bacterial pathogens of humans, fishes, and invertebrates," *Science (80-.).*, vol. 355, no. 6326, pp. 731–733 (2017). 10.1126/science.aal1956

7 F. Tarquinio, G. A. Hyndes, B. Laverock, A. Koenders, and C. Säwström, "The seagrass holobiont: understanding seagrass-bacteria interactions and their role in seagrass ecosystem functioning," *FEMS Microbiol. Lett.*, vol. 366, no. 6, p. fnz057 (2019). https://doi.org/10.1093/femsle/fnz057

8 C. L. Ettinger, S. L. Williams, J. M. Abbott, J. J. Stachowicz, and J. A. Eisen, "Microbiome succession during ammonification in eelgrass bed sediments," *PeerJ*, vol. 5, p. e3674 (2017). 10.7717/peerj.3674

9 H. I. Yeoh, R. Izzatty, G. Furusawa, A. A. A. Amirul, A. C. Shu-Chien, and Y. Y. Sung, "The Vibrio-predatory filamentous bacteria effectively removed acute hepatopancreatic necrosis disease (AHPND) causative Vibrio parahaemolyticus in vitro," Aquac. Reports, vol. 21, no. August, p. 100910 (2021) 10.1016/j.aqrep.2021.100910

10 B. D. Eyre and A. J. P. Ferguson, "Comparison of carbon production and decomposition, benthic nutrient fluxes and denitrification in seagrass, phytoplankton, benthic microalgae- and macroalgae-dominated warm-temperate Australian lagoons,"

Mar. Ecol. Prog. Ser., vol. 229, pp. 43–59 (2002). 10.3354/meps229043

11 K. E. Brodersen, K. Koren, M. Moßhammer, P. J. Ralph, M. Kühl, and J. Santner, "Seagrass-Mediated Phosphorus and Iron Solubilization in Tropical Sediments," *Environ. Sci. Technol.*, vol. 51, no. 24, pp. 14155–14163 (2017). 10.1021/acs.est.7b03878

12 Y. Nakashima *et al.*, "Microbial Detoxification of Sediments Underpins Persistence of Zostera marina Meadows," *Int. J. Mol. Sci.*, vol. 25, no. 10, p. 5442 (2024). https://doi.org/10.3390/ijms25105442

13 T. Aires, T. M. Stuij, G. Muyzer, E. A. Serrão, and A. H. Engelen, "Characterization and Comparison of Bacterial Communities of an Invasive and Two Native Caribbean Seagrass Species Sheds Light on the Possible Influence of the Microbiome on Invasive Mechanisms," *Front. Microbiol.*, vol. 12, no. August, pp. 1–19 (2021). 10.3389/fmicb.2021.653998

14 F. Short, T. Carruthers, W. Dennison, and M. Waycott, "Global seagrass distribution and diversity: A bioregional model," *J. Exp. Mar. Bio. Ecol.*, vol. 350, no. 1–2, pp. 3–20 (2007). 10.1016/j.jembe.2007.06.012

Morphometric insights of least concern to vulnerable Elasmobranch in Tuban Waters, East Java: baseline data and fisheries implications

Akhmad Nurhijayat[1,3], *Zairion* Zairion[1,2,3*], *Hollie* L. Booth[3,4], *Luky* Adrianto[2,3], *Mennofatria* Boer[2], *Muhammad* Ghozali Salim[5], *Yon* Vitner[2,3], *Marandi* Eko Wibowo[6], *Firmansyah* Tawang[5], *Antika* Milata Rizka[5], *Dimas* Ardiansah[5]

[1] Coastal and Marine Resources Management Study Program, Department of Aquatic Resources Management, Faculty of Fisheries and Marine Sciences, IPB University, 16680 Bogor, Indonesia
[2] Department of Aquatic Resources Management, Faculty of Fisheries and Marine Sciences, IPB University, 16680 Bogor, Indonesia
[3] Center for Coastal and Marine Resources Studies – International research Institute for Maritime, Ocean and Fisheries - IPB University, Bogor 16143, Indonesia
[4] Department of Biology, University of Oxford, Oxford-OX1, United Kingdom
[5] Mobula Project Indonesia, Pure 01/09, Krajan, Tembokrejo, Muncar, 68472 Banyuwangi, Indonesia
[6] Department of Marine Science - Ronggolawe University, Manunggal Street No. 61, Wire, Kutorejo - Tuban, East Java 62391, Indonesia

Abstract. Indonesia is a mega biodiverse archipelagic nation with a high diversity of elasmobranch species (sharks and rays), yet it faces major challenges in their management. The northern coast of East Java, particularly Tuban Regency within Fisheries Management Area (FMA) 712, is a key landing site for these species. This study provides baseline morphometric data for elasmobranchs categorized as Least Concern (LC), Near Threatened (NT) and Vulnerable (VU) according to the IUCN Red List, and assesses their implications for sustainable fisheries management. Data were collected through daily observations of landings at Palang Fish Landing Base / Fish Landing Unit (PPI Palang) from May 2024 to April 2025. Shark landings consisted of 45.5% LC, 27.3% NT, and 27.3% VU species. Rays were composed of 38.5% LC, 30.8% NT, and 30.8% VU. Most individuals were below the known mature length (Lm), indicating that immature individuals were being harvested. These findings highlight the risk of silent collapse, where population declines progress unnoticed until a critical threshold is reached. Integrating biological considerations into fisheries management frameworks and fostering stronger collaboration between scientific and policy stakeholders is crucial to supporting the long-term sustainability of vulnerable species.

1 Introduction

Sharks and rays (elasmobranchs) have biological characteristics such as slow growth, long lifespan, late reproductive maturity, and low fecundity [1]. These characteristics make these species highly vulnerable to intensive fisheries exploitation [2]. Vulnerability is increasing due to global demand for meat, fins, and other by-products, so that most species are under pressure from over-exploitation [1].

* Corresponding author: zairion@apps.ipb.ac.id

Globally, a recent IUCN Red List assessment indicates that elasmobranchs face unprecedented levels of conservation risk, with more than one-third of species classified as Vulnerable, Endangered, or Critically Endangered. Many populations continue to show long-term declining trends, reflecting cumulative pressures from overfishing, habitat degradation, and other anthropogenic pressures. This pattern highlights the increasing risks across shark and ray groups and underscores the urgency of strengthening science-based management and conservation strategies [3]. Indonesia is widely recognized as one of the world's largest contributors to elasmobranch landings, with a significant share derived from small-scale coastal fisheries [4]. However, basic biological information, particularly morphometric data, remains scarce in most Indonesian coastal areas, limiting the availability of comparable large-scale datasets needed to support effective fisheries management [5]. Standard morphometric measurements, including total length, fork length, precaudal length, and disc width, are key biological variables used to characterize elasmobranch population structure, assess size dynamics, and estimate growth and reproductive traits essential for effective stock assessment and management [6].

A number of studies conducted in Indonesia have provided initial insights into the size structure of elasmobranch catches as part of efforts to understand the dynamics of exploitation. However, the available information is still limited to specific locations and observation periods. In certain waters, catch reports show a relatively high proportion of small and juvenile individuals, indicating an interaction between fishing activity and juvenile age groups. This condition must be interpreted with caution due to the multispecies nature of tropical fisheries, where spatial and temporal variations in catch composition can lead to substantial differences in size structure. This issue is particularly important for elasmobranchs, as they are generally characterized by slow growth rates and low reproductive productivity. Such life history traits make elasmobranchs more vulnerable to fishing pressure than many teleost species [7,8].

Size structure of catches can be used as an early indicator to assess population status and exploitation intensity. Size-based approaches are known as relatively simple yet informative evaluation methods, especially in fisheries with limited data [9]. This approach is increasingly important in the context of data-limited fisheries, where simple biological indicators can provide early signals of stock changes [10]. Species morphometry plays a central role in revealing the biological characteristics of elasmobranchs, including growth patterns, size structure, and population variation between regions. This information forms the scientific basis for understanding population dynamics and designing more adaptive management strategies. However, the availability of morphometric data in Indonesia still largely depends on landing-based records and has not been systematically integrated. This data gap has the potential to hamper the development of evidence-based conservation strategies, particularly in the context of data-limited fisheries. Therefore, strengthening the local morphometric database is an important requirement for improving the quality of management decision-making [11]. Within this framework, morphometric studies not only provide an overview of the condition of shark and ray fisheries, but also serve as an important basis for monitoring changes in population structure over time. In line with these objectives, this study aims to present basic morphometric data on elasmobranch species with conservation statuses ranging from Least Concern to Vulnerable that are landed in the waters of Tuban, East Java. The findings of this study can serve as a baseline for establishing reference values for population size and structure. Overall, the results are expected to enhance our understanding of size structure variability and to contribute scientifically to the development of more adaptive and conservation-oriented fisheries management strategies in the northern coastal region of Java.

2 Method

2.1 Study Area

This research was conducted at the Palang Fish Landing Base / Fish Landing Unit (PPI), Tuban Regency, East Java, Indonesia (Fig. 1). This location serves as one of the primary landing centers along the northern coast of Java and is recognized as an important local landing site for elasmobranch species.

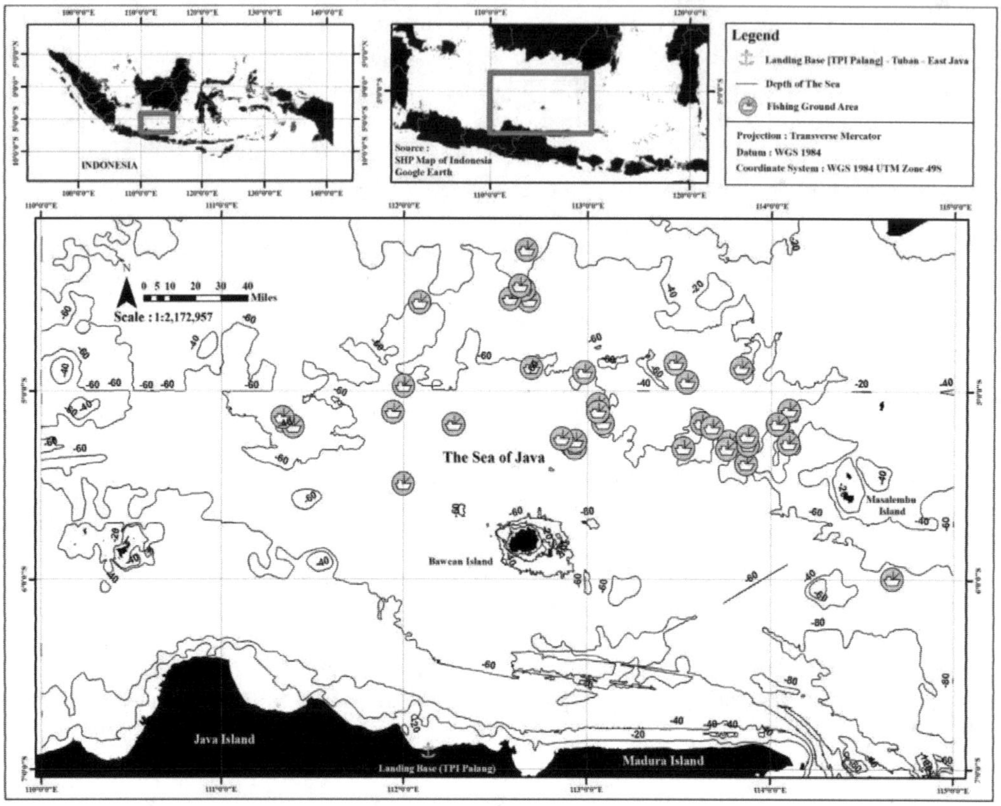

Fig. 1. Research location: Palang Fish Landing Base / Fish Landing Unit (PPI Palang), Tuban Regency, East Java, Indonesia.

2.2 Data collection

The data used in this study were morphometric data of elasmobranchs (sharks and rays) landed at the Palang Fish Landing Base / Fish Landing Unit (PPI), Tuban Regency, East Java. The morphometric data collected included total length (TL) for sharks and disc width (DW) for rays, recorded in centimeters (cm). In addition, species data, number of individuals, and conservation status according to the IUCN Red List were also documented for each specimen. Data collection was conducted directly in the field through observations during the landing of the catch. Each identified individual was measured using a measuring tape with an accuracy of 0.1 cm, then recorded by species. The recording process took place continuously from May 2024 to April 2025, providing a more comprehensive temporal picture of the variation in elasmobranch catches at the study site.

2.3 Data analysis

Data analysis was conducted descriptively and exploratively using R version 4.5.1 with the *ggplot2*, *dplyr* and *ggpubr* packages. Data analysis encompassed the composition of shark and ray catches, the proportional representation of species based on their IUCN Red List conservation status, and the comparative distribution of size classes among species. Size distributions were derived from total length (TL) for sharks and disc width (DW) for rays, while length at maturity (Lm) was obtained from secondary data sourced from (https://www.fishbase.org/home) and IUCN Red List (https://www.iucnredlist.org/en). The results of this analysis provide a morphometric overview as well as important information to support ecological and conservation studies of elasmobranchs.

3 Results

3.1 Composition of ray and shark catches

There are 11 identified shark species, with the dominant *Carcharhinus tjutjot* (212 individuals, VU) and *Chiloscyllium punctatum* (203, NT). Other relatively abundant species are *Hemipristis elongata* (86, VU), *Chiloscyllium plagiosum* (57, NT), and *Hemigaleus microstoma* (53, VU). Overall, the majority of shark catches were from VU category species (Fig. 2). A total of 12 ray species were identified. The catch was dominated by *Neotrygon orientalis* (146 individuals, LC) and *Pateobatis uarnacoides* (139 individuals, VU). In general, the VU category dominates in terms of the number of species, however many LC individuals were also found at the study site (Fig. 3).

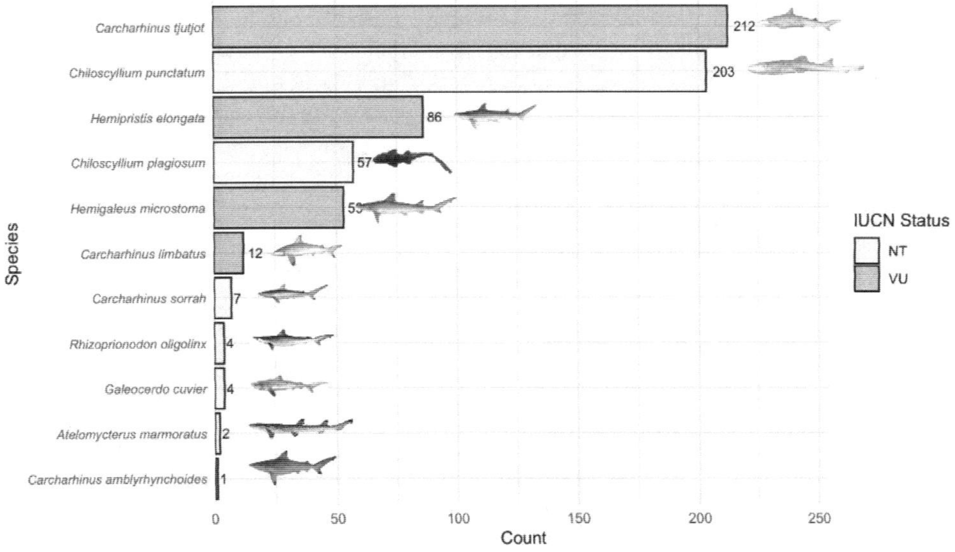

Fig. 2. Species composition of shark catches based on number of individuals.

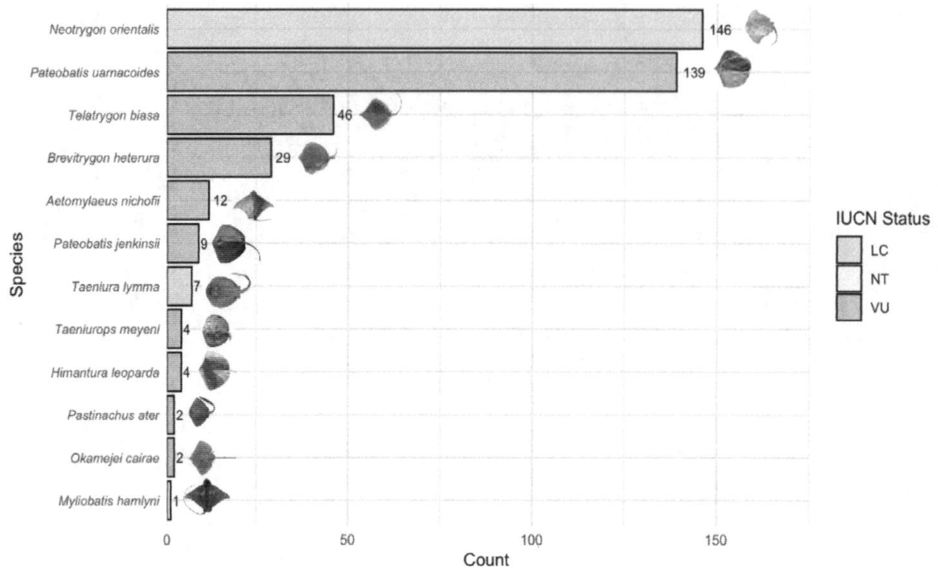

Fig. 3. Species composition of ray's catches based on number of individuals.

3.2 Proportion of elasmobranch conservation status

The composition of conservation status based on the IUCN Red List shows clear differences in the proportion of sharks and rays landed at the Palang Fish Landing Base (PPI Palang), Tuban Regency, East Java, Indonesia. In the shark group, species with the Vulnerable (VU) category dominate with a proportion of 57.9%, while the Near Threatened (NT) category is recorded at 42.1%, and no species are in the Least Concern (LC) category. In the Rays group The Vulnerable (VU) category also dominates with a higher percentage, namely 61.6%, followed by Least Concern (LC) at 38.2%, while the Near Threatened (NT) category is only represented in very small numbers (0.1%) (Fig. 4). The dominance of the Vulnerable category in both sharks and rays indicates that these species tend to be caught more frequently.

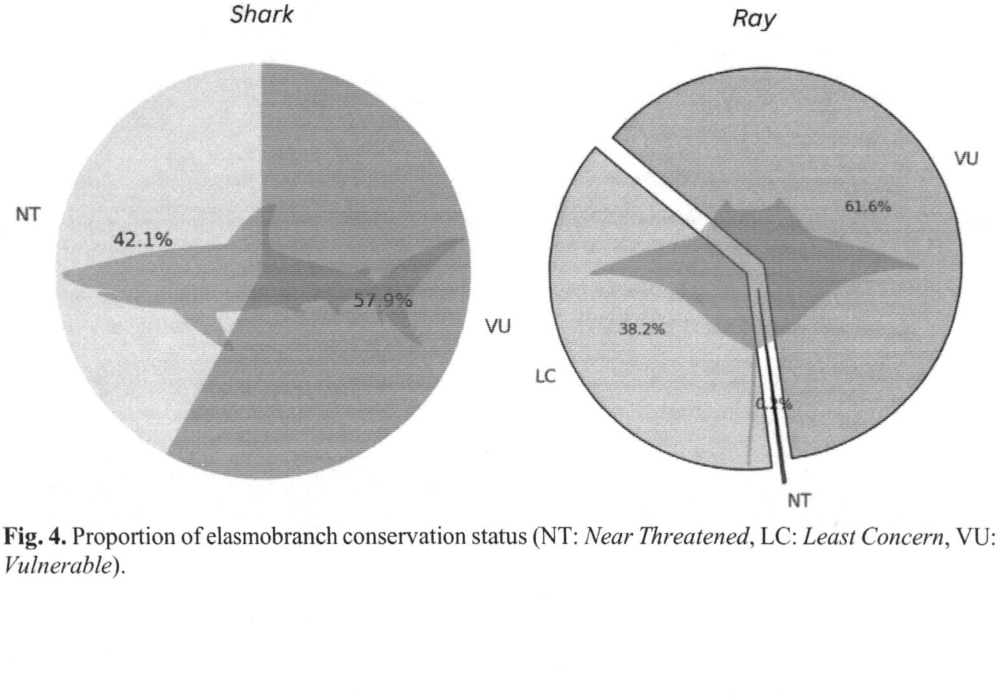

Fig. 4. Proportion of elasmobranch conservation status (NT: *Near Threatened*, LC: *Least Concern*, VU: *Vulnerable*).

3.3 Morphometric size distribution: Total Length (TL) in sharks and Disc Width (DW) in rays

The TL distribution graph shows the size variation among the shark species landed. Some species, such as *Galeocerdo cuvier* and *Hemipristis elongata*, has individuals with sizes approaching or exceeding Lm, although their numbers are relatively limited. As for species such as *Chiloscyllium punctatum*, *Hemigaleus microstoma*, and *Rhizoprionodon oligolinx* dominated by individuals with a size below Lm (Fig. 5). The distribution of Disc Width (DW) in ray species also shows a similar pattern. Several large species such as *Pateobatis uarnacoides*, *Himantura leoparda* and *Taeniurops meyeni* shows some individuals that have passed the mature gonadal size. However, the majority of species such as *Neotrygon orientalis*, *Telatrygon biasa*, and *Brevitrygon heterura* caught in sizes below Lm (Fig.6). This condition confirms the high level of exploitation of juvenile groups in the ray in the research area.

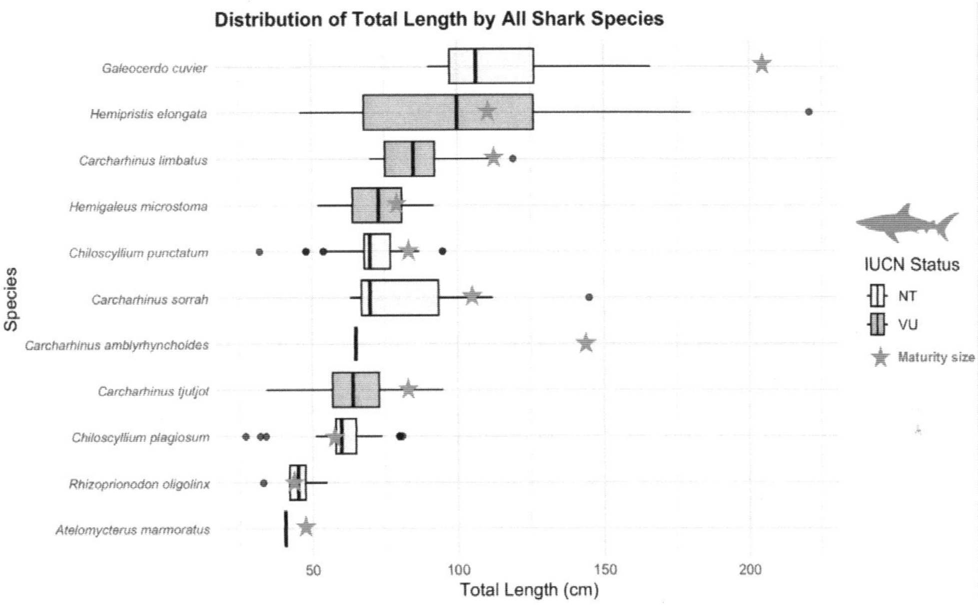

Fig. 5. Total length distribution of shark species.

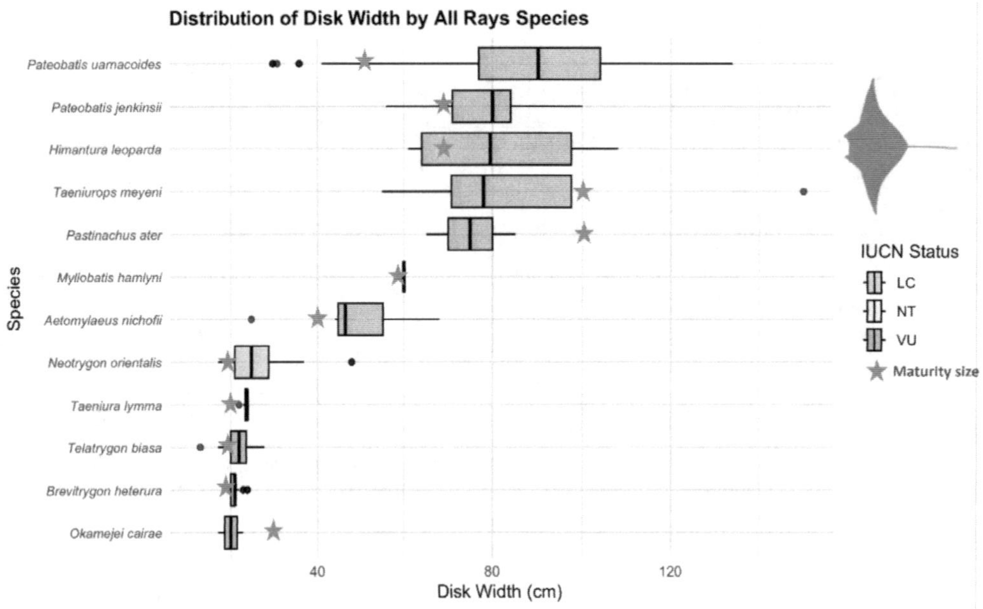

Fig. 6. Disk width distribution of ray species.

4 Discussion

The study results show that the proportion of elasmobranchs in the Vulnerable (VU) category is quite high, both in the shark (57.9%) and ray (61.6%) groups. This finding is in line with global conditions, where most elasmobranch species, especially in Southeast Asia and Indonesia, are recorded as threatened or experiencing population decline, with an estimated 59% of species being threatened and 72.5% showing a declining trend [12]. In the shark group, the species that dominate the catch include: *Carcharhinus tjutjot, Hemipristis elongata* and *Hemigaleus microstoma.*

The observed length frequency distributions indicate a fishing pattern that disproportionately captures immature individuals. Several shark species, including *Chiloscyllium punctatum* and *Rhizoprionodon oligolinx*, are frequently recorded below their sexually mature length (Lm). A similar trend is seen in rays such as *Neotrygon orientalis*, *Telatrygon ordinary* and *Brevitrygon heterura*. The predominance of subadult individuals in the catch suggests that some resources are exploited before reproduction [13]. Exploitation pressure on vulnerable elasmobranch species generates not only ecological concerns but also important socio-economic implications. In the study area, shark and ray landings are actively traded in local markets and partly integrated into domestic supply chains, suggesting that market demand plays a meaningful role in shaping exploitation intensity. The strong dependence of coastal communities on these commodities illustrates that elasmobranch fisheries operate within a closely linked socio-ecological system, where ecological sustainability and economic well-being are inseparable [14]. Contemporary fisheries governance increasingly recognizes that biological indicators must be interpreted alongside socio-economic realities. In elasmobranch fisheries, effective management requires balancing conservation objectives with the livelihood needs of fishing communities. The implementation of minimum landing sizes represents a practical tool to reduce juvenile exploitation while preserving the reproductive structure of stocks. Beyond ecological benefits, protecting immature cohorts can contribute to long-term socio-economic stability by improving stock productivity, enhancing catch quality, and reducing vulnerability to boom-and-bust cycles associated with overexploitation. Integrating conservation targets with livelihood-sensitive approaches is therefore essential to minimize trade-offs and support more equitable transitions in small-scale fisheries [15]. In addition to scientific approaches based on catch size-structure assessments, governance strategies that emphasize Community-Based Fisheries Management (CBFM) provide an important complementary pathway. CBFM functions not only as a monitoring mechanism but also as a platform for local empowerment, strengthening the role of fishers through active participation in rule-making, surveillance, and catch reporting. Such engagement fosters a sense of stewardship, increases social legitimacy, and reduces implementation costs because regulations emerge from deliberative processes perceived as fair by the community. This perspective is particularly relevant to the study area, where conservation approaches that rely solely on restrictive measures—without offering viable economic alternatives—may generate resistance and social tension. More integrative strategies that combine conservation initiatives with the strengthening of local economic opportunities are therefore more likely to produce durable outcomes. By reinforcing both ecological sustainability and socio-economic resilience, these approaches can promote fisheries governance that is adaptive, inclusive, and socially equitable, while supporting long-term resource stewardship in coastal communities.

5 Conclusion

The findings indicate that elasmobranch catches in the Tuban waters are dominated by species classified as Vulnerable (VU), with the majority of individuals measuring below their length at maturity (Lm). This pattern suggests fishing pressure that disproportionately affects smaller or younger size classes. The morphometric data collected provide important baseline information on size variation and population structure of elasmobranchs in the northern coastal waters of Java. These insights can inform future biological research and offer a scientific foundation for developing more adaptive and conservation-oriented fisheries management strategies that support both resource sustainability and the socio-economic resilience of coastal communities.

Acknowledgements are extended to the Center for Coastal and Marine Resources Studies (CCMRS) IPB, the Darwin Initiative, and the University of Oxford for providing financial support for this research. Appreciation is also given to the Mobula Project Indonesia for their valuable collaboration.

References

1 N.K. Dulvy, S.L. Fowler, J.A. Musick, R.D. Cavanagh, P.M. Kyne, L.R. Harrison, J.K. Carlson, L.N.K. Davidson, S.V. Fordham, M.P. Francis, C.M. Pollock, C.A. Simpfendorfer, G.H. Burgess, K.E. Carpenter, L.J.V. Compagno, D.A. Ebert, C. Gibson, M.R. Heupel, S.R. Livingstone, J.C. Sanciangco, J.D. Stevens, S. Valenti, W.T. White. Extinction risk and conservation of the world's sharks and rays. eLife (2014). https://doi.org/10.7554/eLife.00590

2 A.S. Ferreira, M.A. Naré, J.I. Robalo. Evaluating techniques for determining elasmobranch body size: A review of current methodologies. PeerJ 12, e16743 (2024). https://doi.org/10.7717/peerj.16743

3 D. Cardeñosa, S.K. Shea, H. Zhang, G.A. Fischer, C.A. Simpfendorfer, D.D. Chapman. Two thirds of species in a global shark fin trade hub are threatened with extinction: Conservation potential of international trade regulations for coastal sharks. Conserv. Lett. 15, e12910 (2022). https://doi.org/10.1111/conl.12910

4 R. Okes, G. Sant, An overview of major shark traders, catchers and species. TRAFFIC International, Cambridge (2019)

5 N. Hikmah, H. Nufus, S. Syahrial, R. Ezraneti, R. Astuti, Sharks and rays at Pelabuhan Perikanan Samudera (PPS) Lampulo, Banda Aceh, Indonesia: Morphometric characteristics and differentiators based on multivariate analysis. Journal of Marine Studies 1(2), 1204 (2024). https://doi.org/10.29103/joms.v1i2.17630

6 J. C. Carrier, C. A. Simpfendorfer, M. R. Heithaus, K. E. Yopak (eds.), Biology of Sharks and Their Relatives, 3rd ed., CRC Press, Boca Raton (2022).

7 F. Ferretti, B. Worm, G.L. Britten, M.R. Heithaus, H.K. Lotze, Patterns and ecosystem consequences of shark declines in the ocean. Ecol. Lett. 13, 1055–1071 (2010). https://onlinelibrary.wiley.com/doi/epdf/10.1111/j.1461-0248.2010.01489.x

8 J.D. Stevens, R. Bonfil, N.K. Dulvy, P.A. Walker, The effects of fishing on sharks, rays, and chimaeras. ICES J. Mar. Sci. 57, 476–494 (2000). https://doi.org/10.1006/jmsc.2000.0724

9 M.J.N. Bergman, S.M. Ubels, G.C.A. Duineveld, E.W.G. Meesters, Effects of a 5-year trawling ban on the local benthic community in a wind farm in the Dutch coastal zone. ICES J. Mar. Sci. 72, 962–972 (2015). https://doi.org/10.1093/icesjms/fsu193

10 J. Prince, A. Hordyk, S.R. Valencia, N. Loneragan, K. Sainsbury, Revisiting the concept of Beverton–Holt life-history invariants with the aim of informing data-poor fisheries assessment. ICES J. Mar. Sci. 72, 194–203 (2015). doi:10.1093/icesjms/fsu011

11 S. Oktaviyani, Wanwan Kurniawan, Fahmi Fahmi. Species composition and size distribution of sharks and rays caught in Bali Strait and its surrounding area and its relation to fisheries management. Jurnal Iktiologi Indonesia 20(1), 509 (2019). https://doi.org/10.32491/jii.v20i1.509

12 N. Clark-Shen, A. Chin, S. Arunrugstichai, J. Labaja, M. Mizrahi, B. Simeon, N. Hutchinson, Status of Southeast Asia's marine sharks and rays. Conserv. Biol. 37, e13962 (2022). https://doi.org/10.1111/cobi.13962

13 M.R. Heupel, C.A. Simpfendorfer, Estuarine nursery areas provide a low-mortality environment for young bull sharks Carcharhinus leucas. Mar. Ecol. Prog. Ser. 433, 237–244 (2011). https://doi.org/10.3354/meps09191

14 F. Berkes, J. Colding, C. Folke, Navigating social-ecological systems: Building resilience for complexity and change (Cambridge University Press, Cambridge, 2016)

15 H. Booth, D. Squires, E.J. Milner-Gulland, The mitigation hierarchy for sharks: a risk-based framework for reconciling trade-offs between shark conservation and fisheries objectives. Fish Fish. 21, 269–289 (2020). https://doi.org/10.1111/faf.12429

Macrozoobenthic community structure in the mangrove ecosystem of Teluk Bogam Village, Central Kalimantan

Farenta Indira Alma[1], *Sulistiono* Sulistiono[1*], *Ayu* Ervinia[1], *Dudi* Muhammad Wildan[1], *Endang* Sunarwati Srimariana[2], *Ahmad* Shofy Mubarak[3], *Ismail* Ismail[4], and *Abdul* Wahab Radjab[5]

[1] Departement of Aquatic Resources Management, Faculty of Fisheries and Marine Sciences, IPB University, Bogor, 16680, Indonesia.
[2] Departement of Marine Sciences and Technology, Faculty of Fisheries and Marine Sciences, IPB University, Bogor, 16680, Indonesia.
[3] Department of Marine Science, Faculty of Fisheries and Marine Sciences, Airlangga University, Surabaya, 60115, Indonesia.
[4] Marine and Fisheries Polytechnic of Sorong, Ministry of Marine Affairs and Fisheries, Sorong, 98411, Indonesia.
[5] Research Center for Deep Sea, National Research and Innovation Agency, Jl. Pasir Putih I, Ancol Timur, Pademangan, Jakarta Utara, 14430, Indonesia.

Abstract. The mangrove ecosystem in Teluk Bogam Village plays an essential role in maintaining the balance of the coastal environment. Variations in the utilization of mangrove areas have the potential to cause differences in water quality conditions. To date, no study has specifically examined the macrozoobenthic community structure in Teluk Bogam Village. This study aims to analyze the macrozoobenthic community structure within the mangrove ecosystem of Teluk Bogam Village, Central Kalimantan. Sampling of macrozoobenthos and water quality observations were conducted in March, June, and September 2024, as well as in January 2025, at three observation stations. Data analyses included assessments of macrozoobenthic community structure and Principal Component Analysis (PCA). Macrozoobenthos in the study area consisted of the classes Gastropoda, Bivalvia, and Polychaeta, with the genus *Cerithium* being the most dominant. The community structure was categorized as stable, characterized by moderate diversity (H'=2.04–2.06), high evenness (E= 0.92–0.97), and low dominance (D = 0.13–0.14). The PCA results indicated that macrozoobenthic density exhibited a strong association with water quality parameters, particularly turbidity and total suspended solids (TSS).

1 Introduction

The coastal area of Teluk Bogam is located in Teluk Bogam Village, Kumai District, West Kotawaringin Regency, Central Kalimantan Province, and directly borders the Java Sea. Among the key ecosystems along the Teluk Bogam coastline is the mangrove ecosystem. Mangroves are one of the most important coastal ecosystems, providing essential ecological functions for both humans and aquatic biota. Mangrove forests protect shorelines from erosion, buffer strong winds and ocean waves, trap sediments, and retain fine particles carried

* Corresponding author: onosulistiono@gmail.com

by water flow. Furthermore, mangrove ecosystems serve as critical spawning, nursery, and feeding grounds for aquatic organisms such as fish and macrozoobenthos [1].

Macrozoobenthic organisms are generally sedentary and possess limited mobility. These organisms exhibit varying levels of sensitivity to environmental changes depending on their species. Therefore, the presence of specific macrozoobenthic taxa can be used as bioindicators to assess the ecological condition of aquatic environments. According to [2], the mangrove ecosystem in Teluk Bogam Village is located adjacent to residential areas, fishing activities, and tourism sites, which may lead to spatial variation in environmental conditions. Such variation can influence the structure of macrozoobenthic communities. Understanding the macrozoobenthic community structure in this mangrove ecosystem is thus essential for evaluating ecological balance and the quality of benthic habitats.

Previous studies on macrozoobenthic communities have been conducted in various coastal areas of Indonesia, including the Margagiri–Grenyang waters of Banten Bay [3] and the mangrove ecosystem of Pejarakan Village, Bali [4]. However, to date, no study has specifically examined the macrozoobenthic community structure within the mangrove ecosystem of Teluk Bogam Village. Therefore, this study aims to analyze the macrozoobenthic community structure in the mangrove ecosystem of Teluk Bogam Village, Central Kalimantan, as a basis for assessing aquatic environmental conditions and as reference data to support the sustainable management of mangrove ecosystems.

2 Research method

2.1 Time and location

The study was conducted in March, June, and September 2024, and January 2025 within the mangrove ecosystem of Teluk Bogam Village, Kumai District, West Kotawaringin Regency, Central Kalimantan (Figure 1). Three observation stations were established to represent the variation in environmental conditions and the local utilization of the mangrove area by the surrounding community. Station 1 was located near the coastal lodging area, Station 2 was situated around the fishing boat landing site and silvofishery area, and Station 3 was positioned adjacent to residential settlements and coastal tourism sites. Water quality parameters were analyzed both in the field and at the Environmental Laboratory, while macrozoobenthic analyses were carried out at the Macrobiology Laboratory I, Faculty of Fisheries and Marine Science, IPB University.

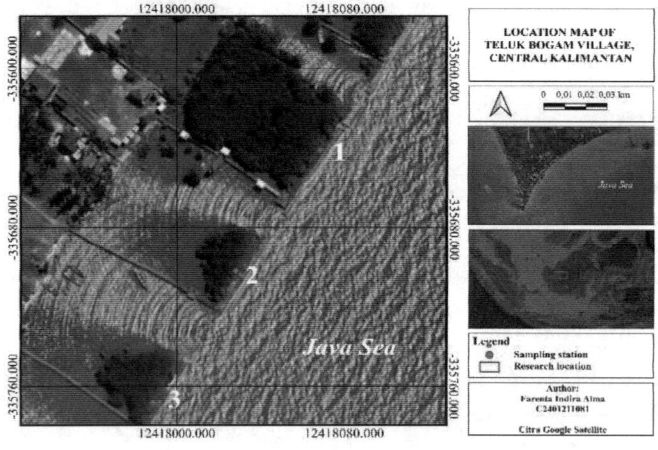

Fig. 1. Research location in Teluk Bogam Village, Central Kalimantan.

2.2 Data collection

Water quality sampling and measurements were conducted with four replications at each station, following the method outlined by APHA 2017 (Table 1). Sediment samples were collected simultaneously with macrozoobenthos sampling using 1 × 1 m quadrat transects and a Van Veen Grab sampler. Sediment grain size classification followed [5], which categorizes sediment fractions from gravel to clay based on particle diameter (mesh size). Macrozoobenthos samples were sorted and identified using the book *Recent and Fossil Indonesian Shells* by Dharma 2005 and the World Register of Marine Species (WoRMS: www.marinespecies.org), in the Macrobiology Laboratory I, Faculty of Fisheries and Marine Sciences, IPB University. Water quality parameters were analyzed both *in situ* (in the field) and *ex situ* (in the laboratory) at the Environmental Laboratory. The physical, chemical, and biological parameters of the water, along with the analytical methods and instruments used, are presented in Table 1.

Table 1. Parameters analyzed and observed.

Parameters	Unit	Instrument	Method
Physical			
Substrate	%	*Sieve Shaker*	Dry sieving
Temperature	°C	Multiparameter	*Electrometric method*
Depth	m	*Secchi disk*	Direct measurement
Turbidity	NTU	Multiparameter	*Nephelometric method*
Total Suspended Solid (TSS)	mg/L	Gravimetric scale	*Gravimetric method*
Total Dissolved Solid (TDS)	mg/L	Multiparameter	*Electrometric method*
Chemical			
pH	-	Multiparameter	*Electrometric method*
Dissolved Oxygen (DO)	mg/L	Multiparameter	*Electrometric method*
Salinity	ppt	Refractometer	Direct measurement
Biological			
Macrozoobenthos	-	*Van Veen Grab*	Ecological indices, PCA

2.3 Data analysis

2.3.1 Macrozoobenthos density

The formula used to calculate the density of macrozoobenthos at a given location is as follows [6]:

$$K_i = \frac{10000 \times N_i}{A} \quad (1)$$

Where, Ki = Density of macrozoobenthos species i (ind/m^2), Ni = Number of individuals of macrozoobenthos species i (ind), A = Total sampling area for macrozoobenthos (m^2), The value 10,000 is a conversion factor from cm^2 to m^2.

2.3.2 Ecological index

The Shannon-Wiener diversity index was calculated using the following formula [7]:

$$H' = -\sum_{i=1}^{n} p_i \ln p_i \quad (2)$$

Where, *H'* = Shannon-Wiener diversity index, *pi* = Proportion of individuals of species i relative to the total number of individuals (ni/N), *ni* = Number of individuals of species i. *N* = Total number of individuals across all species.
The evenness index ranges from 0 to 1 and is calculated based on the diversity index. The formula for the evenness index is as follows [8]:

$$E = \frac{H'}{H'_{max}} \quad (3)$$

Where, *E* = Evenness index, *H'* = Shannon-Wiener diversity index, *s* = Total number of macrozoobenthos species. $H'_{max} = ln\ S$

The dominance index ranges from 0 to 1 and is calculated using the Simpson's Dominance Index formula [8]:

$$D = \sum_{i=1}^{n} \left(\frac{n_i}{N}\right)^2 \quad (4)$$

Where, *D* = Dominance index, *ni* = Number of individuals of macrozoobenthos species i, *N* = Total number of individuals of all macrozoobenthos species

2.3.3 Physicochemical water parameters

The values of water quality parameters were compared with water quality standards as stipulated in [9], specifically regarding water quality standards for marine biota, as well as [10], and supporting literature.

2.3.4 Sediment fraction classification

The total weight of filtered sediment was identified based on the Wentworth scale (1922) [5]. The dry sieving method was used to separate sediment into seven fractions: gravel (4 mm), granule gravel (2 mm), very coarse sand (1 mm), coarse sand (0.5 mm), medium sand (0.25 mm), fine sand (0.063 mm), and silt (<0.063 mm).

2.3.5 Relationship between macrozoobenthos density and physico-chemical water parameters

The relationship between macrozoobenthos density and water quality parameters was analyzed using Principal Component Analysis (PCA) with the assistance of XLSTAT2021 software. The relationship between the macrozoobenthos community and the physical and chemical parameters was further analyzed using Pearson correlation analysis. The correlation coefficient values range from 0.00 to 1.00, indicating a relationship that varies from very weak to very strong. The Pearson correlation coefficient can be calculated using the following formula [11].

$$r_{xy} = \frac{n\sum xy - \sum x \sum y}{\sqrt{n\sum x^2 - (\sum x)^2}\sqrt{n\sum y^2 - (\sum y)^2}} \quad (5)$$

Where, *r* = Pearson correlation coefficient between variables *x* and *y*; *x, y* = Correlated variables; *n* = Number of data points used.

3 Results and discussion

3.1 Results

3.1.1 Macrozoobenthos composition and density

The macrozoobenthos composition found in the mangrove ecosystem of Teluk Bogam Village consisted of three classes and 33 genera, including 7 genera from the class Bivalvia, 20 genera from Gastropoda, and 6 genera from Polychaeta (Table 2).

Table 2. Macrozoobenthos genera found in the Mangrove Ecosystem of Teluk Bogam Village.

Class	Genera				
Bivalvia	*Anadara*	*Donax*	*Lithophaga*	*Vasticardium*	
	Antigona	*Fimbria*	*Tellina*		
Gastropoda	*Aliculastrum*	*Conus*	*Nassaria*	*Pila*	*Terebra*
	Cerithidea	*Coralliophila*	*Nassarius*	*Pirenella*	*Terebralia*
	Cerithium	*Hemifusus*	*Nerita*	*Planaxis*	*Trochus*
	Clypeomorus	*Littoraria*	*Oliva*	*Strombus*	*Turitella*
Polychaeta	*Glycera*	*Nereis*	*Paraonis*		
	Lumbrineris	*Notomastus*	*Scoloplos*		

The composition of macrozoobenthic classes identified in the Teluk Bogam mangrove ecosystem, from the highest to the lowest percentage, was as follows: Gastropoda (71.90%), Polychaeta (16.12%), and Bivalvia (11.98%) (Figure 2). The dominance of Gastropoda indicates that the water conditions remain favorable for organisms with high tolerance to environmental fluctuations and elevated organic matter availability. Seasonal differences during sampling periods did not show a significant effect on the composition and distribution of macrozoobenthos within the Teluk Bogam mangrove ecosystem.

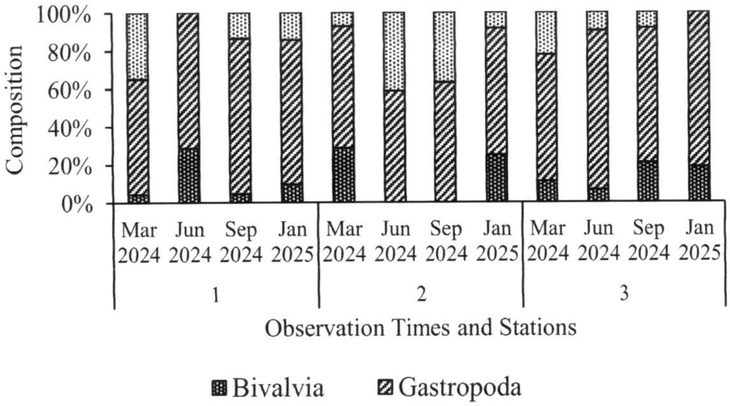

Fig. 2. Composition of macrozoobenthos found in the Mangrove Ecosystem of Teluk Bogam Village.

The density of macrozoobenthos across the three stations and four observation periods exhibited considerable variation. The highest density was recorded at Station 1 in January 2025, reaching 279 individuals/m^2. The most abundant genus identified was *Cerithium*. The dominant composition and density of macrozoobenthos reflect the adaptive capacity of benthic organisms to environmental conditions, as well as the interactions between environmental factors and anthropogenic activities within the ecosystem [12]. The variation in macrozoobenthos density is illustrated in Figure 3.

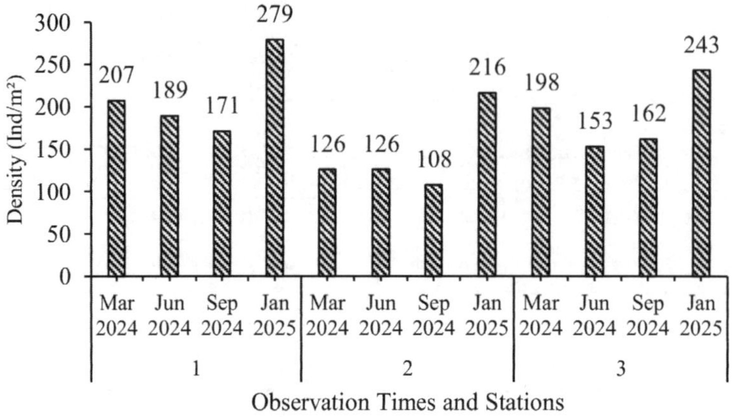

Fig. 3. Macrozoobenthos density in the Mangrove Ecosystem of Teluk Bogam Village.

3.1.2 Ecological index

An ecological index analysis, comprising diversity (H'), evenness (E), and dominance (D), was conducted to describe the structure of the macrozoobenthic community and to assess the stability level of the mangrove ecosystem in Teluk Bogam Village. The results of these index calculations are presented in Table 3.

Table 3. Diversity index (H'), evenness index (E), and dominance index (D) of macrozoobenthos in the mangrove ecosystem of Teluk Bogam Village.

Stations	Observation Periods	H'	E	D
1	March 2024	2,09	0,95	0,13
	June 2024	2,10	0,95	0,13
	September 2024	2,09	0,95	0,14
	January 2025	2,02	0,92	0,15
2	March 2024	2,01	0,97	0,14
	June 2024	2,11	0,96	0,13
	September 2024	2,02	0,97	0,14
	January 2025	2,13	0,92	0,14
3	March 2024	2,03	0,97	0,14
	June 2024	1,99	0,96	0,15
	September 2024	2,03	0,98	0,14
	January 2025	2,02	0,92	0,15

Based on Table 3, the diversity index (H') of macrozoobenthos in the mangrove ecosystem of Teluk Bogam Village ranged from 1.99 to 2.13 across the three stations during the four observation periods. The diversity of macrozoobenthos showed moderate variation, although it was lower than that reported by [4] in the mangrove ecosystem of Pejarakan Village, Bali. A moderate diversity index indicates that the ecosystem is in a relatively balanced ecological condition with a moderate level of environmental stress. The evenness index (E) of macrozoobenthos in the Teluk Bogam mangrove waters ranged from 0.92 to 0.97, which is categorized as high since the calculated values approached 1. This suggests that no single species dominated the community, and the distribution of individuals among species was relatively even [7].

The dominance index (D) of macrozoobenthos in the mangrove ecosystem of Teluk Bogam ranged from 0.13 to 0.15, indicating a low level of dominance, as the calculated values were close to 0. The dominance index reflects the tolerance or adaptive capacity of macrozoobenthos to dynamic environmental conditions within the aquatic ecosystem. Overall, the ecological index analysis revealed moderate diversity, high evenness, and low dominance. These findings are consistent with those reported by [13] in the mangrove ecosystem of Serangan Island, Bali, which exhibited similar ecological index values. Such conditions indicate that the macrozoobenthic community structure within the Teluk Bogam mangrove ecosystem is ecologically stable.

3.1.3 Physicochemical water parameters

The measurements of physicochemical parameters in the mangrove ecosystem waters of Teluk Bogam Village revealed that the Total Suspended Solids (TSS) and turbidity values at all three stations exceeded the established water quality standards. This condition indicates an increase in suspended particulate matter, which may be attributed to tidal activities and sediment resuspension within the mangrove area.Other parameters such as temperature, pH, dissolved oxygen (DO), and salinity remained within the acceptable range of marine water quality standards for aquatic organisms (Table 4). These findings suggest that the aquatic environment is still capable of supporting marine life, despite relatively high levels of turbidity and suspended solids. Turbidity can be influenced by various environmental factors, including rainfall, seasonal variation, and tidal dynamics. Strong wave action and rapid water currents can enhance sediment resuspension from the seabed, leading to an increase in suspended particle concentration within the water column.

Table 4. Mean and standard deviation of physical and chemical water parameters in the Mangrove Ecosystem of Teluk Bogam Village.

Parameters	**units**	**Stations**			**Water Quality Standard**
		1	**2**	**3**	
Physical					
Temperature	°C	28,58 ± 2,32	28,45 ±1,79	28,93 ± 1,66	28-32*
Depth	m	0,51 ± 0,08	0,56 ± 0,06	0,55 ± 0,08	Natural
Turbidity	NTU	158,75 ± 85,06	85,23 ± 49,36	90 ± 59,78	<5**
TSS	mg/L	339 ± 287,26	222 ± 120,17	284 ± 120,83	80*
TDS	mg/L	15,90 ± 3,60	16,10 ± 4,10	15,32 ± 3,82	80***
Chemical					
pH	-	7,82 ± 0,21	7,84 ± 0,32	7,81 ± 0,19	7 - 8,5*
DO	mg/l	5,68 ± 0,33	5,43 ± 0,22	5,43 ± 0,29	>5*
Salinity	ppt	21,60 ± 1,23	21,25 ± 0,90	21,38 ± 0,95	up to 34*

Source: *[9]; **[10]; ***[14].

3.1.4 Substrate characteristics

Differences in the percentage composition of sediment fractions were observed among the three sampling stations. Based on the analysis results, all stations were classified as having silty sand substrate types (Table 5). The Gastropoda group was the most abundant class across all stations. Silty sand substrates provide suitable habitats for the growth and survival of gastropods, thereby supporting higher levels of diversity and abundance within this group.

Table 5. Substrate composition of the Mangrove Ecosystem Waters in Teluk Bogam Village

Stations	**Fraction Composition (%)**							**Substrate Types**
	Gravel	Granule gravel	Very coarse sand	Coarse sand	Medium sand	Fine sand	Silt	
1	0,17	0,19	2,93	15,56	22,02	42,95	16,19	Silty sand
2	0,17	0,87	6,71	19,01	22,37	36,70	14,17	Silty sand
3	0,25	0,87	9,83	22,08	21,11	31,93	13,93	Silty sand

3.1.5 Relationship between macrozoobenthos and physico-chemical water parameters

The relationship between macrozoobenthos and the physico-chemical parameters of the water was analyzed using Principal Component Analysis (PCA). The results of the PCA are presented in the biplot shown in Figure 4.

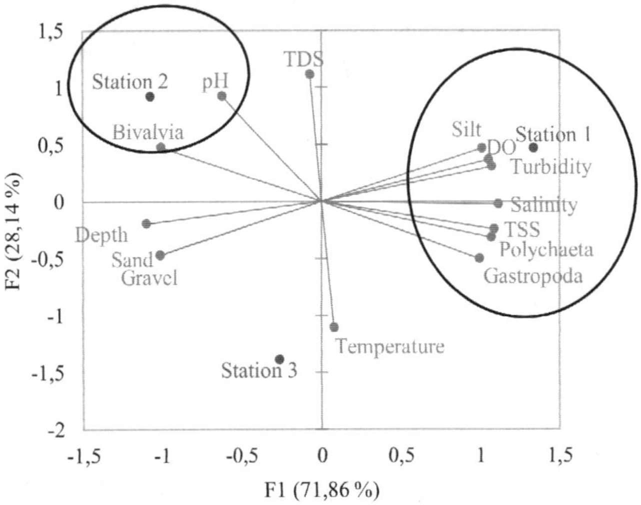

Fig. 4. Biplot of the Relationship Between Macrozoobenthos and Physico-Chemical Water Parameters.

Based on the results of the Principal Component Analysis (PCA), the parameters TSS, salinity, turbidity, dissolved oxygen (DO), and mud substrate were found to be associated with the Gastropoda and Polychaeta groups. The relationships among these parameters exhibited strong to very strong positive correlations according to Pearson's correlation analysis. This condition indicates that macrozoobenthos tend to be more abundant in waters with higher TSS and turbidity values, which may be attributed to greater availability of organic matter and nutrients, particularly in muddy substrates that are commonly found in mangrove ecosystems. These findings are consistent with those of [15], who reported that total suspended solids have a strong influence on macrozoobenthos density. High TSS and turbidity values can act as limiting factors for species that are sensitive to suspended sediments; however, they still support the presence of taxa with high environmental tolerance, such as Gastropoda and Polychaeta. The positive correlation between salinity and these two groups further indicates their strong adaptive capability to varying salinity levels. In addition, the positive correlation between DO and the abundance of Gastropoda and Polychaeta suggests that these organisms thrive under conditions with sufficient oxygen availability. Dissolved oxygen is an essential factor for the metabolic processes and survival of macrozoobenthos.

Meanwhile, Bivalvia exhibited a strong positive correlation with pH, indicating that this group tends to develop optimally within stable pH ranges. The pH parameter itself is an important indicator of water quality, as it influences the physiological processes of aquatic organisms and the availability of essential nutrients. The presence and abundance of macrozoobenthos may also be influenced by mangrove vegetation density. High mangrove density plays a significant role in enhancing litter production, which subsequently enriches sediments with organic matter. The accumulation and decomposition of litter contribute to the formation of nutrients and detritus, serving as a primary food source for macrozoobenthos. Consequently, mangrove ecosystems with dense vegetation tend to support higher abundances of benthic organisms [11].

The dominant mangrove species found in the Teluk Bogam mangrove ecosystem is *Rhizophora mucronata*, which typically grows in estuarine areas influenced by tidal fluctuations. Sandy to silty substrates support the adaptability of *R. mucronata* to its environment, facilitating optimal propagule dispersal and growth.

4 Conclusion

Based on a four-month study conducted in the mangrove ecosystem of Teluk Bogam Village, macrozoobenthos belonging to the classes Gastropoda, Bivalvia, and Polychaeta were identified across the three observation stations. The most dominant genus recorded was *Cerithium*. The highest macrozoobenthos density was observed in January 2025, reaching 279 individuals/m^2. The community structure of macrozoobenthos was categorized as stable, characterized by moderate diversity, high evenness, and low dominance. The results of the Principal Component Analysis (PCA) indicated that turbidity and total suspended solids (TSS) were the primary environmental factors influencing the distribution of macrozoobenthos, particularly Gastropoda, within the Teluk Bogam mangrove ecosystem. These findings provide valuable insights and serve as a scientific basis for the sustainable management and conservation of mangrove ecosystems.

References

1. E. Hilmi, L.K. Sari, T.N. Cahyo, R. Dewi, T. Winanto, The structure communities of gastropods in the permanently inundated mangrove forest on the north coast of Jakarta, Indonesia. J. Biol. Divers. **23**, 2699–2710 (2022)
2. R. Nirmalasari, Keanekaragaman bivalvia di Pantai Teluk Bogam, Kecamatan Kumai, Kabupaten Kotawaringin Barat, Kalimantan Tengah. J. Ilmu Alam Lingkung. **10**, 9–23 (2019)
3. Khasanah S., Riani E., Zairion, Sulistiono, Community structure of macrozoobenthos as a water quality indicator at Margagiri-Grenyang coastal waters, Banten Bay, in Proceedings of the International Seminar on Demersal and Crustacean Fisheries Management 2023, IOP Conf. Ser.: Earth Environ. Sci., IPB International Convention Center, Bogor, Indonesia, August 2–3 (2023), **1400**. 10.1088/1755-1315/1400/1/012012
4. A.F. Zafirah, A.M. Samosir, Sulistiono, I. Solihin, T. Nugroho, T. Hestirianoto, E.S. Srimariana, R.E. Arhatin, Rastina, A.S. Mubarak, Ismail, Macrozoobenthic community structure in mangrove ecosystems of Pejarakan Village, Bali. BIO Web Conf. **156**, 01002 (2024). https://doi.org/10.1051/bioconf/202515601002
5. C.K. Wentworth, A scale of grade and class terms for clastic sediments. J. Geol. **30**, 377–392 (1922)

6. J.E. Brower, J.H. Zar, C.N.V. Ende, Field and Laboratory Methods for General Ecology (Wm. C. Brown Publisher, Dubuque, 1990)
7. C.J. Krebs, Ecology: The Experimental Analysis of Distribution and Abundance (Harper and Row, New York, 1989)
8. M.C. Dash, S.P. Dash, Fundamentals of Ecology, 3rd ed. (Tata McGraw-Hill Education, New Delhi, 2009)
9. [PPRI] Pemerintah Republik Indonesia, Peraturan Pemerintah Republik Indonesia Nomor 22 Tahun 2021 tentang Penyelenggaraan Perlindungan dan Pengelolaan Lingkungan Hidup (Jakarta, 2021)
10. [KLH] Kementerian Lingkungan Hidup, Keputusan Menteri Negara Lingkungan Hidup Nomor 51 Tahun 2004 tentang Baku Mutu Air Laut (Jakarta, 2004)
11. B.C. Bayudana, I. Riyantini, Sunarto, S. Zallesa, Asosiasi dan korelasi makrozoobentos dengan kondisi ekosistem mangrove di Pulau Pari, Kepulauan Seribu. Bull. Mar. Oceanogr. **11**, 1–10 (2022)
12. M.S. Nadaa, S.P.J.N. Taufiq, S. Redjeki, Kondisi makrozoobentos (Gastropoda dan Bivalvia) pada ekosistem mangrove Pulau Pari, Kepulauan Seribu, Jakarta. Bull. Mar. Oceanogr. **10**, 33–41 (2021). https://doi.org/10.14710/buloma.v10i1.26095
13. I.K.D. Ariawan, I.G.B.S. Dharma, E. Faiqoh, Struktur komunitas makrozoobentos di ekosistem mangrove Pulau Serangan Bali. J. Mar. Aquat. Sci. **7**, 224–231 (2021)
14. S.L. Undap, H. Pangkey, N.P.L. Pangemanan, Analisis fisika-kimia kualitas air perairan Bahoi Kecamatan Likupang Barat, Sulawesi Utara. Budidaya Perairan. **6**, 38–44 (2018)
15. A. Mustofa, Pengaruh total padatan tersuspensi terhadap biodiversitas makrozoobentos di Pantai Telukawur Kabupaten Jepara. J. Disprotek. **9**, 37–45 (2018). https://doi.org/10.34001/jdpt.v9i1.656

BIO Web of Conferences **221**, 03007 (2026) https://doi.org/10.1051/bioconf/202622103007
7th EMBRIO International Symposium (EIS 2025)

Reproductive biology of the mud crab (*Scylla serrata*) in the mangrove area of Teluk Bogam Village, Central Kalimantan

Resti Putri Andria[1], *Zairion* Zairion[1], *Sulistiono* Sulistiono[1*], *Intan* Rabiyanti[1], *Nurfajar* Nurfajar[1], *Thomas* Nugroho[2], *Gondo* Puspito[2], *Fadly Y* Tantu[3]

[1]Department of Aquatic Resources Management, Faculty of Fisheries and Marine Science, Bogor Agricultural University, Bogor, 16680, West Java, Indonesia
[2]Department of Fisheries Resources Utilization, Faculty of Fisheries and Marine Science, Bogor Agricultural University, Bogor, 16680, West Java, Indonesia
[3]Aquatic Resources Study Program, Faculty of Animal Husbandry and Fisheries, Tadulako University, Palu, 94119, Central Sulawesi, Indonesia

Abstract. The mud crab (*Scylla serrata*) is one of the fisheries resources in Teluk Bogam Village Central Kalimantan, and is utilized by the local community as a livelihood; however, information on its biology is very limited. This study aims to analyze the reproductive biology of the mud crab (*Scylla serrata*) in the mangrove area of Teluk Bogam Village as a basis for fisheries management. The study was conducted from August 2024 to January 2025, with a total of 215 individuals (93 males and 122 females), caught by fishermen using folding traps. The parameters observed included length frequency distribution, carapace width and weight relationships, condition factors, sex ratio, gonad maturity stages (GMS), first gonadal maturity size, gonado-somatic index, fecundity, and oocyte diameter. The study results showed a sex ratio of 1: 0.76 for mud crabs, with a dominance of females. The peak spawning season for mud crabs is estimated to occur in November. The size at first maturity for males was 113,13 mm, while for females was 116,67 mm. Fecundity ranged from 820,858 to 2,419,923 oocytes. The spawning type of mud crabs is a total spawner, meaning all fertilized oocytes are released simultaneously.

1. Introduction

Teluk Bogam Village is one of the villages located in Kumai Subdistrict, Kotawaringin Barat Regency, Central Kalimantan. This area has significant natural resource potential, particularly in the fisheries sector, especially mud crabs (*Scylla serrata*). This commodity is utilized by the local community as one of their primary sources of income. *S. Serrata* has the widest distribution compared to other mud crab species. This species is widely distributed throughout the western Indo-Pacific region and is closely associated with mangrove ecosystems. *S. serrata* makes a significant contribution to both industrial and traditional fisheries across its range, with catch and landing levels increasing significantly over the past few decades. This indicates that *S. serrata* is not only ecologically important but also economically important [1].

The high potential presence of *S. serrata* in mangrove ecosystems and its economic value make this species an important resource that needs to be conserved. High pressure on

* Corresponding author: onosulistiono@gmail.com

mangrove ecosystem degradation due to human activities, such as residential development and fishing activities in the waters of Bogam Bay, can lead to a decline in habitat quality and threaten the *S. serrata* population. Additionally, uncontrolled fishing without proper fisheries management also harms the sustainability of this species.

One important factor influencing the sustainability of mud crabs is reproduction. Reproduction plays an important role in sustainable management, particularly in developing catch strategies to maintain the availability of exploited populations and evaluating the effectiveness of fisheries regulations [2]. To date, the biological aspects of crab reproduction in the mangrove areas of Teluk Bogam Village have not been studied. Therefore, it is important to research the biological aspects of reproduction in this region. This study aims to analyze the reproductive biology of mud crabs (*S. Serrata*) in the mangrove areas of Teluk Bogam Village as a basis for effective fisheries management.

2 Research methods

2.1. Time and location

Crab sampling was conducted monthly, for 6 months (August 2024 - January 2025) in Teluk Bogam Village. Sample analysis was conducted at the Macro Biology Laboratory, Department of Aquatic Resource Management, Faculty of Fisheries and Marine Science, Bogor Agricultural University.

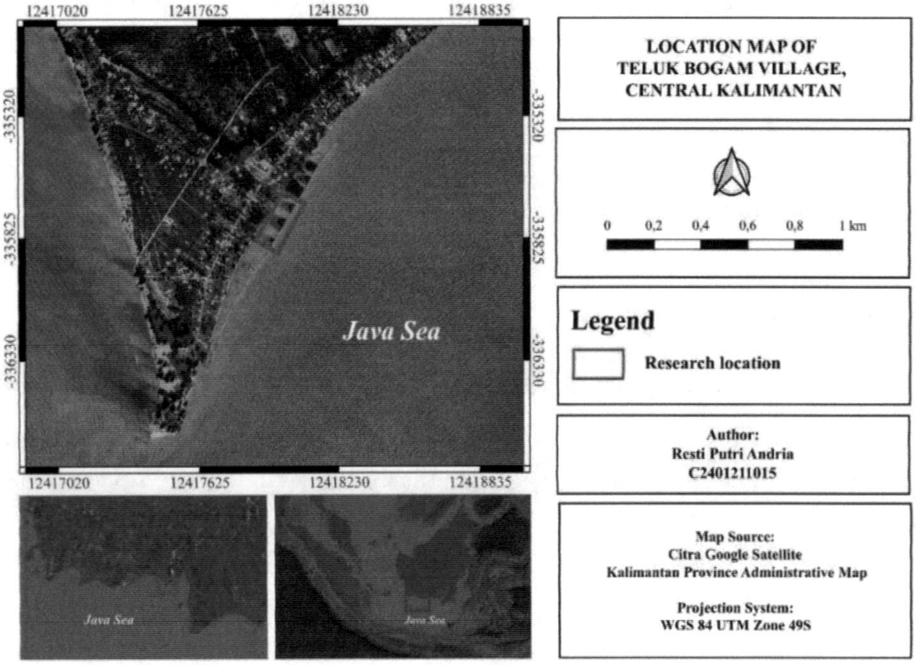

Fig. 1. Location of the mud crab (*S. serrata*) research in the mangrove area of Teluk Bogam Village, Kumai District, Kotawaringin Barat Regency, Central Kalimantan.

2.2. Data collection

The data used in this study is primary data. The sample was obtained from the catch of fishermen in Teluk Bogam Village using folding crab traps, amounting to 215 crabs (122 females and 93 males). Weighing was performed using a digital scale (accuracy 1 gram). Observation of crab gonads was conducted by dissecting the crab carapace and classifying the gonadal maturity level based on [3]. The calculation of oocyte number (fertility) was performed using a combined gravimetric and volumetric method on females that had reached maturity and had gonads that were used. Oocyte diameter measurements were performed using a composite microscope equipped with a micrometer and connected to a computer with 4x10 magnification, assisted by the *Beta View* 3.7 application.

2.3. Data analysis

2.3.1. Carapace width distribution frequency

The frequency distribution of carapace width is determined through class intervals, class midpoints, and frequencies in each carapace width group. The frequency distribution of carapace width is presented in a bar graph between carapace width and crab weight, which is used to observe the normal distribution. According to [4], the frequency distribution of carapace width can be calculated using the following formula:

$$\text{Number of classes} = 1 + 3{,}32 \text{ Log n} \tag{1}$$

$$\text{Class Width} = \frac{(Max - Min)}{Number\ of\ classes} \tag{2}$$

$$\text{Class distance} = \text{Class width} + \text{nst} \tag{3}$$

Description:
n : Number of data observed
nst : Smallest unit value

2.3.2. Sex ratio

Sex ratio was calculated by comparing the number of male crabs and the number of female crabs using the formula of [4]:

$$\text{Sex ratio} = \frac{Number\ of\ male\ crabs}{Number\ of\ famale\ crabs} \tag{4}$$

The relationship between male and female crabs in determining the similarities or differences in portions between male and female mud crabs can be known by analyzing the sex ratio using the *Chi-square* test, using the following formula:

$$X^2 = \frac{\Sigma(o_i - e_i)x^2}{o_i} \tag{5}$$

Description:
x^2 count : *Chi-square* test
o_i : Observed frequency of male and female mud crabs
e_i : Expected frequencies of male and female mud crabs

2.3.2. Gonad maturity stage (GMS)

The level of maturity of crab gonads was determined by comparing several factors, such as color, size, and volume of gonads and egg grains, and looking at changes in gonad morphology based on the classification of gonad maturity levels [3].

2.3.3. Size at first gonad maturity stage (CW_{m50})

The value of the size of the first maturing gonads was determined at a percentage of 50% using the [5] equation:

$$P = \frac{1}{1+e^{[-r(CW-CWm)]}} \quad (6)$$

Description:

P : Proportion of gonadally mature crabs in each carapace width class interval.
CW_m : A measure of the width of the carapace of gonadally mature crabs.
r : *Slope* of the curve
a : Intercept

2.3.4 Gonado-somatic maturity (GSI)

The gonado-somatic index is the relative ratio between the weight of the gonads and the body weight of the crab. The gonadal maturity index value is determined using formula [4]:

$$GSI = \frac{\text{Gonad weight}}{\text{Crab body weight}} \times 100 \quad (7)$$

2.3.5 Fecundity

Fecundity values can be used to determine the number of oocytes in the gonads of mature and spent crabs. Calculation of the number of oocytes was done using a combined method of gravimetry, volumetry, and counting. The following formula is used to calculate fecundity [4]:

$$F = \frac{G\ x\ VxX}{Q} \quad (8)$$

Description

F : Fecundity
G : Total gonad weight (g)
V : Dilution content (cc)
X : Number of oocytes per cc
Q : Weight of sample egg (gram)

2.3.6. Egg diameter

Egg diameter measurements were taken under a compound microscope with the aid of a micrometer connected to a computer. The results of the analysis were presented in the form of a histogram to show the distribution of egg sizes.

3 Results and discussion

3.1 Results

3.1.1 Frequency distribution of carapace width

Carapace width frequency distribution is the distribution of the frequency of individuals or the number of individuals in various size groups of a particular carapace width. Frequency distribution diagram of carapace width of mud crabs from August 2024 to January 2025. The frequency distribution of male mud crab length was mostly found in the class interval 102.25-109.25 mm (with a total of 32 individuals), while in the class interval 134.25-141.25 mm, no male mud crabs were found. The frequency distribution of the length of female mud crabs was mostly found in the 102.25-109.25 mm class interval (with a total of 25 individuals), while the lowest frequency distribution of female mud crab length was found in the 70.25-77.25 mm and 134.25-141.25 mm class intervals, with a total of 2 individuals.

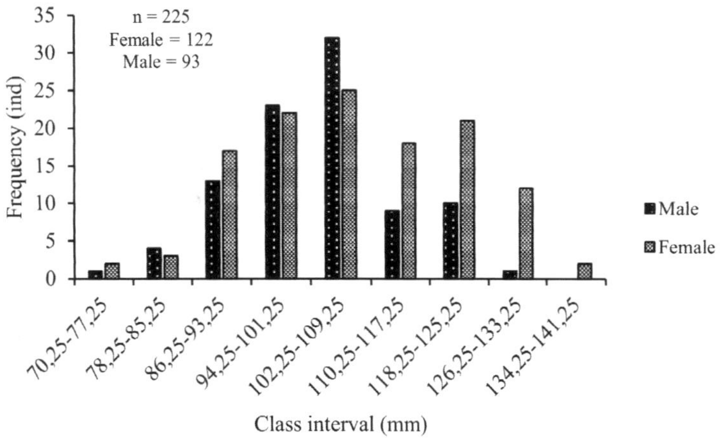

Fig. 2. Frequency distribution of the carapace width of mangrove crabs (*S. serrata*) in Teluk Bogam Village.

3.1.2 Sex ratio

The sex ratio is a parameter used to determine the balance between the male and female sexes in the population. The sex ratio of mud crabs during the observation period from August to January is presented in Figure 2. The sex ratio of male to female mud crabs (*S. serrata*) based on Figure 2 shows the highest value in January, with a value of 1.47, and the lowest sex ratio in October, with a value of 0.29. Based on the chi-square test, the total sex ratio of mud crabs in Teluk Bogam Village is 1: 0.76 with a predominance of female crabs. The sex ratio of mud crabs (*S. serrata*) based on shows that the dominant size of female mud crabs is found in the 86.25-93.25 mm size class, while male mud crabs are found in the 78.25-85.25 mm size class.

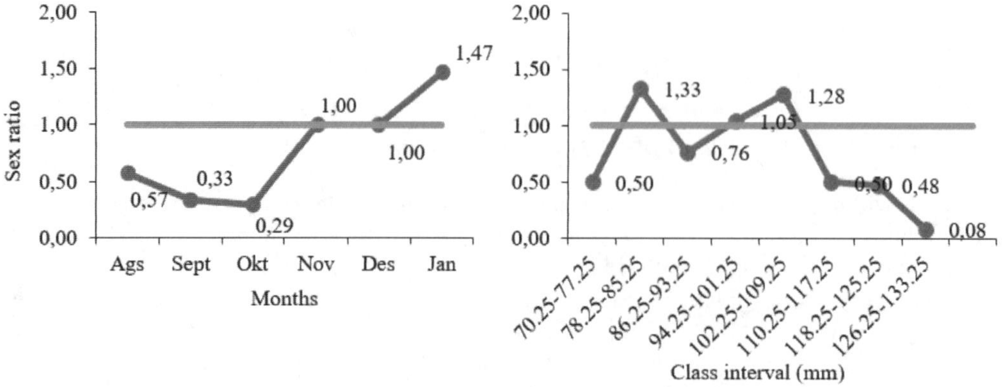

Fig. 3. Sex ratio of mud crabs (*S. serrata*) based on observation time and size in Teluk Bogam Village, Central Kalimantan.

3.1.3 Gonad maturity stage (GMS)

Determination of the level of gonadal maturity in reproduction is used to determine whether an individual has matured gonads with immature gonads in stock populations in the waters [4]. Based on morphological observations, the level of gonadal maturity in mud crabs is classified into five. The results of morphological identification of mud crab GMS during observations from August to January are presented in Figure 3.

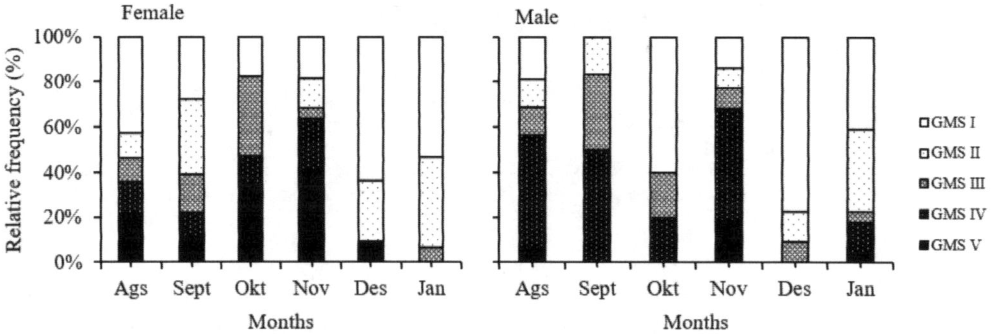

Fig. 4. Gonad maturity stage of mud crab (*S. serrata*) in Teluk Bogam Village, Central Kalimantan.

The morphological maturity level of mud crab (*S. serrata*) gonads during the observation period from August to January, as shown in Figure 3, indicates that most of the mud crabs caught had not yet reached gonadal maturity. The percentage of female mud crabs that were not yet gonadally mature was 68.9%, while that of male mud crabs was 65.6%. Meanwhile, mature mud crabs accounted for 31.1% of female mud crabs and 34.4% of male mud crabs, with the highest gonadal maturity occurring in November.

3.1.4 First gonadal maturity size (CWm50)

The first gonadal maturity size is the size at which 50% of the observed species population has reached gonadal maturity, denoted by the notation L50 or Lm (CWm50 for crabs). The first gonadal maturity size of female and male mud crabs is presented in Figure 4.

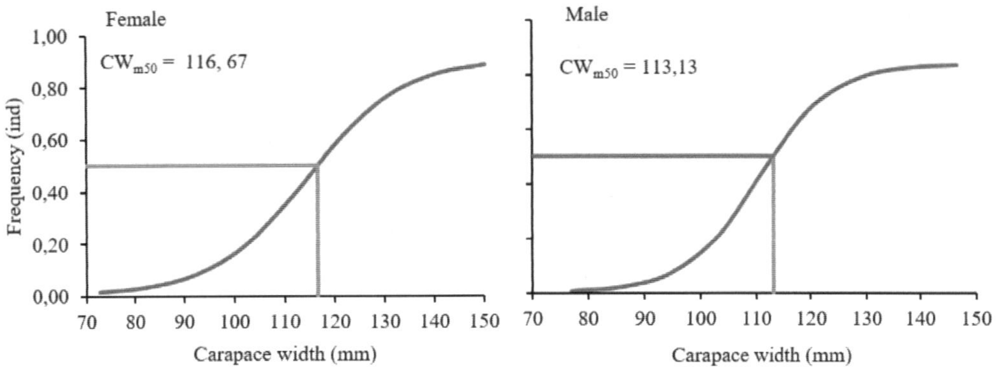

Fig. 5. Size of gonads at first maturity in mud crabs (*S. serrata*) in Teluk Bogam Village, Central Kalimantan.

The results of the analysis (CWm50) in Figure 10 show that the first gonadal maturity value for female mud crabs is found at a carapace width of 116.67 mm. Meanwhile, the first gonadal maturity value for male mud crabs is found at a carapace width of 113.13 mm.

3.1.5 Gonado-somatic index (GSI)

The gonad maturity index can be used to determine the reproductive ability of crabs, as well as a reference to predict crabs that are ready to spawn and show the percentage of weight in producing mature gonad. The percentage of gonad maturity level will increase or reach the maximum limit at the time of spawning. After the spawning process is complete, the level of gonad maturity will decrease again. The results of the gonado-somatic index (GSI) every month during the observation are presented in Figure 5.

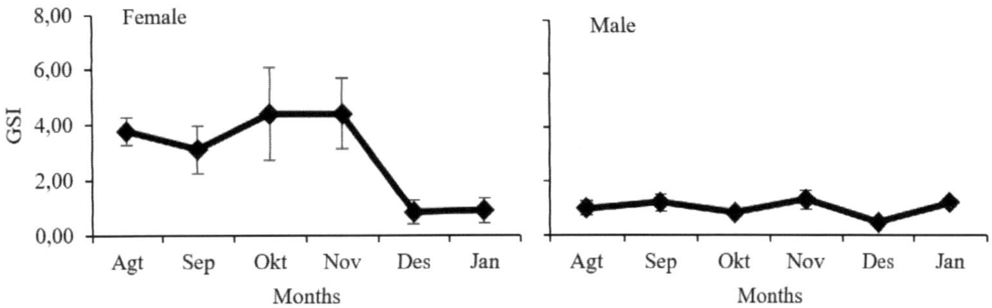

Fig. 6. Gonad maturity index of mud crab (*S. serrata*) in Teluk Bogam Village, Central Kalimantan.

The results of the gonadal maturity index analysis based on Figure 5 show that female mud crabs have a higher average GSI value than male mud crabs. The highest GSI values for both female and male mud crabs were recorded in November, with values of 4.4166±1.2270 for female mud crabs and 1.3048±0.3505 for male mud crabs. The lowest GSI values for both female and male mud crabs were recorded in December, with values of 0.8330±0.4421 for female mud crabs and 0.4586±0.1428 for male mud crabs.

3.1.6 Fecundity

Fecundity in reproductive analysis is used to predict the number of individuals that will be produced and to determine the number of individuals in a particular age class [4]. The fecundity observed in female mud crabs is that of female mud crabs that have reached gonadal maturity. The fecundity value of mud crabs ranged from 820,858 to 2,419,923 oocytes. The mud crabs used were gonadally mature females with a carapace width of 86.25-136.25 mm (Figure 6).

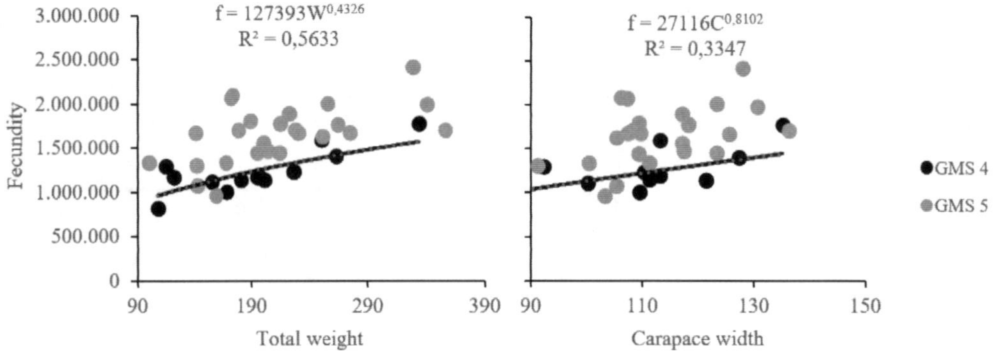

Fig. 7. Relationship of fecundity with carapace width and weight of mud crab (*S. serrata*) in Teluk Bogam Village, Central Kalimantan.

The relationship between fecundity and body weight and carapace width, based on Figure 6, shows a positive correlation, where the greater the weight and carapace width, the greater the number of oocytes produced. The relationship between fecundity and total weight is shown by the equation $F = 127393W^{0.4326}$ with a coefficient of determination value of $R^2 = 0.5633$. The relationship between fecundity and carapace width is shown by the equation $F = 24689C^{0.3347}$, with a coefficient of determination value of $R^2 = 0.3347$.

3.1.7 Oocyte diameter

The diameter of an egg is the centre line or length of the egg, measured using a calibrated micrometer on a microscope. The diameter of oocytes in the ovary increases as the gonads mature [4]. The distribution of mud crab egg diameter is presented in Figure 7. The results of egg diameter measurements conducted on female mud crabs based on Figure 7 obtained an average diameter of 0.09 ± 0.01 mm for GMS I, 0.13 ± 0.02 mm for GMS II, 0.19 ± 0.02 mm for GMS III, 0.33 ± 0.02 mm for GMS IV, and 0.36 ± 0.02 mm for GMS V.

3.2 Discussion

The frequency distribution of carapace width can be used to group carapace sizes from smallest to largest, which is useful in determining age groups. Additionally, this frequency distribution is used to observe shifts in carapace size, which reflect the number of age groups (cohorts) present in a population [6]. Crab length is measured based on carapace width, because during growth, carapace width increases more significantly than length.

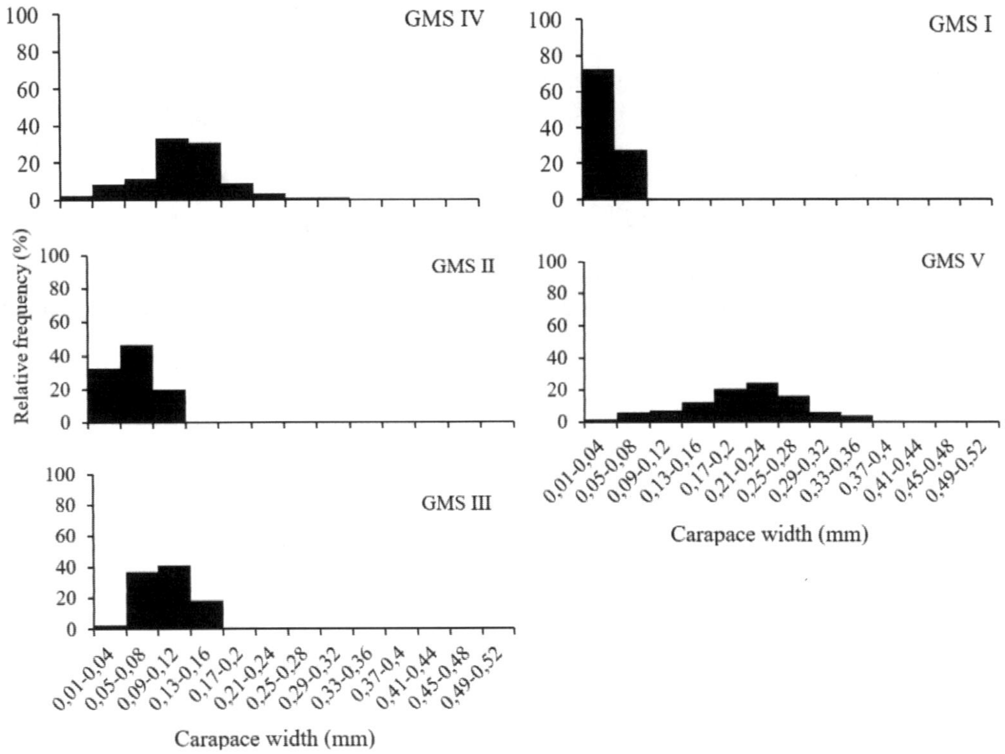

Fig. 8. Oocyte diameter of mud crab (*S. serrata*) oocytes in Teluk Bogam Village, Central Kalimantan.

The results of the carapace width frequency distribution based on Figure 3 show that the peak frequency of mud crabs is at 102.25-109.25 mm for both males and females. Most of the crabs caught in this study still do not meet the minimum catch size (Minimum legal size) as regulated in Ministry of Marine Affairs and Fisheries Regulation No. 7 of 2024 on the Management of Lobsters (*Panulirus* spp.), crabs (*Scylla* spp.), and mud crabs (*Portunus* spp.), which stipulates that the minimum legal size for crabs is a carapace width of over 12 cm or 120 mm per individual. According to [7], the carapace width of mud crabs (*S. serrata*) is classified into three phases: juvenile, young, and adult. In the juvenile phase, the carapace width is <70 mm, in the young phase it is 70–120 mm, and in the adult phase it is >120 mm.

The frequency distribution results show that the mud crabs caught in Teluk Bogam Village are dominated by young crabs, accounting for 89% of females and 98% of males. Adult crabs caught consisted of 11% females and 2% males, while juvenile crabs were not found. This finding suggests that the fishing gear used by fishermen remains insufficiently selective, potentially disrupting recruitment processes and contributing to overfishing or overexploitation. This condition is related to the developmental phase at the megalopa stage, where crabs begin to migrate to the muddy seabed and then move towards the coast. This process begins with movement towards the river mouth, which is then continued towards the mud ecosystem to return to the mating process [8].

In comparison, research conducted by [7] shows that crab catches using bubu traps in Lawele Bay, Southeast Sulawesi, consisted of 9.5% juvenile crabs, 90.5% adult crabs, and no juveniles. These results indicate that the bubu fishing gear used to catch mudcrabs in Lawele Bay is highly selective. Conversely, the fishing gear used in Teluk Bogam Village is less selective and therefore needs to be designed or modified to be able to catch crabs that have reached adulthood.

In addition to the use of less selective fishing gear, another possible reason for the crab catch in Teluk Bogam Village not meeting the requirements of Ministry of Marine Affairs and Fisheries Regulation No. 7 of 2024 and the classification of carapace width is that the average size of adult mud crabs in Teluk Bogam Village does not reach the minimum size permitted for capture according to current regulations. This condition is due to the provisions in the Ministry of Marine Affairs and Fisheries Regulation No. 7 of 2024 regarding the minimum size of mud crabs that can be caught, which is universal across all locations and for all types of crabs. However, the size and biological characteristics of crabs vary by location due to differences in environmental conditions.

The distribution of mud crabs in the waters is influenced by several factors, such as oceanography, physical and chemical conditions of the water, food availability, behaviour and life cycle. The distribution of crab carapace width is influenced by several factors, including environmental conditions and exploitation levels. A supportive environment accelerates mud crab growth, and low exploitation levels provide opportunities for crabs to reach optimal size. In addition, crab feeding habits, such as their response to bait on fishing gear and irregular feeding patterns, also influence size distribution. Crabs prefer to hide in holes to avoid predators and engage in burrowing activities during their life cycle, which also contributes to carapace size variation [9].

The sex ratio of mud crabs in Teluk Bogam Village is unbalanced, at 1: 0.76, with a predominance of female crabs. These results are in line with the findings of a study conducted by [10], which found an imbalance in the sex ratio of male and female crabs, with female crabs dominating. This imbalance is thought to be related to food availability and the life cycle of mud crabs, especially during the reproductive period. In contrast to the results of [11], male crabs were more dominant than females. The difference in sex ratio is suspected to be influenced by seasonal variations, location, and crab body size.

The sex ratio composition tends to change according to the spawning season [10]. In this study, the spawning season occurred in November, when crabs began to migrate to the sea to spawn. Before migrating, mud crabs usually settle first in mangrove or coastal areas to forage, mate, and accelerate gonadal maturation. This is indicated by the dominance of female crabs before November. Female mudcrabs generally seek water bodies with suitable conditions for the spawning process, particularly regarding water temperature and salinity parameters. According to [12], the reproductive season or spawning season for mud crabs occurs throughout the year, but the peak varies in each water body. According to [13], the level of gonadal maturity in crabs is influenced by environmental factors. The main factors affecting gonadal development include temperature, food, light period, and season. During the reproductive phase, mud crabs require sufficient nutrient intake, both in terms of quality and quantity.

The first gonadal maturity size in a species occurs at a certain age, when available energy is no longer used for body metabolism but is instead focused on gonadal maturation. The first gonadal maturation size (CWm50) in male mud crabs is 113.13 mm, smaller than that of female mud crabs, which have a first gonadal maturation size of 116.67 mm. These results are consistent with the research by [10], which showed that the first gonadal maturity size in male mud crabs is smaller than that of female crabs, with a size of 80.13 mm for males and 82.54 mm for females. Each species has a different first gonadal maturity size, and even the same species can experience size differences depending on environmental conditions and geographical location. Species of the same size and age do not always reach gonadal maturity at the same age and length.

Fecundity is one of the important parameters that needs to be studied in reproductive biology analysis, which refers to the number of oocytes produced by female mud crabs to determine their reproductive potential. Mud crabs that have reached gonadal maturity have varying levels of fecundity. This difference is closely related to the size of the female parent, where the number of oocytes increases as the parent's body size increases. Fertility increases with body weight and carapace width, reaching its peak at a certain size and age. This increase is influenced by the fertility cycle within the organism's body. Additionally, fertility is influenced by species, age, individual size, diet, and living environment [4]. The results of this study are in line with the findings of [14], with fertility ranging from 2,116,514 ± 505,669 oocytes.

Egg diameter increases with gonadal maturity. Egg size varies between species, possibly due to differences in morphological characteristics. Additionally, it is known that mud crab diameter ranges from 0.36 ± 0.02 mm. Based on these measurements, it is known that mud crabs have a total spawning type, meaning all fertilised oocytes are released at once. This finding is consistent with the results of a study by [15]. According to [4], as the gonads develop, the egg diameter also increases.

The lack of information on the reproductive biology of crabs (*S. serrata*) in Teluk Bogam Village, combined with increasing exploitation intensity, necessitates appropriate management efforts to ensure the crab population remains stable. Based on this study, one form of management that can be applied is seasonal management, particularly by reducing catches during the spawning season in November. This aims to provide mud crabs with the opportunity to reproduce optimally. Another effort that can be made is to avoid catching crabs below the first gonadal maturity size, which is 116.67 mm. Catching crabs below this size prevents them from having the opportunity to lay oocytes or hatch their oocytes. Additionally, the implementation of environmentally friendly aquaculture practices, such as silvofishery, can also be a solution in maintaining the sustainability of mud crabs. Silvofishery is an aquaculture activity conducted alongside mangrove ecosystem conservation. This activity creates a stable environment that supports the natural reproduction of mud crabs.

4 Conclusion and recommendation

4.1 Conclusion

Mud crabs (*Scylla serrata*) in the mangrove area of Teluk Bogam Village still do not meet the minimum catch size and show an unbalanced sex ratio with a predominance of females. The peak spawning period for female and male mud crabs is estimated to occur in November, with fecundity ranging from 820,858 to 2,419,923 oocytes, and the spawning type classified as total spawner.

4.2 Recommendation

Further research is needed on the reproductive biology of mud crabs (*Scylla serrata*) in the mangrove area of Teluk Bogam Village, with a larger sample size and a longer observation period (one year), to obtain more representative results and determine the spawning frequency over a year. Additionally, research on the gonads of crabs carrying oocytes is needed to confirm that the crabs are total spawners, enabling the development of sustainable mud crab management strategies based on a thorough understanding of reproductive biology and optimal management approaches.

References

1. H. A. Hubatsch, S.Y. Lee, J.O. Meynecke, K. Diele, I Nordhaus, M Wolff, Life-history, movement, and habitat use of *Scylla serrata* (Decapoda, Portunidae): Current knowledge and future challenges. Journal of Hydrobiologia. **763**, 1 (2015).
2. W. Kantun, S. Wulandari, H. Anggreni, Nisbah kelamin dan ukuran pertama kali matang gonad kepiting bakau, *Scylla serrata* (Forskal, 1775) di Perairan Sungai Sanrangang Kabupaten Takalar. Bawal Widya Riset Perikanan Tangkap. **14,** 2 (2022).
3. M. S. Islam, K. Kodama, H. Kurokura, Ovarian development of the mud crab *Scylla paramamosain* in a Tropical Mangrove Swamps, Thailand. Journal of Scientific Research. **2**, 2 (2010).
4. M. I. Effendie, Metoda Biologi Perikanan (Yayasan Dewi Sri, Bogor,1979).
5. M. King, Fisheries Biology: Assessment and Management. Second Edition (Blackwell Science Ltd Oxford, 2007).
6. M. I. Effendie, Biologi perikanan (Yayasan Pustaka Nusatama, Yogyakarta, 2006).
7. L. Sara, Study on the size structure and population parameters of mud crab *Scylla serrata* in Lawele Bay, Southeast Sulawesi, Indonesia. Journal of Coastal Development. **13**, 2 (2010).
8. Sunarto, Sulistiono, I. Setyobudiandi, Hubungan jenis kepiting bakau (*Scylla* Spp.) dengan mangrove dan substrat di Tambak Silvofishery Eretan, Indramayu. Marine Fisheries. **6**, 1 (2015).
9. Sulistiono, N. M. Yahya, E. Riani, Distribusi *Scylla* spp. di perairan estuari Sungai Donan Segara Anakan Bagian Timur, Cilacap. Habitus Aquatica J. **2**, 1 (2021).
10. E. Tiurlan, A. Djunaedi, E. Supriyantini, Analisis aspek reproduksi kepiting bakau (*Scylla* sp.) di Perairan Kendal, Jawa Tengah. Journal of Tropical Marine Science. **2**, 1 (2019).
11. I. P. Sanur, Sulistiono, Yonvitner, D. M. Wildan, A. Ervinia, Pendugaan Pertumbuhan Kepiting Bakau (*Scylla serrata* Forskal) di Perairan Karangsong, Indramayu, Provinsi Jawa Barat. Jurnal Teknologi Perikanan Dan Kelautan. **12**, 1 (2021).
12. K. Adnam, Budidaya Kepiting Bakau Dan Biologi Ringkas. Jakarta Bhatara 1991, Jakarta, (1996).
13. I. S. B. Ramandey, S. Mulyani, N. A. Umar, Sebaran ukuran panjang berat dan tingkat kematangan gonad kepiting bakau *Scylla* sp. di Perairan Teluk Youtefa Jayapura. Journal of Aquaculture and Environment. **6,** 1 (2023).
14. S. Koolkalya, U. Matchakuea, T. Jutagate. Comparison of catch sizes and reproductive biology of mud crab species of the genus *Scylla*, coexisting populations in the eastern Gulf of Thailand. International Journal of Agricultural Technology. **12**, 7 (2016).
15. H Asmara, E Riani, A Susanto, Analisis beberapa aspek reproduksi kepiting bakau (*Scylla serrata*) di perairan Segara Anakan, Kabupaten Cilacap, Jawa Tengah. Jurnal Matematika Sains dan Teknologi. **12**, 1 2011.

Fish assemblage patterns influenced by nutrient levels and thermal effluent from power plants in the lower reaches of Berung and Wadas Rivers, Serang Regency, Banten, Indonesia

Muhammad Irfan Afif[1]*, *Firsta Kusuma* Yudha[1], and *Intan* Rabiyanti[1]

[1] Department of Aquatic Resources Management, Faculty of Fisheries and Marine Sciences, IPB University, Bogor, West Java 16680, Indonesia

Abstract. The downstream of the Berung and Wadas Rivers in Serang Regency, Banten Province, Indonesia, possesses economically significant fishery resources. However, these waters face environmental threats, notably thermal effluent from a Steam Power Plant. This study aimed to analyze fish species diversity and habitat conditions by examining the relationship between fish assemblages and key physicochemical factors, specifically temperature and nutrient levels. The study was conducted in January 2020 at six sampling stations representing riverine, estuarine, and marine zones. Fish samples were collected using gillnets, while physicochemical parameters (including temperature, nutrients, DO, salinity, and pH) were measured concurrently. Fish diversity was evaluated using the Shannon index, and the relationship between fish assemblages and environmental variables was analyzed using Principal Component Analysis (PCA) and Canonical Correspondence Analysis (CCA). A total of 63 individual fishes, comprising 11 species from 10 families, were recorded. The squaretail mullet (*Ellochelon vaigiensis*) was the dominant species (67%). Fish diversity varied significantly across stations. Stations 1, 2, and 6, located in productive estuarine areas, exhibited higher fish diversity. Conversely, Station 5, situated near the power plant's thermal discharge, recorded elevated temperatures (37.5°C) and low fish diversity, indicating negative impacts from thermal pollution. The findings confirm that the thermal effluent acts as a pollutant, creating environmental instability for the aquatic biota. Therefore, improved waste management strategies are necessary to mitigate the ecological impacts in this area.

1 Introduction

The waters surrounding Bojonegara, especially in the downstream area of Berung and Wadas Rivers, Serang Regency, Banten Province, Indonesia, possess economically significant fishery resources. These resources are exploited not only by local fishermen but also by the broader Banten community, as evidenced by the Karangantu Fish Auction Place. The sustainability of fish stocks in the downstream area of the Berung and Wadas River is inextricably linked to the environmental quality of the marine habitat.

An ecobiological approach can be employed to analyze the relationship between aquatic environmental conditions and the status of fishery resources. This approach examines the environmental interconnections (abiotic-biotic interactions) that can either support or hinder

* Corresponding author: irfanafif@apps.ipb.ac.id

the sustainability of fish stocks. Currently, the sustainability of these fishery resources is under considerable pressure from anthropogenic activities. A demonstrated impact of such pressures is the decline of fish resources in Indonesia's coastal waters due to environmental degradation and pollution [1].

The coastal area is currently undergoing construction and development for a steam power plant, leading to its continuous transformation through new constructions, including land reclamation, port facilities, and jetty bridges. During electricity generation, the steam power plant utilizes water in its cooling system, subsequently discharging heated effluent into the marine environment. This process involves releasing large volumes of thermal effluent into the surrounding waters, leading to an increase in sea surface temperature [2]. Elevated water temperatures accelerate biological and chemical processes and reduce the solubility of dissolved oxygen in the water. These alterations can profoundly affect the fish resources at the downstream parts of the Wadas and Berung Rivers. This is shown by a study of the Tanjung Jati Coal Power Plant in Jepara, Central Java, which has been the subject of environmental monitoring since the early phases of its expansion. The research found a noticeable diversity decline in taxa and species abundance over the study period (January 2021 – December 2022), with species ranging from 79-197 species (19-23 per sample). Critically, the Shannon diversity index declined from 2.67 to 2.43, while dominance indices increased from 0.11 to 0.51, indicating ecological stress and a shift toward species dominance patterns typical of moderately polluted systems.

A study is therefore required to elucidate the diversity and distribution structure of the fish community at the Bojonegara coast, particularly in response to the increased sea surface temperatures caused by the steam power plant. Furthermore, the influence of organic pollution from various riverine inputs on these fish resources will also be assessed. This study aims to investigate the species composition and the relationship between fish assemblages, nutrients, and thermal effluents.

2 Methods

2.1 Time and location

The study was conducted in the downstream area of the Berung and Wadas Rivers, Serang Regency, Banten Province. A single sampling was carried out in January 2020. Samples were collected from six stations, selected based on their distinct aquatic characteristics (Fig. 1). Stations 1, 2 and 3 were for Wadas River, while 6, 5 and 4 were for Berung River. Stations 1 and 6 were located in the riverine zone, Stations 2 and 5 in the estuarine zone, and Stations 3 and 4 in the marine zone. Subsequent sample analysis was performed at the Bio-Macro Laboratory, Department of Aquatic Resources Management, Faculty of Fisheries and Marine Sciences, IPB University.

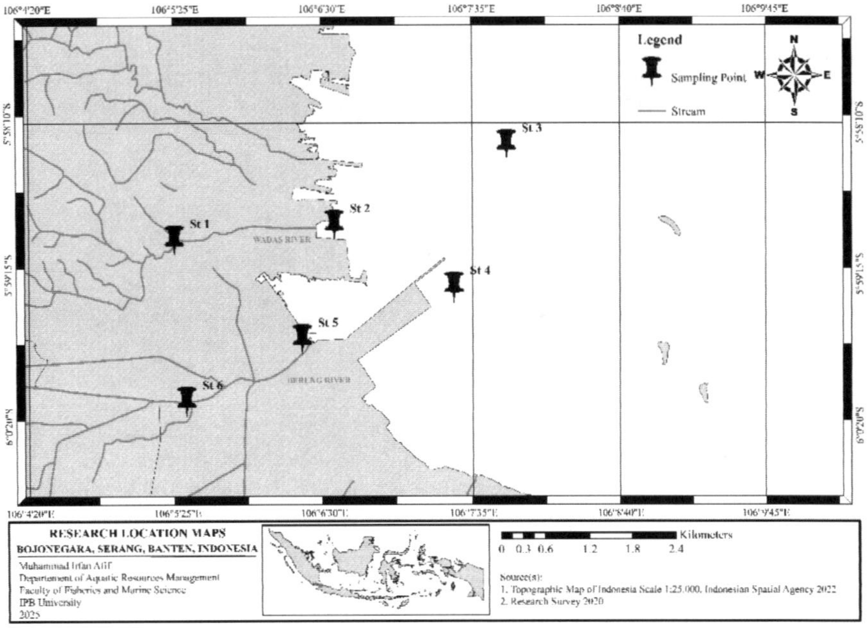

Fig. 1. Map showing the research sites in the Berung and Wadas Rivers and the Bojonegara coastal area.

2.2 Materials

The equipment used for sample collection included a gillnet (1-3 inches), field notebooks, a Global Positioning System (GPS) unit, DO meter, pH meter, refractometer, a measuring tape, bottles for nutrient samples, syringes, and a turbidity meter. A 10% formalin solution was used as a preservative for the collected samples.

2.3 Data collection

Fish samples were obtained from the gill net which was left to soak for about 30 minutes to 1 hour. Fish samples were sorted and preserved with 10% formalin and brought to the Bio-Macro Laboratory for identification. On the other hand, the measurement of water quality was obtained by collecting the physicochemical parameters as presented in Table 1.

Table 1. The physicochemical parameters that have been measured.

Parameters	Units	Principles of analysis
Temperature	°C	instrument
Turbidity	mg/L	laboratory analysis
Total suspended solid (TSS)	mg/L	laboratory analysis
pH		instrument
Salinity	‰	instrument
Dissolved oxygen (DO)	mg/L	instrument
Total Phosphate	mg/L	laboratory analysis
Nitrate	mg/L	laboratory analysis
Nitrite	mg/L	laboratory analysis

2.4 Data analysis

The frequency of occurrence illustrates the temporal distribution of fish throughout the research period and was calculated using the following formula [3]:

$$\left[Fk = \frac{P_i}{P} x\ 100\right] \quad (1)$$

where:

Fk: Frequency of occurrence (%)

Pi: Number of observations for species *i*

P: Total number of observations

The Shannon–Wiener Diversity Index (H′) was used to measure species diversity by considering both species richness (number of species) and evenness (relative abundance among species). The index is calculated as:

$$H' = -\sum_{i=1}^{S} p_i ln\ (p_i) \quad (2)$$

where:

S: total number of species

p_i: proportion of individuals belonging to the *i-th* species ($p_i = \frac{n_i}{N}$)

n_i: number of individuals of species *i*

N: total number of individuals across all species

The relationship between environmental factors and fish abundance was analyzed using Principal Component Analysis (PCA) and Canonical Correspondence Analysis (CCA). The XLSTAT plugin was used in conjunction with Excel to examine the relationship of sampling stations, nutrient parameters, and fish assemblages. The results of these analyses were interpreted and discussed descriptively based on the relevant literature and additional environmental data, such as temperature, TSS, turbidity, and water productivity.

3 Results and discussion

3.1 Fish assemblages

During the sampling period, a total of 11 fish species from 10 families and 5 orders were identified across the six stations. The identified species are presented in Table 2. The lowest diversity was observed at stations 2 and 5. A total of 63 individuals were recorded, with the squaretail mullet (*Ellochelon vaigiensis*) from the family Mugilidae being the dominant species, accounting for 67% of the total individuals. The calculation results of the Shannon-Wiener index, classified according to Hashemi et al. [15], indicated that only station 4 possessed moderate diversity, while the other stations exhibited low diversity. These results reflected a situation where certain points contained a high abundance of fish, albeit belonging to only a single species.

The family Mugilidae was commonly found. This finding is consistent with Huang et al. [4], who stated that this family has one of the widest distributions in estuarine areas. Furthermore, these species tolerate a wide range of salinity, as they can inhabit brackish and marine environments. They are catadromous species that typically form large schools over sand or mud substrates.

Ellochelon vaigiensis and *Scomberoides commersonianus* exhibited the same frequency of occurrence, as both species were present at two of the sampling stations. Frequency of occurrence indicates the extent of a species' temporal distribution among stations, as observed from the frequency of fish caught during the observation period. Although they shared the same frequency of occurrence, their total abundance differed. This difference can be attributed to the schooling behavior of *E. vaigiensis*, whereas *S. commersonianus* tends to be more solitary [5].

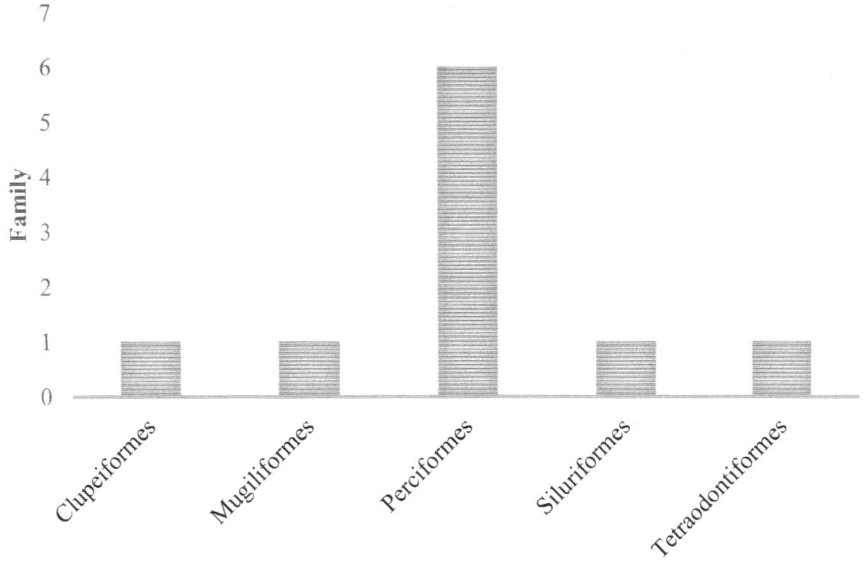

Fig. 2. This family-level classification in the downstream areas of the Wadas and Berung Rivers.

In addition to identifying the most abundant species, it is also necessary to examine the diversity of the catch. Grouping the species by family facilitates the interpretation of diversity and its relationship with environmental conditions. This family-level classification is illustrated in Fig. 2. A closer examination of the data reveals that the order Perciformes contained the highest number of families. The high representation of the order Perciformes aligns with its well-established status as the largest and most taxonomically diverse order of teleosts. This order is a dominant and characteristic feature of virtually all global marine ecosystems.

3.2 Physicochemical parameters of water

The measurement of water physicochemistry parameters yielded distinct results for each station (Table 3). Several parameters, specifically salinity, pH, and Dissolved Oxygen (DO), were sampled twice to assess potential differences between surface and bottom water layers. Additionally, nutrient levels were evaluated through the analysis of nitrate, nitrite, and total phosphate concentrations.

Table 2. Fish distribution in the downstream areas of the Berung and Wadas Rivers

No	Ordo	Family	Species	Station						Total	% Total	FK
				1	2	3	4	5	6			
1.	Clupeiformes	Engraulidae	*Thryssa hamiltoni*			3				3	5%	17%
2.	Mugiliformes	Mugilidae	*Ellochelon vaigiensis*	17					25	42	67%	33%
3.	Perciformes	Anabantidae	*Anabas testudineus*						7	7	11%	17%
4.	Perciformes	Carangidae	*Scomberoides commersonnianus*		1		1			2	3%	33%
5.	Perciformes	Cichlidae	*Oreochromis mossambicus*						3	3	5%	17%
6.	Perciformes	Cichlidae	*Oreochromis niloticus*						1	1	2%	17%
7.	Perciformes	Osphronemidae	*Trichopodus trichopterus*						1	1	2%	17%
8.	Perciformes	Polynemidae	*Eleutheronema tetradactylum*				1			1	2%	17%
9.	Perciformes	Sciaenidae	*Johnius trachycephalus*				1			1	2%	17%
10.	Siluriformes	Ariidae	*Arius arius*				1			1	2%	17%
11.	Tetraodontiformes	Tetraodontidae	*Lagocephalus lunaris*					1		1	2%	17%
Total				17	1	3	4	1	37	63	100%	
Shannon Wiener Index				0	0	0	1.4	0	1			

The chemical water quality varied depending on the sampling location. Compared to the seawater quality standards for marine biota established by the Government Regulation of Republic Indonesia No. 22 of 2021 (PP RI No. 22/2021), several parameters exceeded the permissible limits. These include the surface pH at station 5; surface DO at stations 3 and 5; bottom DO at stations 1, 2, and 3; nitrate concentrations at all stations; and total phosphate concentrations at stations 1, 2, and 6 (Table 3).

The pH value at station 5, which fell outside the standard range, is likely influenced by the thermal discharge from a nearby power plant. According to Menon et al. [6], elevated water temperatures can alter numerous physicochemical parameters, including causing a decrease in pH. Low DO levels were observed near the thermal discharge area and at the bottom of the river mouth. The reduced DO near the discharge area is attributable to the inverse relationship between water temperature and dissolved oxygen solubility. Meanwhile, the low DO at the river mouth bottom is likely due to high concentrations of suspended particles, which increase turbidity and reduce light penetration. This limited light availability inhibits photosynthesis by primary producers [7].

Table 3. The measurement of physicochemical parameters in the downstream area of the Wadas and Berung Rivers. Asteriks denote values exceeding the Indonesian national seawater quality standards for marine biota.

Parameters	**Units**	**Sample location**	**Station**					
			1	**2**	**3**	**4**	**5**	**6**
Salinity	‰	Surface	0	23	24	23	18	0
		Seabed	8	23	24	25	25	0
pH	-	Surface	7.17	7.87	7.88	7.65	6.63*	7.1
		Seabed	7.17	7.8	7.86	7.85	7.78	7.1
DO	mg/l	Surface	6.5	5.1	4*	5	4.3*	5.4
		Seabed	0.3*	4.1*	4.6*	5.6	5.2	5.4
Nitrate	mg/L	Surface	0.867*	0.794*	0.773*	0.762*	0.777*	0.873*
Nitrite	mg/L	Surface	0.128	0.064	0.062	0.065	0.067	0.064
Total phosphate	mg/L	Surface	0.094*	0.072*	< 0.001	< 0.001	< 0.001	0.321*
Temperature	°C	Surface	28.4	30.4	30.4	34.5	37.5	29.5
Turbidity	NTU	Surface	111	28.0	8.6	5.5	10.6	380
TSS	mg/L	Surface	57.6	66.4	79.0	21.0	11.0	108

Nitrate concentrations were observed to be relatively low across the sampling sites, with all stations registering values below the 0.1 mg/L threshold typically associated with eutrophic waters [8]. However, spatial variations in nutrient loading were evident from the phosphate data. The highest phosphate concentrations were detected specifically in the river mouth areas (Stations 1 and 6). This finding aligns with Colborne et al. [9], who identified riverine input as a primary source of phosphate in marine environments. This pattern suggests a significant contribution of organic materials derived from terrestrial anthropogenic activities, such as domestic wastewater or agricultural runoff, entering the coastal ecosystem via the rivers.

3.3 Relationship between fish assemblages and water quality

The clustering of the sampling stations provides an overview of the defining parameters for each location. A Principal Component Analysis (PCA) revealed significant correlations between environmental parameters and the stations (Fig. 3), as well as between fish species and the stations (Fig. 4). The resulting PCA biplots clearly illustrate these groupings.

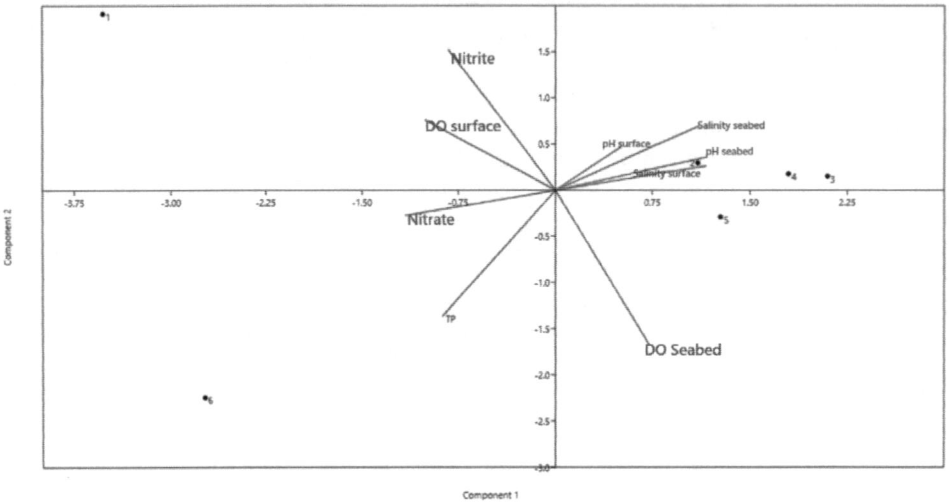

Fig. 3. Biplot diagram of PCA between sampling station points and water quality parameters.

The conditions at stations 2, 3, and 4 were predominantly influenced by salinity and pH (Fig. 3). This is attributed to their location at the confluence of freshwater and seawater, making salinity a dominant environmental factor. The pH of freshwater in an estuarine environment tends to be more acidic, an effect potentially exacerbated by the nearby power plant, which can alter the pH of the receiving waters [10]. In contrast, stations 1 and 6 were more strongly influenced by nutrient concentrations, characteristic of their riverine locations. Rivers typically exhibit high nutrient levels resulting from land erosion and anthropogenic waste discharge. Fig. 3 also suggests an inverse relationship between nutrient concentrations and salinity. This indicates that as the water body extends towards the sea, nutrient levels tend to decrease while salinity increases.

The distribution of fishes across the stations is illustrated in Fig. 4. The analysis shows that several freshwater species, namely *Oreochromis niloticus*, *Trichopodus trichopterus*, *Anabas testudineus*, and *Oreochromis mossambicus*, were primarily found at station 6. This distribution is consistent with the low salinity at this station (Table 3), which provides a suitable habitat for these species. The estuarine fish *Ellochelon vaigiensis* was also found within this group; this species typically inhabits near-shore areas and enters river mouths to forage [6]. A second distinct group consisted of strictly marine species, including *Arius arius*, *Scomberoides commersonnianus*, *Eleutheronema tetradactylum*, and *Johnius trachycephalus* [11]. As expected, these species were absent from the freshwater environments of stations 1 and 6.

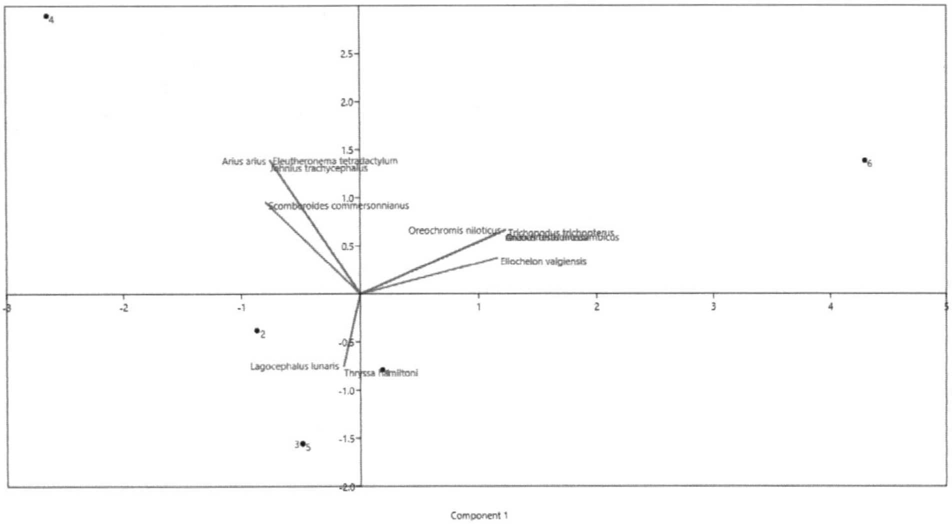

Fig. 4. Biplot diagram of PCA based on fish species and sampling station points.

Canonical Correspondence Analysis (CCA) was employed to visualize the relationships between stations, water chemistry parameters, and fish species. CCA is a direct gradient multivariate method used to explain the relationship between a species community and its environmental variables. It is particularly effective for assessing the influence of environmental parameters on fish abundance. Prior to the analysis, all water chemistry and fish abundance data were log-transformed (log(x+1)) to achieve normality.

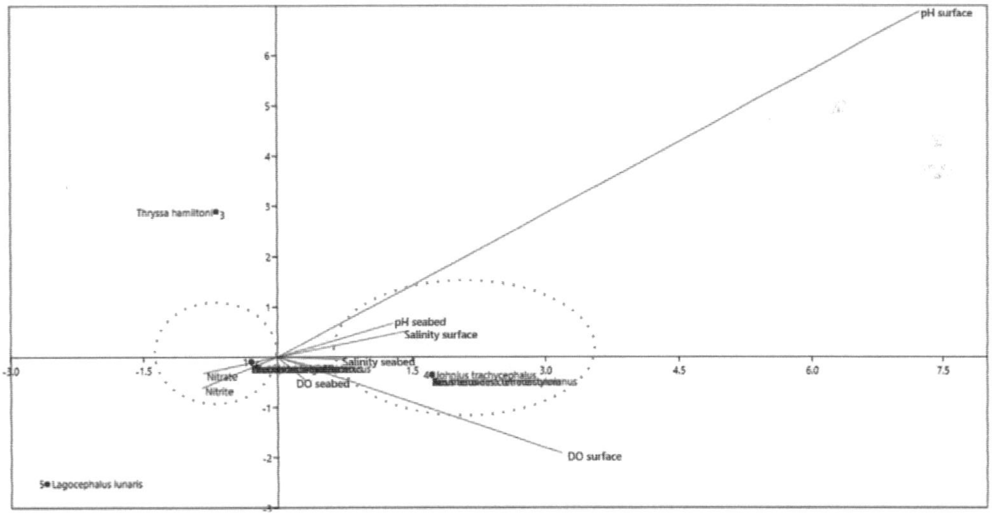

Fig. 5. Biplot diagram of CCA showing a relationship between fish species and water quality parameters.

The CCA biplot revealed two distinct clusters based on the relationships between stations, fish species, and water chemistry (Fig. 5). The first cluster was characterized by a strong association with nitrate and nitrite, which were the defining parameters for station 6. This station served as a congregation point for species such as *Ellochelon vaigiensis*, *Anabas testudineus*, *Oreochromis niloticus*, *Oreochromis mossambicus*, and *Trichopodus trichopterus*. These species likely gather at station 6 for foraging. As omnivores, *A. testudineus*, *O. niloticus*, *O. mossambicus*, and *T. trichopterus* can exploit the abundant food resources available in estuaries [12]. Similarly, *E. vaigiensis* probably utilizes this area to feed on detritus, which is highly available in river mouths.

In contrast, the second cluster was characterized by higher pH, salinity, and dissolved oxygen (DO), which were representative of station 2. This station was predominantly inhabited by *Johnius trachycephalus*, *Scomberoides commersonnianus*, and *Eleutheronema tetradactylum*. The area around station 2 features estuarine waters with extensive mangrove vegetation. Mangrove ecosystems are known for their high productivity and serve as crucial feeding grounds for numerous marine fish species.

Overall, in the downstream areas of both the Wadas and Berung river, the sampling stations are located within an estuarine region that is naturally expected to be highly productive. However, this ecosystem is highly susceptible to disturbances from various anthropogenic activities. This is particularly evident at station 5, the site of thermal effluent discharge from a power plant, which exhibited impoverished fish diversity. This low diversity strongly suggests that the thermal discharge has a significant negative impact on the aquatic biota. Such impacts are well-documented; for instance, research by Michie et al. [13] indicates that thermal effluent can lead to decreased primary productivity, fish recruitment failure, direct fish mortality, and alterations in fish migration patterns.

The results of this study are corroborated by findings from Ronauli et al. [14], which also indicate a relationship between nutrient dynamics, thermal effluent from power plants, and aquatic productivity. The discharge of heated water and high nutrient loads were found to cause phytoplankton blooms, particularly dominated by diatom groups such as *Chaetoceros* sp. and *Thalassiothrix* sp. This condition enhances primary productivity, which in turn supports a higher abundance of zooplankton, theoretically leading to a more active aquatic food web.

However, the dominance of specific phytoplankton species and the suppression of zooplankton diversity due to elevated temperatures suggest an ecosystem imbalance. This ecological pressure could potentially reduce the carrying capacity of the aquatic environment for higher trophic level organisms, including fish, which are highly dependent on a stable food chain. Therefore, although thermal effluent may locally increase primary and secondary productivity, the decline in habitat quality and the observed fluctuations in the plankton community indicate a significant risk of reduced fishery potential at stations located near the power plant's thermal discharge points.

4 Conclusion

The waters surrounding the Berung and Wadas river estuaries in Serang Regency, Banten Province, support high-value economic fishery resources. However, this environment faces threats, such as thermal waste from a steam power plant. In this study, fish diversity varied among the sampling stations, each possessing distinct characteristics. Stations 1, 2, and 6 exhibited high fish diversity, attributed to their locations within the estuary, which is rich in organic matter and high in productivity, serving as a critical feeding ground for fish. In contrast, Station 5 (Berung River), the site of the power plant's thermal waste discharge, recorded low fish diversity. The thermal effluent acts as a pollutant, causing environmental instability for the aquatic biota.

References

1 Y. Rustandi, A. Fahrudin, M.N. Arkham, A. Zahid, Hamdani, The decline of fish resources in the coastal waters of Karawang Regency, West Java, Indonesia. AACL Bioflux **11**, 1825-1835 (2018). https://doi.org/10.1088/1755-1315/241/1/012003

2 I. Suyatna, A.S. Sidik, I.F. Almadi, S. Rizal, K. Sukarti, Fish community structure in high water temperature around Bontang Industrial Estate, East Kalimantan, Indonesia. Biodiversitas **17**, 558-564 (2016). https://doi.org/10.13057/biodiv/d170216

3 F. Shuai, X. Li, Y. Li, J. Li, J. Yang, S. Lek, Temporal patterns of larval fish occurrence in a large subtropical river. PLoS One **11**, e0146441 (2016). https://doi.org/10.1371/journal.pone.0146441

4 C.W. Huang, W.N. Tzeng, Estuarine use and movement patterns of seven sympatric Mugilidae fishes in the Tatu Creek estuary, central western Taiwan. Estuar. Coast. Shelf *Sci.* **68**, 62-74 (2006). https://doi.org/10.4194/1303-2712-v16_2_10

5 M. Eighani, S.Y. Paighambari, S.M. Bayse, Comparing handline and trolling fishing methods in the recreational pelagic fishery in the Gulf of Oman. Sci. Mar. **83**, 215-222 (2019). https://doi.org/10.3989/scimar.04864.10A

6 N.N. Menon, A.N. Balchand, N.R. Menon, Hydrobiology of the Cochin backwater system—a review. Hydrobiologia **430**, 149-183 (2000). https://doi.org/10.1023/A:1004033400255

7 J.D. Allan, M.M. Castillo, K.A. Capps, Primary producers, in Stream Ecology: Structure and function of running waters (Springer International Publishing, Cham, 2021), pp. 141-176. https://doi.org/10.1007/978-1-4020-5583-6_6

8 A. Ménesguen, X. Desmit, V. Dulière, G. Lacroix, B. Thouvenin, V. Thieu, M. Dussauze, How to avoid eutrophication in coastal seas? A new approach to derive river-specific combined nitrate and phosphate maximum concentrations. Sci. Total Environ. **628**, 400-414 (2018). https://doi.org/10.1016/j.scitotenv.2018.02.025

9 S. Colborne, T. Maguire, B. Mayer, M. Nightingale, G. Enns, A. Fisk, K. Drouillard, M. Mohamed, C. Weisener, S. Mundle, Water and sediment as sources of phosphate in aquatic ecosystems: The Detroit River and its role in the Laurentian Great Lakes. Sci. Total Environ. **647**, 1594-1603 (2019). https://doi.org/10.1016/j.scitotenv.2018.08.029

10 A.B. Larin, B.M. Larin, Improvement of chemical monitoring of water-chemistry conditions at thermal power stations based on electric conductivity and pH measurements. Therm. Eng. **63**, 374-378 (2016). https://doi.org/10.1134/S0040601516030058

11 N.U. Saher, R.M. Niazi, A.H. Narejo, N. Hawa, A.H. Baloch, M. Tabish, M. ul Ain, F. Razi, N.A. Qureshi, Feeding Diversity of Finfish in Different Wild Habitat, in Pertinent and Traditional Approaches Towards Fishery (IntechOpen, 2023). https://doi.org/10.5772/intechopen.110113

12 A. Zahid, C.P. Simanjuntak, M.F. Rahardjo, S. Sulistiono, Iktiofauna ekosistem estuari Mayangan, Jawa Barat [Ichthyofauna of Mayangan estuary, West Java]. J. Iktiologi Indones. **11**, 77-85 (2011). http://dx.doi.org/10.22146/jfs.31498

13 L. Michie, Thermal pollution in Australian freshwater ecosystems: ecological consequences for fish larvae and juveniles. Ph.D. thesis, School of Life Sciences, University of Technology Sydney (2020). http://hdl.handle.net/10453/147345

14 E.C. Ronauli, N.T.M. Pertiwi, H. Effendi, A phytoplankton biodiversity and pollution bioindicator in Bojonegara coastal waters, Banten bay, Indonesia. J. Biospecies **15**, 64-77 (2022). https://doi.org/10.22437/biospecies.v15i1.12491

15 S.A. Hashemi, R. Ghorbani, F. Kymaram, S.A. Hossini, G. Eskandari, A. Hedayati, Fish species composition, distribution and abundance in Shadegan Wetland. Fisher. Aquac. J., **6**(2), 1000128 (2015). https://doi.org/10.4172/2150-3508.1000128

Nutritional composition and antioxidant potential of brown seaweed (*Padina minor*) from Bengkulu, Indonesia

Nurlaila Ervina Herliany[1*], *Maya* Angraini Fajar Utami[2], *Mukti* Dono Wilopo[2], *Zamdial*[2], *Nurul* Permata Sari[2], and *Emely* Chantika[2]

[1]Departement of Fisheries Science, Faculty of Agriculture, University of Bengkulu, Jl. WR Supratman, Bengkulu 38371, Bengkulu, Indonesia
[2]Departement of Marine Science, Faculty of Agriculture, University of Bengkulu, Jl. WR Supratman, Bengkulu 38371, Bengkulu, Indonesia

Abstract. Seaweed is a popular food that has potential as a new source of bioactive compounds for humans. Bengkulu Province has abundant natural seaweed resources, one of which is *Padina minor*. Despite that, no research has been conducted on its nutritional composition and antioxidant potential thus far. Understanding this information is critical for utilizing *P. minor*, particularly in the food industry. This research was carried out to examine the nutritional composition and antioxidant activity of *P. minor* to provide useful information, especially for coastal communities. The seaweed powder were characterized for their nutritional composition (water, ash, protein, lipid, crude fiber and carbohydrates), while the fresh seaweeds were extracted using methanol. The methanol extract was analyzed to determine its antioxidant activity using DPPH solution. The results showed that the largest nutritional components in *P. minor* are water, carbohydrates and ash. The IC_{50} value of *P. minor* was 11.26 ppm, exhibited very strong antioxidants, similar with ascorbic acid (IC_{50} = 3.27 ppm). Based on this, *P. minor* from Bengkulu has the potential to be developed as an alternative source of minerals, dietary food and a source of antioxidant.

1 Introduction

High health concerns, particularly those associated with degenerative diseases, raise consumer awareness of lifestyle and healthy eating habits. This impacts the establishment of food consumption habits that are beneficial to health. Food with added health benefits, such as antioxidants, anti-cholesterol, and antidiabetic properties, is known as functional food. High demand for functional food generates a wide range of food products, one of which is derived from marine products.

Seaweed is an abundant fishery and marine commodity in Indonesia, both naturally and cultivated. According to certain studies, seaweed has numerous health benefits. Seaweed is rich in polysaccharides, proteins with complete essential amino acid content, unsaturated lipidty acids, and various vitamins and minerals [1]. Nowadays, seaweed gets a lot of attention because of its bioactive compounds, which have diverse biological and chemical activities like antioxidant, anti-inflammatory, anti-dyslipidemic, antitumor, and anti-microbial and can reduce the risk of cardiovascular disease.

Bengkulu Province has a relatively high diversity of marine resources, including seaweed. A preliminary study showed that Bengkulu has 20 species of seaweed; one of them was *P.*

* Corresponding author: vivien.unib@gmail.com

minor. Unfortunately, this species is not utilized by local communities due to the lack of scientific data regarding its nutritional content. There is a knowledge gap on the nutritional composition of *P. minor* from the coastal waters of Bengkulu. As a result, the purpose of this study was to gather such information to favour potential development of *P. minor* utilization in Bengkulu. The present study is the first published data on the proximate composition and antioxidant activity of *P. minor* from Bengkulu. Based on these preliminary findings, we investigated the proximate composition and antioxidant activity of *P. minor* collected from Bengkulu to provide scientific information, particularly for coastal communities.

Futhermore, the biochemical composition of seaweed vary according to species, location, season, and growth. Since the Bengkulu shoreline is oriented directly towards the Indian Ocean, it leads to relatively strong currents. Besides that, Bengkulu coastal water exhibits semidiurnal tides, with the lowest low tide reaching up to 50 meters from the shoreline. This exposes the seaweed to direct sunlight, causing it to dry out. This combination of environmental conditions can affect its nutritional content and antioxidant activity, so we hypothesized that *P. minor* from Bengkulu exhibits distinct nutritional and antioxidant profiles due to its unique local environment.

2 Materials and Method

2.1 Sampling procedures

Seaweed was collected from Teluk Sepang Beach, Bengkulu (3° 54' 57.23" S, 102° 16' 5.20" E) (Fig. 1). The psychochemical characteristic of Teluk Sepang Beach could be seen in Table 1. Random sampling method was carried out at a single location. Fresh *P. minor* was collected during low tide by hand picking and then identified in the Laboratory of Fisheries, Faculty of Agriculture, Bengkulu University. The *P. minor* samples was cleaned with water and shade-drying at room temperature then oven dried at 60°C for 12 hours. After drying, the samples were pulverized in the grinder to obtain seaweed powder. Seaweed powder was kept in plastic at room temperature for further analysis [2].

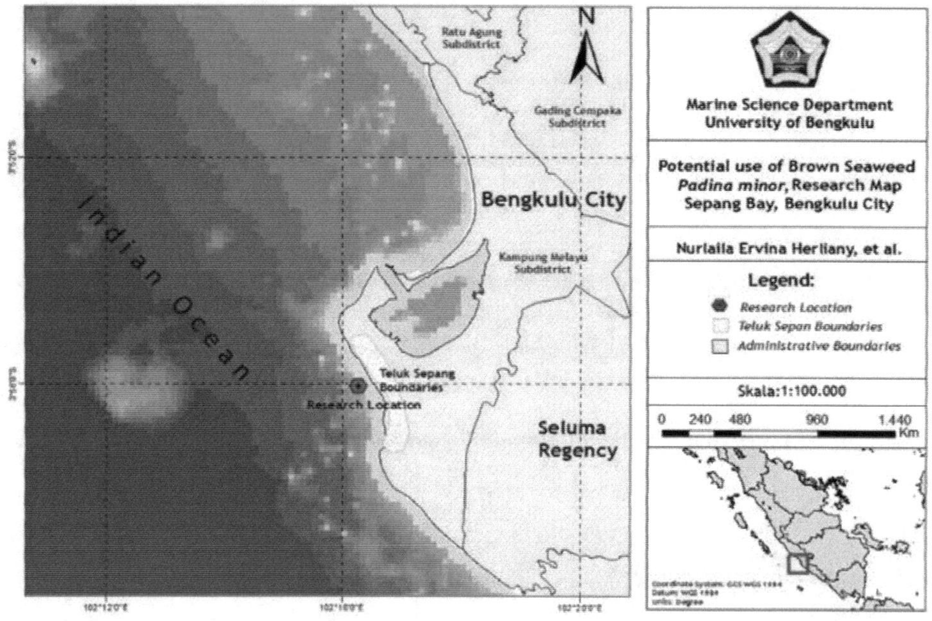

Fig. 1. The sampling site in Teluk Sepang Beach, Bengkulu.

Table 1. Psychochemical characteristic of Teluk Sepang Beach.

Parameters	Content
Salinity (ppt)	34.67
Temperature (°C)	35.67
pH	8.29
Current (m/s)	0.53
Nitrate (mg/L)	0.06
Phosphate (mg/L)	0.004

2.2 Proximate analysis

Seaweed powder was analyzed for its proximate composition following by procedur of [3], including moisture content, ash content, lipid content, protein content, and crude fiber. The carbohydrate content was obtained by difference method following the equation below:

$$\% \text{ Carbohydrate} = 100 - (\% \text{ moisture} + \% \text{ ash} + \% \text{ protein} + \% \text{ lipid}) \quad (1)$$

The analysis was repeated two times (duplo).

2.3 Seaweeds extraction

Extraction was carried out using the maceration method using a single solvent, methanol. Fresh *P. minor* (150 g) was cut into small pieces and soaked in methanol solvents for 72 hours at room temperature, with the ratio of seaweed: methanol of 1:4. The solution was filtered using Whatman filter paper No. 42, then evaporated with a rotary evaporator at 40 °C to obtaine paste. The paste was then washed using ethanol to remove salts (desalting). After washing, the paste then centrifuged at 4,200 rpm for 10 minutes. The supernatant was evaporated using the rotary evaporator at 40 °C to obtain methanol extract. Then, a stock solution with a concentration of 100 mg/ml was made using ethanol solvent and stored in the refrigerator for further analysis [2].

2.4 Determination of yield

The yield is obtained from the ratio of the weight of the seaweed extract to the weight of the seaweed powder. The yield value is calculated using the following formula:

$$Yield\ (\%) = \frac{Weight\ of\ extract}{Weight\ of\ seaweed\ powder}\ x\ 100\% \quad (2)$$

2.5 2,2-diphenyl-1-picrylhydrazyl (DPPH) radical scavenging assay

The scavenging activity of *P. minor* extract was analyzed using DPPH solution, following the procedure of [3]. Briefly, a total of 2 ml of sample solution at various concentrations (50, 100, 150, 200, and 250 ppm) was placed in a test tube and added with 2 ml DPPH 0.1 mM solution. The solution was mixed for 5 seconds using a vortex before being stored in a dark room for 30 minutes at room temperature. The absorbance of the solution was read using specthophotometer at wavelength of 515 nm. As a control solution, the absorbance of a blank solution containing 2 mL DPPH and 2 mL ethanol was read. Ascorbic acid was used as a positive control. The percentage of reduced DPPH color can be used to calculate free radical scavenging activities using the following formula :

$$\text{DPPH Inhibition (\%)} = (\text{blank abs} - \text{sample abs})/(\text{blank abs}) \times 100\ \% \quad (3)$$

Where blank abs is absorbance of control and sample abs is absorbance of sample.

Antioxidant activity is expressed in IC_{50} (Inhibition Concentration 50), the concentration of antioxidants needed to inhibit 50% free radicals. The IC_{50} value was calculated from the regression equation, y = a + bx, where Y is the percent inhibition and X is sample concentration (ppm).

2.6 Data analysis

The proximate data are analyzed descriptively and presented as means ± standard deviations of two assay replicates. The regression analysis was analyzed using MS Office Excel 2013.

3 Results

3.1 Morphological of *P. minor*

P. minor is a brown seaweed with a fan-like thalii and a yellowish brown to white in color (Fig. 2.). The blades surface has concentric lines that are evenly spaced. Blades up to 10 cm in height and 4 cm broad. This seaweed is found attached to rocks and dead coral in the intertidal zone of Teluk Sepang Beach, Bengkulu.

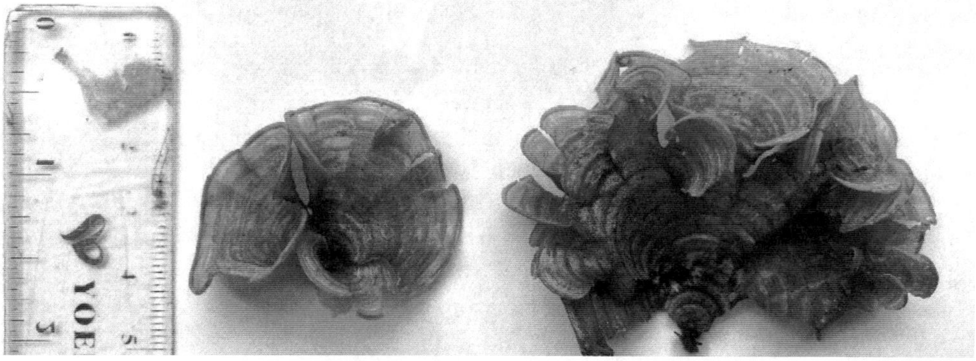

Fig. 2. Morphological of *P. minor* from Teluk Sepang Beach.

3.2 Nutritional composition of *P. minor*

Based on this research, the composition of *P. minor* from the highest to the smallest was moisture content, carbohydrate content, ash content, crude fiber content, protein content, and lipid content, respectively (Table 2.).

Table 2. Nutritional composition of *P. minor*.

Chemical composition	Content
Moisture (%, wet basis)	68.00±0.00
Ash (%, dry basis)	42.97±0.16
Crude Protein (%, dry basis)	1.95±0.15
Crude Lipid (%, dry basis)	1.79±0.72
Crude Fiber (%, dry basis)	6.01±0.27
Carbohydrate (%, dry basis)	49.27±0.24

*Values are expressed as mean ± SD (n = 2)

3.3 Extract yield of *P. minor*

The extraction process is used to obtain active compounds from raw materials using specific solvents. To determine the effectiveness of the solvent used, it is necessary to measure the extract yield. In this study, the yield of *P. minor* extract was 0.91±0.04%.

3.4 2,2-diphenyl-1-picrylhydrazyl (DPPH) Radical Scavenging Assay

Antioxidants are substances that have the ability to neutralize the harmful effects of free radicals. The antioxidant activity of *P. minor* ethanol extract was assessed by DPPH tests. Antioxidant activity is measured by the IC_{50} value (inhibition concentration of 50%), which is the concentration of samples that can inhibit 50% free radicals. The IC_{50} value of fresh *P. minor* extract was 11.26 ppm exhibited very strong antioxidant, calculated using the regression equation y = 0.1881x + 47.882 (Fig. 3.). Meanwhile, ascorbic acid as a control positive has an IC_{50} value of 3.27 ppm.

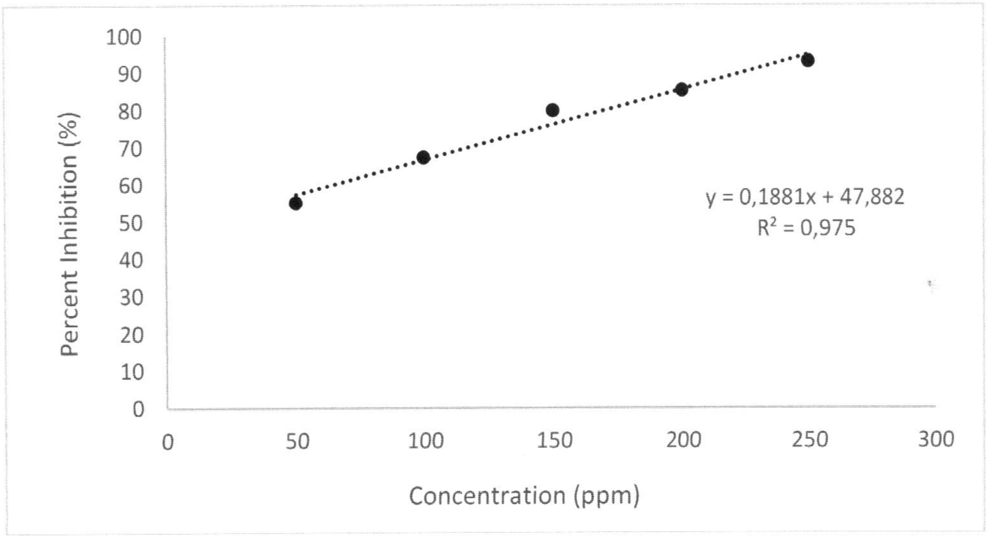

Fig. 3. The sample concentration and percent inhibition correlation.

4 Discussions

4.1 Morphological of *P. minor*

P. minor had yellowish brown to light brown or slightly whitish thallii due to light calcification. *P. minor* is characterized by its 2-layered thallus, with hairs restricted to the ventral surface. This species widely distributed in intertidal to subtidal zones, grows in a solid substrate including rocky or sandy bottom [4] and in many tropical areas, including Indonesia.

4.2 Nutritional composition of *P. minor*

The results showed that *P. minor* contained high amounts of water (68 ± 0.00 % wb). Vijay *et al.* [5] reported that seaweed contain 80-90% of water. Marine algae's polysaccharides increase water holding capacity, leading to increased initial moisture content. Hydrogen bonds between polysaccharides form extrinsic moisture, while tight junction proteins regulate intrinsic moisture.

P. minor also contained high carbohydrates, which was 49.27 ± 0.24 %. Meanwhile, [6] stated that carbohydrate content of *P. minor* from Puntondo Coast, Takalar, Indonesia, which was 48.06%. This variation can be influenced by environmental factors such as seasonal differences and light intensity. Seaweed carbohydrate levels are generally higher in the dry season because increased sunlight intensity accelerates photosynthesis, causing the algae to form and store more sugars and polysaccharides [7]. Kamal *et al.* [8] reported that the carbohydrate content of all macroalga studied in Egyptian Red Sea Shore peaked in summer compared to winter. High water temperatures indicate increased sunlight intensity, which also increases the rate of photosynthesis.

Furthermore, a study by El-Manawy et al [7] stated that seaweed polysaccharide content varies depending on the reproductive stage; the more mature the seaweed, the higher the polysaccharide content. During the juvenile phase, a large amount of energy is allocated to the synthesis of structural and reserve polysaccharides. Because the algae retain a lot of growth resources during this period, polysaccharide synthesis is decreased. Some of the stored polysaccharides are released during the vegetative phase to supply the energy required for the development of mature reproductive structures and spores. During this stage, yields are comparatively stable and start to rise as the algae mature. When algae reach the reproductive phase, some of the accumulated polysaccharides are mobilized to support the development of reproductive structures and spore formation. Because reproduction uses a lot of energy, polysaccharide stores are used up at this stage. As a result, polysaccharide levels rise as algae grow.

Carbohydrates in seaweed have health benefits, including antioxidants, anticoagulants, anti-inflammatory, anti-arthritic, immunomudulatory, and immunostimulant properties. Carbohydrates in brown algae are found in the form of fucoids, laminarin, cellulose, and alginate. One component of carbohydrates is crude fiber. The crude fiber of *P. minor* was 6.01 ± 0.27%. Seaweed contains low crude fiber, less than 10% [5]. The variation in crude fiber content could be related to variability in species, photosynthetic activity, growth stage, and seasonality, which are caused by environmental changes that affect photosynthesis and nutrient absorption [6]. The crude fiber content in seaweed is very good for digestive health, can prevent colon cancer, and is good for diet.

Seaweed has a greater ash concentration than terrestrial plants or animal products. Seaweed grows in a seawater environment rich in dissolved minerals, causing the algal cells to continuously absorb inorganic ions across the entire thallus surface, not just through the roots, resulting in more effective nutrient uptake. As a result, seaweed has a high ash content [5]. Brown seaweed generally has a higher ash content compared to red and green seaweed due to its alginate and sulfated polysaccharide content, which is rich in polygalacturonic acid units that act as anionic carboxylic binding sites that can interact with metallic cations.

In this study, the ash content of *P. minor* was 42.97 ± 0.16%, higher than in the research of Khadijah et al [6] from Puntondo coast, Takalar, Indonesia, and Salosso et al [9] from Kelapa Lima coastal waters, Kupang Bay, Indonesia which was 38.02% and 34.58%, respectively. The ash content of seaweed varies according to several factors, including age, species, geographical conditions, physiology, and mineralization methods [6]. Moreover, salinity is also suspected to be a factor influencing seaweed ash content. Excessive salinity levels can cause cells water content to rapidly drop, which raises the salt concentration inside the cells. During the study, Teluk Sepang Beach had a relatively high salinity of 34.67 ppt. This condition is thought to have contributed to the high ash content of *P. minor*.

Seaweed ash content is also influenced by the mineral content of the water. Mineral content differences between locations can result in varied seaweed ash content [6]. Unfortunately, this study did not measure ash or mineral content at Teluk Sepang Beach, so we cannot confirm whether the ash content of *P. minor* is influenced by the mineral amount of its environment. Besides that residual salt adhering to the seaweed during sample also influences ash content. Improper seaweed washing can leave salt residue on the surface of the thallus, resulting in a high ash content.

Protein and lipid content in *P. minor* was 1.95 ± 0.55% and 1.79 ± 0.72%, respectively. Seaweed has low protein and lipid content, namely 1-44% and 1-6% [5], consecutively. The protein content of *P. minor* in this study were lower compared to the study by Khadijah et al [6] on *Padina* sp. from Puntondo Coast, Takalar, Indonesia, which reported a protein content of 12.33%, but almost have the same lipid content (1.60%).

Differences in seaweed protein content are influenced by variations in species, habitat, season, and environmental conditions such as water temperature, salinity, light intensity, current strength, and nutrient supply [5]. According to Kamal et al [8], the protein content of seaweed is generally higher during winter and spring compared to summer, when rapid growth occurs due to increased photosynthetic rates. This statement is consistent with the present study, where protein levels were low because the sample was collected during the dry season. Furthermore, El Manawy et al [7] added that the protein content have a positive correlation with nitrate and phosphate levels in the water. According to Darmawan et al [10], the optimal nitrate : phosphate ratio for seaweed growth is 30 : 1. Meanwhile, the research location has a low nitrate : phosphate ratio of 12 : 1, which impacts the low protein content of *P. minor*.

Variations in lipid content are influenced by several factors, including species, season, and environmental conditions such as salinity and water temperature. According to Xiong et al [11], high salinity leads to ion stress, osmotic stress, and secondary stress. Ion stress results in Na+ poisoning. Excessive Na ions on the thallus surface can inhibit K+ absorption from the environment. Osmotic stress can disrupt water and nutrient absorption because of high environmental osmotic pressure. Ion stress and osmotic stress due to high salinity will cause secondary stress, specifically damage to cell structures and macromolecules such as lipids. Salinity at the research location was relatively high, resulting in low lipid content.

Higher temperatures generally result in lower lipid content. Vijay et al. [5] added that lipid content in tropical seaweed is lower than in temperate seaweed. Under stress conditions due to high temperatures, seaweed adjust their lipid metabolic pathways to synthesize and accumulate neutral lipids in the form of triacylglycerols. It provides a carbon and energy storage function in cells, allowing algae to tolerate high temperature. The combination of these various factors can explain the difference in lipid content of *P. minor* compared to previous studies.

Although seaweed contains relatively low levels of protein and lipid, the amino acids and fatty acids present contribute positively to human health. According to Salosso et al [9], *P. australis* is rich in aspartic and glutamic acid, which can act as an immune system, maintaining digestion and generate neurotransmission in the nerves of the brain and muscles. Proteins derived from macroalgae are also important as a source of peptides and amino acid extracts, especially after enzymatic digestion, which enhances their solubility in water, allowing them to be utilized in numerous industries [8]. Low lipid content provides health benefits for people with obesity. However, seaweed has an essential amino acid that is quite complete and rich in unsaturated fatty acids that are good for brain intelligence, such as EPA and DHA. Corsetto et al. [1] stated that seaweed fatty acids perform important functions in nutrition and cell membrane development, including essential-linolenic fatty acids that cannot be produced by mammals, whereas terrestrial plants can only produce it in small quantities. Based on this, seaweed has the potential to be an alternative supplement for unsaturated fatty acids, essential amino acid, and natural flavourings.

4.3 Extract yield and of *P. minor*

The yield obtained in this study was lower than that reported by Afrin et al [12] for *P. tetrastromatica* from Saint Martin's Island, Bangladesh, which was 14.26±0.85%. Herliany et al[2] suggests that variations in yield may result from differences in seaweed species, specifically due to the polarity of active compounds between species. Typically, extracts are most efficient when the solvent and active compound have similar polarity. When a seaweed species contains a higher concentration of polar active compounds, extraction with a polar solvent typically results in higher yield. Moreover, extraction yield may vary depending on the extraction method, environmental conditions, and season [2, 13].

4.4 2,2-diphenyl-1-picrylhydrazyl (DPPH) radical scavenging assay

There are different IC_{50} value of same genus from several coastal waters. Gazali *et al.* [14] reported that fresh *P. australis* from West Aceh, Indonesia, had an IC_{50} value of 265 µg/ml, while [14] found that dried *Padina* sp. from Selayar Island, South Sulawesi, had an IC_{50} value of 116.83 ppm. Various factors can influence antioxidant activity, including environmental conditions, seasons, harvest age, species, drying processes, and extraction procedures. According to Herliany *et al* [2], season influences the condition of the aquatic environment, which can impact the content of seaweed. The bioactive content in seaweed is influenced by seasonal variables that cause environmental changes, such as direct exposure to sunlight, nutrients, salinity, pH, and water temperature. The extraction method affects the antioxidant properties of seaweed because the suitable solvent allows for the optimal extraction of components from plants [13]. The polarity of seaweed's compounds will determine the appropriate solvent for extraction. Polar compounds dissolve in polar solvents (methanol, ethanol, water, and butanol), while non-polar compounds dissolve in non-polar solvents (n-hexane, ether, and chloroform).

This study found that *P. minor* from Bengkulu has strong antioxidant activity, similar to ascorbic acid. Because of this, *P. minor* has potential as an antioxidant supplement. Antioxidants are essential for counteracting free radicals, which contribute to degenerative diseases. Antioxidant needs can be met by consuming external sources. Further research should examine how processing methods affect *P. minor's* antioxidant activity to identify optimal methods that preserve its efficacy.

Conclusion

P. minor has high nutritional value, making it a potential functional food. The nutritional composition from the highest to the lowest is water, carbohydrates, ash, crude fiber, protein and lipid, respectively. The high antioxidant activity of *P. minor* suggests its potential application in the development of functional foods or nutraceuticals; however, further purification and compound identification are required.

Acknowledgments

We thank Faculty of Agriculture, Bengkulu University for providing this research funds in 2022 and 2023, with grant number 6009/UN30.11/LT/2022 and 5222 /UN30.11/PG/2023.

References

1. P. A. Corsetto, G. Montorfano, S. Zava, I. Colombo, B. Ingadottir, R. Jonsdottir, K. Sveinsdottir, A. M. Rizzo, Characterization of antioxidant potential of seaweed extracts for enrichment of convenience food. Antioxidants. **9**, 249 (2020). doi: 10.3390/antiox9030249.
2. N. E. Herliany, M. A. F. Utami, M. D. Wilopo, F. I. Wardani, Antioxidant activity of green seaweed *Boergesenia forbesii* from Teluk Sepang Beach Bengkulu, in Proceedings of the 6th EMBRIO International Symposium, Bogor, Indonesia, November 6-7 (2023), BIO Web of Conferences, 112, 06003. https://doi.org/10.1051/bioconf/202411206003.
3. AOAC [Association of Official Agricultural Chemists], Official methods of analysis of analytical chemist. 20th ed., (AOAC International, Rockville, USA, 2016).
4. F. Faradilla, A. D. Putri, G. A. Agustin, L. Nurkaromah, M. W. Febrianti, M. A. Budhiman, U. Salamah, R. Chasani, F. Nikmah, Macroalgae diversity at Porok Beach, Gunungkidul, Yogyakarta, Indonesia. J. of Agriculture and Appl. Biol. **3**, 1 (2022). doi: 10.11594/jaab.03.01.06.
5. K. Vijay, S. Balasundari, R. Jeyashakila, P. Velayathum, K. Masilan, R. Reshma, Proximate and mineral composition of brown seaweed from Gulf of Mannar. Inter. J. of Fish. and Aquatic Studies. **5**, 5 (2017).
6. K. Khadijah, N. H. Soekamto, F. Firdaus, S. M. T. Chalid, Y. M. Syah, Chemical composition, phytochemical constituent, and toxicity of methanol extract of brown algae *(Padina* sp.) from Puntondo Coast, Takalar (Indonesia). J. of Food Quality and Hazards Control. **8** (2021) : 178-185. doi: 10.18502/jfqhc.8.4.8259.
7. I. M. El-Manawy, M. Z. Nassar, S. H. Rashedy, Spatial and temporal variations in the nutritional composition of some seaweeds from the Red Sea, Egypt. Egyptian J. of Phycol. **20** (2019).
8. M. Kamal, N. Abdel-Raouf, K. Alwutayd, H. AbdElgawad, M. S. Abdelhameed, O. Hammouda, K. N. M. Elsayed, Seasonal changes in the biochemical composition of dominant macroalgal species along the Egyptian Red Sea Shore. Biology. **12**, 411 (2023). https://doi.org/ 10.3390/biology12030411
9. Y. Salosso, S. Aisiah, L. N. L. Toruan, W. Pasaribu, Nutrient content, active compound, and antibacterial activity of *Padina australis* against *Aeromonas hydropillar*. Pharmacogn J. **12**, 4 (2020).
10. M. Darmawan, N. P. Zamani, H. E. Irianto, H. Madduppa, Diversity and abundance of green seaweed *Caulerpa* (chlorophyta) across Indonesian coastal waters with different nutrient levels: Bintan Island, Jepara, and Osi Island. J. Ilmu dan Teknologi Kelautan Tropis. **14**, 2 (2022).
11. I. Xiong, J.K. Zhu, Salt tolerance in the Arabidopsis, (American Society of Plant Biologi, Rockville, Maryland, 2002)
12. F. Afrin, T. Ahsan, M. N. Mondal, M. G. Rasul, M. Afrin, A. A. Silva, C. Yuan, A. K. M. A. Shah, Evaluation of antioxidant and antibacterial activities of some selected seaweeds from Saint Martin's Island of Bangladesh. Food Chem. Advances. **3**, 100393 (2023)
13. A. Akbar, N. H. Soekamto, Firdaus, Bahrun, Antioxidant of n-hexane, ethyl acetate and methanol extracts of *Padina* sp. with DPPH method, in Proceedings of International Conference on Sustainable Utilization of Natural Resources, Ambon, Indonesia, November 28 (2020). doi:10.1088/1755-1315/800/1/012019 1 1-6.

14. M. Gazali, M. Nursid, A. Husni, Nurjanah, N. P. Zamani , Zuriat, C. D. Febrina, N. Zurba, H. Husin, D. Hidayat, The effect of different drying methods on antioxidant compounds and fucoxanthin content of brown seaweed from the intertidal zone of West Aceh, Indonesia. AACL Bioflux. **15**, 6 (2022).

The impact evaluation of fisherman assistance policies on the economic resilience and coastal communities' health: a case study of West Sulawesi

Tedjaprasadja Leonardo[1]*, *Puspita* Ni Putu Rekha[2], *Satyafebrianti* Kendra Callista[3]

[1]Department of Public Health, Faculty of Medicine, Universitas Pembangunan Nasional Veteran Jawa Timur, Indonesia
[2]Department of Economics Development, Faculty of Economics and Business, Universitas Pembangunan Nasional Veteran Jawa Timur, Indonesia
[3]Faculty of Medicine, Universitas Pembangunan Nasional Veteran Jawa Timur, Indonesia

Abstract. West Sulawesi's fisheries sector ranks second in the province's Gross Regional Domestic Product, yet it continues to record among the highest maternal and child mortality rates in Indonesia. This paradox raises concerns about the broader impact of government-supported fisherman assistance program beyond economic metrics. This study evaluates the influence of these programs on the economic resilience and health outcomes of coastal communities, focusing on maternal mortality rate (MMR), infant mortality rate (IMR), under-five mortality rate (U5MR), nutritional access, and educational participation. A descriptive analytical method was applied using secondary data from BPS Indonesia and provincial health profiles. The analysis revealed significant inter-district disparities: Polewali Mandar and Center Mamuju exhibited elevated MMR and IMR, while Mamasa recorded the lowest per capita fish consumption. Secondary school enrolment was also lowest in Polewali Mandar and Center Mamuju. These findings suggest uneven program benefits, shaped by geographic isolation, infrastructure gaps, and socioeconomic inequality. The study underscores the need for district-specific, multisectoral policy interventions that integrate health, nutrition, and education strategies. Aligning these efforts with Sustainable Development Goals (SDGs 2, 3, and 4) is essential to promote inclusive and equitable development in Indonesia's coastal regions.

1 Introduction

The fisheries sector plays a pivotal role in Indonesia's regional economic development, particularly in coastal provinces where marine resources are abundant. In West Sulawesi Province, the Gross Domestic Regional Product (GDRP) of the fishing sector ranks second only to agriculture (BPS, 2025), underscoring its strategic importance in local livelihoods and food systems[1,2]. Despite this economic prominence, the province continues to grapple with alarming public health indicators, including elevated maternal mortality, infant mortality, and child mortality rates[3]. Based on The Long Form Survey conducted by BPS Indonesia on Table 1, the maternal mortality rate (MMR) in West Sulawesi ranks fourth after Papua, West Papua, and West Nusa Tenggara. On the other side, the infant mortality rate

* Corresponding author: leonardo.tedjaprasadja.fk@upnjatim.ac.id

(IMR) ranks fifth after Papua, West Papua, Maluku and Gorontalo. Similarly, the under-five mortality rate (U5MR) ranks fourth after Papua, West Papua, and Maluku. West Sulawesi, with its vast coastline and rich marine resources, stands as one of Indonesia's emerging fisheries provinces where the strength of the blue economy is deeply intertwined with community health and well-being. The fisheries sector not only sustains the province's economic growth but also serves as a vital pillar for nutrition, food security, and household resilience among coastal populations. This paradox-economic strength in fisheries juxtaposed with poor health outcomes-raises critical questions about the effectiveness of government assistance programs aimed at improving the welfare of coastal communities.

Table 1. The Long Form Survey of Maternal, Infant, and Under-Five Mortality in Indonesia SP2020

No.	Province	MMR	IMR	U5MR
1	Aceh	201	19,41	22,88
2	Sumatera Utara	195	18,28	21,37
3	Sumatera Barat	178	16,35	19,19
4	Riau	158	15,69	18,24
5	Jambi	177	16,99	19,98
6	Sumatera Selatan	175	16,78	19,62
7	Bengkulu	179	19,73	23,38
8	Lampung	192	15,69	18,3
9	Kepulauan Bangka Belitung	198	16,75	19,67
10	Kepulauan Riau	142	13,31	15,43
11	DKI Jakarta	48	10,38	12,02
12	Jawa Barat	187	13,56	15,91
13	Jawa Tengah	183	12,77	14,81
14	DI Yogyakarta	58	10,9	12,62
15	Jawa Timur	184	13,49	15,64
16	Banten	127	13,83	16,15
17	Bali	85	13,26	15,37
18	Nusa Tenggara Barat	257	24,64	29,37
19	Nusa Tenggara Timur	316	25,67	30,72
20	Kalimantan Barat	246	17,47	20,44
21	Kalimantan Tengah	200	17,95	21,19
22	Kalimantan Selatan	224	17,22	20,1
23	Kalimantan Timur	177	15,51	18,04
24	Kalimantan Utara	194	16,65	19,43
25	Sulawesi Utara	230	17,23	20,4
26	Sulawesi Tengah	264	27,72	33,44
27	Sulawesi Selatan	192	18,2	21,49
28	Sulawesi Tenggara	226	23,29	27,93
29	Gorontalo	266	29,47	35,85
30	Sulawesi Barat	274	29,21	35,49
31	Maluku	261	29,82	36,54
32	Maluku Utara	255	28,61	34,58
33	Papua Barat	343	37,06	47,23
34	Papua	565	38,17	49,04

Source: BPS Indonesia (2025)

Government interventions in the fisheries sector are designed not only to enhance productivity and income but also to foster nutritional resilience through increased access to seafood. Fish and other marine products are rich in essential nutrients such as protein, omega-3 fatty acids, and micronutrients that are vital for maternal and child health[1,4]. However, the extent to which these assistance programs translate into tangible improvements in health and economic indicators remains underexplored.

In coastal communities, economic vulnerability often intersects with health risks, particularly in regions heavily dependent on marine resources. Previous Research using Data Envelopment Analysis (DEA) revealed that provinces with high marine dependency also face significant health challenges, including infectious diseases and climate-related risks[5]. These findings suggest that economic interventions must be integrated with public health strategies to achieve sustainable development outcomes. Moreover, the nutritional potential of seafood is frequently underutilized due to limited purchasing power, inadequate infrastructure, and gaps in health education[6].

The Indonesia government has implemented various assistance schemes for fishermen, including insurance programs, subsidies, and capacity-building initiatives. While these programs aim to reduce economic precarity and improve productivity, their impact on health outcomes-particularly among vulnerable groups such as mothers and children- has not been systematically assessed[7,8]. Studies from the Journal of Fisheries Social Economics and the Indonesian Fisheries Policy Journal emphasize the need for cross-sectoral policy evaluation frameworks that consider both economic and health dimensions [9].

This study employs a descriptive analysis approach to evaluate the effectiveness of fisher assistance programs in enhancing the welfare of coastal communities, with a specific focus on economic conditions and key health indicators. Particular attention is given to maternal, infant, and under-five mortality rates as critical measures of public health. While the analysis offers valuable insights, it is constrained by the limited availability of region-specific data, which may affect the generalizability of the findings. Nonetheless, the research seeks to inform the development of more responsive and equitable coastal development policies.

2 Methods

This study employs a descriptive analysis approach to evaluate the effectiveness of government assistance for fishermen in improving the economic resilience and health outcomes of coastal communities in West Sulawesi. Descriptive analysis is particularly suited for identifying patterns, trends, and relationships among variables without manipulating the research environment. The selected indicators include maternal mortality rate, infant mortality rate, under-five mortality rate, full educational participation, number of maternal deaths, and per capita fish consumption. These indicators are chosen based on their relevance to both economic and health dimensions of coastal welfare, as supported by previous studies such as on maternal mortality determinants in Indonesia and on infant mortality in East Java[10,11].

Data for this research are sourced from official government databases, including the BPS Indonesia, and regional health profiles. The maternal and child mortality indicators are analysed to assess the health status of coastal populations, while educational participation serves as a proxy for socio-economic empowerment. Fish consumption data are included to evaluate nutritional access and dietary behaviour, which are critical for maternal and child health. The use of descriptive statistics allows for the identification of disparities and trends across districts within West Sulawesi Province. This method has been validated in similar studies, such as the spatial analysis of infant mortality[12]and the consumption behaviour analysis[13].

The analytical framework involves tabulating and visualizing the selected indicators to reveal geographic and demographic patterns. Mortality rates are expressed per 100.000 live births (for MMR) and per 1000 live births (for IMR and U5MR), while fish consumption is measured in kilograms per capita per year. Educational participation is quantified as the percentage of school-age children enrolled in formal education. This methodology not only supports the research objectives but also contributes to the broader discourse on integrated development strategies, as emphasized in their systematic review of maternal mortality trends in Indonesia[14].

3 Results and Discussion

3.1 District disparities in West Sulawesi in health status, maternal and child mortality rates

For the first selected indicators for health status across West Sulawesi, the districts show marked variation in maternal and child mortality. Coastal or remote districts like Mamasa and Polewali Mandar tend to report the highest maternal (MMR) and under-five mortality, reflecting more limited access to health services. In contrast, the provincial capital Mamuju and Mamuju Tengah generally exhibit lower mortality rates, likely due to better facilities. These patterns align with national analyses: Sulawesi ranks among the regions with the highest MMR in Indonesia, with dramatic interprovincial gaps [14]. Polewali Mandar also tend to report the highest infant mortality. Spatial-health studies in Indonesia similarly document clear geographic clusters of infant/child deaths, validating our district-level approach[12].

Table 2. Maternal mortality rate in West Sulawesi in 2023

District	Pregnancy Death	Labor Death	Post Labor Death	Total
Majene	2	1	3	6
Polewali Mandar	6	0	8	14
Mamasa	1	2	2	5
Mamuju	2	5	2	9
Pasangkayu	2	2	4	8
Central Mamuju	1	0	4	5
West Sulawesi	14	10	23	47

Source: BPS Indonesia (2024)

Table 3. Infant mortality rate in West Sulawesi in 2019 - 2022

		2019			2020		
No	**District**	**Infant Death**	**Infant Birth**	**Rate**	**Infant Death**	**Infant Birth**	**Rate**
1	Majene	69	3.850	17,9	45	3.679	12,2
2	Polewali Mandar	98	8.584	11,4	98	8.294	11,8
3	Mamasa	10	2.557	3,9	24	2.520	9,5
4	Mamuju	70	5.228	13,4	33	5.676	5,8
5	Pasangkayu	47	2.843	16,5	67	2.879	23,3
6	Central Mamuju	29	2.507	11,6	36	2.591	13,9
West Sulawesi		323	25.589	12,6	303	25.639	11,8

		2021			2022		
No	**District**	**Infant Death**	**Infant Birth**	**Rate**	**Infant Death**	**Infant Birth**	**Rate**
1	Majene	19	3.575	5,3	40	3.817	10,5
2	Polewali Mandar	86	8.323	10,3	78	8.144	9,6
3	Mamasa	26	2.497	10,4	23	2.584	8,9
4	Mamuju	36	4.646	7,7	50	5.568	9,0
5	Pasangkayu	59	2.925	20,2	62	3.223	19,2
6	Central Mamuju	35	2.746	12,7	49	2.622	18,7
West Sulawesi		261	24.712	10,6	302	25.938	11,6

Source: BPS Indonesia (2024)

3.2 Fish consumption patterns

Fish consumption per capita varies by district, largely mirroring coastal access. For example, Majene (a fishing area) has a relatively low intake – about 43.8 kg per person per year, below the national average (~46.5 kg). This suggests even "fish-rich" areas may under-consume due to socioeconomic factors. By contrast, larger fishing hubs (e.g. Polewali Mandar) are expected to have higher per-capita consumption (though exact district data are seldom published). In general, districts with strong fisheries see higher fish diets, whereas interior or poorer districts consume.

Table 4. Food consumption in West Sulawesi in 2022-2023

District	**Cattle Meat (kg)**		**Avian Meat (kg)**		**Eggs (kg)**		**Milk (liter)**		**Fish (kg)**	
	2022	**2023**	**2022**	**2023**	**2022**	**2023**	**2022**	**2023**	**2022**	**2023**
Majene	0,30	0,40	1,70	1,10	3,90	4,10	1,70	1,30	39,30	31,80
Polewali Mandar	1,10	9,50	1,70	17,70	4,40	31,30	1,10	34,10	30,70	136,90
Mamasa	4,10	1,80	2,80	2,90	4,20	6,90	0,80	0,00	23,20	25,60
Mamuju	0,40	0,70	2,80	2,50	4,50	3,80	1,90	2,20	36,00	33,70
Pasangkayu	1,40	1,90	3,00	2,50	6,30	5,40	2,30	2,20	34,20	31,60
Central Mamuju	0,70	1,40	3,60	2,00	5,30	39,00	2,00	12,80	35,10	184,30

District	Cattle Meat (kg)		Avian Meat (kg)		Eggs (kg)		Milk (liter)		Fish (kg)	
	2022	2023	2022	2023	2022	2023	2022	2023	2022	2023
West Sulawesi	1,30	1,40	2,50	2,20	4,70	4,10	1,50	1,70	33,20	30,50

Source: BPS Indonesia (2024)

3.3 Educational participation (school enrollment)

Educational participation (school enrollment rates) is generally high in Sulawesi Barat but still shows district variation. Urban centers like Mamuju have the highest school attendance, whereas more remote, rural Mamasa often reports lower enrollment, reflecting infrastructure and economic disparities. These differences matter because education is a key development factor: improving infrastructure and schooling opportunities is an explicitly noted strategy for reducing maternal and child mortality[15]. In practice, districts with better school access tend to also have stronger socioeconomic empowerment, which correlates with better health outcomes.

Table 5. West Sulawesi's educational profile in 2023

District	Total students by educational level			Crude Participation Rate		
	Elementary School	Middle School	High School	Elementary School	Middle School	High School
Majene	19.621	7.076	8.885	88	62	96
Polewali Mandar	43.593	15.497	16.380	81	56	56
Mamasa	16.249	8.664	8.379	94	94	83
Mamuju	31.859	10.504	11.022	94	60	61
Pasangkayu	18.764	6.951	6.227	88	62	48
Central Mamuju	13.976	4.394	4.432	85	57	52
West Sulawesi	144.962	53.086	55.325	87	63	63

Source: BPS Indonesia (2024)

Table 6. West Sulawesi's educated population by age in 2023

District	Age (Years Old)		
	7 - 12	13 - 15	16 - 18
Majene	22.318	11.379	9.266
Polewali Mandar	53.890	27.794	29.357
Mamasa	17.235	9.238	10.042
Mamuju	33.990	17.417	17.973
Pasangkayu	21.407	11.295	12.875
Central Mamuju	16.386	7.721	8.491
West Sulawesi	165.226	84.844	88.004

Source: BPS Indonesia (2024)

4 Conclusion

The analysis of multi-sectoral indicators across the six districts of West Sulawesi reveals significant intra-provincial disparities in health outcomes, nutritional access, and educational participation, suggesting the need for targeted and localized policy interventions.

1. Infant and Maternal Mortality (IMR)
   Disparities between 2019 and 2022 exhibit substantial fluctuation across districts. In 2022, Pasangkayu (19.2/1000) and Mamuju Tengah (18.7/1000) recorded the highest IMRs, indicating persistent vulnerabilities in maternal-child health systems. Conversely, Mamasa (8.9/1000) and Polewali Mandar (9.6/1000) showed lower rates, possibly reflecting better primary health service accessibility. Similarly, maternal mortality in 2022 was disproportionately concentrated in Polewali Mandar (14 maternal deaths), compared to Majene and Mamuju Tengah (5–6 deaths), highlighting potential weaknesses in obstetric and postnatal care systems in certain areas.
2. Nutritional Access and Fish Consumption
   Despite being a coastal province, fish consumption remains uneven. In 2023, Polewali Mandar (136.9 kg/capita) and Mamuju Tengah (184.3 kg/capita) demonstrated the highest per capita fish intake, suggesting adequate food availability and cultural dietary integration. However, Mamasa (25.6 kg/capita)—an inland district—showed the lowest level, reflecting geographical barriers and possibly socioeconomic constraints to accessing protein-rich marine foods. Year-on-year trends (2022 to 2023) indicate a slight decline in fish consumption in some districts like Majene, which dropped from 39.3 to 31.8 kg/capita, implying instability in food systems or economic purchasing power.
3. Educational Participation as a Proxy for Socioeconomic Empowerment
   Enrollment data shows a stark contrast in educational participation. While Majene (SMA/SMK APK: 96%) and Mamasa (83%) maintain high school-age participation, districts like Polewali Mandar (56%) and Mamuju Tengah (52%) lag significantly, particularly in secondary education. This indicates systemic challenges in access, retention, or affordability of higher education, which could perpetuate intergenerational poverty and health vulnerability. Integrated Patterns of Disadvantage districts such as Pasangkayu and Mamuju Tengah emerge as high-risk areas across multiple indicators: elevated IMR, high maternal mortality, low education participation (especially in Mamuju Tengah), and mixed nutritional access. These overlapping vulnerabilities point to entrenched structural inequalities, demanding multi-sectoral intervention strategies that integrate health services, nutrition programs, and educational reforms.
4. Policy Implementation
   The findings of this study highlight the critical need for district-specific, evidence-based policy interventions in West Sulawesi Province to address persistent disparities in health, nutrition, and education outcomes. A prioritized policy agenda should include targeted strengthening of maternal and child health services in districts with disproportionately high mortality rates, such as Mamasa and Polewali Mandar, where indicators suggest systemic limitations in antenatal, delivery, and

neonatal care capacity. Improving service coverage, referral systems, and skilled birth attendance in these areas is vital to reduce preventable deaths.

Simultaneously, efforts to enhance food security and improve the distribution and accessibility of nutrient-dense foods, particularly fish, are essential—especially in inland and under-resourced districts such as Mamasa, where per capita fish consumption remains significantly below both provincial and national averages. Interventions may include infrastructure development for cold-chain logistics, nutrition education, and locally adapted fish-based food programs to support dietary diversification and address undernutrition.

Addressing these interlinked challenges in a coordinated, district-sensitive manner is imperative not only for improving individual and community well-being, but also for fulfilling broader development commitments. These include the Sustainable Development Goals (SDGs), specifically SDG 2 (Zero Hunger), SDG 3 (Good Health and Well-being), and SDG 4 (Quality Education). As such, a multisectoral and spatially targeted approach is essential to advancing equitable human development in West Sulawesi and similar coastal, decentralized regions of Indonesia.

References

1. A. Bashir, Z. Ishak, I. Asngari, Mukhlis, P. Atiyatna, I. Hamidi, The performance and strategy of Indonesian fisheries: a descriptive review. *Int. J. Econ. Financ. Issues* **9**, 31–36 (2019). https://doi.org/10.32479/ijefi.7188
2. Hasani Q., Julian D., Damai A. A., Yudha I. G., Diantari R., Yuliana D., Reza M., Caesario R., Putriani R. B., 2024 Priority strategy in the development of sustainable capture fisheries in the Marine Protected Area of Kiluan Bay, Lampung, Indonesia. AACL Bioflux 17(2):764-774.
3. S.M.B. Bangun, N.I. Lipoeto, N. Nazir, R. Rasyid, Y. Yonariza, H. Hardisman, M. Masrul, F. Firdawati, Differences in food security based on family income and its impact on stunting among children under five in Padang Lawas Regency, North Sumatera Province. *Contagion* **7**, 37 (2025). https://doi.org/10.30829/contagion.v7i1.23728
4. Shellock RJ, Fullbrook L, McKinley E, Cvitanovic C, Kelly R, Martin V. The nature and use of ocean literacy in achieving sustainable ocean futures: a systematic map. *Ocean Coast Manag.* 257, 107325 (2024). https://doi.org/10.1016/j.ocecoaman.2024.107325
5. D.R. Wati, Sea surface salinity and precipitation patterns as indicators of climate change-induced health risks in Indonesian coastal villages: a data envelopment analysis approach. *J. Manaj. Kesehat. Indones.* **12**, 307–320 (2024). https://doi.org/10.14710/jmki.12.3.2024.307-320
6. N. Nurbaya, N. Najdah, Z. Irwan, Y. Yudianti, N. Nurliah, R. Rasmaniar, E. Nurlaela, Exploration of adolescent health services and utilization in Mamuju District, West Sulawesi. *J. Public Health Trop. Coast. Reg.* **8**, 51–58 (2025). https://doi.org/10.14710/jphtcr.v8i1.25518
7. F. Riza, Legal guarantees for fishermen's work risks. *De Lega Lata* **8**, 245–252 (2023). https://doi.org/10.30596/dll.v8i2.15113
8. A.R. Septiana, M.M. Reginaldis, R. Tohopi, Policy implementation: strengthening communication in the fishermen insurance government program (BPAN) in

Tanggetada District, Kolaka Regency. *J. Ilm. Ilmu Adm. Publ.* **10**, 508 (2019). https://doi.org/10.26858/jiap.v10i2.16198

9. E.N. Manafi, Multitemporal evaluation of coral health index in Pieh Island Conservation Area and surrounding sea (2021–2024): biophysical analysis and its implications for ecosystem-based management. **17**, 59–75 (2025).
10. Cameron L, Contreras Suarez D, Cornwell K (2019) Understanding the determinants of maternal mortality: An observational study using the Indonesian Population Census. PLOS ONE 14(6): e0217386. https://doi.org/10.1371/journal.pone.0217386
11. Mulugeta SS, Wassihun SG. Determinant of infant mortality in Ethiopia: demographic, socio economic, maternal and environmental factors. MOJ Women's Health. 11(2):49–57 (2022). https://doi.org/10.15406/mojwh.2022.11.00305
12. F.D. Cahyoko, H.F.A. Subratha, Identification of infant mortality rate factors using spatial autoregressive moving average. *J. Public Health Trop. Coast. Reg.* **7**, 235–248 (2024). https://doi.org/10.14710/jphtcr.v7i3.24856
13. A.E. Tiffany, A.K. Mudzakir, B.A. Wibowo, Analisis tingkat konsumsi ikan laut dan faktor-faktor yang mempengaruhi. *J. Fish. Resour. Util. Manag. Technol.* **9**, 25–34 (2020).
14. M. Syairaji, D.S. Nurdiati, B.S. Wiratama, Z.D. Prüst, K.W.M. Bloemenkamp, K.J.C. Verschueren, Trends and causes of maternal mortality in Indonesia: a systematic review. *BMC Pregnancy Childbirth*. **24**, 1–14 (2024). https://doi.org/10.1186/s12884-024-06687-6
15. X.R. Guo, Y.L. Ji, S.Y. Yan, T. Shi, K. Chamroonsawasdi, J. Liu, H.J. Wang, Impact of scaling up health intervention coverage on reducing maternal mortality in 26 low- and middle-income countries: a modelling study. *J. Glob. Health*. **14**, 04221 (2024). https://doi.org/10.7189/jogh.14.04221

The impact of mini trawl (arad) fisheries based on fish and social resources in Cilincing, Jakarta Bay

Roza Yusfiandayani[1*], *Domu* Simbolon[1], *Tri Dewi K* Pribadi[2], *Mercy* Patanda[3], *Rachmad* Caesario[4], *Bertoka Fajar Surya Perwira* Negara[5], and *Susanti* Rahayu[1]

[1] Department of Fisheries Resources Utilization, IPB University, Bogor, West Java, 16680. Indonesia
[2] Department of Biology, Universitas Padjadjaran, West Java, 45363. Indonesia
[3] Department of Fisheries Resources Utilization, Satya Negara Indonesia University, Jakarta Capital Special Region, 12240, Indonesia
[4] Department of Marine and Fisheries, Lampung University, Lampung, 35141. Indonesia
[5] Student of Department of Marine Biotechnology and Resources, National Sun Yat-Sen University, Kaohsiung 80424, Taiwan

Abstract. Mini trawl fisheries is a fishing method typically used in small to medium-scale fisheries. It involves using a net with meshes large enough to prevent fish from getting trapped. However, using mini trawl can significantly impact the aquatic environment, including the waters in Cilincing, North Jakarta, Indonesia. This study has two objectives: (1) Determine the level of sustainability of mini trawl in Cilincing waters and (2) identify fishermen's perceptions of using mini trawl. The analysis method used in this research is by conducting interviews and the EAFM (*Ecosystem Approach to Fisheries Management*) approach. The ecosystem approach to fisheries management ensures the sustainability of fisheries. The results of this study show that the sustainability of mini trawl use in Cilincing is moderate. This result was obtained using the EAFM approach. The value obtained from the composite analysis of each domain in the EAFM indicator is 60.67, which is classified as moderate. Most (50%) of the set net fishermen in Cilincing waters showed a positive perception (in favor) of using set nets for fishing, and the remaining 40% were neutral, and 10% disagreed. The impact are fishing using mini trawls in Cilincing waters still maintains sustainable fish resources.

1 Introduction

One of the government's efforts to maintain the sustainability of marine resources is to issue Ministerial regulations governing the use of trawls as stipulated in the Minister of Marine Affairs and Fisheries Regulation No. 18 of 2021 [1]. The definition of *trawls* based on the Indonesian National Standard 01-7233-2006 is a fishing gear consisting of two trawl wing, the body, and the trawl bag.

*Corresponding author: ocha_roza@apps.ipb.ac.id

Various types of fishing gear, such as lift nets, set nets, traps, gillnets, seine nets, and mini trawls, are operated in Jakarta Bay [2]. In 2022, fish production landed at the Cilincing fish auction was 2,927,046kg and valued at Rp 88,218,904,500 [3]. Mini trawl fishing in Cilincing, North Jakarta, is one of the fishing activities commonly conducted by local fishermen. This fishing practice is primarily carried out to meet the daily needs of fishermen and local communities. However, as in many other places, the mini trawl in Cilincing also has the potential to cause negative impacts if not appropriately managed. The first aspect to consider regarding the mini trawl fishery in Cilincing is the sustainability of fish resources. Regarding this, the monitoring and management of fish resources are important to prevent overfishing. These measures include setting fishing quotas, minimum sizes for catchable fish, and appropriate fishing seasons. Selective nets can reduce the impact of bycatch or capture of non-target fish. Therefore, fishermen can optimize the net designs to reduce the capture of unwanted fish. The second aspect to be considered is marine habitat management. As mini trawling can damage seabed habitats, measures must be taken to minimize this damage, such as avoiding vulnerable areas and protecting coral reefs and seaweeds. This issue can be addressed by encouraging the diversification of local fishers' livelihoods, which could be a good solution. This could include training fishers in other sustainable fishing practices or developing sustainable marine resource-based economic ventures.

According to the latest Minister of Marine Affairs and Fisheries Regulation number 36 of 2023 [4] concerning the placement of fishing gear and fishing aids, mini trawls are not categorised as prohibited fishing gear in Indonesia. However, it was banned in 2015 through the Minister of Marine Affairs and Fisheries Regulation number 2 of 2015 [5], and starting in 2016, the fishing gear are again allowed in all fisheries management areas of the Republic of Indonesia. The permit to operate mini trawl is limited to the fishing vessel with a maximum tonnage of 5 GT.

The development of capture fisheries activities in the Cilincing area, characterized by the increasing number of fishermen, which give pressure on capture fisheries in Cilincing waters due to the large number of fishing operations in the Cilincing area [6]. Increased pressure on capture fisheries may have a negative impact on existing fisheries resources. If left unchecked, the sustainability of capture fisheries in the Cilincing area will be at risk. Therefore, assessing the sustainability of the existing capture fisheries activities in Cilincing waters is necessary before the fish resources are overfished. In addition, data limitations are also common in small-scale fisheries. The lack of data recording for small-scale fisheries statistics, due to the absence of enumerators in Cilincing, also warrants attention. Hence, the utilization of small-scale fisheries statistics has yet to be maximized as initial information in sustainable fisheries management. With the study of the sustainability of fisheries activities in the research location, it is expected that information related to the level of sustainability of capture fisheries activities in Cilincing waters will be obtained to be used as a reference in making policies for future fisheries management. This study aims to determine the sustainability level of mini trawl fishery activities in Cilincing waters and identify fishers' perceptions of using mini trawl.

2 Material and method

2.1 Research location and times

The research was conducted from January to May 2024 at the Cilincing fish auction site, North Jakarta (Figure 1). Data was collected covering two domains: the fish resource domain and the fishing technique domain.

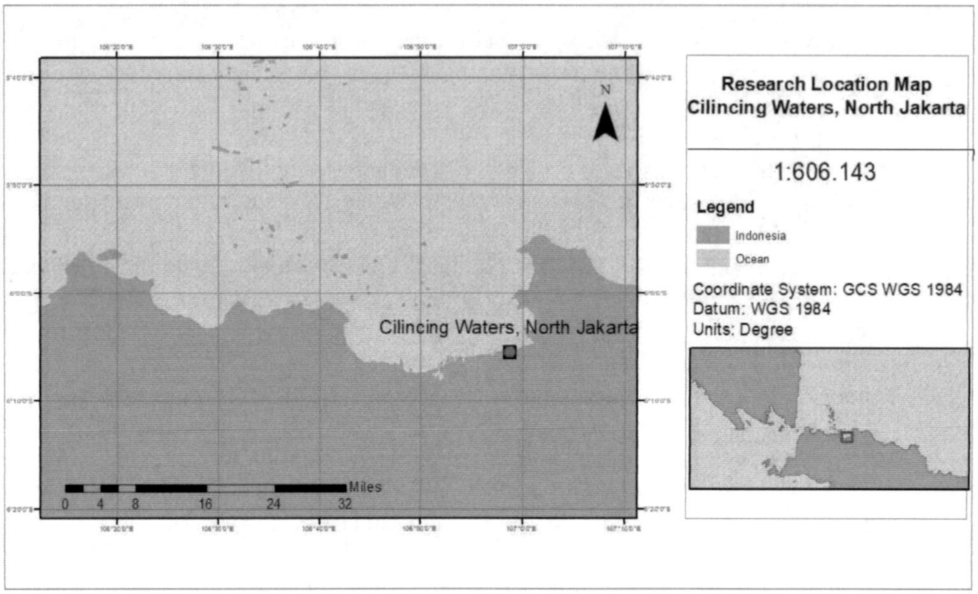

Fig. 1. Research Location in Cilincing, North Jakarta.

Each domain has indicators used to assess its sustainability level. Information collection through interviews was conducted using questionnaire. A total of 15 fishermen were selected as key informants in this study using a purposive sampling technique. Further data collection was obtained from the Marine and Fisheries Agency. The types of data used to assess the sustainability level of mini trawl fisheries in Cilincing using the EAFM approach can be seen in Table 1.

Table 1. Types of data and collection methods used to assess the sustainability of the mini trawl fishery in Cilincing.

Domain/Aspect Fish resources			
	Indicator	**Definition / Explanation**	**Data Collection**
1	CPUE	It is the catch per unit of fishing effort.	*Logbooks*, Observers for at least 2 years from the fishery unit under study.
2	Proportion of juvenile fish caught	The percentage of fish caught before reaching *maturity*.	Sampling
3	Species composition of the catch	Utilized and unutilized target species	*Logbook,* survey, and monitoring
4	*Range collapse of* fish resources	Increasingly distant fishing grounds	Survey and monitoring, *logbook*, observation,

			and interview
Domain/Aspect of Fishing Techniques			
	Indicator	**Explanation**	**Collection**
1	Effect of fishing gear operation on the environment	More than 50% of the water bottom is damaged 25-50% of the water bottom is damaged <25% water bottom is damaged	Interview
2	Modification of fishing gear and fishing aids	Use of fishing gear and aids that harm fish resources	Observer, sampling of target fish size/ dominant fish, Lm size based on literature study
3	Capture selectivity	Fishing activity is associated with a diversity of catches from the use of existing fishing gear.	Capture fisheries statistics, *logbooks*, and surveys

2.2 Data Analysis

Analysis of the level of sustainability of mini trawl fisheries activities in Cilincing waters was conducted using the EAFM (*Ecosystem Approach to Fisheries Management*) approach. The composite approach to be used is a simple composite approach using the MCA (*multi-criteria analysis*) approach in which a criterion is developed to analyse the behavior of fisheries management through the development of a composite index with several stages referring to Budiarto [7]. The assessment of the fish resource domain includes four indicators: CPUE, the proportion of juvenile fish caught, species composition of the catch, and *range collapse of* fish resources. The fishing technique domain assessment includes three indicators: operation of fishing gear against the environment, modification of fishing gear, and selectivity of fishing [8]. The assessment criteria and weights for each indicator can be seen in Table 2.

Table 2. Assessment criteria, scores, and weights of fish resource and fishing technique domain indicators [9].

Domain of Fisheries resource				
No	Indicator	Criteria	Score	Weight
1.	CPUE	Sharp decline (average decline >25% per month Slight decline (average decline <25% per month) Stable or increasing	1 2 3	40
2.	Proportion of juvenile fish caught	A lot (>60%) Many (30-60%) Little (<30%)	1 2 3	15
3.	Species composition of the catch	Less proportion target (< 15% of total volume) Proportion of targets equal to non-targets (16-30% of total volume) Greater proportion	1 2 3	10
4.	*Range collapse* of fish resources	Increasingly difficult depending on the target species Relatively fixed depending on target species Getting easier depending on the target species	1 2 3	10
Domain of fishing techniques				

Domain of Fisheries resource				
1.	Operation of fishing gear on the environment	More than 50% of the water bottom is damaged 25-50% of the water bottom is damaged <25% water bottom damaged	1 2 3	30
2.	Modification of fishing gear ang fishing aids	More than 50% of the target species size <Lm 25-50% target species size <Lm <25% target species size <Lm	1 2 3	25
3.	Capture selectivity	Low (75% of fish caught are non-target) Medium (50-70% of fish caught are non-target) High (<50% of fish caught are non-target)	1 2 3	15

The value is obtained by multiplying the score by the weight. The indicator with the highest weight indicates that it is the *main indicator* in a domain. In contrast, the indicator with the smallest weight indicates that it has a low level of importance in assessing a domain. Indicator values are divided into five categories that describe the level of fisheries management in each domain (Table 3).

Tabel 3. Domain value score boundaries.

Value Range			Description
≥ 33.33	-	≤ 46,17	Bad
≥ 46.67	-	≤ 59.5	Less
≥ 6	-	≤ 72.83	Medium
≥ 73,33	-	≤ 86.17	Good
≥ 86,67	-	≤ 100	Very

2.3 Perception analysis

Information on fishermen's perceptions of mini trawl was collected through several questions covering productivity, sustainability, and conflict (Table 4). Response options for perception were agreed (A), neutral (N), and disagree (DS). The score for agree (A) is 3, neutral (N) is 2, and disagree (DS) is 1.

Table 4. Fishermen's perception questions on the use of mini trawl.

No	Question
1	Use of mini trawl nets to catch fish
2	Fishermen's perception of mini trawl nets as environmentally friendly fishing gear
3	Fishermen's perceptions of catch productivity are influenced by the availability of fisheries resources in a body of water.
4	Fishermen's perception of continuous fishing of undersized fish, which will reduce the fish population in a body of water.
5	Fishermen's perception of continuous fishing of undersized fish, which will reduce the fish population in a body of water.
6	Fishermen's perception of the use of non-selective mini trawl will affect the number (volume) of catches.
7	Fishermen's perception of undersized fish, which should be released, because they may reduce the fish population.
8	Fishermen's perceptions of differences in catchability between mini trawl fishermen and other fishermen will lead to social conflicts between fishermen.

3 Results

3.1 Sustainability level of mini trawl fishery in Cilincing waters

The type of vessel is a typical 10 GT wooden boat with an *inboard* engine. The size of the vessels is generally the same, between 7-9 meters long, 2.5-3 meters wide, and 1-2 meters deep. The vessels are typically powered by a Dongfeng 24 HP motor. Each of them has one fish hold, with a capacity of 1 ton.

Indicators of the fish resources domain consist of CPUE, the proportion of juvenile fish caught, species composition of catches, and range collapses of fish resources. In contrast, the fish resources domain consists of the effects of fishing gear operation on the environment, modifications to fishing gear and fishing aids, and selectivity in fishing. The indicator values for each domain are shown in Table 5.

Table 5. Criteria for assessing indicators in the fish resource domain.

Domain of Fisheries resource					
No	Indicator	Criteria	Score	Weight	Indicator Value (score x weight)
1.	CPUE	Sharp decline (average decline >25% per month Slight decline (average decline <25% per month) Stable or increasing	1 2 3	40	40
2.	Proportion of juvenile fish caught	A lot (>60%) Many (30-60%) Little (<30%)	1 2 3	15	45
3.	Species compositi on of the catch	Less proportion target (< 15% of total volume) Proportion of targets equal to non-targets (16-30% of total volume) Greater proportion	1 2 3	10	10
4.	*Range collapse* of fish resources	Increasingly difficult, depending on the target species Relatively fixed, depending on the target species Getting easier depending on the target species	1 2 3	10	10
Domain of fishing techniques					
No	Indicator	Criteria	Score	Weight	Indicator Value (score x weight)
1.	Operation of fishing gear on the environment	More than 50% of the water bottom is damaged 25-50% of the water bottom is damaged <25% water bottom damaged	1 2 3	30	90
2.	Modificatio n of fishing gear and	More than 50% of the target species size <Lm 25-50% target species size <Lm	1 2 3	25	50

	fishing aids	<25% target species size <Lm			
3.	Capture selectivity	Low (75% of fish caught are non-target) Medium (50-70% of fish caught are non-target) High (<50% of fish caught are non-target)	1 2 3	15	50

Overall, the sustainability level of mini trawl fisheries in Cilincing waters was moderate, with a composite analysis score of 60.67 across the two existing domains using an ecosystem approach (Table 6). The average value of each domain is in the medium category (60.67).

Table 6. Value of each domain of EAFM of mini trawl in Cilincing waters.

Domain	Domain value	Description
Fish resources	51,11	Less
Fishing techniques	70,24	Medium
Average aggregate	60,67	Medium

3.2 Perceptions of mini trawl fishermen in Cilincing waters

Small-scale fisheries in Jakarta Bay generally use mini trawls. One of the main problems in mini trawl fisheries is the low level of selectivity, which results in catches dominated by bycatch rather than targeted fish. This will undoubtedly have an impact on fish resources. Fishermen, who are the main stakeholders, need to understand the various aspects related to the use of mini trawl. For this reason, fishermen's perceptions were identified through eight questions covering the effectiveness of the mini trawl, resource sustainability, and social conflicts among fishermen (Figure 2).

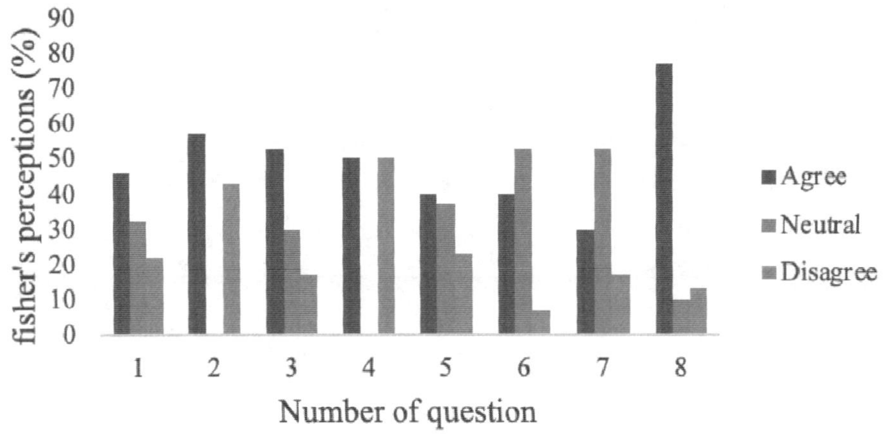

Fig. 2. Cilincing fishermen's perception of the mini trawl fishery in Cilincing waters, North Jakarta.

In total, 46% of 15 respondents stated that mini trawl are effective to catch the fish in Cilincing water, 32% stated that they are neutral towards the use of mini trawl for catching fish, and the remaining 22% disagreed that the use of mini trawl is effective for catching fish (Question 1). As many as 57% of fishermen stated that mini trawl are environmentally friendly fishing gear, as long as it is not operated near the coast. The rest (43%) disagreed that using mini trawl is environmentally friendly because the gear can damage the habitat (Question 2). According to productivity (Question 3), 53% of fishermen reported that using mini trawls can produce more catches, thereby increasing catch productivity. Meanwhile, the other 30% of fishermen believe that the catch fluctuation is influenced by fish stock condition, which is highly dependent on weather and seasons. Good weather conditions (conducive) make it possible to catch fish in large quantities, while bad weather can make it difficult to catch fish in small quantities. The remaining 17% of fishermen did not agree that using mini trawls could increase catch productivity because the catches obtained so far could not be determined in size or quantity. Question 4 collected fishermen's perceptions regarding the effect of continuous catch of undersized catch on fish populations. Approximately, 50% of fishermen agree that mini trawl fishing gear affects fish population when undersized fish are found. Meanwhile, another 50% of respondents disagreed. Those who disagree believe that there are still many fish in a body of water because fish will never run out in the sea. Question 5 addressed fishers' perceptions of the effect of mini trawl use on catch volume. Results show that 40% of fishers perceived mini trawls as very effective in increasing catch volume, as a large number of fish can be captured in a single haul. In contrast, 37% of respondents expressed a neutral perception, acknowledging that catch volume is not determined solely by the operation of mini trawls but is also influenced by fish availability, weather conditions, and seasonal factors. Meanwhile, 23% of fishers disagreed that the use of non-selective mini trawls significantly affects catch volume. These respondents emphasized that different fish species exhibit distinct behavioral patterns; under certain conditions, fish may only pass through a fishing area or remain briefly, reducing the likelihood of being caught and, consequently, affecting total catch volume. Question 6 explored fishers' perceptions of government regulations on mesh size in mini trawl fisheries and their role in achieving sustainable fisheries management. The results indicate that 40% of fishers agreed that larger mesh sizes are necessary to reduce conflicts among fishers and to support the conservation of fish resources by allowing fish to grow and reproduce. More than half of the respondents (53%) expressed a neutral perception, reflecting uncertainty about the effectiveness of mesh size regulations in promoting sustainability. The remaining 7% disagreed that mesh size regulations contribute to sustainable fisheries management. Question 7 examined fishers' perceptions of releasing catches that do not meet the minimum catchable size. A proportion of fishers (30%) agreed with releasing undersized fish, as they believed this practice supports sustainable fisheries management and helps maintain fish stocks with higher economic value, which may ultimately increase fishers' income. However, the majority of respondents (53%) expressed a neutral perception, primarily due to limited knowledge and information regarding size regulations, which makes it difficult for them to determine whether a caught fish meets the legal catch size. The remaining 17% of fishers disagreed with the practice of releasing undersized catches. Question 8 focused on fishers' perceptions of whether differences in catchability between mini trawl fishers and other fishers could lead to social conflict. Most respondents (77%) believed that differences in catch levels caused by varying catchability between mini trawls and other fishing gears could potentially lead to social conflict among fishers. Meanwhile, 10% of fishers expressed a neutral perception, suggesting that conflict may arise from social jealousy toward fishers using more advanced technology, depending on how individual fishers respond to declining catches. If such situations are managed appropriately and in accordance with existing regulations, conflicts may be avoided. The remaining 13% disagreed with the likelihood of social conflict, arguing that fishers generally accept technological differences as long as fishing practices comply with established rules and that catch levels largely depend on individual fishing effort.

The categorical value of perceptions regarding the use of mini trawl was obtained from respondents' responses using three categories (agree, neutral, and disagree). The qualitative answers were then quantified by giving a score value. Respondents who answered agree, neutral, and disagree were given a score of 3, 2, and 1, respectively. All answers were summed up based on each respondent. The fishermen's perception of the use of mini trawl based on the grouping of scores concluded that most of the fishermen respondents had a high level of perception with a percentage of 50% with nine respondents, a medium level of 5 people (40%) and a low level of 2 people (10%). The medium category is in the interval 20-22. Perception categories based on respondents' answers can be seen in Table 7.

Table 7. Perception categories based on respondents' answers.

Perception categories	Number of respondents	Interval	Percentage (%)
Low	2	<20	10%
Medium	5	20-22	40%
High	8	>22	50%
Total	15		100%

4 Discussion

Jakarta bay waters are included in the state fisheries management area of the Republic of Indonesia 712. Most of the main fish resource commodities in WPPNRI 712 are in fully exploited status (reef fish, panaeid shrimp, lobster, crab, and squid), some fish groups are at overexploited (large pelagic fish and demersal fish) and moderate (small pelagic fish) utilization levels [10]. Fishery management focuses solely on the target species, without considering interactions or relationships among the various aspects of the ecosystem.

The perception of fishermen towards fisheries sustainability needs special attention because most fishermen feel that the existence of fish in the ocean is a gift from God that will not run out as long as the ocean still exists. This perception needs to be changed gradually through community leaders' approaches and education for fishermen and their descendants. In addition, fishermen also need to learn the size of fish that is worth catching and not worth catching so that fish obtained in a dead condition is impossible to throw into the sea so that it is used for sale and daily consumption.

For this reason, it is necessary to conduct education by related parties so that fishermen know the size of catchable and non-catchable fish. Fishermen generally agree with government regulations regarding the size of mini trawls. However, there are still fishermen who need to learn the regulations on this matter, so the regulations need to be applied more effectively through education for fishermen. This activity will give insight into the science for fishermen so that the possibility of problems in fisheries management can be overcome and the condition of the aquatic environment is well maintained. This socialization activity is expected to increase fishermen's compliance with regulations. The application of this regulation is also an evaluation material for the government and related stakeholders in improving sustainable fisheries management. Concerning sustainable small-scale fisheries, we should provide more evidence on the direction of the causal arrows: from nature to resource to the individual to culture to community, or in the opposite direction [11]. The government's lack of an education program will affect the pattern of community life, making it prone to social conflict. Weak government supervision ultimately leads to a lack of trust in the fishing community in government performance.

Counseling related to sustainable fisheries needs to be provided to build fishermen's knowledge and awareness and to ensure that fishery resources are available and well-maintained. Fishermen must know that undersized fish should not be caught and, if caught, must be released into the water. Therefore, the government should provide information on fishing seasons. Policy strategies for managing legal-size fishing can be implemented by regulating the fishing season based on the emergence of legal-size category fish and setting size quotas for legal-size category fish in each fishing area [7]. Based on these strategies, fishermen's skills and awareness level in conducting fishing operations can also be built. This can help fishermen conduct fishing operations so that the time used is effective and efficient. Increased coaching and counseling will increase the insight and knowledge of fishermen so that the quality of fishermen's human resources is improving, which in turn can increase income and get out of poverty.

Conflict between fishermen can occur due to strong social jealousy; for example, differences in the use of fishing technology between traditional and modern fishermen. This difference in the use of fishing gear can significantly affect catch rates. This is consistent with the findings that stated that disparities in fishing technology often trigger social conflicts between fishermen [12]. Traditional fishers usually want access to more advanced fishing technology but are limited by their limited capital compared to modern fishers. Based on this, traditional and modern fishermen compete to obtain fish resources, triggering social jealousy and stress due to decreased catches, creating social conflict between fishermen. Differences in fishing technology have the opportunity for social conflict between fishermen [13]. This is caused by social jealousy over the number of catches that eventually leads to protests manifested in physical attacks and the destruction of fishing boats. Therefore, the government must manage conflict with relevant parties, including the fisheries agency, fishing communities, community leaders, religious leaders, and local area managers.

5 Conclusion

The sustainability level of mini trawl fisheries activities in the Cilincing area was rated as moderate, with a score of 60.67. Most mini trawl fishers in this area (50%) support using mini trawl for fishing, while 40% are neutral, and 10% disagree. The impact are fishing using mini trawls in Cilincing waters still maintains sustainable fish resources.

Acknowledgements

This paper was supported by the Directorate of Research and Innovation, IPB University with National Research (Ri-Na) program activities with the title Seine Net fishing and Its impact in Cilincing waters, Jakarta based on cooperation agreement letter No.537/IT3.D10/PT.01.03/P/B/ 2023 dated 24th November 2023. The author would also like to thank the Sub Directorate of Food Security, Marine Affairs and Agriculture, Cilincing, and the mini trawl fishermen in Cilincing, North Jakarta.

References

1. Ministry of Marine Affairs and Fisheries of Republic Indonesia, Placement of fishing gear and fishing aids in the fisheries management area of the Republic of Indonesia and the high seas as well as the arrangement of fishing zones, Pub. L. No. 18/PERMEN-KP/2021 (2021).
2. B. Nugraha, Fisheries status and habitat conditions of Jakarta Bay waters. Jakarta Research Journal **13**, 17–28 (2020). https://doi.org/10.37439/jurnaldrd.v13i1.17
3. Statistics Indonesia of North Jakarta Municipality, Production and value of fish landed at TPI locations 2019–2022 (Statistics Indonesia, Jakarta, 2024)
4. Ministry of Marine Affairs and Fisheries of Republic Indonesia, Placement of fishing gear and fishing aids in measured fishing zones and fisheries management areas of the Republic of Indonesia in inland waters, Pub. L. No. 36/PERMEN-KP/2023 (2023)
5. Ministry of Marine Affairs and Fisheries of Republic Indonesia, Prohibition of the use of trawls and seine nets in the fisheries management area of the Republic of Indonesia, Pub. L. No. 2/PERMEN-KP/2015 (2015)
6. Ministry of Marine Affairs and Fisheries of Republic Indonesia, Estimation of potential fish resources, allowable fish catch, and utilization rate of fish resources in the fisheries management area of the Republic of Indonesia, Pub. L. No. 19/KEPMEN-KP/2022 (2022)
7. T. D. K. Pribadi, R. Yusfiandayani, D. Simbolon, M. Patanda, R. Caesario, E. J. Sayidina, S. R. Violita, A comparison of diversity and catch composition of mini trawl (arad) and lift net (bagan) in Cilincing, Jakarta Bay, Indonesia. Biodiversitas. **26**, 265–277 (2025). https://doi.org/10.13057/biodiv/d260127
8. Budiarto, Blue swimming crab fisheries management with ecosystem approach in the Java Sea waters (FMA 712), Thesis, IPB University, Bogor (2015)
9. Food and Agriculture Organization of the United Nations, Ecosystem approach to fisheries (FAO, Rome, 2003)
10. L. Adrianto, A. Habibi, A. Fahrudin, A. Azizy, A. H. Susanto, I. Musthofa, M. M. Kamal, S. H. Wisudo, Y. Wardiatno, P. Raharjo, Z. Nasution, Y. Vitner, Indicators for fisheries management using an ecosystem approach to fisheries management (Ministry of Marine Affairs and Fisheries of Republic Indonesia, 2014)
11. S. Jentoft, Life above water: small-scale fisheries as a human experience. Maritime Studies. **19**, 389–397 (2020). https://doi.org/10.1007/s40152-020-00203-0
12. L. Daris, M. Yusuf, M. S. S. Ali, Wahyuti, Priority strategies for conflict resolution of traditional fishermen and mini trawl in Maros District, South Sulawesi Province, Indonesia. AACL Bioflux. **13**, 496–502 (2020)
13. D. Simbolon, Fishing ground planning, degradation and planning (IPB Press, Bogor, 2019)

BIO Web of Conferences **221**, 04004 (2026) https://doi.org/10.1051/bioconf/202622104004
7th EMBRIO International Symposium (EIS 2025)

Fisheries co-management: small-scale fisher's perception of the function and impact of FCMC (*fisheries co-management committee*) on efforts to improve small-scale tuna fisheries in Buru and Central Maluku regencies

Adjie Dharmasatya[1*], *I* Putu Agus Widi Pranata[1], *Karel* Yerusa[1], *I* Gusti Ayu Kaisha Ari Komala[2]

[1]Masyarakat dan Perikanan Indonesia Foundation, Denpasar 80237, Indonesia
[2]Department of Aquatic Resources Management, Udayana University, Jimbaran 80361, Indonesia

Abstract. Co-management in small-scale fisheries serves as a strategic approach to bridge the needs of various stakeholders, especially in the complex and high-value management of Tuna fisheries. One form of co-management facilitated by Masyarakat dan Perikanan Indonesia (MDPI) is the Fisheries Co-Management Committee (FCMC). This study aims to explore fisher's perceptions of the impact of FCMC on efforts to improve Tuna fisheries in Buru and Central Maluku Regencies, Maluku Province. A qualitative phenomenological approach was chosen for this research. Data was collected through in-depth interviews with 30% representation of the total Tuna fisher involved in FCMC. The unit of analysis, fisher's perceptions, was examined using the Advocacy Coalition Framework, revealing that fisher interpret FCMC as a vital and necessary dialogue forum between fisher and relevant stakeholders in Tuna fisheries management. Overall, FCMC is considered to have a positive function as a bridge between fishers and authorities; however, there is a need for strengthening in terms of forum sustainability, consistency of policy implementation, and the genuine involvement of government parties in the field.

1 Introduction

Fisher who are classified as small-scale fisheries are very vulnerable to the influence of various factors, namely biology, technology, and the environment [1]. As an effort to tackle the challenges faced by fishers, multi-stakeholders need to be involved in the process of solving these challenges, one of the things the government does is to accommodate fishers groups in the administrative function to get a fair assistance program [2]. However, not all rules are made by involving various stakeholders, especially in this case small-scale fishery.

As an effort to bridge and facilitate the participation of various stakeholders, collaborative management or co-management is carried out, as a form of management that is carried out jointly between stakeholders from various sectors. Co-management as a form of management partnership between the government, industry, and fishers that can strengthen management and produce a good impact [3].

*Corresponding author: adjie.dharmasatya@mdpi.or.id

Fisheries Co-Management Committee (FCMC) Tuna is an example of collaborative management involving various stakeholders in it. It FCMC Tuna was formed based on the mandate of Law No. 31 of 2004 juncto No. 45 of 2009 that fisheries management is carried out based on the principles of justice, partnership, openness and sustainable development. Thus, FCMC functions as a forum for stakeholder consultation in the fisheries sector to synergize in the development of a more resilient fisheries sector to achieve fisheries management goals which include: improving the standard of living of fisher, expanding and creating employment opportunities, ensuring the availability and consumption of fish protein sources, optimizing the utilization of fish resources, and ensuring the sustainability of fish resources [4]. FCMC is designed as a platform for stakeholders to participate in tuna fisheries management. The participation in question is a joint arrangement between stakeholders to share responsibility and authority in fisheries management practices through consultation and negotiation, including conflict resolution, decision-making, strengthening partnerships, and efficiency in tuna fisheries management [5]. This was done to reach a mutual agreement.

The potential of Maluku's fisheries is one of the highest in Indonesia [6]. Meanwhile, the largest contributor to tuna fisheries comes from small-scale tuna fisher [7]. With the exclusivity that has been attached to small-scale tuna fisher, the MDPI Foundation tries to raise not only data concerns but also issues that occur in coastal communities into a forum attended by multi-stakeholders which in this context is FCMC. The Maluku provincial FCMC has been facilitated by MDPI since 2014 and continues to run every year until now. With the consistency of the implementation, we need to see if there has been an impact perceived by small-scale tuna fisher since the FCMC was implemented. The study that needs to be carried out is an assessment of the impact of FCMC on the implementation of tuna management, especially in the Maluku area.

Research Questions

1. What are the impacts and changes felt by fishers participating in FCMC based on the Advocacy Coalition Framework?
2. What is the relevance of the impact or change for fishers to FCMC's initial goals and design?

2 Framework

The Advocacy Coalition Framework was chosen as the main framework to analyze the findings of the Respondents in this study. ACF has proven to be able to show cross-level stakeholder mapping and coalitions between actors (government, NGOs, academics, and fisher) in discussing marine protected area policies. ACF is used to identify belief systems, resources from all actors, strategies, and policy discussions [8]. The research became the basis for using ACF in identifying each dimension of belief system, coalition, policy learning, and stable parameters to describe the perception of FCMC's impact on improving small-scale tuna fisheries management. The Advocacy Coalition Framework was established with the aim of understanding collective decision-making and governance, utilizing frameworks and theories to identify what is important (and not important) in policy-building, implementation, and public policy revision.

3 Research design

Qualitative research is an approach with the aim of exploring and understanding the meaning that individuals or social groups consider to come from social or humanitarian conditions. The qualitative research process is carried out by involving the collection of natural setting data such as interviews, observations, or doctoral analysis. This study uses a qualitative paradigm with the framework analysis method as a technical data analysis. Framework analysis is a qualitative data analysis technique that is systematic and flexible. This method is used in applied research such as policy, and health services. Framework analysis was developed to encourage data analysis in a transparent, traceable analysis process, rigid based on the framework, but still flexible if other themes are found outside the framework [9,10]. This research focuses on describing the Tuna SSF's perceptions of the FCMC's impact on improving tuna management practices. The unit of analysis is small-scale fishers who serve as local champions representing other fishers in the FCMC forum. The context studied includes belief systems, policy learning, coalition direction, and stable parameters within the policy subsystem—the FCMC itself. The entire context of this research is summarized within a single framework, the Advocacy Coalition Framework (ACF), which served as the basic reference for gathering information for this study.

3.1 Participants

This research was conducted in Indonesia, specifically in Maluku Province, in Buru and Central Maluku Regencies, with a sample of 5 fisher representing a total of 30% of the participants who participated in the FCMC during 2022-2024. All participants were tuna fisher living in coastal areas, members of, and former members of, the Fair-Trade Tuna ecolabel-certified fishers in Indonesia. Given these sample characteristics, this research was conducted using purposive sampling. This sampling was chosen considering the potential for high contributions from the selected participants to understanding the phenomenon under study [11].

3.2 Data collection and instrument

In-depth interviews served as the data collection instrument, while data documentation was conducted verbatim based on audio recordings of the interviews. A questionnaire was prepared, adopting the Advocacy Coalition Framework, prior to data collection. However, during the interview process, not all questions were understood, requiring modifications to obtain equally valuable information. Due to distance constraints and unfavourable weather conditions, interviews were conducted by telephone and recorded using a mobile phone recording app. This aimed to minimize the possibility of forgetting or losing the essence of the interview process.

3.3 Research procedures and methodology

The research process begins with identifying the problem considered as the main theme of the research. This objective can be achieved by understanding individuals who have similar experiences related to meetings in the Fisheries Co-Management Committee (FCMC) forum. The individuals referred to in this study are specifically small-scale tuna fisher who have attended the FCMC either as participants or speakers. After the sample is determined and the data is obtained, all data from respondents in the form of interview results, audio recordings of interviews, and verbatim are collected. Using framework analysis, the collected data will then go through a process of familiarization, thematic identification based on the accepted framework, indexing or coding, charting, and finally mapping and interpretation [9,10]. The use of framework analysis was chosen because of its structured, flexible characteristics, the ability to combine deductive and inductive coding, and use matrix outputs in accordance with the framework used [9]. This can reduce research bias by structuring the interview guide and discussion of results but does not rule out the possibility of including new inductive information as new theme code in the research findings.

This research was primarily conducted to understand how tuna fisher in coastal Buru Regency and Central Maluku, Maluku, perceive the impact of FCMC on improving small-scale tuna fisheries management. The following is a diagram of the research procedures:

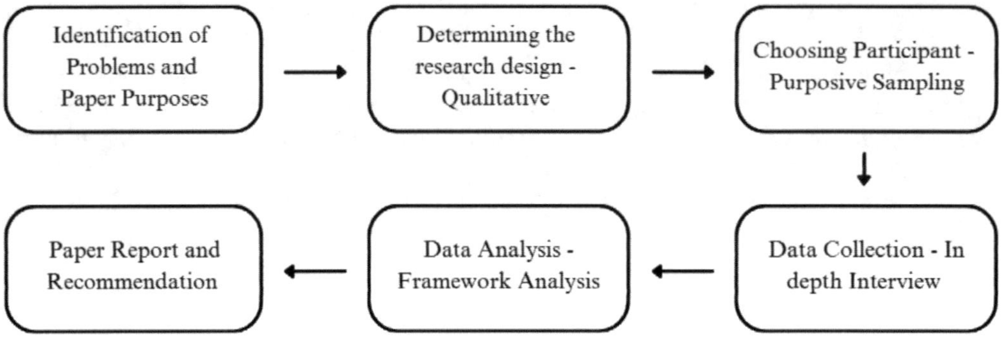

Fig. 1. Research procedures.

3.4 Data analysis

The data collected in this qualitative research were analyzed using framework analysis, which means that this research is based on a framework used as a reference for deductive indexing and provides an opportunity for inductive indexing of new themes that emerged from the respondents. The steps taken are familiarisation (reading the results of the data transcript - verbatim results of audio recordings), thematic framework (compiling themes based on the specified framework of thought and marking themes that emerge from the framework and themes that are outside the framework), indexing (coding verbatim data), charting (compiling a matrix based on the results of indexing), mapping and interpretation (compiling diagrams and mind maps of the whole charting), and validation (done with *cross-checking* Other fisher's answers about the data findings [9].

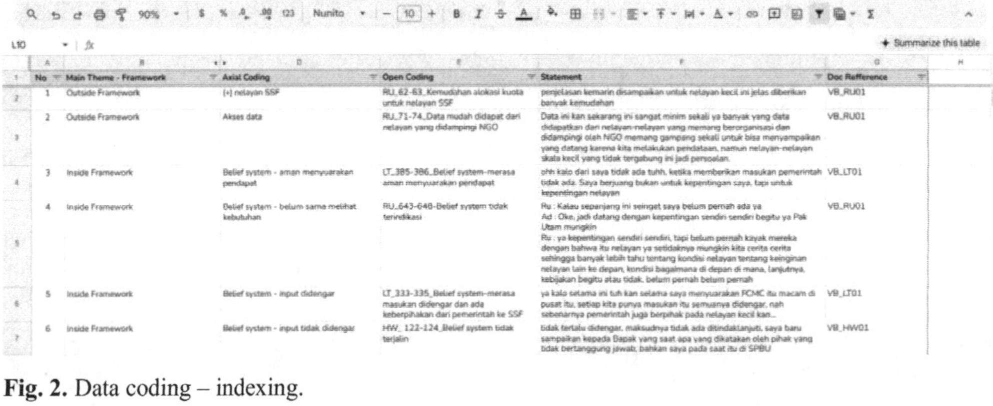

No	Main Theme - Framework	Axial Coding	Open Coding	Statement	Doc Refference
1	Outside Framework	[+] nelayan SSF	RU_62-63_Kemudahan alokasi kuota untuk nelayan SSF	penjelasan kemarin disampaikan untuk nelayan kecil ini jelas diberikan banyak kemudahan	VB_RU01
2	Outside Framework	Akses data	RU_71-74_Data mudah didapat dari nelayan yang didampingi NGO	Data ini kan sekarang ini sangat minim sekali ya banyak yang data didapatkan dari nelayan-nelayan yang memang berorganisasi dan didampingi oleh NGO memang gampang sekali untuk bisa menyampaikan yang datang karena kita melakukan pendataan, namun nelayan-nelayan skala kecil yang tidak tergabung ini jadi persoalan.	VB_RU01
3	Inside Framework	Belief system - aman menyuarakan pendapat	LT_385-386_Belief system-merasa aman menyuarakan pendapat	ohh kalo dari saya tidak ada tuhh, ketika memberikan masukan pemerintah tidak ada. Saya berjuang bukan untuk kepentingan saya, tapi untuk kepentingan nelayan	VB_LT01
4	Inside Framework	Belief system - belum sama melihat kebutuhan	RU_643-648-Belief system tidak terindikasi	Ru : Kalau sepanjang ini seingat saya belum pernah ada ya Ad : Oke, jadi datang dengan kepentingan sendiri sendiri begitu ya Pak Utam mungkin Ru : ya kepentingan sendiri sendiri, tapi belum pernah kayak mereka dengan bahwa itu nelayan ya setidaknya mungkin kita cerita cerita sehingga banyak lebih tahu tentang kondisi nelayan tentang keinginan nelayan lain ke depan, kondisi bagaimana di depan di mana, lanjutnya, kebijakan begitu atau tidak, belum pernah belum pernah	VB_RU01
5	Inside Framework	Belief system - input didengar	LT_333-335_Belief system-merasa masukan didengar dan ada keberpihakan dari pemerintah ke SSF	ya kalo selama ini tuh kan selama saya menyuarakan FCMC itu macam di pusat itu, setiap kita punya masukan itu semuanya didengar, nah sebenarnya pemerintah juga berpihak pada nelayan kecil kan...	VB_LT01
6	Inside Framework	Belief system - input tidak didengar	HW_ 122-124_Belief system tidak terjalin	tidak terlalu didengar, maksudnya tidak ada ditindaklanjuti, saya baru sampaikan kepada Bapak yang saat apa yang dikatakan oleh pihak yang tidak bertanggung jawab; bahkan saya pada saat itu di SPBU	VB_HW01

Fig. 2. Data coding – indexing.

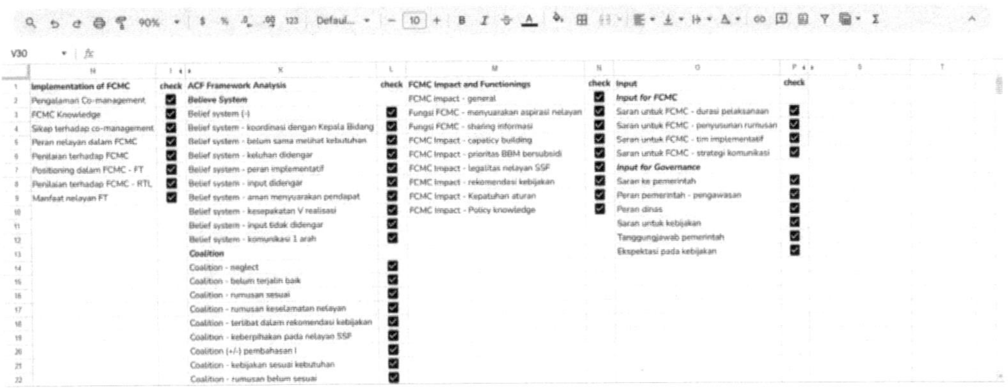

Fig. 3. Axial coding - charting process.

After going through the process of digging data through interviewing the sources, we made interview transcripts and open coding. Open coding was our first step to map the themes that emerge from the respondents. The process continued by mapping open coding into axial coding as shown in Fig.2, we group open coding into a larger, specific theme – closer to the Advocacy Coalition Framework theme. Next, we identified themes in axial coding into the charting process with the same goal, namely, to group sub-sub-themes into larger themes and can include related sub-themes.

4 Findings

As is known in the charting process, the grouping of sub-themes into main themes is carried out with the aim of mapping further results. We identified at least 22 sub-themes spread across 6 major themes, namely the challenges and conditions of fisheries faced by SSF tuna fisher, the role of NGOs-MDPI, the implementation of FCMC, ACF mapping in FCMC impact (main framework), FCMC impact and suggestions for FCMC and stakeholders involved. However, to be more focused on the research questions, here is a mind map for the 2 main themes that will be discussed, namely ACF mapping in FCMC impact (main framework), FCMC impact and suggestions for FCMC and stakeholders involved.

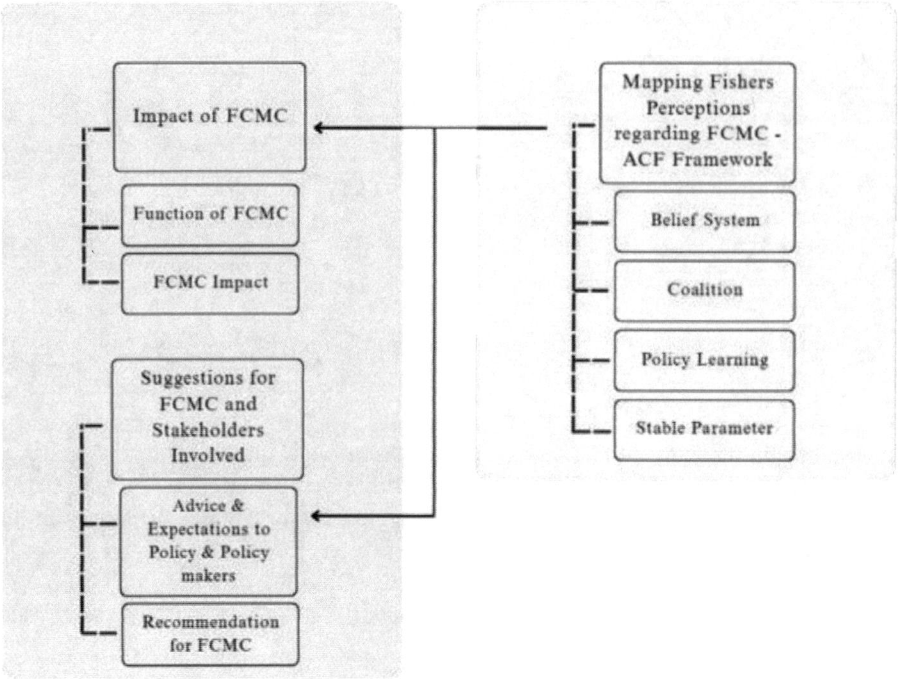

Fig. 4. Research findings diagrams.

4.1 Advocacy coalition framework – mapping fishers' perception of FCMC impact

We map the advocacy coalition framework based on its dimensions, namely belief system, coalition, policy learning, and stable parameters. In this study, we found that each respondent-fisher had quite varied and not a few contradictory answers in reflecting their knowledge and experience in each dimension, especially to view the connections, trusts, and coordination in the FCMC policy sub-system.

Table 1. Sub-theme cluster and indicators in ACF framework.

Sub-Theme Cluster	Indicators
1. Belief System - Indicated to Strengthen	Direct coordination with policy makers, inputs and complaints heard, feel safe to voice opinions
2. Belief System - Weakening Indications	There is no similarity in seeing policy needs in the field, a minimal role of policy implementers, agreements that are not entirely in accordance with the phenomenon in the field, one-way communication, and input that is not heard
3. Coalition - Strong	There is a formulation in the FCMC according to the needs of fishers (safety), fishers are involved in the formulation of policy recommendations, the alignment of policy makers with fishers
4. Coalition - Weak	There is a policy formulation that is not in accordance with the needs of fisher, the ignorance of policy makers on the needs of

	fisher, and the relationship between actors that has not been well established.
5. Policy Learning - Adaptive	Formal and non-formal coordination between actors in FCMC forums
6. Policy Learning - Resistance	Non-formal coordination between actors in FCMC forums
7. Stable Parameter	Neutral perception

4.1.1 Belief System – Strengthening and Weakening Indications

The points that we are trying to explore in the dimension of the belief system are adjusted to the accompanying indicators, namely the process of sharing trust between actors over the formulation of problems and solutions to certain policies in the FCMC forum. In this study, we found that there are indications that strengthen and weaken the belief system. As listed in Table 1, there are several sub-themes that show indications of strengthening the belief system built in the FCMC forum and felt directly by fishers, one of which is feeling that their inputs and complaints are heard during the FCMC forum. The following is one of the respondents' statements that explains that the process of sharing beliefs – belief system in FCMC is correctly felt by respondents:

yes, so far, as long as I have voiced FCMC, it's like at the centre, every time we have input, everything is heard, now actually the government is also on the side of small fisher, right... (Respondents 2)

If the office that was yesterday had a chance to see the head of the department yesterday just tell the story about the similarity, what is the name of this quota, then this data as we talk about is only not at the place of activity, but talking about it at the place of eating, in this stall in the hotel (Respondent 2)

However, each respondent also stated that not only did it strengthen the belief system felt, not a few told of unpleasant experiences related to the process of sharing trust between actors, especially between fisher and policy makers. One of them is respondents who have never been invited to discuss before the forum started or was completed. In this case, the respondent assessed that each actor brought the interests and needs of each representative in FCMC.

Never discussed or approached first, yes, at least the beta first talked to mainly about the problem, one of which is the fuel problem (Respondent 3)

Yes, their own interests, but they have never been like them with that they are fisher, yes at least we can tell a story so that we know more about the condition of fisher about the wishes of other fisher in the future, what conditions are in the future where, he continued, the policy is so or not, never (Respondent 2)

The time of the activity is (not) given time. Because maybe time is too short. And then there's the fisher who don't have the time. (Respondents 4)

4.1.2 Coalition – strong vs weak

The coalition built in FCMC is indicated by the formulation of recommendations or suggestions for policies based on needs that target the welfare of small-scale fishers and apply the opposite, namely when policies can be socialized and implemented by fisher – as part of the context of compliance with fisheries management rules. Thus, the coalition dimension is diversified into two, namely strong coalition and weak coalition. Each of these types of coalitions is indicated by the sub-themes that have been identified. Strong coalition is characterized by the conformity of policies and recommendations submitted, involving various actors in formulation and decision-making, especially small-scale fisher in recommendations for decisions or related policies. Meanwhile, the weak coalition is characterized by decision-making and recommendations dominated by certain actors in the FCMC forum.

The strong coalition identified in this study is based on the respondents' statements regarding several decisions such as the recommendation for the sea distance for the use of ships under 1 GT is below 12 nautical miles. This is realized by the respondents as a step of concern for the safety of fisher by policy makers. Furthermore, the involvement of small-scale fishers in developing recommendations is also an indication of the establishment of positive relationships between actors (policy makers – small-scale fishers). The involvement is characterized by a sense of ownership of the recommendations discussed at the end of the FCMC session and a sense of equality with the actors and policymakers, as they feel the proposals are accepted in the forum. This is directly proportional to the perception of fisher in policy makers, for their alignment with small-scale fisher.

> *Yes, yesterday it was like a discussion, then we put in ahh that's the secretary of the ministry made points that must be discussed if to make his work report, right, he made it in the form of points, there we also have fisher involved, right, so it means that in this case the results of our votes are included in yesterday's work program for the future, then what things can we make in these results, the government can also know that oh... This is the voice of a small fisher, he has a will like this. Ah that means we are involved in it because our proposal is accepted.* (Respondents 1)

On the other hand, the respondents also showed indications of a weak coalition during their participation in the FCMC forum. Some of them are neglect and the formation of formulas that are not in accordance with needs, in this context fisher feel ignored the suggestions and challenges experienced by fisher, especially regarding the condition of fisheries on Buru Island with obstacles such as the operation of many mini-purse saine fishing nets around the Buru sea. One of the respondents explained that the challenges presented in the FCMC forum were considered not in accordance with the conditions in the field, this was conveyed by the respondent who did not accept it because the respondents considered that they understood the conditions of fisher in the field better.

> *The problems that we often convey to the relevant agencies, this is only a hope.* (Respondents 2)

> *It is not yet appropriate, because the time given to us is not clear the first. Secondly, we have aspirations or suggestions that we get in the field are not used. It is as if the government is making up* (Respondent 3)

In general, respondents who have experience in co-management explained that there are some recommendations that are in accordance with the needs of fisher and indicate a strong coalition, but there are still recommendations or policies discussed in FCMC that are still not in accordance with the needs of fisher. Some of the decisions or activities after the implementation of FCMC that are in accordance with the needs of fishers are the issuance of letters of recommendation to prioritize getting fuel subsidies for fishers, good fish handling training, acceleration of the legality of small-scale fishing boats, and fishers' safety support programs – the issuance of fishers' proficiency letters. While one of the policies discussed in FCMC that has not been felt is the legality of FADs, which is due to the difficulty of managing the legality of FADs, the respondents are still registering through NGOs.

> *Because it was a bit of a hassle, if Mr. Karim came along, it was very troublesome. Because the letter came out for too long, or the FAD had fallen, the letter was not there* yet (Respondent 4)

4.1.3 Policy learning – adaptive or resistance learning

In the policy learning dimension, we found that results have been categorized into adaptive and resistance learning sub-themes. Policy learning is driven by the exchange of information in the FCMC forum; adaptive learning, in this case, is characterized by the respondent's presentation of how the learning process unfolds through clear and formal coordination. For example, one of the respondents felt that his needs could be met if the opinion expressed in the forum was written in the results of the FCMC meeting recommendations.

> *He made it in the form of points there as well as we have fisher involved, right.. So it means that in this case the results of our votes are included in yesterday's work program for the future, then what things can we do in this result, the government can also know that oh... This is the voice of a small fisher, he has a will like this. Ah that means we are involved in it because our proposal is accepted... Then the discussion of the related results that participated in it was also attached, from the Maluku Provincial Office, from PSDKP and from NGOs, from small fisher, so that's it* (Respondent 1)

Not only formal coordination, but also non-formal coordination experienced by respondents with policymakers was also felt by the fishers as a form of relationship established several times during FCMC activities. One of the respondents gave a testimony about the good relationship with policymakers after participating in FCMC several times. This is explained in the form of mutually beneficial relationships that seek to be built in non-formal coordination through the availability of data and possible recommendations for increasing catch quotas for small-scale fishers who report tuna catches.

> *Yesterday I had a chance to talk to the head of the department as well, please sir, if we can help together. So the data of fisher, from fisher, fisher to groups, can then be deposited to him, so that they can know later.* (2 Respondents)

> *Yes, the coordination was very good yesterday* (Respondent 2)

> *In fact, I have often discussed with the fisheries service and the head of the district.* (2 Respondents)

Respondents – small-scale fishers who participated in co-management forums such as FCMC indicated that they had increased their capacity as they began to realize the importance of recording the results of discussions when discussing certain issues. This is reflected from the respondents' view of non-formal coordination, which can be interpreted as unclear coordination, and followed by uncertain decisions regarding how to follow up on the results of discussions conducted in the FCMC forum itself.

Why, the FCMC activities that I said that we have often done, but all we are talking about is that yesterday it was a problem that for next year that comes next year how in every FCMC activity we immediately decide and determine there and there must be changes there (Respondent 2)

The government wants this thing, right, but it actually has to be written, right. He wasn't just talking like that. (Respondents 4)

Yes, indeed, anywhere, if we want, if, I am from a child, from the bottom, yes, if in a group, if in this group there is a meeting, then, there is a making something, a plan, it is finished, after the meeting goes home, if the plan is not evaluated, it must be finished, it is gone. Then, at FCMC, if it is, sure, it's big, there are a lot of opinions, a lot of these suggestions, a lot of this, that's if, not implemented, evaluation too, every 3 months, or 1 month, it's the same. (Respondents 5)

4.1.4 Stable parameter – neutral perception

Stable parameters are dimensions in the ACF that are defined as external variables of the FCMC sub-system itself. They are mainly related to factors outside the FCMC that can affect the direction of the discussion in the forum – either providing an opportunity or hindering the main purpose of the discussion. We try to dig up simple information about the changes that occur during the FCMC activities, especially if there are changes in the structuring of the policy makers involved in the forum. In this study, we found that respondents tend to respond neutrally, suggesting that there are not many significant changes felt by fishers towards FCMC or in the implementation of action plans after changes in structuring, leadership, or new information introduced into the forum.

But if we talk about what is called direct change in the field, if I see it, it is not too obvious (Respondent 2)

There is no change at all, the region has no power for its own area, only the help of bodies and machines and then the fish do not want to be used, what is it? there is no change in the permits of FAD and nets, in fact it continues to increase in this Buru (Respondent 3)

One of the respondents gave an interesting answer, that there was a change felt in the discussion of the FCMC forum, after the change in the structure of the Ministry of Maritime Affairs and Fisheries, namely with the ease of carrying out one of the stages of the legality of FADs.

If it is from the ministry of KKP, now we see about the change, yes it is for the issue of ship legality, the legality of the FAD explanation is indeed not how the eee that the owner of this FAD ship can do, passing through that stage. (Respondents 2)

4.2 Impact of FCMC

The impact felt by small-scale fishers from the implementation of FCMC is categorized based on the sub-theme of FCMC function and the impact of the FCMC forum itself. This categorization is based on the context of the answers that appear inductively or deductively – based on the ACF framework.

Table 2. Sub-Theme Cluster and Indicators of FCMC Impact

Sub-Theme Cluster	Indicators
1. Function of FCMC	Voice the aspirations of fishers, a platform for sharing information.
2. FCMC Impact	Capacity building, issuance of subsidized fuel recommendations, legality of small-scale fishers, involvement and implementation of policy recommendations, rule compliance, relationship with governance, and policy knowledge.

4.2.1 Function of FCMC

Respondents in this study agreed that FCMC plays an important role and function for small-scale fishers. One of them is as a connector of aspirations and a forum to convey the challenges of small-scale fishers to policymakers. One of the respondents believes that the actors involved in the FCMC forum are committed to alleviating the challenges faced by small-scale fishers.

The function is very good, there is a symbol of fisher. The provincial government, the district government, actively voiced, actively aspired to the fisher there. I was there shouting that I voiced my aspirations, I am also one of the fisher still feeling what I feel. That voice is what I have to fight for (Respondent 1)

I think it is still necessary, first the fisher have complaints, there is still a container, and at least they are there to listen first. So that it can be solved from obstacles or problems, this is important.. (3 Respondents)

If my hope is that, FCMC should not be stopped, it should be continued. Because in my opinion, FCMC is in fact an aspiration for FCMC. If there is no FCMC, it means that fisher are a bit difficult. Because FCMC has that aspiration, he can achieve it. From within the forum, everyone is committed (Respondent 5)

Another interesting thing found in larger forums, such as regional FCMCs, is the exchange of information among fishers. Fishing methods, experience running a business with fisher groups, and motivation to be independent of the shackles of the supply chain were shared in the forum. One of the respondents stated that the benefits of sharing fishing methods can increase the effectiveness of fishing by eliminating the need to carry weights tied to the fishing hook. He obtained this knowledge from fisher from other provinces.

Then yesterday we FCMC we have a different method from other regions. Well, including my discussion with the team from Gorontalo. He didn't use a knife, but he used a knife. (Respondents 1)

4.2.2 FCMC impact

Small-scale tuna fishers perceive FCMC as an activity that presents all stakeholders in fisheries with the aim of conveying obstacles, problems, and developments that occur in their respective fisheries areas. In the FCMC forum, fishers fought for their interests and tried to find solutions to the challenges they felt. In the FCMC forum, they felt that there was active collaboration, not only from fishers, but also from policymakers to review fishers' challenges carefully and comprehensively.

The needs that exist in this provincial area are all discussed at FCMC. So, we collaborate there, collaborate to talk about the problems that occur, and find solutions there and find an agreement (Respondent 1)

Based on the perception of FCMC from the fishers, we review how the impact of this co-management forum can be explicitly explained through the experience of the respondents in running FCMC. We do this to complement the findings of mapping fishers' perceptions and experiences of FCMC through the ACF framework. In the FCMC impact sub-theme, fishers' responses are quite varied, some feel that there have been no significant changes after the existence of FCMC related to fisheries management. Meanwhile, some others felt the positive impact of this forum, especially in combating IUU fishing by accelerating the legality of fishing boats and various capacity increases. Some of them also involve establishing relationships with policymakers since participating in FCMC activities.

Sing (not) exists in my opinion. Anyway, nothing has been fulfilled in my opinion. There has been no decision, or any kind or suggestion has not been reached, not yet (Respondent 5)

but he can also talk, he can do this... can sit together with people in office. Let's say you have the resources (Respondent 2)

New experiences. This means that there may be policies from the government or there are new changes, maybe I can know. Add new experiences (Respondent 4)

Yesterday was a small fisher's document, after all, that's it. This subsidized fuel has not been felt, but this subsidized fuel is ,,,,but before that, where is it... On average, there is none (Respondent 1)

The change yes,,, the change in the rules that we feel is the first one, our boat is legal. Our boats, small fisher who are currently included in NGOs, have been legalized, this is a good thing. Then the second, we as fisher also get strengthening through the SKN program. Then there is a program program that is saved by fisher in the sea, then get a program what is called a fish handling activity, many things and many benefits that we get through FCMC activities, these are some of the benefits that we can get (Respondent 2)

5 Discussion

This study aims to discuss how the impact and changes felt by small-scale tuna fisher of FCMC participants based on the analysis of the ACF framework and what is the relevance of impacts or changes from fishers if reflected on the purpose and initial design of FCMC itself. Using ACF as a framework to analyse perceptions of the impact of co-management, we found that there are at least two main themes, namely mapping fishers' perceptions of FCMC impacts, impacts, and suggestions for FCMC. In this discussion section, we divide the discussion based on the themes found.

On the theme of mapping fisher perception, we categorize respondent information into 7 sub-theme clusters. The first is the experience and understanding of fisher which leads to the strengthening of the belief system, which in this study is mentioned as the process of sharing information and trust between actors in the FCMC sub-system. In this sub-theme, fishers stated that they have felt safe in conveying challenges and suggestions and confirmed that there was no pressure from any party while participating in FCMC activities. The sense of security expressed by fishers is a good indication to strengthen the process of sharing beliefs with other actors involved in the forum. This is in line with the results of previous research that public involvement in democratic decision-making can increase the perception of legitimacy in related policy processes [12].

In contrast to the previous sub-themes that strengthen the belief system, the sub-theme belief system – weakening indication contains the experiences and views of fishers that have the potential to weaken the belief system or the process of sharing trust in the co-management forum. One of the indicators put forward by fishers is the lack of a policy-enforcement role for FCMC recommendations, as perceived by fishers. In this case, fishers feel that they are not involved in the policy supervision system, especially its implementation in the field. This has the potential to reduce the level of trust between actors, especially fishers, to policymakers. This finding is in line with the results of previous research, namely that public policy is lacking or meaningless in the absence of public participation. Furthermore, if the public feels that their participation is not responded to or has no impact, then trust in policy makers is more likely to decline [13].

The next sub-theme in the cluster mapped in this study is coalition. The coalition in this study is explored in the context of whether a policy recommendation formed at FCMC is in accordance with the needs of each actor, but in this case, what is emphasized is a policy that is in accordance with the needs of small-scale fishers. As with the sub-theme of the cluster belief system, the fisher's response is quite diversified into two sub-themes, namely strong coalition and weak coalition. The strong coalition felt by fishers is the formulation of FCMC recommendations that are tailored to the context of fishers' safety through recommendations for distance to sea and the direct involvement of fisher in the formulation of recommendations at the forum. This can foster a strong coalition with a sense of participation from each actor, especially actors who are directly involved with policies, namely small-scale fishers. These findings are in line with previous research, namely that management or policies prepared through the active involvement of fishers in co-design produce more relevant solutions, minimize conflicts, and strengthen coalitions between fishers and stakeholders. Participatory forums and reflective networks build trust, facilitate information sharing, and create collective commitment to policy implementation. This mechanism is effective because it combines local needs with policy objectives, creating a real incentive for all parties to work together [14].

In addition to the belief system and coalition, one of the most crucial things in mapping fishers perception of the impact of FCMC is on the policy learning dimension. The extraction of data or information about fishers experiences in the context of policy learning is reflected in the existence of formal and non-formal coordination between actors and its implications for the adaptive learning and resistance learning process. In this study, there are fishers who experience direct and formal coordination with policymakers, such as a sense of pride when their name is on the list of names of policy recommendation compilers carried out in the FCMC forum and we also found fishers who feel non-formal coordination as an effort to fulfil data compliance. This process can improve adaptive learning because trust is established based on written and formal coordination. This is in line with the results of other studies that say adaptive policy learning is most effective in the form of a formal mechanism, in accordance with written rules combined with non-formal coordination through the practice of sharing quick information outside the forum [15].

This study found that there are impacts that have been mapped using the ACF framework, especially in the context of direct coordination with policymakers and feeling safe and heard in the delivery of opinions, which is reflected as a resource for relationships built between fishers and policymakers. Another impact that is mapped from fishers perception of FCMC is in the dimension of strong coalition, which is indicated by the involvement of fisher in policy formulation and the acceptance of fishers input. This is in accordance with the impact conveyed by fishers after the implementation of FCMC, such as the issuance of recommendations for fuel priorities for small-scale fishers. The next impact felt by fishers can also be quite identifiable from the policy learning dimension, namely the growing sense of pride in the inclusion of fishers' identities in the FCMC recommendation document which is also in line with the formal coordination shown in the FCMC forum.

The results of this study also show that the FCMC functions as intended to be formed, namely as a collaborative management forum that allows various stakeholders to share responsibility and authority in tuna fisheries management practices. Fishers' perceptions confirm that FCMC has facilitated more open consultation and aspiration, as well as providing space for them to be involved in the decision-making process and formulation of policy recommendations. In addition, the forum strengthens partnerships between actors, particularly between fishers, governments, and non-governmental organizations, which is reflected through formal and informal coordination. FCMC also plays a role in opening up a space for negotiation and conflict resolution related to fisheries management dynamics, although its effectiveness still faces limitations in the aspect of policy follow-up. In addition, FCMC contributes to increasing the efficiency and capacity of fishers through a joint learning process, both in terms of technical fishing and knowledge related to regulations. These findings confirm that FCMC is an important instrument in strengthening fair, participatory, and sustainable tuna fisheries governance.

6 Conclusion

This study shows that the FCMC plays an important role as a forum for dialogue and communication between small-scale tuna fishers and policymakers in fisheries management in Buru and Central Maluku Regencies. Through an analysis of the Advocacy Coalition Framework (ACF), it was found that fishers assessed that FCMC provides a positive function in the form of conveying aspirations, capacity building, accelerating the legality of fishing vessels, and access to policy recommendations such as fuel subsidies and safety programs. However, the effectiveness of this forum still faces challenges in the form of weak policy implementation in the field, one-way communication, and limited role of the government in following up on the results of the forum. Therefore, the sustainability of FCMC requires strengthening the aspects of consistency of implementation, policy socialization, policy supervision, and increased attention to the socio-economic conditions of smallholder fishers so that the initial goal of co-management, namely fair, participatory, and sustainable fisheries governance, can be achieved.

7 Recommendation

The findings of this study highlight the importance of delivering knowledge to fishers in a logical and adaptive manner to enhance their understanding of the objectives behind action plans and policy implementation. This process requires direct engagement from policymakers and support from non-governmental organizations (NGOs), academics, and industry actors. Strengthening fisher capacity based on their interests—including improving communication skills—is essential to foster participation and build confidence in expressing opinions. Fishers emphasize that policy implementation must be effectively carried out, with policymakers actively involved in field-level execution. They also recommend that the authority for managing marine areas within 20 nautical miles be returned to local governments to facilitate FAD licensing processes, alongside assistance in determining coordinate points. Furthermore, comprehensive policy outreach is needed, along with strict supervision of Fisheries Fuel Stations and transparent communication regarding the availability of subsidized fuel. Increased attention to the socio-economic conditions of small-scale fishers is also necessary, given their significant role in food security and the coastal economy.

7.1 Limitation

In this study, we applied a framework analysis to understand the perspective of small-scale fishers on the impact of FCMC for improving small-scale fisheries management, and the participants involved were Tuna fishers in Buru and Central Maluku Regency who were Fair-Trade certified and had experience directly participating in co-management activities both as resource persons and participants in activities. However, this study has several limitations that should be considered when interpreting the findings. First, we conducted the data collection process by interviewing via telephone media, this is because the distance of the respondents is far, the time of data collection coincides with the activities of the respondents to build cooperatives, the catch season, and one of them is currently switching professions to become delivery workers in the mining complex so that it is not possible to present each respondent to conduct face-to-face interviews. The limitation of using telephone media for interviews is that we were not fully able to understand the real experiences contained in the respondents' nonverbal expressions and did not describe their experiences in a complete and more in-depth way.

Reference

1. M. S. Baskoro and D. R. Monintja, Analysis of Production Factors of Small-Scale Fisheries using Arad Nets in Tegal City , Indonesia, Dev. Ctry. Stud. **5**, 98 (2015).
2. A. Zulham and R. Pramoda, Pengorganisasian Nelayan Skala Kecil di Zona Penangkapan Ikan Perikanan Industri dalam Mendukung Rencana Kebijakan Penangkapan Ikan Terukur Organizing Small Scale Fishers for Catching Fish in The Industrial Fishing Zone to Support A Measurable Fishing Pol, Bul. Ilm. Mar. Sos. Ekon. Kelaut. Dan Perikanan, Ejournal Balitbang 89 (2021).
3. P. Raakjær Nielsen, Jesper; Degnbol, Fisheries Co-Management - An Institutional Innovation, Iifet 1 (2002).
4. MDPI, Panduan pelaksanaan pengelolaan bersama perikanan tuna 2.0 di provinsi maluku, (2021).
5. R. L. De Moura, C. V. Minte-Vera, I. B. Curado, R. B. Francini-Filho, H. D. C. L. Rodrigues, G. F. Dutra, D. C. Alves, and F. J. B. Souto, Challenges and prospects of fisheries co-management under a marine extractive reserve framework in northeastern Brazil, Coast. Manag. **37**, 617 (2009).
6. W. Waileruny, T. Kesaulya, and Y. M, Analisis Usaha Perikanan Pancing Tuna Di Kecamatan Amahai Kabupaten Maluku Tengah, Trit. J. Manaj. Sumberd. Perair. **18**, 38 (2022).
7. E. Hoshino, S. Pascoe, I. Van Putten, B. P. Resosudarmo, F. Satria, and L. Sadiyah, Survey-based approach to generate regional multipliers for the Indonesian tropical tuna fisheries, Environ. Dev. Econ. 405 (2024).
8. C. M. Weible, An advocacy coalition framework approach to stakeholder analysis: Understanding the political context of California marine protected area policy, J.

Public Adm. Res. Theory **17**, 95 (2007).

9. N. K. Gale, G. Heath, E. Cameron, S. Rashid, and S. Redwood, Using the Framework Method for the Analysis of Qualitative Data in Multi-Disciplinary Health Research, 2013.
10. J. Ritchie and L. Spencer, *Qualitative Data Analysis for Applied Policy Research*, in *Analyzing Qualitative Data* (routledge, 2002), pp. 173–194.
11. M. Q. Patton, *Qualitative Research & Evaluation Methods: Integrating Theory and Practice* (Sage publications, 2014).
12. M. Germann, Mini-Publics, (Lack of) Representativeness, and Legitimacy Beliefs, Br. J. Polit. Sci. **55**, 1 (2025).
13. A. N. M. Abdi, A. I. Nageye, and H. O. Sabriye, Examining the effects of public participation on residents' trust in local government: The mediating effect of perceived responsiveness, PLoS One **20**, 1 (2025).
14. W. C. Lewin, M. E. Pierce, R. Arlinghaus, M. S. Weltersbach, and H. V. Strehlow, Stakeholder workshops on western Baltic cod fisheries–conflict and consensus in the face of a highly dynamic ecosystem, Biol. Conserv. **302**, (2025).
15. S. Akther and J. Evans, Emerging attributes of adaptive governance in the global south, Front. Environ. Sci. **12**, 1 (2024)

Strengthening Indonesia's role in the ASEAN blue economy through policy and roadmap alignment

Imadadienan Fadlin Jaida[1*], *Marwa* Irfan Hanif[2], *Maulanie* Makhardani[1]

[1]School of Environmental Science and Engineering, Nanjing University of Information Science and Technology, Pukou District, Nanjing, China, 211544
[2]Department of Marine Sciences, Faculty of Fisheries and Marine Science, Diponegoro University, Semarang 50275, Indonesia

Abstract. Indonesia's expansive maritime domain and geostrategic position in Southeast Asia underscore its critical role in advancing sustainable ocean governance through the blue economy paradigm. This study examines the alignment between Indonesia's *Blue Economy Roadmap (2023–2045)* and the *ASEAN Blue Economy Framework (2023)*, intending to assess policy coherence, institutional coordination, and implementation synergies at both national and regional levels. Employing a descriptive qualitative methodology and a meta-synthesis of policy documents, institutional reports, and scholarly literature. The analysis evaluates four dimensions of alignment: vision and objectives, institutional arrangements, implementation mechanisms, and stakeholder engagement. The findings demonstrate substantial convergence in sustainability-oriented priorities, inclusive economic growth, and ecosystem stewardship, particularly across five strategic pillars: blue bioeconomy, blue financing, ecosystem restoration, governance coordination, and marine data systems. Both frameworks articulate a shared commitment to integrating ecological resilience with economic transformation. Nonetheless, persistent challenges remain, including fragmented marine data management, limited inter-ministerial coordination, and insufficient mechanisms for local community participation. By strategically aligning its national roadmap with regional priorities, Indonesia can catalyze cohesive and resilient maritime governance. More broadly, the study contributes to the literature on regional ocean governance by positioning ASEAN as a potential global exemplar of sustainable and inclusive blue economic transformation.

1 Introduction

The blue economy is a significant concept that involves harnessing ocean resources to help the economy grow, improve people's lives, and create jobs while protecting ocean ecosystems. It covers a wide range of traditional and emerging sectors, such as fishing, tourism, shipping, oil and gas production, marine biotechnology, ocean waste management, and renewable energy. In the blue economy, many things happen on the ocean and along the coast [1]. The goal is to balance profit-making, environmental protection, and ensuring everyone has the same rights. The blue economy is excellent in the long term because it keeps ocean ecosystems healthy and encourages individuals to use ocean resources to make money and get jobs. It has a lot in common with the United Nations' Sustainable Development Goals

* Corresponding author: imadadienan@gmail.com

(SDGs), especially SDG 14, which calls for the sustainable use and protection of seas, oceans, and marine resources. The blue economy also helps other SDGs by promising to end hunger, poverty, and inequity [2].

Indonesia, which has the largest maritime area in Southeast Asia, spans more than 17,000 islands located at the strategic crossroads of the Indian and Pacific Oceans. Indonesia plays a crucial role in trade and maritime security as it controls key waterways essential for international commerce. Indonesia has a long coastline and abundant marine resources. These things might help it build a thriving blue economy [3]. Indonesia has officially launched its *Blue Economy Roadmap* (2023–2045), with the overarching objective of improving the welfare of coastal communities, advancing sustainable economic development, and ensuring the prudent management of the country's maritime resources. The plan outlines strategies to address ocean degradation, inadequate infrastructure, and limited public awareness of marine stewardship. It shows the importance of legal, political, social, technological, environmental, and economic tools to ensure the blue economy model works [4].

Evaluating the coherence between Indonesia's *Blue Economy Roadmap (2023–2045)* and the *ASEAN Blue Economy Framework (2023)* is a crucial step in understanding how Indonesia's national strategies can be effectively aligned with ASEAN's regional priorities to promote sustainable and integrated maritime governance. This comparative assessment seeks to identify areas of alignment and divergence, assess the feasibility of joint implementation, and determine Indonesia's contribution to ASEAN's overarching blue economy vision. Analyzing Indonesia's strategic pillars, including blue bioeconomy, blue financing, ecosystem restoration, governance coordination, and marine data systems, the research underscores how policy coherence can reinforce collective governance, enhance regional resource management, and support ASEAN's pursuit of sustainable, inclusive, and equitable maritime development. In doing so, it positions Indonesia's leadership and policy innovation as catalysts for a more integrated and resilient regional blue economy.

The ASEAN Blue Economy Framework was formed in 2023 to promote cooperative management of maritime resources and environmental sustainability among member nations. This framework clarifies the necessity for coordination among ASEAN member states to protect the environment while efficiently exploiting their marine resources. It also addresses the gaps between human and technological resources [5]. Indonesia plays a pivotal role within ASEAN and has made significant progress in advancing the blue economy agenda. By sharing knowledge, best practices, and regulatory frameworks with other member states, Indonesia not only strengthens regional policy coherence but also facilitates the implementation of sustainable ocean governance across Southeast Asia. Its leadership helps align national initiatives with ASEAN priorities, ensuring that regional collaboration is both effective and adaptive to environmental, social, and economic challenges [6].

This study evaluates the alignment between Indonesia's Blue Economy Roadmap (2023–2045)[4] and the ASEAN Blue Economy Framework (2023) [6]. It aims to assess the coherence of their visions, strategic priorities, governance structures, and implementation mechanisms in advancing sustainable ocean development and regional cooperation in Southeast Asia.

2 Methods

This study examines Indonesia's role in the ASEAN Blue Economy through a descriptive qualitative methodology. The qualitative approach in the literature review is used to synthesize and summarize the results of descriptive qualitative research. This method of synthesizing and summarizing qualitative research results is called meta-synthesis. Meta-synthesis is a qualitative research methodology that integrates findings from multiple qualitative studies to generate comprehensive insights and enhance the applicability of research outcomes in practical settings [7].

2.1 Data sources

This study relies on secondary data collected from official policy documents, institutional reports, and academic literature. The materials were selected based on their relevance to Indonesia's and ASEAN's Blue Economy policy framework, credibility, and publication recency, as listed in Table 1.

Table 1. Summary of data sources.

Type of Document	Source / Title	Description	Publisher / Institution	References
National Policy Document	*Indonesia's Blue Economy Roadmap (2023–2045)*	Outlines Indonesia's long-term vision, strategic goals, and implementation mechanisms for sustainable marine and coastal development.	Ministry of National Development Planning/ National Development Planning Agency (Bappenas)	[4]
Regional Policy Framework	*ASEAN Blue Economy Framework (2023)*	Provides regional principles, strategies, and cooperation mechanisms for member states in promoting the blue economy.	ASEAN Secretariat	[6]
Government Report	*Annual Report on Marine and Fisheries Performance*	Presents data on Indonesia's marine sector achievements, institutional initiatives, and challenges related to blue economy programs.	Ministry of Marine Affairs and Fisheries (KKP)	[8-,9]
Academic Literature	*Selected peer-reviewed journal articles and policy evaluations*	Provide theoretical foundations and analytical insights into blue economy governance, sustainability, and regional cooperation.	Various academic publishers (Scopus/Google Scholar)	[3, 5, 10-,11]

2.2 Evaluation criteria

The alignment between Indonesia's Blue Economy Roadmap (2023–2045) and the ASEAN Blue Economy Framework (2023) is assessed using four criteria. Each criterion evaluates specific aspects of policy coherence, governance, implementation, and stakeholder engagement. These four criteria were chosen because they represent the key dimensions of effective policy alignment between national and regional frameworks, as summarized in Table 2.

Table 2. Evaluation criteria.

Criterion	Description	Analytical Focus
Alignment of Vision and Objectives	Assesses how well Indonesia's roadmap goals, principles, and strategic priorities align with those outlined in the ASEAN Blue Economy Framework.	Identification of shared priorities, targets, and expected outcomes.
Institutional Arrangements	Examines the governance structures and institutional mechanisms outlined in both documents.	Evaluation of stakeholder roles and responsibilities, including government agencies, private sector actors, and local communities.
Implementation Mechanisms	Reviews the instruments, strategies, and action plans proposed for realizing policy goals.	Analysis of policy feasibility, resource allocation, timelines, and monitoring processes.
Stakeholder Engagement	Evaluates the inclusiveness and participatory nature of the policy development process.	Assessment of multi-actor participation, public involvement, and cross-sector collaboration.

2.3 Data analysis

The analysis process in this study was conducted manually using a qualitative descriptive approach. Each document listed in Table 1 was carefully reviewed and coded to identify key themes, objectives, and implementation mechanisms relevant to Indonesia's and ASEAN's Blue Economy agendas. The review focused on extracting and comparing policy language, institutional structures, and strategic directions outlined in the selected materials.

A content analysis technique was applied to interpret patterns, similarities, and differences across the documents. The analysis emphasized how Indonesia's national roadmap aligns with the regional framework in terms of vision, governance structure, implementation approach, and stakeholder participation, as outlined in Table 2. To ensure analytical rigor, the documents were cross-checked for consistency and triangulated with supporting academic literature and government reports. Manual analysis was considered appropriate due to the limited number of papers and the interpretive nature of the comparative policy analysis. Manual review allowed for a more nuanced understanding of policy intent, contextual differences, and institutional dynamics that might not be fully captured by automated tools [7].

3 Result and discussion

3.1 Alignment of vision and objects

The alignment between Indonesia's Blue Economy Roadmap (2023–2045) and the ASEAN Blue Economy Framework (2023) matters because it strengthens regional cooperation and policy coherence in managing shared marine resources. This alignment allows Indonesia and ASEAN to coordinate strategies for sustainable ocean governance, promote trade and investment in blue sectors, and address transboundary challenges such as marine pollution and overfishing more effectively. Moreover, by aligning national and regional priorities, Indonesia enhances its leadership role in shaping ASEAN's blue economy agenda while ensuring that national initiatives contribute directly to regional sustainability and resilience goals [3, 4-6].

At a deeper level, this synergy underscores the interdependence between national ambitions and regional integration efforts. Both frameworks emphasize sustainability, innovation, and inclusiveness as key drivers of ocean-based growth. Such alignment ensures policy coherence across scales, strengthens institutional collaboration, and enables the harmonization of marine data systems, spatial planning, and regulatory standards. For Indonesia, convergence with ASEAN's framework enhances its legitimacy as a regional leader in blue governance, positioning it as a bridge between local implementation and regional cooperation. The shared commitment also allows both Indonesia and ASEAN to respond collectively to transboundary issues like illegal, unreported, and unregulated (IUU) fishing, plastic pollution, and ecosystem degradation that transcend national boundaries and require multilateral coordination [5].

As illustrated in Table 3, the visions and objectives of both frameworks align closely. Indonesia's roadmap envisions sustainable and knowledge-based marine resource management to create socio-economic prosperity, while ASEAN's framework promotes inclusive and environmentally responsible growth through marine and freshwater resource utilization. Both highlight sustainability and prosperity as the core of blue economy transformation. This alignment is not merely conceptual, but it also operationalizes shared objectives through measurable targets, such as increasing marine protected areas, expanding the maritime sector's GDP contribution, and enhancing maritime employment. These goals demonstrate Indonesia's dedication to advancing ecological restoration, economic diversification, and social welfare three pillars that directly reinforce ASEAN's aspiration for integrated regional growth [5, 10-11].

Furthermore, the alignment enhances Indonesia's strategic contribution to ASEAN's blue economy by enabling harmonized policy design and collective progress monitoring. For example, Indonesia's target to expand marine protected areas to 30% of national waters complements ASEAN's goal of developing regional actions to conserve ecosystems and combat IUU fishing. Similarly, Indonesia's ambition to raise the maritime sector's share of GDP and employment strengthens ASEAN's economic integration and inclusivity agenda. Through this shared vision, Indonesia not only supports ASEAN's sustainability narrative but also leverages its vast marine potential to position Southeast Asia as a global leader in blue economic transformation.

However, despite the overall coherence, differences remain in the pace of governance and in the implementation mechanisms. Indonesia's growing economic and political influence enables it to move beyond ASEAN's consensus-based "ASEAN Way," sometimes leading to a faster national response than regional coordination. While this divergence can create temporary policy gaps, it also introduces opportunities for institutional learning and adaptive governance within ASEAN. Indonesia's proactive diplomacy such as its Indo-Pacific framework, extends ASEAN's blue economy discourse beyond its traditional focus, fostering cross-sectoral collaboration and expanding the regional scope of sustainable maritime development [10].

In essence, the alignment of vision and objectives between Indonesia's roadmap and the ASEAN framework is not only symbolic of shared aspirations but also functional in advancing coordinated action. It deepens Indonesia's contribution to ASEAN's blue economy by aligning ecological protection, economic innovation, and social inclusion under a unified policy direction, thereby enhancing regional resilience, integration, and collective sustainability [3].

Table 3. Comparison of the Indonesia's blue economy roadmap and ASEAN blue economy framework.

Aspect	**Indonesia's Blue Economy Roadmap (2023–2045)**	**ASEAN's Blue Economy Framework (2023)**	**Alignment Level**
Vision	*"Our diverse coastal and marine resources are sustainably managed through a knowledge-led Blue Economy to create socio-economic prosperity, ensure a healthy marine environment and strengthen resilience for the benefit of current and future generations."*	*"ASEAN's Blue Economy Vision emphasises value creation by advancing sustainable practices that promote sustainable and inclusive economic and social development related to marine and freshwater activities and livelihoods, contributing to ASEAN economic integration efforts, economic growth, and sustainable development. By leveraging the potential of the blue economy, ASEAN envisions a new engine for growth that is socially, economically, and environmentally sustainable."*	**High** – Both emphasize sustainability, inclusiveness, and prosperity through ocean-based growth.
Target / Objective 1	Increase Marine Protected Areas to **30%** (97.5 million ha) of national waters by 2045 to preserve biodiversity, restore fisheries, and protect key ecosystems (mangroves, seagrass, coral reefs).	Develop regional actions to protect marine and coastal ecosystems, combat IUU fishing, and promote sustainable aquaculture and fishing practices.	**High** – Strong alignment in ecosystem conservation and sustainable fisheries.
Target / Objective 2	Increase maritime sector GDP contribution to **15%** of national GDP by 2045.	Promote marine industrial development, trade and connectivity, blue governance, biotechnology, and innovation for sustainable growth.	**High** – Both target economic growth from sustainable marine industries.
Target / Objective 3	Increase maritime employment to **12%** of total national employment by 2045 to improve social well-being.	Support inclusiveness, livelihood improvement, and climate change adaptation in coastal and freshwater communities.	**High** – Shared emphasis on socio-economic benefits for communities.
Cross-cutting	–	Ensure a cross-sectoral and cross-stakeholder approach, create a value chain, support inclusiveness, and ensure sustainability as a new engine for ASEAN's future economic growth. Serve as a non-binding reference for ongoing engagement among AMS and with partners.	**Medium–High** – Indonesia emphasizes national integration and sectoral targets, ASEAN adds a stronger regional collaboration dimension.

3.2 Strategic pillars and implementation mechanisms

To strengthen Indonesia's role in the ASEAN Blue Economy, it is essential to examine the strategic pillars and implementation mechanisms shared between Indonesia's national roadmap and the ASEAN framework. These pillars of blue economy: blue bioeconomy, blue financing, ecosystem restoration, coordination and governance, and marine data systems reflect both parties' commitment to sustainable ocean governance and outline practical steps for achieving their shared vision, as illustrated in Fig 1. The blue bioeconomy is vital as it promotes innovation-based growth and sustainable resource use across sectors such as aquaculture, marine biotechnology, and bioproducts, thereby enhancing Indonesia's contribution to ASEAN's collective economic resilience. Blue financing supports these initiatives by mobilizing investments through sustainable funding mechanisms and partnerships, enabling both Indonesia and ASEAN to expand environmentally responsible maritime industries. Ecosystem restoration, exemplified by initiatives like the Coral Triangle Initiative (CTI), strengthens biodiversity conservation, coastal protection, and community livelihoods, ensuring that ecological sustainability underpins economic progress. Coordination and governance are equally crucial, as Indonesia's integrated approach linking ministries, industries, and local communities provides a model for regional collaboration and institutional synergy within ASEAN. Lastly, marine data and information systems enhance transparency, monitoring, and evidence-based policymaking; Indonesia's Maritime Intelligent Platform demonstrates how digital integration can support efficient marine resource management. Together, these strategic pillars reinforce Indonesia's leadership and accelerate ASEAN's progress toward a sustainable, inclusive, and resilient blue economy [7-9].

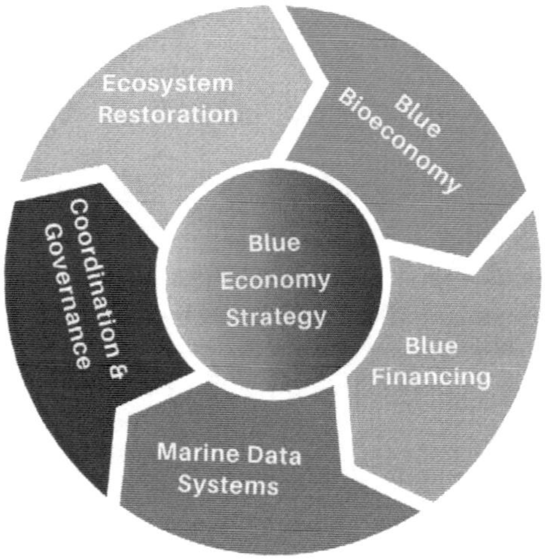

Fig. 1. Strategic pillars of blue economy.

3.2.1 *Blue bioeconomy*

The blue bioeconomy serves as the cornerstone of sustainable ocean-based development by integrating biotechnology, renewable marine resources, and innovation to generate high-value products. For Indonesia, this pillar is essential because it promotes economic diversification beyond traditional fisheries and supports environmentally responsible industries such as aquaculture, marine bioproducts, and pharmaceuticals. By advancing the blue bioeconomy, Indonesia strengthens ASEAN's collective capacity to achieve sustainable growth, create jobs in coastal communities, and reduce reliance on extractive marine activities. This aligns directly with ASEAN's goal of balancing economic expansion with ecological conservation [4-6].

In Indonesia's Blue Economy Framework, in Phase II (2025–2029), Indonesia focuses on increasing the added value of the blue economy through investment, technology adoption, quality assurance, and promotion of emerging sectors, including biotechnology and marine conservation, supported by robust governance and ecosystem stewardship. Phase III (2030–2034) emphasizes expansion through diversification, increasing investment and product variety in established and emerging sectors, improving feasibility and accessibility, and creating quality jobs while integrating research, innovation, and blue food initiatives. Phase IV (2035–2039) advances competitiveness, commercialization of high-value products, integration into global supply chains, and development of regional blue economy centres. These phases ensure that Indonesia's blue bioeconomy not only strengthens national economic growth and coastal livelihoods but also contributes significantly to ASEAN's agenda for a resilient, innovation-driven, and ecologically sustainable blue economy [4, 8-9].

3.2.2 Blue financing

Blue financing provides the financial infrastructure necessary to turn sustainable marine projects into reality. It includes mechanisms such as green bonds, blended finance, and public–private partnerships to mobilize investment in marine conservation, renewable energy, and low-carbon maritime transport. For Indonesia, enhancing blue financing means improving access to international funding sources and promoting innovative investment models that can be replicated across ASEAN. Thereby, strengthening regional financial resilience and supporting a shared commitment to sustainable development [2].

A landmark achievement in this regard is Indonesia's issuance of its first Blue Bond in May 2023, with 7- and 10-year tenors totalling JPY 20.7 billion through the Samurai Bond market. This issuance, prepared using lessons from Indonesia's first green Islamic bond and SDGs bond, demonstrates strong national commitment and provides a positive precedent for advancing the blue economy. The process was guided by comprehensive frameworks, including the Republic of Indonesia Green Bond and Green Sukuk Framework (2018), updated in 2021 to the SDGs Government Securities Framework, which identifies eligible social, green, and blue sectors for thematic bonds. Additionally, the Blue Financing Strategic Document (UNDP and Coordinating Ministry of Maritime Affairs and Investment, 2021) provides guidelines on financing key sectors of the blue economy, estimating that annual investments between IDR 3.64 to 1,392.22 trillion are needed to achieve a 15% contribution of the blue economy to GDP by 2045. Supporting policies, such as the Blue Finance Policy Note (ICCTF-Bappenas with World Bank PROBLUE), recommend strengthening the enabling environment for finance mobilization and aligning government spending with sustainable blue projects. These efforts collectively guide the preparation of Blue Finance Instruments Development Guidelines, which define eligible projects, issuance processes, and expected benefits. Beyond bonds, Indonesia has the potential to explore diverse financing schemes including debt-based instruments, other bonds, Sharia financing, trust funds, impact investment, blended finance, and ecological fiscal transfers. By championing blue financing, Indonesia not only ensures funding for sustainable projects but also reinforces its leadership role in ASEAN, promoting a financially resilient, environmentally responsible, and inclusive regional blue economy [2, 4, 8-12].

3.2.3 Ecosystem restoration

Healthy ecosystems are the foundation of a sustainable blue economy. This pillar emphasizes the restoration of coral reefs, mangroves, and seagrass meadows that provide critical ecosystem services such as carbon sequestration, coastal protection, and fisheries productivity. Indonesia's participation in initiatives like the Coral Triangle Initiative demonstrates its leadership in marine biodiversity protection and its ability to scale restoration projects regionally. Through this pillar, Indonesia helps ASEAN enhance resilience to climate change and strengthen the ecological base necessary for long-term ocean economy growth [3].

Indonesia's conservation efforts have yielded measurable benefits, with the Ocean Health Index (OHI) rising from 57.08 in 2012 to 65.8 in 2020, reflecting improvements in marine environmental quality and surrounding social conditions. However, certain components, particularly food provision, which encompasses wild-caught fisheries and aquaculture declined over the same period, indicating unsustainable fish harvesting and aquaculture practices. The Marine Trophic Index (MTI) in 2020 further highlights challenges, scoring only 16.3 out of 100 (ranked 56th globally), reflecting declines in high-trophic-level fish populations relative to lower-trophic species and signalling potential disruptions in ecosystem function and food chains. Currently, 28.91 million ha of Indonesian waters are designated as Marine Protected Areas (MPAs), and under its commitments to the Convention on Biological Diversity (CBD) and SDG 14, Indonesia aims to increase this to 32.5 million ha by 2030. Despite rapid progress in designation, the effectiveness and management of MPAs remain limited, and as marine sectors expand, competition for space will place additional pressure on ecosystems. Without adequate planning and preservation, ecosystem degradation could undermine fisheries, marine tourism, and other blue economy sectors, as well as the protective services these ecosystems provide to coastal communities. Achieving SDG 14 and ensuring a sustainable blue economy will therefore require a careful balance between economic utilization, ecosystem preservation, and long-term environmental stewardship [4, 8, 9].

3.2.4 Coordination and governance

Effective coordination and governance ensure that blue economy policies are coherent, inclusive, and enforceable. Indonesia's multi-sectoral approach linking ministries, local governments, industries, and communities provides a governance model that can inspire ASEAN-wide cooperation. Strengthening institutional coordination reduces policy overlap, enhances compliance, and promotes integrated coastal and marine planning across borders. By sharing governance best practices, Indonesia supports ASEAN's efforts to establish a harmonized regional framework for sustainable ocean management. At the national level, coordination is led by the Coordinating Ministry for Maritime Affairs and Investment, which oversees ministries with maritime competencies, including Marine Affairs and Fisheries, Energy and Mineral Resources, Transportation, Environment and Forestry, Public Works and Housing, Tourism, Investment (BKPM), and others. This structure helps streamline communication, reduce policy overlap, and facilitate integrated marine spatial and economic planning. Each ministry contributes its sectoral expertise from fisheries management, marine conservation, and renewable energy development to investment facilitation, tourism promotion, and human resource development ensuring that blue economy initiatives are supported by competent institutions, harmonized regulations, and cross-sectoral collaboration [1,4, 8].

Indonesia's decentralized governance system, with 38 provinces, 416 regencies, and 98 municipalities, adds complexity to policy implementation. While local governments have autonomy over sectors such as public works, education, and environmental management, jurisdictional fragmentation can lead to overlapping regulations and conflicts. The Blue Economy Roadmap addresses this challenge through coordinated mechanisms and legislative support, including the establishment of the Coordinating Ministry of Maritime Affairs and Investment and the revisions of key laws such as Law No. 26/2006 on Spatial Planning and Law No. 32 on the Sea. These reforms aim to integrate land and marine spatial planning, clarify institutional roles across national, provincial, and local levels, and strengthen policy coherence. The recent Law No. 6/2023 on Job Creation further streamlines the division of labor among ministries and local governments, supporting the effective delivery of programs, policies, and funding. Together, these institutional and legal measures enhance Indonesia's capacity to implement the Blue Economy Roadmap effectively, serving as a governance model that can inspire ASEAN regional coordination, harmonize ocean management practices, and ensure sustainable development of marine resources [4, 6].

3.2.5 Marine data and information systems

Accurate and accessible marine data systems are essential for informed decision-making, monitoring, and adaptive management. Indonesia's development of platforms such as the Maritime Intelligent Platform illustrates its commitment to digital transformation in ocean governance. By improving transparency, data sharing, and cross-border collaboration, this pillar enhances ASEAN's collective ability to monitor marine health, track pollution sources, and evaluate the impacts of blue economy initiatives. Indonesia's leadership in data integration helps establish a regional knowledge base that supports both policy coherence and scientific innovation [5-6].

Key actions include improving marine data repositories and open-access platforms, streamlining permits and licensing, facilitating investment in sustainable marine sectors, and developing marine research and innovation hubs. Multiple ministries and agencies including the Coordinating Ministry for Maritime Affairs and Investment, Ministry of Marine Affairs and Fisheries, Ministry of Environment and Forestry, Ministry of Finance, and National Research and Innovation Agency are involved to ensure effective governance. By enhancing data infrastructure and scientific capacity, Indonesia bolsters national marine management while contributing to ASEAN's ability to monitor marine health, track pollution, and evaluate blue economy initiatives, supporting sustainable and resilient ocean growth [4].

3.3 Key challenges in indonesia

While Indonesia has shown strong commitment to advancing the Blue Economy both nationally and regionally, several structural and operational challenges still hinder its effectiveness. Fragmented marine data systems without a unified national platform hinder comprehensive resource assessment and reflect broader institutional coordination issues. Weak collaboration among ministries leads to policy overlap and inefficiency, while limited community capacity and financial constraints prevent coastal populations from fully engaging in Blue Economy initiatives. Strengthening data integration, institutional coordination, community empowerment, and financial access is therefore essential to achieving sustainable and inclusive outcomes. Addressing these issues is critical to ensuring that policy alignment with ASEAN translates into tangible and sustainable outcomes.

3.3.1 Marine data and information systems

One of the most significant challenges in implementing Indonesia's Blue Economy agenda lies in the fragmented nature of marine data management. Currently, data on fisheries, ocean health, and maritime activities are collected by different ministries and agencies without a unified national database or standardized data-sharing protocols. This institutional fragmentation hampers integrated decision-making and weakens Indonesia's ability to develop evidence-based marine policies. Without a centralized and interoperable marine data platform, it becomes difficult to monitor ocean conditions, evaluate policy outcomes, or measure the progress of blue economy initiatives. The policy consequence of this fragmentation is the reduced effectiveness of cross-sectoral coordination and the inability to design adaptive management strategies that respond dynamically to environmental and economic changes. Establishing a centralized marine data system, such as a national Blue Economy dashboard, would enhance transparency, improve monitoring efficiency, and align Indonesia's practices more closely with ASEAN's data integration goals [4].

3.3.2 Limited institutional coordination

The complexity of Indonesia's ocean governance structure poses another major challenge to the effective implementation of blue economy policies. Multiple ministries such as the Ministry of Marine Affairs and Fisheries (KKP), the Ministry of Environment and Forestry (KLHK), and the Ministry of National Development Planning (Bappenas) share overlapping mandates, leading to fragmented decision-making and duplicated efforts. This lack of institutional synergy weakens policy coherence and slows the execution of integrated marine programs. The policy consequence of weak coordination is inefficiency in implementing multi-sectoral initiatives, which undermines Indonesia's ability to present a unified approach within ASEAN. Moreover, it risks creating conflicting regulations and inconsistent enforcement across national and regional levels. Strengthening institutional coordination through a centralized inter-ministerial framework or "Blue Economy Task Force" could help Indonesia streamline its governance processes and foster more cohesive cooperation within the ASEAN framework [4].

3.3.3 Limited community participation and capacity

Coastal communities, as the primary beneficiaries and stakeholders of the blue economy, often face barriers to participation due to limited financial resources, technical capacity, and institutional support. Many local communities lack access to training programs, funding, or partnerships needed to engage effectively in sustainable fisheries, eco-tourism, or aquaculture development. As a result, blue economy projects risk becoming top-down initiatives that do not address local needs or knowledge systems. The policy consequence of this gap is the exclusion of vulnerable coastal populations from economic benefits, thereby undermining the blue economy framework's inclusivity and social equity objectives. Empowering communities through capacity-building programs, micro-financing, and participatory planning would not only enhance local ownership but also strengthen Indonesia's contribution to ASEAN's commitment to equitable and inclusive blue growth [4].

3.3.4 Financial and investment barriers

Access to adequate funding remains a critical challenge for Indonesia's blue economy initiatives. While national and regional policies highlight the importance of blue financing, many projects particularly at the local or community level struggle to secure sufficient investment for sustainable marine and coastal development. Limited financial resources restrict the implementation of ecosystem restoration programs, sustainable aquaculture, and marine conservation initiatives. The policy consequence of these financial constraints is the underachievement of strategic objectives outlined in both Indonesia's Blue Economy Roadmap and the ASEAN framework, such as biodiversity protection, livelihood improvement, and innovation in marine industries. Strengthening access to blue financing through public-private partnerships, regional investment funds, and international development assistance would not only support the scaling of projects domestically but also demonstrate Indonesia's leadership in promoting collaborative, resource-efficient strategies across ASEAN [4, 11].

3.4 Opportunities through ASEAN cooperation

In addition to national efforts, Indonesia's engagement within ASEAN opens significant opportunities to advance the blue economy agenda. Regional cooperation not only strengthens collective capacity but also allows member states to share resources, expertise, and data in ways that individual countries could not achieve alone.The ASEAN Blue Economy Data Platform is an excellent way for ASEAN countries to work together in this area because it has all the information about blue economy sectors, including data on the economy and politics. This portal will help find ASEAN's "Blue Economic Posture" and let member countries work together on the proper projects and improve their talents. Indonesia and other ASEAN member countries could collaborate on policy and technical issues to help the blue economy grow [3, 5-6, 10].

As a member of ASEAN, Indonesia has played a big role. It has enabled people to collaborate, share knowledge, and establish robust laws for long-term marine governance. ASEAN can assist in solving problems that impact everyone, such as not having enough skilled labour and technological differences. Overall, this will help enterprises in the blue economy improve. Indonesia is the best country to lead efforts in the region to develop the ocean in a way that is healthy for the environment, given its good location and growing political and economic power. Indonesia can modify the norms and how things are done in the area by utilizing its leadership role to push for the responsible use of marine resources and economic growth. Indonesia is dedicated to long-term growth and plays a significant role in the region's efforts to implement blue economy principles. Some of the things it is doing are the Indonesian Ocean Policy and several blue economy programs [4, 9].

3.5 Policy Recommendation

Getting Indonesia more involved in the ASEAN Blue Economy framework is a big step forward. One way to do this is to make it easier for ministries to work together. Various groups need to work together to solve huge societal and environmental problems, such as climate change and the loss of the ecosystem. A holistic coordination approach that combines governance roles and action scenarios can also fill the gaps where people do not work together. This can be even better with responsive regulation frameworks and network governance tactics that have worked well in the past. For example, the Republic of Macedonia and a few Central Asian countries have proven how crucial it is for ministers to work together on policy if they wish to transform how the government functions [13].

Fixing problems with data is another key goal, and the ASEAN Regional Data Platform is a fantastic method. Ocean data science initiatives (ODSIs) that utilize data to make choices indicate that managing the coast and ocean well requires data from many sources. However, there are significant gaps between national marine databases and what regions need. Because of this, the people in charge of the groups need to revise their plans and adjust. Metadata compilation can help fix these issues by making it easier for professionals from other professions to share data and for people to discover information. The Ocean Data

Interoperability Platform (ODIP) is a good example of how to move data between countries safely. It has information from Canada, the US, Australia, and Europe [14].

Coastal communities are crucial actors in the blue economy, and it is important to find and give them power so that the economy can flourish in a fair and long-lasting fashion. These groups need to be part of the decision-making process so that development plans consider variations in social and economic status and make the area stronger. Programs that give people the skills they need to get jobs, money, and improve the environment through eco-tourism are all examples of empowerment programs. Community leaders need to ensure that policies are grounded in what is happening in the community and meet the needs of the people living there. Communities of Practice (CoPs) are also helpful in engaging local specialists to work together to help the blue bioeconomy grow in a way that stakeholders guide. It is important to include small-scale fishermen and other local stakeholders in formulating policies so that the outcomes are fair and just for everyone [1, 10-13].

Finally, the national roadmap will only function in the long run if it is implemented openly, is adaptable, and includes everyone. Many people need to work together to develop fair and transparent rules, while ensuring open information sharing among all actors. Policies should evolve with social, economic, and environmental changes through adaptive governance supported by proper monitoring and evaluation systems. Indonesia's Integrated Coastal Management (ICM) initiative in places such as Bintan and Bali serves as a strong national case of adaptive governance and multi-stakeholder engagement in marine policy. By applying similar approaches in implementing the Blue Economy Roadmap, Indonesia can maintain policy relevance, integrate new scientific and technological insights, and demonstrate leadership within ASEAN's regional Blue Economy agenda [4, 8-9]

4 Conclusion

Indonesia, with the largest coastal area in Southeast Asia, holds significant marine resources and regional influence, but faces challenges such as fragmented marine data systems, limited coordination among ministries, and insufficient coastal community engagement, which can hinder policy coherence and effective implementation of its Blue Economy Roadmap (2023–2045). To address these gaps, Indonesia leverages strategic pillars including the blue bioeconomy to drive innovation and sustainable livelihoods, blue financing to mobilize investment through instruments like its 2023 Blue Bond, ecosystem restoration to safeguard biodiversity and coastal resilience, coordination and governance to integrate national and local policies across multiple ministries, and marine data and information systems to enhance transparency, monitoring, and evidence-based decision-making. By strengthening these pillars in alignment with the ASEAN Blue Economy Framework, Indonesia not only advances its own sustainable ocean development but also serves as a regional leader, promoting harmonized governance, cross-border collaboration, and resilient, inclusive growth throughout ASEAN.

References

1 S. Li, L. Chen, & Y. Liu, Actors, themes, approaches, and imbalances in Blue Economy cooperation: A systematic review and future prospects. *Ocean and Coastal Management*, **267** (2025). https://doi.org/10.1016/j.ocecoaman.2025.107698.

2 U. Sarangi, Blue economy, blue finance and ocean governance for achieving sustainable development goals. *Natural Resources Forum*, **47** (2023) 3–21. https://doi.org/10.1111/1477-8947.12267.

3 O. Sabrina & R. A. Putra, Navigating the Blue Economy: Indonesia's Regional Efforts in ASEAN to Support Sustainable Practices in Fisheries Sector. *Sustainability (Switzerland)*, **17** (2025). https://doi.org/10.3390/su17156906.

4 L. A. T. Sambodo, D. Damayan, P. Pane, L. Hasni Pertamawa, I. Thofiq Firdaus, T. Wikapuspita, S. Kharis Harianto, S. Collin, T. Muhammad, W. Mugijayani, C. Mangunsong, M. Suardhin, D. Narjoko, A. Fauzi, S. Mayas, T. Apita Nugrahaeni, M. Thufail Mukhairiq, M. Ariq Rifqi Alfarisy, R. M. Aldo Manullang, R. Riyanto, F. Nurhadi, R. Tamadhika, & D. Kusuma Andhika, *Indonesia Blue Economy Roadmap* (Ministry of Natonal Development Planning/Natonal Development Planning Agency (BAPPENAS), 2023).

5 F. M. M. Aziz, A Review on ASEAN Blue Economy Framework: Instilling Blue Science, Technology, and Innovation for Sustainable Economic Growth. (2025), pp. 214–224. https://doi.org/10.2991/978-94-6463-791-5_24.

6 ASEAN Secretariat, *ASEAN BLUE ECONOMY FRAMEWORK* (2023).

7 M. A. Mohammed, R. J. Moles, & T. F. Chen, Meta-synthesis of qualitative research: the challenges and opportunities. *International Journal of Clinical Pharmacy*, **38** (2016) 695–704. https://doi.org/10.1007/s11096-016-0289-2.

8 KKP (Kementrian Kelautan dan Perikanan), *Bijak Mengelola Laut untuk Ekonomi Biru* (2024).

9 KKP (Kementrian Kelautan dan Perikanan), Berlayar dengan Ekonomi Biru untuk Indonesia Emas, Kinerja Kementrian Kelautan dan Perikanan 2020-2024. (2024).

10 R. N. Gamage, Blue economy in Southeast Asia: Oceans as the new frontier of economic development. *Maritime Affairs*, **12** (2016) 1–15. https://doi.org/10.1080/09733159.2016.1244361.

11 B. Endarto, D. E. Indriastuty, & F. Mardiana, Legal transplantation of blue bond regulation in Indonesia. *Environmental Development*, **53** (2025). https://doi.org/10.1016/j.envdev.2024.101118.

12 J. Elston, H. Pinto, & C. Nogueira, Tides of Change for a Sustainable Blue Economy: A Systematic Literature Review of Innovation in Maritime Activities. *Sustainability (Switzerland)*, **16** (2024). https://doi.org/10.3390/su162411141.

13 J. A. Reis-Filho, V. Hatje, & F. Barros, Navigating blue justice: Policy gaps and conflicts in coastal development from small-scale fisher perspectives. *One Earth*, **7** (2024) 1786–1802. https://doi.org/10.1016/j.oneear.2024.07.014.

14 J. Pearlman, D. Schaap, & H. Glaves, Ocean Data Interoperability Platform (ODIP): addressing key challenges for marine data management on a global scale. *OCEANS 2016 MTS/IEEE Monterey* (2016), pp. 1–7. https://doi.org/10.1109/OCEANS.2016.7761406.

Estimation of carbon footprint resulting from tuna fishing activities using different fishing gear

Faza Kamila[1], *Vita Rumanti* Kurniawati[1*], *Yopi* Novita[1], *Dwi Putra* Yuwandana[1], *Prihatin Ika* Wahyuningrum[1], and *Eko Sri* Wiyono[1]

[1] Departement of Fisheries Resources Utilization, Faculty of Fisheries and Marine Science, IPB University, Dramaga, Bogor, 16680, Indonesia

Abstract. Tuna fishing is a fishing activity with high fuel use intensity (FUI) that contributes significantly to greenhouse gas emissions. In order to support Indonesia's commitment to the Paris Agreement, efforts to reduce emissions are needed across all sectors, including fisheries. Mapping the carbon footprint (CFP) of tuna fishing activities is essential in identifying key emission points and formulating targeted strategies. This research aims to calculate and compare FUI and CFP from tuna fishing using handline, purse seine, troll line, and longline in several Indonesian fishing ports. Data were obtained through an inventory of previous studies and port statistics. Results showed that FUI ranged from 0.18-2.32 kg/kg, with the highest value for tuna longlines at Palabuhanratu Fishing Port and the lowest for purse seines at Kutaraja Fishing Port. Estimated CFP of tuna ranged from 0.06-1.72 kgCO_2eq/kg, with the highest value for tuna longlines at Nizam Zachman Fishing Port and the lowest for purse seines at Kutaraja Fishing Port. Based on each vessel's gear type, size, or location, no specific pattern was found for either FUI or CFP. This research confirms that FUI is influenced by fuel consumption and catch, while tuna CFP is primarily influenced by FUI and composition of the tuna catch.

1 Introduction

Indonesia agreed to commit to reducing greenhouse gas (GHG) emissions through the Paris Agreement, which was later renewed at the 2021 United Nations Climate Change Conference-COP26 in Glasgow, Scotland. In 2022, Indonesia joined the United Nations Framework Convention on Climate Change (UNFCCC). It submitted a plan to reduce emissions by 31.89% through its own efforts and 43.20% with international support through an Enhanced Nationally Determined Contribution, with a vision of achieving net zero emissions by 2060 or sooner.

Energy transition policy experts emphasise that failure to meet the Paris Agreement commitments could undermine international confidence and hamper green investment. In addition, uncontrolled GHG emissions will exacerbate the climate crisis and impact vital economic sectors. This underscores the importance of reducing GHG emissions in Indonesia.

GHG emissions come from various human activities and natural processes. The energy and transport sectors produce GHG emissions mainly through the combustion of fossil fuels. GHG gases can be defined as a combination of carbon dioxide (CO_2), methane (CH_4), nitrous oxide (N_2O), and other fluorinated gases. Among these gases, carbon dioxide (CO_2) is the most dominant in the atmosphere, accounting for about 76% of global GHG emissions. While other GHGs have a higher warming potential, carbon's much larger volume of emissions and

* Corresponding author: vitarumanti@apps.ipb.ac.id

long lifetime in the atmosphere make its impact on climate change more significant. Therefore, GHG emissions reduction is often focused on reducing carbon emissions, also known as the carbon footprint.

Carbon footprint measures the total carbon dioxide (CO_2) emissions released directly or indirectly caused by an activity. Direct emissions come from sources controlled or owned by the activity, such as emissions from burning fuel oil. Meanwhile, indirect emissions are produced due to the activity, but occur outside of direct control, such as from ship construction, provision of equipment and feed, fuel production, refrigeration, and so on.

One of the fishing activities that produce a high carbon footprint is tuna fishing. Tuna fisheries have a relatively high fuel use intensity (FUI) compared to other fisheries [1]. On the other hand, the number of tuna fishing vessels in Indonesia is relatively increasing annually [2].

Tuna fishing activities in Indonesia involve the use of a variety of fishing gear, such as handlines, purse seines, troll lines, and long lines. Handline is a passive fishing gear because fishermen wait for fish to eat the bait that is stretched manually without any active movement of the boat [3]. Meanwhile, purse seine is an active fishing gear operated by encircling schools of fish using a large net stretched by the fishing vessel, then the bottom of the net is pulled closed, forming a bowl-like shape. A troll line is an active fishing gear pulled horizontally by a boat while waiting for fish to eat the bait. Longline is a passive fishing gear operated through setting, immersing, and hauling the fishing line. Due to differences in the method and scale of operation, the carbon footprint generated from fishing using these gears will also be different.

To date, carbon footprint reduction efforts have been conducted through studies focusing on improving fuel efficiency mapping of various fishing activities. However, the unavailability of carbon footprint mapping of different types of tuna fishing gear means that efforts to reduce carbon emissions in the tuna capture fisheries sector cannot be accurately measured. In addition, the absence of data on the performance of tuna fishing gear in terms of the carbon footprint risks causing government policies to be less targeted, because they are not based on identifying efficiency targets and specific reduction strategies.

Studies on the carbon footprint generated by tuna fishing activities using different fishing gears in Indonesian waters have been conducted by several researchers. For example, longline at Palabuhanratu Fishing Port [4], handline at Pondokdadap Fishing Port [5], and handline and purse seine at Kutaraja Fishing Port [6] have previously been assessed. Considering the results of existing research, this study is intended to compare the findings of the previous studies and provide comprehensive information on the performance of different tuna fishing gear in terms of the carbon footprint generated in various regions of Indonesian waters.

The benefits of this study are to provide information related to tuna fishing gear that has more efficient performance in terms of the carbon footprint, help businesses to compare the performance of fishing gear based on the carbon footprint, and provide a basis for further research on technological innovations in fishing gear that are more energy efficient and environmentally friendly.

2 Methods

The research was conducted in March-June 2025. The study was conducted by collecting data on fuel oil (BBM) consumption and the amount and composition of catch per trip per vessel for one year or more than one year from tuna fishing units. The data was obtained by reviewing research results. The data obtained were from handline fishing units at Pondokdadap Fishing Port, Malang, East Java; handline and purse seine at Kutaraja Fishing Port, Banda Aceh, Aceh; and troll line and longline at Palabuhanratu Fishing Port, Sukabumi,

West Java. Data were also collected from the literature for Nizam Zachman Fishing Port, Jakarta, Special Capital Region of Jakarta, covering handline, purse seine, and longline fishing units (annual report of Nizam Zachman Fishing Port).

The research began by conducting a literature study on fuel consumption data and the amount and composition of catch per trip per vessel for one year or more than one year in handline, purse seine, troll line, and longline fishing units. Selected studies from the search were first reviewed to ensure data availability. Data validation was conducted by checking the credibility of the sources, as well as the completeness and clarity of the data. Field data were also collected as harbour statistics for handline, purse seine, and longline fishing units. Data processing was carried out so that the data can be analysed. Each data from each source had various data processing techniques. However, in general, what was done included grouping fishing vessels data based on the GT classification according to Indonesian Fisheries Statistics, namely motorboats <5 GT, 5-10 GT, 10-20 GT, 20-30 GT, 30-50 GT, 50-100 GT, 100-200 GT, and>200 GT. Furthermore, the units in the data were converted to consistent units, namely using kilograms. This unit was also chosen to adjust the carbon footprint, where the unit of fuel is kilograms.

Data analysis was carried out using the FUI calculation formula as follows.

$$FUI_{fish} = \frac{\sum FC\ (kg)}{\sum HT\ (kg)} \quad (1)$$

Description of formula (1):

$\sum FC$ = Fuel consumption per vessel per trip (kg)

$\sum HT$ = Total catch per vessel per trip (kg)

Based on FUI, data analysis was then carried out using a formula for calculating the estimated carbon footprint for the total catch, tuna, and non-tuna. The estimated carbon footprint is expressed in kilograms of carbon dioxide equivalent per kilogram of fish ($kgCO_2eq/kg$). The formula for calculating the carbon footprint per kilogram of fish caught was as follows.

$$CO_2\ emission = \frac{FUI \times NCV \times EF}{10^6} \quad (2)$$

Description of formula (2):

NCV = Net Calorific Value (MJ/kg)

EF = CO_2 Emission Factor (kg CO_2/TJ)

10^6 = conversion of MJ to TJ (1 TJ = 10^6 MJ)

Fishing vessels in Indonesian waters mostly use diesel fuel, which has an EF value of 74,100 kg CO_2/TJ and an NCV value of 43.00 MJ/kg.

Furthermore, numerical and descriptive comparative analyses were conducted to determine information on the performance of fishing gear in terms of the carbon footprint generated.

3 Result and discussion

Based on the results of data collection related to fuel oil (BBM) consumption as well as the number and composition of catches per trip per vessel for one year or more than one year from four types of tuna fishing gear, information from a total of eight tuna fishing units was obtained spread across four different fishing ports, namely Nizam Zachman, Kutaraja, Palabuhanratu, and Pondokdadap. The eight fishing units consisted of three units of handline, two units of purse seine, one unit of troll line, and two units of longline. A detailed summary of the data obtained is presented in Table 1.

Table 1. Details of data acquisition regarding the analysed tuna fishing gear.

Source	Type of fishing gear	Vessel size (GT)	Number of vessels (Unit)	Number of trips per year	Average trip length (days/trip)	Range of data taken
Statistics Nizam Zachman Fishing Port	Handline	30-50	2	1	85	2024
		50-100	16	1-3	113	
		100-200	2	1	236	
	Purse Seine	50-100	1	1	201	
		100-200	224	1-2	244	
		>200	9	1-2	287	
	Longline	30-50	5	1-3	121	
		50-100	47	1-5	178	
		100-200	21	1-5	148	
Kutaraja Fishing Port [6]	Handline	5-10	1	15	3	2023
		10-20	8		5	
		20-30	5		9	
	Purse Seine	20-30	6	12	6	
		30-50	7		11	
		50-100	4		9	
Palabuhanratu Fishing Port [7]	Troll Line	5-10	15	24	11	2023
Pondokdadap Fishing Port [5]	Handline	5-10	2	6	10	2022
		10-20	15	7	10	
		20-30	15	9	10	
Palabuhanratu Fishing Port [4]	Longline	20-30	128	12	-	2015-2021

Table 1 shows that these four types of fishing gear are used in various sizes, from small-scale 5-10 GT (Gross Tonnage), such as troll line vessels, to industrial-scale>200 GT purse seine vessels. The variation in vessel size indicates the diverse scale of operations in tuna fishing activities in Indonesia, from traditional to large industrial scales.

The number of vessels sampled also varied between locations, from 1 vessel to 224 boats, depending on the completeness of data from each source. Fishing activity, as reflected by the number of trips and trip length, also varied depending on the scale of operation and gear type. Large vessels tended to make few trips, with 1-5 trips per year and long fishing durations ranging from 85 to 287 days per trip. In contrast, small to medium-sized vessels usually made more trips, reaching up to 24 trips per year with a duration of fishing between 5 and 15 days per trip. This shows the capacity and cruising power difference between small, medium, and large vessels.

Data from Nizam Zachman Fishing Port shows that tuna are caught by handline, purse seine, and longline, with the most significant number of vessels being 224 purse seine vessels of 100-200 GT and 47 longline vessels of 50-100 GT. This reflects the character of that harbour as the base of the large-scale tuna fishing industry. Meanwhile, data from Pondokdadap Fishing Port [5] and from Palabuhanratu Fishing Port [7] reflect the condition of small and medium-scale fisheries, with smaller vessel sizes. The variation in gear from different locations provides a strong basis for comparing gear performance.

3.1 Fuel consumption

Fuel consumption is one of the main factors affecting the carbon footprint of tuna fishing activities. A common finding in fisheries life cycle assessment studies is that fuel inputs dominate GHG emissions, with 60-90% of life cycle GHG emissions resulting from direct fuel combustion during fishing [1]. In the fish supply chain, fuel combustion of tuna fishing vessels accounts for about 70-90% of total GHG emissions up to the point of landing. This exceeds the combative energy use and GHG emissions from processing, packaging, and product transport [8]. Fuel consumption per trip for each fishing gear is presented in Table 2.

Table 2. Fuel consumption of each tuna fishing gear.

Type of fishing gear	Name of fishing port	Vessel size (GT)	Average trip length (days/trip)	Average fuel consumption per day (kg)	Total fuel consumption per trip (kg)
Handline	Nizam Zachman Fishing Port	30-50	85	34.79	2,940.00
		50-100	113	93.79	10,225.39
		100-200	236	142.37	33,600.00
	Kutaraja Fishing Port	5-10	3	107.67	323.00
		10-20	5	123.90	632.65
		20-30	9	157.74	1,335.52
	Pondokdadap Fishing Port	5-10	10	100.84	617.01
		10-20	10	125.96	895.62
		20-30	10	141.25	1,259.03
Purse Seine	Nizam Zachman Fishing Port	100-200	243	133.36	29,008.91
		>200	287	152.12	42,702.54
	Kutaraja Fishing Port	20-30	6	241.53	1,380.95
		30-50	11	365.92	3,969.85
		50-100	9	573.81	4,644.42
Troll Line	Palabuhanratu Fishing Port	5-10	11	48.21	530.32
Longline	Nizam Zachman Fishing Port	30-50	121	112.61	9,711.77
		50-100	178	102.73	14,198.38
		100-200	148	291.90	22,486.61
	Palabuhanratu Fishing Port	20-30	-	-	-

In addition to generating GHG emissions, vessel fuel consumption has also been identified as a key driver of environmental toxicity, acidification, eutrophication, particulate matter emission, and abiotic resource depletion, further emphasising the impact of vessel fuel consumption as a life cycle ecological hotspot in capture fisheries [9].

The amount of fuel consumption is influenced by the size of the vessel, the distance to the fishing grounds, the type of fishing gear, the engine used, and the fuel used [5]. In addition, natural conditions, such as weather, wave height, wave period, and wind, also affect the amount of fuel consumption [10].

In general, the larger the vessel size and the longer the duration of fishing, the higher the fuel consumption per trip. Longline fishing gear's fuel consumption varied by port location and vessel size. Purse seine recorded the highest fuel consumption, especially at Nizam Zachman Fishing Port, with vessels >200 GT consuming up to 42,702 kg of fuel/trip over

287 days. Handline and longline also showed an increasing trend in consumption as vessel size and operating length increased.

There are limitations in determining fuel consumption data, especially at Nizam Zachman Fishing Port. The available data is the fuel data recorded when the vessel prepares to go to sea, so it only reflects the initial amount of fuel before the vessel sails. Based on interviews, vessels generally refuel at sea 4-5 times per trip, which is done when fishermen transship their catch. The port has not recorded this refuelling activity. Therefore, the actual fuel consumption during the trip cannot be known explicitly.

These limitations need to be addressed by the fishing port by starting to implement a comprehensive and continuous fuel consumption recording system, not only during departure. For example, vessels can be required to report their total fuel consumption after completing a trip to sea or to record each time they refill fuel at sea. This can assist the port in calculating the amount of fuel used each trip, so that the amount of tax from fuel purchases that needs to be paid by each boat owner can also be appropriate. In addition, the completeness of the data will also make future research more accurate.

3.2 Number and composition of catches

In addition to fuel consumption, FUI and carbon footprint are also strongly influenced by the amount and composition of the catch obtained in one trip during one year. The higher the catch, the lower the FUI because energy consumption is spread over more output. Therefore, the productivity of fishing gear is an essential factor in assessing energy efficiency prior to emission calculations. Knowledge of the proportion of tuna and non-tuna catch is also the basis for calculating FUI and estimating the carbon footprint of fishing activities [4].

Each fishing gear has varying catchability depending on the technical design, operational methods, and biological characteristics of the target species [11]. Differences in these characteristics impact variations in catch, both in terms of total numbers and the proportion of tuna to the overall catch. The number and proportion of tuna catches from each fishing gear sampled in this study are presented in Table 3. Data includes the average catch in one fishing trip that has previously been accumulated for one year from each fishing gear.

Table 3. Number and composition of catches for each tuna fishing gear.

Type of fishing gear	Name of fishing port	Vessel size (GT)	Average trip length (days/trip)	Catch (kg)			Percentage (%)	
				Total	Tuna	Non-tuna	Tuna	Non-tuna
Handline	Nizam Zachman Fishing Port	30-50	85	16,000	4,675	11,325	29	71
		50-100	113	31,440	15,653	15,787	50	50
		100-200	236	35,000	10,000	25,000	29	71
	Kutaraja Fishing Port	5-10	3	1,033	493	540	48	52
		10-20	5	1,800	718	1,082	40	60
		20-30	9	4,026	1,249	2,777	31	69
	Pondokdadap Fishing Port	5-10	10	1,706	853	853	50	50
		10-20	10	2,144	1,072	1,072	50	50
		20-30	10	2,857	1,428	1,428	50	50
Purse Seine	Nizam Zachman Fishing Port	100-200	243	70,442	16,042	54,400	23	77
		>200	287	75,455	16,245	59,209	22	78
	Kutaraja Fishing Port	20-30	6	7,090	707	6,383	10	90
		30-50	11	13,537	1,770	11,767	13	87

Type of fishing gear	Name of fishing port	Vessel size (GT)	Average trip length (days/trip)	Catch (kg)			Percentage (%)	
				Total	Tuna	Non-tuna	Tuna	Non-tuna
		50-100	9	17,228	1,892	15,337	11	89
Troll Line	Palabuhanratu Fishing Port	5-10	12	895	309	586	35	65
Longline	Nizam Zachman Fishing Port	30-50	121	8,500	4,009	4,491	47	53
		50-100	178	22,852	13,459	9,393	59	41
		100-200	148	34,857	28,615	6,241	82	18
	Palabuhanratu Fishing Port	20-30	-	-	-	-	18	82

Handlining at Nizam Zachman Fishing Port and Kutaraja Fishing Port showed moderate to high selectivity, with 29-50% of tuna. At the same time, at Pondokdadap Fishing Port, it was assumed to be 50% due to data limitations. Puse seiners, although focused on skipjack, produced large volumes of tuna but with low proportions (9-23%) and low selectivity. The troll liners produced 35% tuna with catch diversity due to the use of fishing aggregating devices (FADs). Longline showed the highest selectivity, reaching 82% on large vessels at Nizam Zachman Fishing Port, while at Palabuhanratu Fishing Port, it was only 18%.

In general, catches from each gear and location showed variations that reflected differences in operational strategies, vessel scale, and water conditions where operations took place. This variation not only shows the technical and ecological dynamics in the field but also becomes an essential component in calculating fuel use intensity and estimating carbon footprint in a more representative manner.

3.3 Fuel use intensity

Fuel use intensity (FUI) is one of the leading indicators in assessing the energy efficiency of tuna fishing activities. The FUI value is calculated as the ratio between the amount of fuel consumption and the total catch obtained, using formula (1), resulting in units of kg of fuel per kg of fish. This parameter is important because it directly represents how much energy consumption is required to produce one kilogram of caught fish. The smaller the FUI value, the more efficient fuel use and the lower the potential emissions. Table 4 presents the FUI values of each tuna fishing gear.

Table 4. Fuel use intensity of each tuna fishing gear.

Type of fishing gear	Name of fishing port	Vessel size (GT)	Average trip length (days/trip)	Fuel use intensity (kg BBM/kg HT)
Handline	Nizam Zachman Fishing Port	30-50	85	0.18
		50-100	113	0.33
		100-200	236	0.96
	Kutaraja Fishing Port	5-10	3	0.31
		10-20	5	0.35
		20-30	9	0.33
	Pondokdadap Fishing Port	5-10	10	0.36
		10-20	10	0.42
		20-30	10	0.44
Purse Seine		100-200	243	0.41

Type of fishing gear	Name of fishing port	Vessel size (GT)	Average trip length (days/trip)	Fuel use intensity (kg BBM/kg HT)
	Nizam Zachman Fishing Port	>200	287	0.57
	Kutaraja Fishing Port	20-30	6	0.19
		30-50	11	0.29
		50-100	9	0.27
Troll Line	Palabuhanratu Fishing Port	5-10	12	0.59
Longline	Nizam Zachman Fishing Port	30-50	121	1.14
		50-100	178	0.62
		100-200	148	0.65
	Palabuhanratu Fishing Port	20-30	-	2.32

Fuel use efficiency in capture fisheries, interpreted through FUI values, is influenced by various operational and ecological factors. These factors include the type of fishing gear used, the relative abundance of target fish stocks, the distance travelled to the fishing grounds, the skill level of the captain and crew, and the use of FADs or fish aggregating aids [9].

Based on Table 2, FUI values varied significantly between gear types, ports, and vessel sizes. The overall FUI value ranged from 0.18 to 2.32 kg fuel/kg fish. The lowest FUI value was recorded for purse seine gear at Kutaraja Fishing Port, with a range of 0.18-0.29 kg fuel/kg fish, indicating high fuel efficiency. The highest FUI value was found in longline gear at Palabuhanratu Fishing Port at 2.32 kg fuel/kg fish, indicating high fuel consumption for every kilogram of fish produced. This is in line with the results of Tyedmers and Parker's research [1] that tuna caught using purse seines tend to have low FUI or are not too wasteful in terms of energy. In addition, vessels catching tuna by longline have historically consumed more fuel per landing unit than vessels catching tuna by purse seine [1].

Showing more detail, the FUI values of each fishing gear are compared in a graph, as seen in Figure 1.

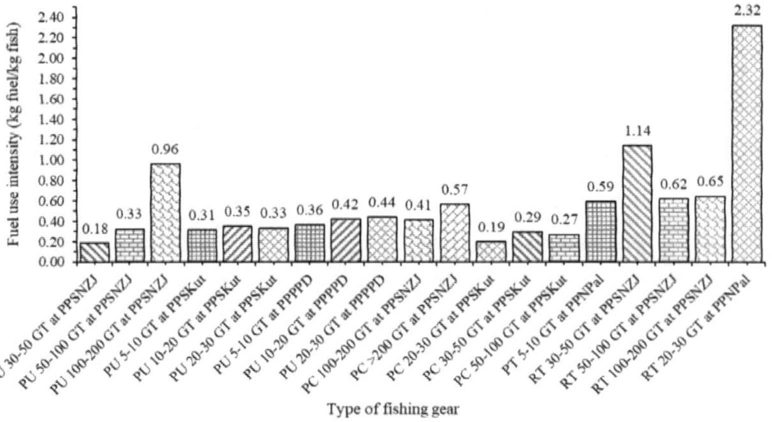

PU	= Handline	PPSNZJ	= Nizam Zachman Fishing Port
PC	= Purse Seine	PPSKut	= Kutaraja Fishing Port
PT	= Troll Line	PPPPD	= Pondokdadap Fishing Port
RT	= Longline	PPNPal	= Palabuhanratu Fishing Port

Fig. 1. Comparison of fuel use intensity in tuna fishing gear.

Fuel use intensity is based on the type of fishing gear in each fishing port, and it can be seen that the handline does not show a consistent pattern between vessel sizes. The FUI value of handline in three fishing ports tends to fluctuate and does not form a particular trend. However, the FUI value at Pondokdadap Fishing Port is slightly higher than for the other two ports, although the difference is insignificant. There is an anomaly in the 100-200 GT handline at Nizam Zachman Fishing Port because it has a significantly different FUI from the other handline vessel sizes.

Furthermore, a slightly more evident pattern can be seen if the comparison is made based on the GT size of the vessel, namely in purse seine and longline fishing gear. In purse seines, the larger the vessel size, the higher the FUI tends to be, except for the 30-50 GT purse seine vessels at Kutaraja Fishing Port, which is higher but not significant, with a difference of 0.10 from the 20-30 GT vessel size and a difference of 0.02 from the 50-100 GT vessel size. Meanwhile, longline displays the opposite pattern, where the larger the vessel size, the lower the FUI tends to be, except for the 100-200 GT longline vessel at Nizam Zachman Fishing Port, which is higher but not significant, with a difference of 0.03 higher than the 50-100 GT vessel size. Handline has no trend, but it is known that the average FUI is moderate. The anomaly with the handline at Nizam Zachman Fishing Port is that the larger vessel size, the higher the FUI. The troll line was not compared because there were no other GT sizes, only 5-10 GT vessels.

Based on the location of the fishing port, at Nizam Zachman Fishing Port, the larger the vessel size for the same gear type, the higher the FUI, whether for handline, purse seine, or longline gear. While at Kutaraja Fishing Port, FUI tends to be equally low, both in handline and purse seine gear, but it appears that purse seine has a lower average FUI compared to handline, even though this may be due to various factors, such as operational support (e.g. the use of FADs), a more organised fishing plan, and a system of mass fishing in a short time. The use of FADs in the waters around Aceh, including Kutaraja, has been shown to increase fishing efficiency [12]. Research demonstrated that production factors such as vessel engine power, number of crew members, and number of lights affect the efficiency of purse seine at Kutaraja Fishing Port [13].

3.4 Carbon footprint

The estimated carbon footprint was calculated by multiplying the FUI values in Table 4 by the emission factor and net calorific value of diesel fuel, according to formula (2). The carbon footprint value is expressed in kilograms of carbon dioxide equivalent per kilogram of fish ($kgCO_2eq/kg$). The tuna carbon footprint column shows how much GHG emissions each kilogram of tuna catch has to "bear". The higher the carbon footprint value, the greater the activity's environmental impact. The estimated carbon footprint resulting from tuna fishing activities with different fishing gear is presented in Table 5.

Table 5. Estimated carbon footprint resulting from tuna fishing activities with different fishing gear.

Type of fishing gear	Name of fishing port	Vessel size (GT)	*Carbon footprint* ($kgCO_2eq/kg$)		
			Total	Tuna	Non-tuna
Handline	Nizam Zachman Fishing Port	30-50	0.59	0.17	0.41
		50-100	1.04	0.52	0.52
		100-200	3.06	0.87	2.18
	Kutaraja Fishing Port	5-10	1.00	0.48	0.52
		10-20	1.12	0.45	0.67
		20-30	1.06	0.33	0.73

Type of fishing gear	Name of fishing port	Vessel size (GT)	*Carbon footprint* (kgCO₂eq/kg)		
			Total	Tuna	Non-tuna
	Pondokdadap Fishing Port	5-10	1.15	0.58	0.58
		10-20	1.33	0.67	0.67
		20-30	1.40	0.70	0.70
Purse Seine	Nizam Zachman Fishing Port	100-200	1.31	0.30	1.01
		>200	1.80	0.39	1.42
	Kutaraja Fishing Port	20-30	0.62	0.06	0.56
		30-50	0.93	0.12	0.81
		50-100	0.86	0.09	0.76
Troll Line	Palabuhanratu Fishing Port	5-10	1.89	0.65	1.24
Longline	Nizam Zachman Fishing Port	30-50	3.64	1.72	1.92
		50-100	1.98	1.17	0.81
		100-200	2.06	1.69	0.37
	Palabuhanratu Fishing Port	20-30	7.40	1.33	6.07

Table 5 shows the variation in carbon footprint values between gear types, vessel sizes, and ports. The fishing gear's overall tuna carbon footprint value ranged from 0.06 to 1.72 $kgCO_2eq/kg$. The highest overall tuna carbon footprint was for longline at Nizam Zachman Fishing Port, with a tuna carbon footprint of 1.72 $kgCO_2eq/kg$ for vessels of 30-50 GT, reflecting a very high level of emissions. The lowest average tuna carbon footprint was for purse seine gear at Kutaraja Fishing Port, with a tuna carbon footprint as low as 0.06 $kgCO_2eq/kg$ for vessels 20-30 GT, reflecting very low emission levels.

This finding aligns with the results of Tyedmers and Parker [1], who also showed that longline gear produces a much higher carbon footprint than purse seine gear. In their study, longline fishing produced emissions of 3.83 $kgCO_2eq/kg$, while purse seine only produced 1.53 $kgCO_2eq/kg$. This difference indicates that, in general, active fishing gear, such as purse seine, tends to be more environmentally friendly in terms of the carbon footprint produced than passive fishing gear, such as longline. Although the carbon footprint values in this study are lower in absolute terms, the comparison pattern between gear types still shows a similar trend. For more detail, the carbon footprint values of each fishing gear are compared in a graph, as seen in Figure 2.

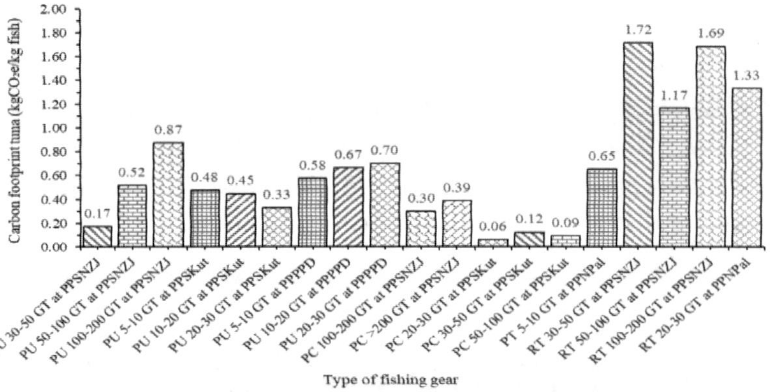

PU = Handline PPSNZJ = Nizam Zachman Fishing Port
PC = Purse Seine PPSKut = Kutaraja Fishing Port
PT = Troll Line PPPPD = Pondokdadap Fishing Port
RT = Longline PPNPal = Palabuhanratu Fishing Port

Fig. 2. Comparison of tuna carbon footprint in tuna fishing gear.

The tuna carbon footprint based on the type of fishing gear at each fishing port does not appear to have a clear pattern. For handliners at Nizam Zachman Fishing Port and at Pondokdadap Fishing Port, the larger vessel size was associated with a higher tuna carbon footprint value. Meanwhile, for handline gear at Kutaraja Fishing Port, the larger vessel size was associated with a lower tuna carbon footprint value. The purse seiners at Nizam Zachman Fishing Port had a higher average tuna carbon footprint than that at Kutaraja Fishing Port. Troll liners had a medium tuna carbon footprint. Meanwhile, longliners at Nizam Zachman Fishing Port fluctuated and had the highest average tuna carbon footprint compared to other gears. This was the same as for the longliners at Palabuhanratu Fishing Port, with a high tuna carbon footprint value compared to other fishing gears.

Furthermore, based on GT size, there was no clear pattern. Handline and longline gear had a tuna carbon footprint that fluctuated significantly. Purse seiners displayed a trend where the larger the vessel size, the higher the tuna carbon footprint value, but there was an anomaly for purse seiners 30-50 GT with a difference that was not significant, only 0.03 higher than purse seiners of size 50-100 GT.

Based on the location of the fishing port at Nizam Zachman Fishing Port for handline and purse seine gear with larger vessel sizes, the tuna carbon footprint was also higher. Meanwhile, longline gear with the smallest size had the largest tuna carbon footprint value. Furthermore, at Kutaraja Fishing Port, the larger vessel size was associated with a lower tuna carbon footprint value, while the purse seine gear showed no trend. At Pondokdadap Fishing Port, the average tuna carbon footprint value was medium, with a trend that the larger the vessel size, the higher the carbon footprint value produced. Fishing ports with their respective gear types that produced the lowest to highest average tuna carbon footprint values are Kutaraja, Pondokdadap, Palabuhanratu, and Nizam Zachman.

The results of this study provide a quantitative overview of the FUI and carbon footprint of different gears in several fishing ports. Findings show significant variations among gear types, vessel sizes, and operational locations, reflecting the field's strategic technical and energy efficiency differences. With this data, efforts to reduce the carbon footprint of the capture fisheries sector can be more measurable because it is based on the value of the fuel consumption and the proportion of catch from each unit.

In addition, this information can serve as the basis for more targeted policy development. The government can prioritize energy efficiency interventions in fishing units with the highest carbon footprint values, and encourage using fishing gear proven more efficient and environmentally friendly. This mapping can also support the development of a gear-based emissions reporting and monitoring system as part of the national commitment to reduce GHG emissions, in accordance with the Paris Agreement targets.

Data limitations mean that measuring the carbon footprint of capture fisheries is still challenging, especially regarding the availability of accurate and detailed fuel consumption data. Currently, most fishing ports only record the amount of fuel when the vessel leaves for sea, with no information on the actual consumption during the trip and refuelling at sea. However, research shows [14] that even for years the data at fishing ports has increased in completeness, it still does not meet the data needs to calculate FUI and carbon footprint more accurately.

Based on this explanation, the government needs to encourage improving the fuel consumption recording system, such as through digital daily logbook filling, fuel flowmeter installation, and port data with the vessel monitoring system (VMS). With the availability of more accurate data, future carbon footprint estimation studies can be more representative and support evidence-based policy making. Investing in a good data collection system is an important foundation for Indonesia's more efficient, low-emission, and sustainable fisheries transformation.

Furthermore, the FUI value of global tuna capture fisheries showed an increasing trend from 2009 to 2023 [9]. However, these findings may be influenced by the limited sample size of the latest study, which was smaller than the sample size of their monitoring in the fisheries sector.

The results of this study, on a national scale, also show that some tuna fishing units produce a high carbon footprint, such as the longliners at Nizam Zachman Fishing Port. This indicates that the issue of energy efficiency in fisheries is a local and globally relevant challenge. Therefore, efforts to improve fuel efficiency must be considered in managing sustainable capture fisheries locally and internationally.

4 Conclusion

The fuel use intensity (FUI) value in tuna fishing units at the research sites ranged from 0.18 to 2.32 kg of fuel/kg of fish. The highest FUI value was recorded for longline gear at Palabuhanratu Fishing Port, while the lowest was for purse seine gear at Kutaraja Fishing Port. Based on these FUI values, the estimated carbon footprint of tuna ranged from 0.06-1.72 kgCO_2eq/kg, with the highest value also found in the 30-50 GT longline gear at Nizam Zachman Fishing Port, and the lowest value found in the 20-30 GT purse seiners, which tend to produce a lower carbon footprint than passive gears such as longlines.

The results of comparisons between gear types, vessel GT sizes, and fishing port locations show no consistent pattern regarding fishing gear performance in terms of the FUI value and the resulting carbon footprint. Nevertheless, it can be confirmed that the FUI value is strongly influenced by fuel consumption and the amount of catch, while the tuna carbon footprint is mainly influenced by the FUI value and the composition of tuna caught by each gear.

References

1 P. Tyedmers, R. Parker, Fuel consumption and greenhouse gas emissions from global tuna fisheries: a preliminary assessment. ISSF Technical Report 2012-03. International Seafood Sustainability Foundation, McLean, Virgina, USA. https://www.iss-foundation.org/about-issf/what-we-publish/issf-documents/issf-technical-report-2012-03-fuel-consumption-and-greenhouse-gas-emissions-from-global-tuna-fisheries-a-preliminary-assessment

2 M. Imron, R. Yusfiandayani, M.S. Baskoro, Produksi dan produktivitas tuna oleh kapal tuna longline yang berbasis di PPN Palabuhanratu. Jurnal Teknologi Perikanan dan Kelautan **10**(2), 173-181 (2019). https://doi.org/10.24319/jtpk. 10.173-181

3 L.A. Tomasila, M. Syamsudin, R. Polhaupessy, Proses penangkapan tuna madidihang (*Thunnus albacares*) dengan alat tangkap pancing ulur (hand line) di Pulau Ambon. TRITON **16**, 2 (2020) https://doi.org/10.30598/TRITONvol16issue2page97-107

4 R.F. Anggawangsa, I.T. Hargiyatno, Suryanto, A.A. Widodo, Estimation of carbon footprint of longline and lines fisheries in the Indonesia FMA 573-Indian Ocean based at Palabuhanratu Fishing Port. IOP Conf. Ser.: Earth Environ. Sci. **1137**, 012049 (2023). https://doi.org/10.1088/1755-1315/1137/1/012049

5 J.Z. Rachman, Konsumsi energi pada operasi kapal penangkapan tuna, cakalang, tongkol di Pelabuhan Perikanan Pantai Pondokdadap, Undergraduate Thesis, IPB University (2023).

6 U. Salsabila, Estimasi carbon footprint pada aktivitas transportasi tuna berbasis di Pelabuhan Perikanan Samudera Kutaraja, Master Thesis, IPB University (2024).

7 A. Mulyana, Efisiensi faktor produksi perbekalan melaut unit penangkapan pancing tonda 6 GT di Pelabuhan Perikanan Nusantara Palabuhanratu, Undergraduate Thesis, IPB University (2024).

8 F. Ziegler, S. Hornborg, BS. Green, O.R. Eigaard, A.K. Farmery, L. Hammar, K. Hartmann, S. Molander, R.W.R. Parker, E.S. Hognes, I. Vázquez-Rowe, A.D.M. Smith, Expanding the concept of sustainable seafood using Life Cycle Assessment. Fish Fisher. **17**(4), 1073-1093 (2016). https://doi.org/10.1111/faf.12159

9 P. Tyedmers, S. Donovan, Fuel consumption and greenhouse gas emissions from global tuna fisheries: 2024 updated report. ISSF Technical Report 2025-03. International Seafood Sustainability Foundation, Pittsburgh, PA, USA (2025). https://www.iss-foundation.org/about-issf/what-we-publish/issf-documents/issf-2025-03-fuel-consumption-and-greenhouse-gas-emissions-from-global-tuna-fisheries-2024-updated-report/

10 J. Prpić-Oršić, R. Vettor, C.G. Soares, O.M. Faltinsen, Influence of ship routes on fuel consumption and CO2 emission, in Maritime Technology and Engineering , eds. C. Guedes Soares & T.A. Santos (CRC Press, London, 2014). https://doi.org/10.1201/b17494

11 A.F. Nasution, M.T. Noor, E. Saraswati, Pengaruh Perbedaan kedalaman setting pada alat tangkap pancing ulur terhadap hasil tangkapan ikan madidihang (*Thunnus albacares*) di perairan Mentawai yang didaratkan di Pelabuhan Perikanan Samudera Bungus. Manfish, Jurnal Ilmiah Perikanan dan Peternakan **3**(1), 109-128 (2025). https://journal.asrihindo.or.id/index.php/Manfish/article/view/129

12 M.A. Chaliluddin, Munzir, E. Miswar, T. Rizwan, A. Rahmah, D. Rianjuanda, I. Rusydi, R. Nellyana, Pengaruh rumpon terhadap hasil tangkapan pukat cincin (purse seine) di perairan utara Aceh. Jurnal Kelautan dan Perikanan Indonesia **2**(1), 51-60 (2022). https://doi.org/10.24815/jkpi.v2i1.27093

13 R.M. Aprilia, Mustaruddin, E.S. Wiyono, N. Zulbainarni, Analisis efisiensi unit penangkapan pukat cincin di Pelabuhan Perikanan Pantai Lampulo Banda Aceh. Jurnal Teknologi Perikanan dan Kelautan **4**(1), 9-20 (2017). https://doi.org/10.24319/jtpk.4.9-20

14 C.A. Syahrin, Estimasi emisi gas buang kapal longline 51-100 GT di Pelabuhan Perikanan samudera Nizam Zachman, Undergraduate Thesis, IPB University (2021).

Formaldehyde detection and quality evaluation of Tambaqui (*Colossoma macropomum*) to ensure food safety in the Coastal Region of Kalianda

Arafah Purnama[1*], *Nurhayati* Tati[1], *Suseno* Sugeng Heri[1], *Ardina* Carissa Mutiara[1] and *Suhaima* Nurafi Razna[2]

[1]Department of Aquatic Products Technology, Faculty of Fisheries and Marine Sciences, IPB University, Bogor, St. Raya Dramaga, 16680, Indonesia
[2]Food Quality Assurance, Vocational School, IPB University, Bogor, St. Kumbang 14, 16128, Indonesia

Abstract. Food safety is a critical component of the national food system, particularly for fishery commodities that are highly perishable and require proper post-harvest handling. In coastal regions such as Kalianda, South Lampung, limited cold chain infrastructure and inadequate post-harvest handling systems can contribute to a decline in fish quality and the potential for substandard practices, including the use of non-recommended additives. This study aimed to analyze the freshness and formaldehyde content of tambaqui (*Colossoma macropomum*) in the fish supply chain in Kalianda. Sampling was conducted from April to June 2025 at the Fish Auction Site (TPI) at Dermaga Bom and traditional markets in the region. The parameters analyzed included organoleptic attributes, pH, Total Volatile Base (TVB), Trimethylamine (TMA), natural formaldehyde (FA), and dimethylamine (DMA). The results indicated that fish obtained from collectors exhibited better quality (average organoleptic score of 7.2) compared to market samples (average score of 5.9). TVB levels in market samples reached 14.12 mg N/100g, higher than the 8.82 mg N/100g found in collector samples; TMA, FA, and DMA values were also higher in market-sourced fish. These findings indicate quality degradation and potential contamination during distribution in traditional markets. Therefore, strengthening the cold supply chain, providing education to stakeholders, and enforcing strict monitoring of hazardous additive usage are essential to ensure the quality and safety of fishery products.

1 Introduction

Food safety plays an integral role in national food systems, particularly concerning fishery commodities, which are highly perishable and demand meticulous post-harvest handling to preserve quality and ensure consumer safety. In coastal regions like Kalianda, South Lampung, the absence of adequate cold chain systems and proper handling procedures may contribute to rapid degradation of fish quality and create conditions that enable the use of improper preservation methods, including potentially hazardous additives.

Fish spoilage is often evaluated using physicochemical and sensory parameters. Total Volatile Base Nitrogen (TVB-N) is widely recognized as a robust spoilage indicator, as it reflects the accumulation of ammonia, trimethylamine (TMA), and other nitrogenous

* Corresponding author: arafah.purnama.r1@alumni.tohoku.ac.jp

compounds during degradation [1]. pH changes also provide valuable information about the freshness status of fish muscle, with increases typically correlating with microbial activity [2]. Trimethylamine oxide (TMAO), abundant in marine fish, can be enzymatically or microbially reduced to dimethylamine (DMA) and TMA, compounds responsible for characteristic off-odors and closely associated with spoilage progression.

Formaldehyde in fish may occur naturally as a result of TMAO degradation during storage, particularly in cold conditions, or may be illegally added as a preservative to prolong shelf life. Natural formaldehyde concentrations in marine species are generally low and vary by species, size, and storage conditions [3]. However, levels exceeding natural thresholds can indicate contamination from external sources, posing a significant health hazard due to its carcinogenic potential.

Tambaqui (*Colossoma macropomum*) is widely consumed in coastal Lampung, yet there is limited information on its post-harvest quality and formaldehyde content across different supply chain nodes. To address this gap, this study evaluates freshness and formaldehyde levels in tambaqui obtained from both fish collectors at Dermaga Bom landing site and local traditional markets. Parameters include organoleptic properties, pH, TVB, TMA, natural formaldehyde (FA), and dimethylamine (DMA). Findings from this investigation will offer insights into the integrity of the fish supply's supply chain, inform food safety protocols, and support policy and infrastructure improvements in coastal fisheries.

2 Methods

2.1 Research materials and equipment

The research was conducted from March to June 2025 , with sample collection carried out at the Fish Auction Site (TPI) Dermaga Bom and traditional markets in Kalianda, South Lampung . Sample preparation was performed at the Way Muli Aquaculture Center, Rajabasa, South Lampung. The sample freezing stage was carried out in cold storage at PT. Iandv Bio Indonesia, Rajabasa, South Lampung. Chemical testings were conducted at the Laboratory of Aquatic Product Raw Material Characteristics and the Biochemistry Laboratory, Department of Aquatic Product Technology, as well as the Fish Nutrition and Feed Technology Laboratory, Department of Aquaculture, Faculty of Fisheries and Marine Sciences, IPB University, Bogor.

The study utilized fresh tambaqui (*Colossoma macropomum*) as the primary material, along with analytical-grade reagents including ammonium acetate, glacial acetic acid, acetylacetone, distilled water, 7.5% TCA solution, Nessler reagent, H_3BO_3 solution, saturated K_2CO_3 (3 N) solution, neutral 10% formaldehyde, 0.0105 N HCl, copper ammonia solution, 5% CS_2–toluene solution, 30% acetic acid, and anhydrous Na_2SO_4 (all Merck or equivalent). The equipment employed included organoleptic score sheets, a pH meter, homogenizer, analytical balance, knives, spatulas, Conway dishes, incubator, ELISA reader, microplates, micropipettes, microtips, glass pipettes, pipette fillers, test tubes, measuring cylinders, glass funnels, filter papers, vortex mixer, ice gels, and styrofoam boxes. These materials and instruments were selected to ensure the precision and reliability of physical and chemical analyses.

2.2 Research stages

The research procedure began with sampling, followed by physical and chemical testing, and concluded with data analysis. Fish samples were collected from two different locations, namely from a collector at the Dermaga Bom fish auction site and a seller at the Kalianda traditional market. The selection of tambaqui from these two sources was considered to represent the actual conditions at each location, so it could be categorized as a representative form of sampling. The samples were then stored in a cool box containing flake ice and ice gel to maintain the quality of the fish until the testing process was carried out.

2.2.1 Sample collection and storage

A total of nine fish were obtained, with three used for organoleptic testing in triplicate and the remaining six processed into homogenized samples for chemical analyses, including pH, TVB, TMA, FA, and DMA. The samples were stored in styrofoam boxes containing ice flakes at a 1:2 fish-to-ice ratio with added ice gel to maintain quality prior to testing. Additionally, interviews with collectors and market vendors were conducted to gather information about fishing and storage practices.

2.2.2 Sample preparation

Morphometric measurements of tambaqui samples, including body height, body width, standard length, and total length, were taken using two precision rulers, and the results were systematically recorded to build a complete dataset. The fish samples were selected based on uniform body proportions to ensure consistency in the analysis. Organoleptic tests were then conducted using human senses to evaluate taste, odor, color, and texture in accordance with the Indonesian National Standard No. 2729 of 2013.

2.2.3 Freezing and transporting samples

The fish samples were transported the next day from Lampung to Bogor during a 10-hour journey, starting at 11:00 a.m. to 9:00 p.m. Western Indonesian Time. The fish samples were transported in a styrofoam box containing flake ice with a ratio of fish to ice of 1:2 and ice gel to maintain the quality of the fish.

2.2.4 Chemical testing

The chemical testing procedure for food involves several systematic steps. pH testing is carried out to determine the acidity or alkalinity of a sample, which affects its quality and safety. The Total Volatile Base (TVB) test measures ammonia and other volatile nitrogen compounds as indicators of freshness, while the TMA test evaluates fish quality by detecting spoilage-related compounds. Natural formaldehyde testing is performed to identify the presence of potentially harmful formaldehyde, and dimethylamine testing is used to further assess product quality, particularly in fish, as it reflects deterioration and spoilage.

2.3 Analysis

2.3.1 Organoleptic assessment

Following the guidelines in Indonesian National Standard No. 2729 of 2013 for fresh fish [4], sensory evaluation was conducted using a score sheet with a nine-point scale. Assessments were based on human sensory perception of several attributes, including the eyes, gills, surface mucus, flesh, odor, and texture. Three trained panelists were engaged to perform the evaluations.

2.3.2 pH measurement

The acidity level (pH) of the fish was determined using a calibrated pH meter applied to three body regions—dorsal, ventral, and caudal—to ensure consistent data. Approximately 5 g of fish flesh was weighed, minced, and homogenized with 45 mL of distilled water using a homogenizer. The pH reading was recorded once the meter displayed a stable value, following the method described by [5].

2.3.3 Total Volatile Base (TVB) analysis

TVB content, which indicates protein decomposition, was measured using the Conway microdiffusion method with Nessler reagent. A 15 g sample was blended with 45 mL of 7.5% TCA, homogenized, and filtered to obtain a clear extract. The extact was placed in the outer chamber of a Conway dish alongside saturated K_2CO_3, while 3% H_3BO_3 was placed in the inner chamber. After sealing, the dish was incubated at 37°C for 2 hours before titrating the boric acid solution with 0.02 N HCl until a pink endpoint appeared. The TVB concentration was calculated using the titration data.

2.3.4 Trimethylamine (TMA) analysis

TMA determination followed the Conway microdiffusion method [7]. In this test, TMA was reduced to TMA, which was then quantified through gas diffusion and titration. The sample extract was combined with neutral formaldehyde and saturated K_2CO_3 in the outer chamber, while 3% H_3BO_3 was placed in the inner chamber. After sealing, the setup was incubated for 60 minutes at 37°C, and the boric acid solution was titrated with 0.02 N HCl until the green color shifted to pink. The TMA concentration was calculated from titration results.

2.3.5 Natural Formaldehyde (NF) analysis

Natural formaldehyde content was measured using the Nash reagent method [8], which involves a solution of ammonium acetate, glacial acetic acid, and acetylacetone. Five grams of fish flesh were homogenized with 15 mL of 7.5% TCA, filtered, and mixed with an equal volume of Nash reagent. The mixture was incubated at 40°C for 30 minutes, cooled to room temperature, and then its absorbance was measured at 412 nm using an ELISA reader. FA levels were calculated based on a standard curve ranging from 0 to 20 ppm.

2.3.6 Dimethylamine (DMA) analysis

DMA quantification, following the method of [9], involved mixing 5 g of sample with 15 mL of 7.5% TCA, then reacting 2 mL of the extract with copper-ammonia solution and CS_2-toluene. The mixture was incubated at 50°C for 2 minutes, combined with 30% acetic acid, and vortexed for 1 minute. The toluene layer was separated into a tube containing anhydrous Na_2SO_4, and absorbance was measured at 440 nm using an ELISA reader. DMA levels were determined using a standard curve (0–19 ppm).

2.4 Data analysis

The procedure for processing physical and chemical test results, including organoleptic test, pH, TVB, TMA, natural formaldehyde, and dimethylamine, were carried out using a parametric approach with Microsoft Excel 2021. This software served as a tool to process data systematically so that more measurable results are obtained. Through this application, the total value, average, and standard deviation of each test parameter was calculated.

3 Result and discussion

3.1 Aquatic Product Supply Chain in Kalianda

The tambaqui fish (*C. macropomum*) belongs to the family *Serrasalmidae*. The tambaqui fish prefers habitats in whitewater, blackwater, and clearwater rivers with flooded forests during part of the year, migrating in large schools during spawning season. tambaqui are commonly caught using gillnets made of multifilament nylon with mesh sizes ranging from 150 to 240 mm stretched. Gillnets account for about 70% of the catches, and other methods include mixed gear (gillnets combined with castnets), drift nets, long lines, and castnets [10].

Morphometric measurements of freshwater pomfret (*C. macropomum*) were recorded to characterize the physical dimensions of the sampled fish. The average total length was 17.50 ± 0.44 cm, with a standard length of 13.93 ± 0.65 cm. The mean fork length was 15.30 ± 0.46 cm, while the average body height was 9.60 ± 0.92 cm. These measurements were used as baseline data to ensure uniformity of the fish samples analyzed in subsequent physicochemical and sensory quality assessments.

Field interviews conducted in the coastal area of Kalianda show a fairly structured distribution pattern in the supply chain for capture fisheries products, which operates on a daily cycle. Fishing activities generally begin when fishermen leave the dock at night to go out to sea. The catch is then brought back to shore in the early morning and immediately distributed to the fish auction site. At the fish auction site, the fresh fish are immediately auctioned to collectors or wholesalers, and the transaction process usually takes place on the same day. This auction process allows fishery products to immediately enter the formal distribution channel, both for local consumption and for supply to other regions, thereby maintaining the freshness and market value of the commodity.

The distribution pattern for fish entering traditional markets is slightly different. These fish generally come from fishing activities conducted by fishermen the previous afternoon. To maintain quality and slow down the degradation process, the catch is immediately stored in insulated containers such as cool boxes equipped with ice as a passive cooling medium. This low-temperature storage aims to suppress the activity of spoilage microorganisms and extend the shelf life of the product until distribution and sale the following day. This storage and distribution pattern highlights the importance of the cold chain in maintaining the quality of fresh seafood throughout the supply chain, as well as illustrating the adaptation of fishermen and traders to fishing times and market demand. An illustration of the supply chain for aquatic products in Kalianda is attached in Figure 1. Documentation of fish handling at the TPI and traditional markets in Kalianda is attached in Figure 2.

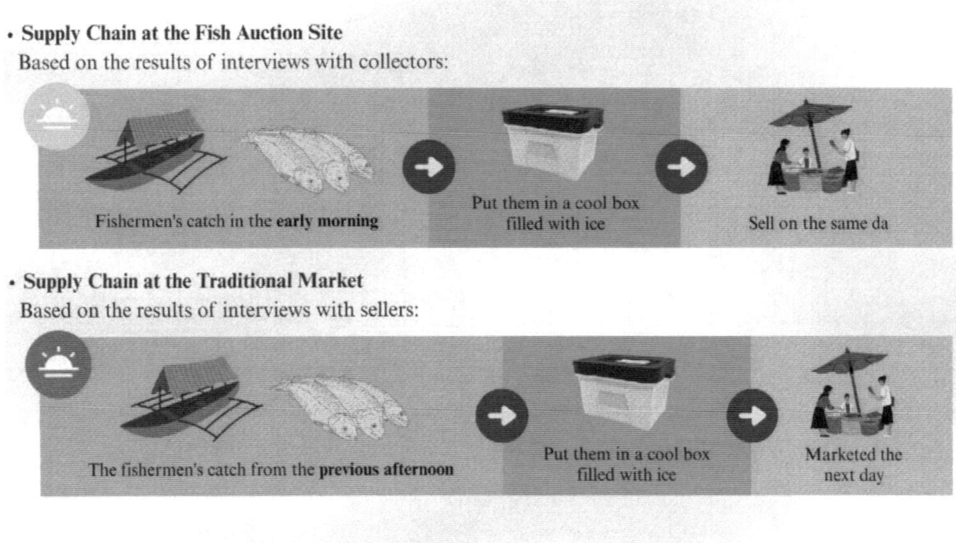

Fig. 1. Aquatic product supply chain in Kalianda.

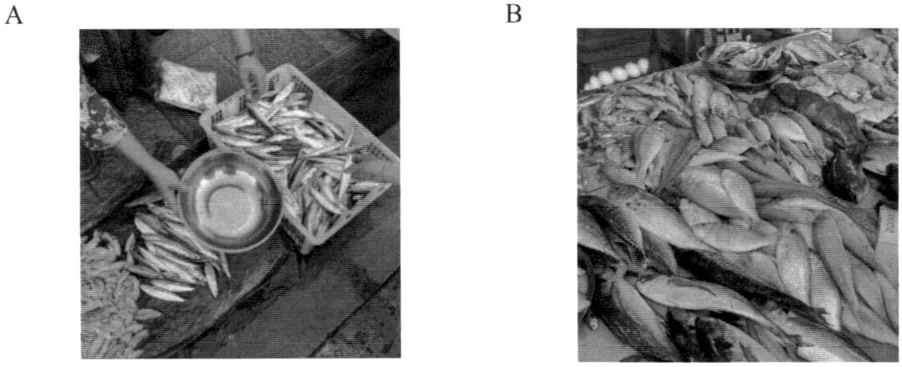

Fig. 2. Fish Handling at the Fish Auction Site and Traditional Market in Kalianda. Description: A = Dermaga Bom Kalianda Fish Auction Site, B = Kalianda Traditional Market.

3.2 Organoleptic Test

Based on field observations, tambaqui (*C. macropomum*) delivered to the Fish Auction Site (TPI) were generally caught by local fishers the night before and promptly placed by collectors into containers filled with water and ice blocks. In comparison, fish sold in traditional markets were typically obtained from collectors the previous afternoon and stored overnight by vendors in insulated boxes with crushed ice. This method of storage is consistent with recommendations, who emphasized that maintaining fish quality begins at the point of acquisition by using crushed ice as a cooling medium and keeping the product in insulated boxes to sustain low temperatures. The differences in organoleptic quality scores between tambaqui (*C. macropomum*) from collectors and those from markets are illustrated in Figure 3.

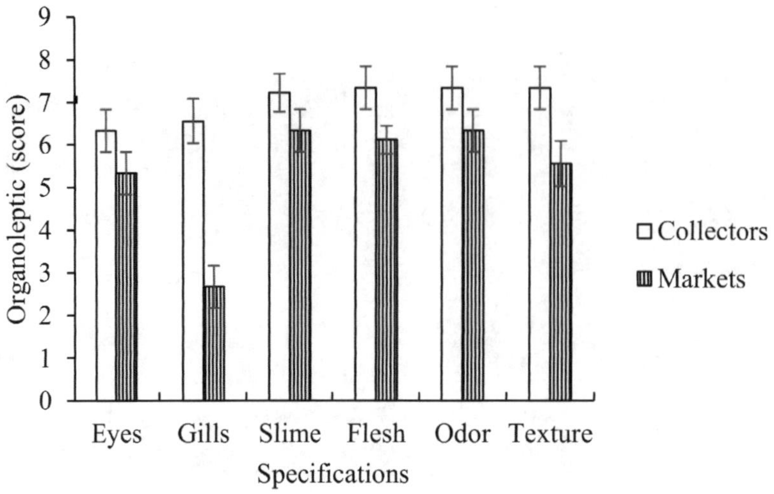

Fig. 3. Sensory quality parameters of Tambaqui (*Colossoma macropomum*). The dotted-line indicates the cut-off value of fresh fish.

Sensory assessments, which examined attributes such as eye brightness, gill coloration, surface mucus, firmness of flesh, odor, and texture, were carried out by three trained panelists and revealed notable quality disparities. Statistical evaluation confirmed that tambaqui obtained directly from collectors at the TPI had significantly higher organoleptic scores than those purchased from market vendors, reflecting greater freshness as a result of more recent harvesting and better handling practices. Similar trends were reported by Rachma *et al.* [11], who found that black tambaqui landed directly by seine net fishers in Kendal Regency retaine d higher freshness levels due to swift post-catch handling and shorter storage durations. Among the sensory indicators, odor was the most decisive factor, as the development of ammonia-like smells was associated with the onset of spoilage. A shorter supply chain proved to be instrumental in preserving the sensory quality of fresh fish products.

3.3 pH

Acidity testing is an instrumental approach to assessing the quality and freshness of fish, carried out using a pH meter after the fish meat has been homogenized with distilled water. This measurement aims to identify the level of acidity or alkalinity of muscle tissue, which directly reflects the post-mortem physiological status of the fish. Fresh fish typically exhibit a pH range of 6.0 to 7.0; deviations from this range indicate quality degradation, with a pH above 7 indicating intensified decomposition processes due to microbial activity, while a pH below 6 indicates quality deterioration due to the accumulation of acidic compounds during storage. The pH values of fish obtained from collectors and market vendors are presented in Figure 4.

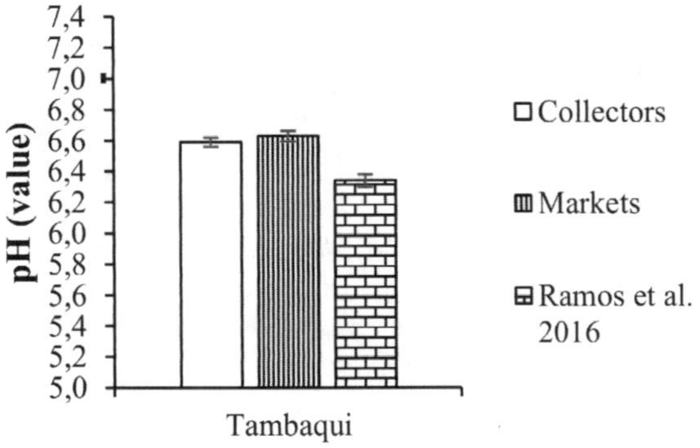

Fig. 4. pH of Tambaqui (*C. macropomum*). The dotted-line indicates the cut-off value of fresh fish.

The pH values of Tambaqui (*C. macropomum*) from collectors (6.59) and markets (6.63) showed only a slight difference, indicating that both sources maintained the fish in relatively fresh condition. Fresh fish muscle generally exhibits a near-neutral pH around 6.0-7.0 (the dotted line), which initially decreases post-mortem due to lactic acid formation before gradually increasing as microbial and enzymatic activity progresses, producing basic compounds such as ammonia and trimethylamine [12]. The observed pH levels are within the range reported for fresh fish and suggest that the market samples, despite undergoing additional handling and storage, have experienced only minimal biochemical changes compared to collector samples. This pattern aligns with previous findings that slight pH elevation is an early indicator of spoilage progression, emphasizing the utility of pH as a rapid and non-destructive freshness assessment parameter in fish quality monitoring.

3.4 Total Volatile Base (TVB)

Total Volatile Base testing is one of the most commonly used parameters in assessing the quality of fishery products, especially seafood. TVB measurement includes the determination of trimethylamine levels produced by spoilage bacteria activity, dimethylamine formed as a result of enzymatic autolysis during frozen storage, and ammonia produced through amino acid deamination and nucleotide catabolism. In addition, TVB also includes other volatile base nitrogen components that are closely related to the spoilage process in fishery products. The TVB levels of fish obtained from collectors and market vendors are presented in Figure 5.

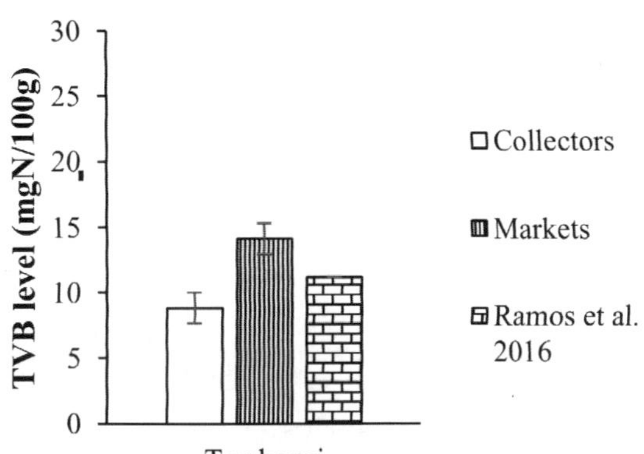

Fig. 5. TVB of Tambaqui (*Colossoma macropomum*). The dotted-line indicates the value of spoiled fish.

Figure 3 illustrates that Tambaqui (*C. macropomum*) samples sourced directly from collectors exhibit notably lower total volatile base (TVB) values (~8 mg N/100 g) compared to those obtained from markets (~14 mg N/100 g), indicating less protein degradation in the former. The dotted line in the figure represents the maximum allowable TVB value of 20 mg N/100 g, as stipulated in SNI 2354.8:2009. Elevated TVB concentrations—comprising ammonia, dimethylamine (DMA), and trimethylamine (TMA)—are linked to microbial and enzymatic spoilage and serve as established indicators of fish freshness, with values above 15 mg N/100 g often signaling the onset of spoilage [13]. The significant disparity between collector and market samples corroborates findings reported before, indicating that TVB-N levels exceeding approximately 35 mg N/100 g reliably reflect advanced spoilage stages. This underscores the necessity of minimizing storage duration and maintaining proper cold chain protocols to preserve sensory and biochemical quality in fish products.

3.5 Trimethylamine (TMA), Natural Formaldehyde (NF), and Dimethylamine (DMA)

The trimethylamine test on fish is used to assess freshness and detect spoilage, as bacteria, particularly in the visceral area, can reduce trimethylamine oxide into trimethylamine, a volatile compound that decreases sensory quality. The natural formaldehyde test identifies and measures endogenous formaldehyde formed during quality degradation, especially under low-temperature or frozen storage. Meanwhile, the dimethylamine test serves as an important molecular parameter to evaluate the decline in quality and freshness of fish, while also helping distinguish between naturally occurring and artificially added formaldehyde. TMA, NF, and DMA values for both sources are shown in Table 1.

Table. 1. Trimethylamine (TMA), Natural Formaldehyde (NF), and Dimethylamine (DMA) of Tambaqui (*Colossoma macropomum*)

Source of Fish	TMA (mgN/100g)	NF (ppm)	DMA (ppm)
Collector	8.24 ± 0.59	0.26 ± 0.02	0.57 ± 0.02
Market	9.22 ± 0.90	0.34 ± 0.03	0.60 ± 0.02

The statistical analysis indicated that the fish from collector had significantly lower TMA, NF, and DMA levels than the one from the market, suggesting superior freshness. Microorganisms use oxygen from TMAO under anaerobic conditions to produce TMA, a process mainly driven by gram-negative bacteria that dominate the microbial community in degrading fish tissue. Natural formation of formaldehyde occurs through the reductive mechanism of trimethylamine oxide catalyzed by the enzyme TMAOase, producing formaldehyde and dimethylamine as by-products [14]. The naturally formed formaldehyde, together with dimethylamine, contributes to protein rigidity, and the extent of protein denaturation is generally proportional to the ratio of natural formaldehyde to dimethylamine [15].

4 Conclusion

The comparative evaluation of *Colossoma macropomum* quality parameters between collectors and market sources revealed consistent trends across physicochemical and sensory indicators. Organoleptic assessment showed significantly higher scores for fish obtained directly from collectors, reflecting superior freshness and minimal sensory deterioration. This finding was supported by pH measurements, where both sources exhibited values within the acceptable freshness range, but collector samples maintained slightly lower pH, indicating reduced microbial and enzymatic activity. Total Volatile Base (TVB) levels were markedly higher in market samples, signifying greater protein degradation, while TMA concentrations followed a similar pattern, with lower values in collector samples corresponding to delayed conversion into volatile amines. Formaldehyde (FA) and dimethylamine (DMA) contents were also significantly elevated in market-sourced fish, suggesting more advanced spoilage and possible risks associated with natural or external contamination. Collectively, these results demonstrate that fish obtained directly from collectors exhibit superior quality and freshness compared to market samples, emphasizing the importance of effective cold chain management, rapid distribution, and strict monitoring to ensure the safety and quality of fishery products.

References

1. C. S. Low, Lai Kim Ng, Determination of chemical properties of meat: Determination of trimethylamine oxide (TMAO-N), trimethylamine (TMA-N), total volatile basic nitrogen (TVB-N) by Conway's method, Laboratory Manual on Analytical Methods and Procedures for Fish and Fish Products, (1987), Accessed: Aug. 15, 2025.https://repository.seafdec.org/handle/20.500.12066/5312
2. F. É. Sylvain, B. Cheaib, M. Llewllyn, T. G. Correia, D. B. Fagundes, A. L. Val, N. Derome, pH drop impacts differentially skin and gut microbiota of the Amazonian fish tambaqui (*Colossoma macropomum*), Sci Reports. **6**, 1 (2016). https:doi.org/10.1038/srep32032.
3. B. K. K. K. Jinadasa, C. Elliott, and G. D. T. M. Jayasinghe, A review of the presence of formaldehyde in fish and seafood, Elsevier Ltd (2022). https://doi.org/10.1016/j.foodcont.2022.108882.
4. BSN, Ikan segar. 2013. [Online]. Available: www.bsn.go.id
5. Apriyantono A, D. Fardiaz, Puspitasari NL, Sedarnawati, and S. Budiyanto, Analisis pangan : petunjuk laboratorium. (1989). Accessed: Aug. 14, 2025. [Online]. Available: https://lib.fkm.ui.ac.id
6. BSN, Cara uji kimia - Bagian 8: Penentuan kadar Total Volatil Base Nitrogen (TVB-N) dan Trimetil Amin Nitrogen (TMA-N) pada produk perikanan. (2009).
7. C. S. Ng, Determination of chemical properties of meat: Determination of trimethylamine oxide (TMAO-N), trimethylamine (TMA-N), total volatile basic nitrogen (TVB-N) by Conway's method, Laboratory Manual on Analytical Methods and Procedures for Fish and Fish Products, p. B-3.1, (1987). Accessed: Aug. 14, 2025. https://repository.seafdec.org/handle/20.500.12066/5312
8. T. Nash, The colorimetric estimation of formaldehyde by means of the Hantzsch reaction. Biochem J. **55**, 3, (1953). https://doi.org:10.1042/BJ0550416,.
9. S. Benjakul, W. Visessanguan, and M. Tanaka, Induced formation of dimethylamine and formaldehyde by lizardfish (*Saurida micropectoralis*) kidney trimethylamine-N-oxide demethylase. Food Chem, **84**, 2, (2004). https://doi.org/10.1016/S0308-8146(03)00214-0.
10. C. P. Campos, R. G. Costa Sousa, M. F. Catarino, G. de Albuquerque Costa, and C. E. C. Freitas, Population dynamics and stock assessment of *Colossoma macropomum* caught in the Manacapuru Lake system (Amazon Basin, Brazil). Fish Manag Ecol, **22**, 5, (2015). https://doi.org/10.1111/FME.12139.
11. H. Rachma, A. Ghofar, S. W. Saputra, Studi Beberapa Aspek Biologi Ikan Bawal Hitam (*Parastromateus*). Management of Aquatic Resources Journal (MAQUARES), **4**, 4, (2015). https://doi.org/ 10.14710/MARJ.V4I4.9762.
12. G. Ólafsdóttir, G. blafsdbttir, E. Martinsdbttir, J. Oehlenschbger, P. Dalgaard, B. Jensen, I. Undeland, I.M. Mackie, G. Henehan, J. Nielsen and H. Nilsen, Methods

to evaluate fish freshness in research and industry, Trends Food Sci Technol. **8**, 8, (1997). https://doi.org/10.1016/S0924-2244(97)01049-2.

13. M. Moosavi-Nasab, S. Khoshnoudi-Nia, Z. Azimifar, and S. Kamyab, Evaluation of the total volatile basic nitrogen (TVB-N) content in fish fillets using hyperspectral imaging coupled with deep learning neural network and meta-analysis. Sci Rep, **11**, 1, (2021). https://doi.org/10.1038/S41598-021-84659.
14. J. Tri Murtini, R. Riyanto, N Priyanto, I Hermana, Pembentukan Formaldehid Alami pada Beberapa Jenis Ikan Laut Selama Penyimpanan dalam Es Curai. JPB Perikanan. **9**, 2 (2014).
15. T. Nurhayati, R. Nugraha, and Nurjanah, Fisiologi, Formasi, dan Degradasi Metabolit Hasil Perairan (2017). Accessed: Oct. 10, 2025. [Online]. Available: https://www.ipbprinting.id/fisiologi-formasi-dan-degradasi-metabolit-hasil-perairan

Mangrove ecosystem services assessment in Tanakeke Island, South Sulawesi

Lorensia Puspita[1*], *Luky* Adrianto[2,3], and *Ario* Damar[2,3]

[1]Study Program of Coastal and Marine Resources Management, Department of Aquatic Resources Management, Faculty of Fisheries and Marine Sciences, IPB University, Kampus IPB Dramaga, Jl. Agatis, Bogor 16680, Indonesia
[2]Department of Aquatic Resources Management, Faculty of Fisheries and Marine Sciences, IPB University, Kampus IPB Dramaga, Jl. Agatis, Bogor 16680, Indonesia
[3]Center for Coastal and Marine Resources Studies, International Research Institute for Maritime, Ocean and Fisheries, IPB University, Kampus IPB Baranangsiang, Jl. Raya Pajajaran No. 1, Bogor 16127, Indonesia

Abstract. Mangrove ecosystems are among the most productive and valuable coastal habitats, providing crucial ecological and socio-economic services. However, intensive exploitation and land conversion have caused serious degradation, including in Indonesia, which holds about one-fifth of the world's mangrove forests. This study aims to assess the mangrove ecosystem services in Minasa Baji Village, Tanakeke Island, South Sulawesi, using the Common International Classification of Ecosystem Services (CICES) Version 5.2 framework. Data collection was carried out through household surveys involving 57 respondents, in-depth interviews, field observations, and secondary data sources. The results show that mangroves contribute significantly to provisioning services, including wood products for charcoal and fishing gear (126,000 kg/year), wild fish catches of 179,165 kg/year, and aquaculture production reaching 250,000 kg/year. In terms of regulating services, mangroves absorb 512.714 tons of CO_2eq, serve as nursery grounds supporting zooplankton abundance of 190 billion individuals, and act as nutrient exporters with nitrate concentrations higher inside mangroves (0.0102–0.0141 mg/L) than outside. Culturally, mangroves are viewed as inherited assets and marriage dowries, reflecting their deep symbolic value within the local Makassarese community. These findings emphasize the critical role of mangroves in maintaining ecological balance, local economies, and cultural identity. Strengthening conservation, rehabilitation, and community-based management is essential to ensure the sustainability of mangrove ecosystem services on Tanakeke Island.

1 Introduction

Mangroves are tropical coastal ecosystems that, along with coral reefs and seagrass beds, provide various ecosystem services and important ecological functions [1]. Ecosystem services refer to the benefits that humans enjoy from an ecosystem [2]. According to the Millennium Ecosystem Assessment (2005), ecosystem services are grouped into four categories, namely regulating, supporting, provisioning, and cultural services. As a provisioning service, mangroves function as a resource for capture fisheries (such as fish, crabs, shrimp, and squid), aquaculture, and a source of wood for charcoal production [3].

* Corresponding author: lorens21puspita@gmail.com

Meanwhile, mangroves as a regulating service play a role in carbon storage, oxygen production, wave breaking, tsunami protection, and seawater intrusion prevention [4]. The supporting services of mangrove ecosystems include their function as nursery areas or feeding grounds for marine life [5]. On the other hand, the cultural services of mangroves generally take the form of non-material benefits that fulfill spiritual and religious needs, such as recreational sites.

Globally, mangrove ecosystems are recognized as one of the most productive and valuable coastal habitats, contributing significantly to climate regulation, shoreline protection, and the livelihoods of millions of people in tropical and subtropical regions. In Indonesia, which holds around one-fifth of the world's mangrove cover, pressures from coastal development, pond expansion, and overexploitation have led to widespread degradation. For example, in Tanakeke Island, South Sulawesi, mangrove cover decreased by about 28.3% (± 337.41 hectares) from 2000 to 2020 [3]. Despite the ecological richness including 11 species from 8 families and moderate diversity indices, health assessments indicate that these forests are in a low to moderate condition, particularly at the seedling level.

The ecological function of mangroves contributes greatly to their economic value. Mangroves are an important resource that supports the economy of the people of Tanakeke Island. The community converts mangroves into ponds and uses them as raw material for making charcoal, firewood, seaweed cultivation poles, and sero/bila fishing poles. According to Setiawan et al. [6], the residents of Tanakeke Island are highly dependent on the mangrove ecosystem to meet their economic needs. This high level of dependence has led to increasingly massive exploitation of mangroves.

The degradation of the mangrove ecosystem on Tanakeke Island has resulted in the loss of various ecosystem benefits, including ecological, physical, and economic aspects. Mangroves, in this context, play a key role in all three aspects. The decline in the quality of the mangrove ecosystem in the region has triggered various negative impacts, such as reduced fish diversity, decreased biological stability and productivity, increased abrasion in a number of villages, effects of extreme weather during the west monsoon (which causes tidal flooding in residential areas), and a decline in fish catches and aquaculture production [7]. According to local communities, such conditions were rare in the past when the mangrove ecosystem was still well preserved.

The evaluation of ecosystem services is a fundamental requirement in environmental management. In current practice, the CICES (Common International Classification of Ecosystem Services) approach has become a widely used standard. This classification system was first developed in 2009, then officially launched in 2013 through version V.4.3, and has continued to undergo refinements until reaching version V.5.2 in 2023. CICES is specifically designed to facilitate the process of measuring, quantifying, and assessing various ecosystem benefits. The calculation method in CICES is detailed with a tiered analysis approach [8]. Sustainable management of mangrove ecosystems in the future can be achieved, and management constraints can be overcome, by paying special attention to key issues encompassing ecological, economic, sociocultural, and institutional aspects. Understanding the valuation of the mangrove ecosystem services directly received by communities can provide a basis for community participation in protecting and preserving this source of income. This study has the specific objective of estimating the value of existing mangrove ecosystem services using CICES Version 5.2, with the study location in Minasa Baji Village, Tanakeke Island, South Sulawesi.

2 Methods

2.1 Time of research and background study of area

This research was conducted from August to November 2024 in Minasa Baji Village, Tanakeke Island, Takalar Regency, South Sulawesi Province. The observation location for data collection was determined using a field survey throughout the Minasa Baji Village area, which is used as a mangrove ecosystem service provider. The research location is shown in Figure 1.

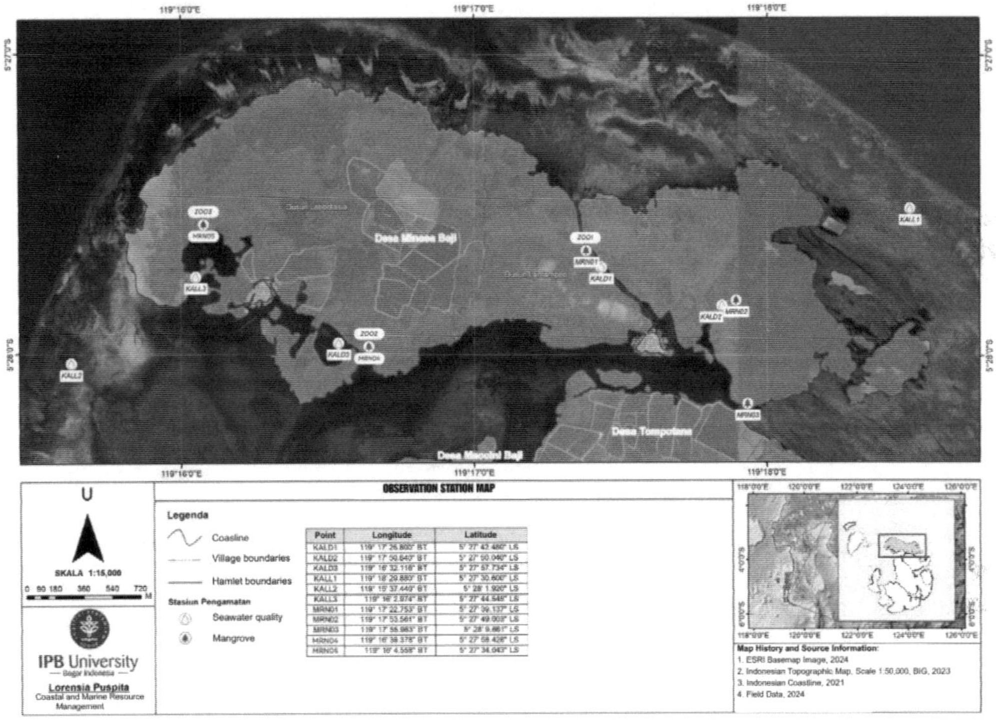

Fig. 1. Research location map. The shaded area denotes land, with the blue line indicating the coastline.

The selection of this location for data collection was based on the existence of the Bangko Tappampang area as a protected mangrove forest area in the region. In this area, there are village regulations governing the utilization and sanctions in mangrove forest management.

2.2 Processing procedure

This study collected data through two main types of sources, namely primary and secondary data. Primary data collection was carried out using three methods: (1) distribution of questionnaires to 57 respondents evenly distributed across two Sub-villages (28 respondents in Lantangpeo and 29 respondents in Labbotalua), (2) in-depth interviews, and (3) simultaneous direct observation in the field. For secondary data, the researcher used various sources including report documents, Excel spreadsheets, accredited scientific journals, peer-reviewed articles, official data from the Takalar Regency Central Statistics Agency, and various other supporting documents relevant to the research topic.

Data collection was conducted through questionnaires and in-depth interviews with community leaders, NGOs, fishermen, fish farmers, and charcoal makers to obtain information on their perceptions, activities, and involvement in the mangrove ecosystem. Data from the questionnaires were analyzed descriptively quantitatively using frequency and percentage tabulations to illustrate patterns of community perceptions and activities. Meanwhile, the results of in-depth interviews and field observations were analyzed qualitatively thematically, namely by identifying, grouping, and interpreting key themes related to the socio-economic interactions of the community and the condition of the mangrove ecosystem.

The questionnaire instrument in this study was developed by adapting the methodology introduced by Finisdore et al. [9]. The classification of mangrove ecosystem services was based on the Common Classification of Ecosystem Services (CICES), which was modified from the versions of Finisdore et al. [9] and Haines-Young [8]. Specifically, provisioning services such as livestock, aquaculture, and wild fish were measured using biomass units. For regulating services, calculations were made using methods of coastline change, zooplankton abundance in waters, and nutrient exports to the open sea. Meanwhile, cultural services were categorized into three main aspects: tourism, education, and research cultural services [8]. The collected data was tabulated, and the ecosystem services of each component were calculated as shown in the table below.

Table 1. Calculation of mangrove ecosystem services.

Ecosystem services	Division	5.2 Code	Description	Calculation	Unit
Mangrove provisioning services					
Wood	Biomass	1.1.5.3	Contribution of mangrove ecosystems to tree growth and wood biomass	Amount of wood used	Kg/ha/year
Wild fish and other natural aquatic biomass	Biomass	1.1.6.1	The contribution of mangrove ecosystems to the growth of fish resources and other aquatic biomass caught in the context of non-cultivated production by economic units for various uses, especially food production.	Catch x catch frequency	Kg/ha/year
Aquaculture provisioning services	Biomass	1.1.2.2	Contribution of mangrove ecosystems to aquaculture biomass harvest.	Harvest yield x harvest frequency x pond area	Kg/ha/year
Mangrove regulating services					
Carbon sequestratrion	Regulation of physical, chemical, biological conditions	2.3.6.1	Contribution of life system types to the number, concentration, or	Carbon value/ha x mangrove area	tonCO_2eq

Ecosystem services	Division	5.2 Code	Description	Calculation	Unit
			climate/atmospheric parameters		
Nursery ground	Regulation of physical, chemical, biological conditions	2.3.2.3	Provision of nursery grounds for fish and shrimp larvae	Abundance of fish and shrimp larvae	Ind larva/m3
Eksportir nutrient ke laut lepas	Regulation of physical, chemical, biological conditions	2.3.2.5	Nutrient provider for the open sea lepas	Increased NO3 concentration in the sea in front of mangroves	NO3 concentration in mangroves and in the sea in front of mangroves
Mangrove cultural services					
Spiritual, artistic and symbolic services	Spiritual, symbolic and other cultural interactions with natural environment	3.4.1.1	The contribution of mangrove ecosystems, particularly through the characteristics and quality of the ecosystem's biophysics, which are recognized by the community for their cultural, historical, aesthetic, sacred, or religious significance	Cultural diversity	Unit

3 Results and discussion

Minasa Baji Village is one of five villages on Tanakeke Island. Minasa Baji Village consists of four Sub-villages, namely Lantangpeo, Labbotalua, Bauluang, and Kampung Beru. The ecosystem shows fluctuations in mangrove area in Labbotalua and Lantangpeo Sub-villages during the period from 2015 to 2024. In Lantangpeo Sub-villages, there was a significant decrease in mangrove area from 96.11 ha in 2015 to 64.73 ha in 2020. This decline is thought to be due to anthropogenic pressures such as land conversion for seaweed cultivation or charcoal production. However, in 2024, the mangrove area increased sharply to 108.08 ha (Figure 2). This increase indicates successful rehabilitation or natural regeneration during that period.

Meanwhile, in Labbotalua Sub-villages, mangrove area experienced a downward trend from 114.58 ha in 2015 to 82.33 ha in 2020. Although there was a slight increase to 84.46 ha in 2024, over all it was still lower than the initial condition in 2015 (Figure 2). This indicates that pressure on the mangrove ecosystem in Labbotalua is relatively high and that it has not yet fully recovered, despite signs of recovery. The mangrove species *Rhizopora stylosa* is the dominant mangrove species in both Sub-villages.

The mangrove forest on Tanakeke Island, Takalar Regency, South Sulawesi, has provided benefits as a sustainable production forest area for the local community. The forest is managed independently by the residents. One of the mangrove areas, Bangko Tapampang, has been designated as a protected zone based on community agreement. To reduce damage caused by tree felling for charcoal production, the village government issued a Village Regulation requiring loggers to leave parent trees to ensure mangrove regeneration. The charcoal raw materials are obtained from mangrove forests, and are either owned by the community or purchased from other residents [10]

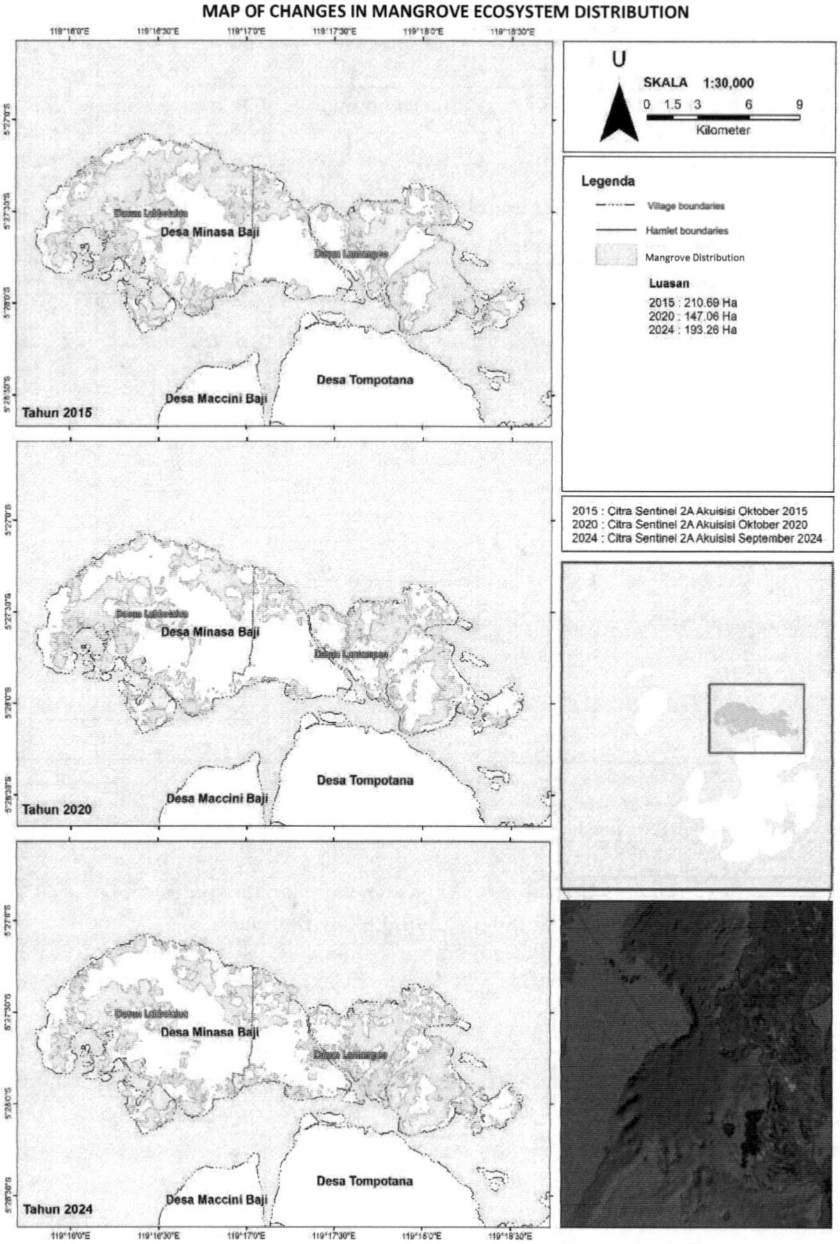

Fig. 2. Distribution of mangroves in Labbotalua and Lantangpeo Sub-Village in the years 2015, 2020 and 2024.

3.1 Provisioning services

3.1.1 Wood

In general, mangrove parts (wood and fruits) in Labbotalua and Lantangpeo Sub-villages are used as raw materials for charcoal, seaweed poles, sero/bila net fishing poles, and paropo fishing gear. Mangrove fruits are used for replanting. Based on interviews, the use of mangrove trunks reaches 126,000 kg/year for charcoal, 28,260 kg/year for seaweed poles, 59,500 kg/year for paropo, and 22,800 kg/year for sero/bila nets (Table 2).

Table 2. Quantity of mangrove ecosystem services in Labbotalua and Lantangpeo Sub-villages.

Utilization of mangrove wood	Area of land cleared (ha)	Weight of wood (kg/ha)	Production kg/ha/year	Yield
Charcoal	1.8	21,000	126,000	1,260 sacks
Seaweed poles	70.65	28,260	28,260	9,420 poles
Paropo	0.0153	59,500	59,500	19,833 poles
Sero/bila poles	4.56	22,800	22,800	7,600 poles

3.1.2 Wild fish and other natural aquatic biomass

The questionnaire results show that fishermen in Lantangpeo and Labbotalua Sub-villages use various fishing gears, including gill nets, sero/bila nets, paropo nets, and fish shooting. Based on these fishing gears, the target catches are fish, crabs, squid, and cuttlefish. The calculated biomass of catches (wild fish and other biota) reached 179,165 kg/year in Lantangpeo and Labbotalua Sub-villages. Details of the catches can be seen in Table 3.

Table 3. Quantity of wild fish and other aquatic biomass in Labbotalua and Lantangpeo Sub-villages.

Type of fishing gear	Area (fish/ha)	kg/ha	Frequency	Production (kg/year)
Sero/bila				
Siganus javus	4.2	1.4	312	433.3
Siganus sp.	4.2	1.4	312	433.3
Cuttlefish	8.3	2.1	312	650.0
Crab	12.5	3.1	312	975.0
Squid	8.3	2.1	312	650.0
Snapper	4.2	2.1	120	250.0
Grouper	4.2	2.1	60	125.0
Caranx sexfasciatus	4.2	2.1	18	37.5
Gill net				
Siganus sp.	571.4	190.5	12	2285.7
Snapper	571.4	285.7	12	3428.6
Grouper	571.4	285.7	12	3428.6
Caranx sexfasciatus	114.3	114.3	12	1371.4
Cuttlefish	2285.7	571.4	12	6857.1
Squid	2285.7	571.4	12	6857.1
***Paropo* (Traditional fishing gear)**				
Siganus sp.	22222.2	7407.4	3	22222.2
Snapper	33333.3	16666.7	3	50000.0
Grouper	33333.3	16666.7	3	50000.0
Siganus javus	33333.3	8333.3	3	25000.0
Dive fishing				

Type of fishing gear	Area (fish/ha)	kg/ha	Frequency	Production (kg/year)
Siganus sp.	200	66.7	12	800
Snapper	200	100	12	1200
Grouper	200	100	12	1200
Caranx sexfasciatus	60	30	12	360
Cuttlefish	100	25	12	300
Squid	100	25	12	300
Total				179,165

3.1.3 Aquaculture

Based on interviews with residents of Labbotalua and Lantangpeo Sub-villages, there were only about four fish farmers, all of whom were located in Labbotalua Sub-villages, with a total pond area of 75 hectares. The harvest per farmer varied between 25,000 and 200,000 fish per harvest, depending on the size of each farm. Harvesting is carried out twice a year. The aquaculture biomass (milkfish) produced in Labbotalua Sub-villages reached 250,000 kg/year, as detailed in Table 4.

Table 4. Aquaculture production in Labbotalua and Lantangpeo Sub-villages.

Pond area (ha)	Yield (fish)	Harvest frequency	Total yield (fish)	Total yield (kg/year)	Total (kg/year)
10 ha	50,000	2	100,000	33333.33	250,000
20 ha	100,000	2	200,000	66666.67	
40 ha	200,000	2	400,000	133333.33	
5 ha	25,000	2	50,000	16666.67	

3.2 Regulating services

3.2.1 Carbon stock estimates

The estimated average carbon stock value of mangroves in Labbotalua and Lantangpeo Sub-villages reached 722,882 tonsC/ha, equivalent to 2.652,98 ton$\boldsymbol{CO_2}$eq/ha. When multiplied by the total mangrove area of 193.26 ha, the total carbon sequestration potential reaches 512.714 tons of CO_2eq (Table 5). This value shows that the mangrove ecosystems in both Sub-villages play an important role as significant carbon sinks.

Table 5. Carbon sequestration in Labbotalua and Lantangpeo Sub-villages.

Carbon stock estimates (tonC/ha)	Estimated conversion of carbon to carbon dioxide (tonCO_2eq/ha)	Mangrove area (ha)	Total (tonCO_2eq)
722,882	2.652,98	193,26	512.714

3.2.2 Nursery ground

Mangroves contribute significantly to the surrounding ecosystem, as shown in the table of zooplankton numbers in the area. Based on field data, the species obtained were unidentified Gastropoda larvae, unidentified Annelida larvae, *Microsetella* sp, and *Balanus* sp. larvae, which are types of zooplankton that serve as food for fish or shrimp larvae. The average zooplankton abundance from the three observation stations ranged from 21,870 ind/L to 52,110 ind/L (Table 6).

Table 6. Zooplankton conditions in Labbotalua and Lantangpeo Sub-villages.

No	Organism	Sample Code								
		Number			Plankter/ml			Abundance/L		
	Species	1	2	3	1	2	3	1	2	3
1	Unidentified Gastropoda larvae	81	108	72	0.081	0.108	0.072	2,430	3,240	2,160
2	Unidentified Annelida larvae	27	0	36	0.027	0	0.036	810	0	1,080
3	*Microsetella* sp.	0	18	27	0	0.018	0.027	0	540	810
4	Larvae of *Balanus* sp.	36	54	54	0.036	0	0.054	1,080	0	1,620
					Average abundance (N)			21,870	44,280	52,110

The results of the calculations in Table 8 show the abundance of zooplankton in the waters of Lantangpeo and Labbotalua Sub-villages, with a total volume of 4,832 m³. Assuming a conversion rate of 1 m³ = 1,000 liters, the total volume of water reaches 4,831,500 liters. Based on the observation results, the average zooplankton abundance is 39,420 individuals per liter. With this value, the total number of zooplankton individuals in the waters is estimated to reach 190,457,730,000 individuals (Table 7). This figure illustrates the enormous and abundant potential of natural food resources in the study area.

Table 7. Zooplankton abundance in Labbotalua and Lantangpeo Sub-villages.

Water area (m3)	Water volume (liter)	Abundance ind/L	Total Ind
4,832	4,831,500	39,420	190,457,730,000

3.2.3 Exporter of nutrients to the open sea

The water quality around mangrove areas is an important indicator in assessing the health of coastal ecosystems. One key parameter to consider is the content of nitrate (NO_3), a compound that acts as an essential nutrient for the growth of phytoplankton and marine vegetation, but which in excessive concentrations can have negative impacts such as eutrophication [11]. Mangrove areas, with their unique physical and biological characteristics, often have complex nutrient dynamics, including nitrate cycles that are influenced by natural processes such as litter decomposition, microbial activity, and interactions with seawater and sediments [12].

Based on the data obtained, nitrate (NO_3) concentrations in the studied mangrove areas showed variations between locations within and outside mangroves. Nitrate values in mangroves ranged from 0.0102 to 0.0141 mg/L, while outside mangroves they ranged from 0.0056 to 0.0087 mg/L (Table 8). This indicates that nitrate concentrations tend to be higher in mangrove ecosystems than in surrounding waters.

High nitrate levels in mangroves can be attributed to the decomposition of organic material from mangrove leaf litter and microbial activity that recycles nutrients [12]. In addition, the complex root structure of mangroves plays a role in trapping sediment and organic matter, thereby increasing the accumulation of dissolved nutrients such as nitrate [13]. Conversely, lower nitrate levels outside mangroves may be due to factors such as dilution by seawater or high nitrate consumption by phytoplankton in open waters [14].

Table 8. Nitrate (NO_3) distribution within and outside mangrove areas

No	Sample Code	Unit	Location	Parameter

				Nitrate (NO_3)-mg/L
1	KALL1	mg/L	Outside the mangrove	0.0079
2	KALD1	mg/L	Inside the mangrove	0.0102
3	KALD2	mg/L	Inside the mangrove	0.0102
4	KALL2	mg/L	Outside the mangrove	0.0087
5	KALL3	mg/L	Outside the mangrove	0.0056
6	KALD3	mg/L	Inside the mangrove	0.0141

The nitrate (NO_3) concentration in the mangrove area was recorded at 0.0041 mg/L with a water volume of 48,315,000 liters. From these calculations, the nitrate content stored in the mangrove area's water system is approximately 198,092 mg (Table 9). This value indicates that although the nitrate concentration is relatively low, the total amount of nutrients in the area is quite large due to the large volume of water.

Table 9. Nutrient levels (NO3) inside and outside mangrove areas

Volume (m^3)	Water volume (Liter)	Nitrate (mg/L)	Total nitrate (mg)
48,315	48,315,000	0.0041	198,092

3.3 Cultural services

3.3.1 Spiritual, artistic, and symbolic services

The population of Tanakeke Island is dominated by indigenous Makassarese people who originated from mainland Sulawesi and have settled permanently on the island. For them, mangrove forests have the same value as land-based plantations and are considered hereditary assets. The majority of mangrove ownership is obtained through inheritance, while a small portion is acquired through purchase. In line with this, Setiawan [15] notes that the community uses mangroves as dowries for marriage and passes them on to the next generation. This view arises because the communities around Tanakeke Island regard the mangroves in their area as private property, so their management is similar to that of land-based plantations.

The total area of mangroves in the two Sub-villages (Labbotalua, Bangko Tapampang, and Lantangpeo) reaches 193.26 ha. The distribution of this area varies, with Labbotalua Sub-villages having the largest mangrove area of 84.46 ha, followed by Lantangpeo with 68.42 ha, and Bangko Tapampang with an area of 40.38 ha. Based on this, the area of mangrove utilization that can be used as cultural value (marriage dowry) is 152.88 hectares.

4 Conclusion

The study concludes that the mangrove ecosystem in Minasa Baji Village, Tanakeke Island, provides crucial ecosystem services that sustain both the environment and local livelihoods. Through the CICES v5.2 framework, the research identified significant contributions across provisioning, regulating, and cultural dimensions. Provisioning services include the production of mangrove wood (used for charcoal, fishing tools, and seaweed poles), wild fish catches totaling 179,165 kg/year, and aquaculture yields of 250,000 kg/year. Regulating services highlight the role of mangroves as major carbon sinks with 512.714 tons CO_2e absorbed, as nursery grounds supporting zooplankton abundance of 190 billion individuals, and as nutrient exporters enhancing marine productivity. Culturally, mangroves function as inherited assets and marriage dowries, reflecting their deep symbolic and social importance within the Makassarese community. Overall, these results emphasize that mangrove sustainability is crucial for maintaining ecological balance, economic stability, and socio-cultural continuity. Therefore, conservation, rehabilitation, and community-based management must be strengthened to ensure the long-term sustainability of mangrove ecosystem services in Tanakeke Island.

References

1 N. Widagti, F. Sidik, N. A. Pradisty, Monitoring Mangrove Untuk Estimasi Potensi Karbon Biru Di Dumai, Riau. JFMR-J. Fisher. Mar. Resour. **5**(2), 459-469 (2021). https://doi.org/10.21776/ub.jfmr.2021.005.02.34

2 R. Costanza, R. d'Arge, R. de Groot, S. Farber, M. Grasso, B. Hannon, K. Limburg, S. Naeem, R.V. O'Neill, J. Paruelo, R.G. Raskin, P. Sutton, M. van den Belt, The Value of the World's Ecosystem Services and Natural Capital. Nature **387**, 253–260 (1997). https://doi.org/10.1038/387253a0

3 Aswin, A. Damar, G. Yulianto, Kondisi Vegetasi dan Perubahan Tutupan Lahan Ekosistem Mangrove Pulau Tanakeke Kabupaten Takalar Provinsi Sulawesi Selatan. Jurnal Ilmu dan Teknologi Kelautan Tropis **13**(2), 305–318, (2021). https://doi.org/10.29244/jitkt.v13i2.33636

4 Iswahyudi, Kebijakan Pengelolaan Ekosistem Hutan Mangrove Berkelanjutan Kota Langsa. Ph.D. thesis, IPB University, Indonesia (2019)

5 I. Nurokhmah, Elastisitas Jasa Ekosistem Mangrove di Teluk Jor, Kabupaten Lombok Timur, Nusa Tenggara Barat. Master Thesis, IPB University, Indonesia (2019)

6 H. Setiawan, R. Purwanti, R. Garsetiasih, Persepsi Dan Sikap Masyarakat Terhadap Konservasi Ekosistem Mangrove Di Pulau Tanakeke Sulawesi Selatan. Jurnal Penelitian Sosial dan Ekonomi Kehutanan **14**(1), 57–70 (2017). https://doi.org/10.20886/jsek

7 Auliansyah, Analisis Ekonomi Pengelolaan Kawasan Ekosistem Mangrove di Pulau Tanakeke Kabupaten Takalar Provinsi Sulawesi Selatan, Master Thesis, IPB University, Indonesia (2018).

8 R. Haines-Young, Common International Classification of Ecosystem Services (CICES) V5.2 Guidance on the Application of the Revised Structure, July. (2023).

9 J. Finisdore, C. Rhodes, R. Haines-Young, S. Maynard, J. Wielgus, A. Dvarskas, J. Houdet, F. Quétier, K.A. Lamothe, H. Ding, F. Soulard, G. Van Houtven, P. Rowcroft, The 18 Benefits of Using Ecosystem Services Classification Systems. Ecosys. Serv. **45**, 101160 (2020). https://doi.org/10.1016/j.ecoser.2020.101160

10 Peraturan Daerah Prov. Sulawesi Selatan No. 4 Tahun 2023 Tentang Pengelolaan dan Pengembangan Ekosistem Mangrove serta Naskah Akademik, (2023).

11 R. W. Howarth R. Marino, Nitrogen as the limiting nutrient for eutrophication in coastal marine ecosystems: Evolving views over three decades, Limnol. Oceanog. **51**(1), 364-376 (2006). https://doi.org/10.4319/lo.2006.51.1_part_2.0364

12 D. M. Alongi, Carbon cycling and storage in mangrove forests. Ann. Rev. Mar. Sci. **6**, 195–219 (2014). https://doi.org/10.1146/annurev-marine-010213-135020

13 E. Kristensen, S. Bouillon, T. Dittmar, C. Marchand, Organic carbon dynamics in mangrove ecosystems: A review, Aquat. Bot. **89**(2), 201–219 (2008). https://doi.org/10.1016/j.aquabot.2007.12.005

14 T. Dittmar, N. Hertkorn, G. Kattner, R. J. Lara, Mangroves, a major source of dissolved organic carbon to the oceans, Glob. Biogeochem. Cycles **20**(1), 1–7 (2006). https://doi.org/10.1029/2005GB002570

15 H. Setiawan, Kajian Etnobotani Mangrove Masyarakat Adat Makassar di Pulau Tanakeke Kabupaten Takalar Provinsi Sulawesi Selatan, in Prosiding Seminar Nasional Perhutanan Sosial Tahun 2016. Banjarbaru: Lambung Mangkurat University Press (2018). pp. 106-113.

BIO Web of Conferences **221**, 04009 (2026) https://doi.org/10.1051/bioconf/202622104009
7th EMBRIO International Symposium (EIS 2025)

Feasibility and productivity analysis of *arad* (mini trawl) fisheries in Cilincing North Jakarta

Rachmad Caesario[1*], *Mercy* Patanda, *Roza* Yusfiandayani[2], *Domu* Simbolon[2], *Tri Dewi Kusumaningrum* Pribadi[3], *Urip* Rahmani[4], *Izza Mahdiana* Apriliani[5,6]

[1]Marine and Fisheries Department, Faculty of Agriculture, Lampung University, Bandar Lampung, Sumatera, Indonesia
[2]Fisheries Resources Utilization Department, Faculty of Fisheries and Marine Sciences, IPB University, Bogor, West Java, Indonesia
[3]Biology Department, Faculty of Natural Science Mathematics, Padjadjaran University, Bandung, West Java, Indonesia
[4]Fisheries Resources Utilization Department, Faculty of Fisheries and Marine Sciences, Satya Negara Indonesia University, South Jakarta, Indonesia
[5]Graduate School of Regional Revitalization Sciences, Utsunomiya University, Utsunomiya City, Tochigi, 321-8585, Japan
[6]Department of Fisheries, Faculty of Fisheries and Marine Science, Padjadjaran University, Sumedang 45363, West Java, Indonesia

Abstract. In 2019, fish production landed at TPI Cilincing weighed 1,476,750 kg, with a value of IDR 47,501,137,000.00. One of the fishing gears used in Cilincing is mini trawl nets. This fishing gear is categorized as illegal by the ministry of marine affairs due to its destructive nature. However, many fishermen still operate this fishing gear and refuse to switch into more environmentally friendly fishing gears. This research aims 1.) to determine the feasibility of the mini trawl in Cilincing, North Jakarta; and 2.) to determine the productivity of mini trawl fishing units and compare it to more environmentally friendly fishing gears. This research was conducted from January - May 2024. The data analyses used were both qualitative and quantitative to provide an overview of mini trawl productivity and financial analysis. The results show that NPV was positive, with B/C >1 (2,21), and an IRR value of 38%. This result shows that mini trawl fishery is beneficial because the return on capital is faster and the profits are higher compared to more environmentally friendly fishing gears. In addition, the productivity value of mini trawl per trip was 0.32 tons/trip and the mini trawl net had a much higher productivity value compared to bottom gillnets (*rampus*) and blue swimming crab gillnets.

1 Introduction

Jakarta Bay boasts abundant fishery resources. This is evidenced by its high capture fishery production. One of the prominent areas contributing to this high production is the Cilincing Fish Auction Site (TPI Cilincing). TPI Cilincing is renowned for its diverse fish resources. In 2019, the fish landed at TPI Cilincing weighed 1,476,750 kg, with a market value of IDR 47,501,137,000. This high production is driven by various fishing activities, including set nets, guiding barriers (*sero*), crab nets/traps, gillnets, bottom seine nets (*dogol*), and mini trawls (*arad*).

* Corresponding author: Rachmad.caesario@fp.unila.ac.id

Trawls are used to catch demersal fish, as well as non-fish species such as clams. Trawls catch economically valuable species such as squid, cuttlefish, mantis shrimp, white pomfret, crabs, and several types of shrimp including banana prawn, jinga shrimp, and giant tiger prawn [1]. Bycatch, which has a lower economic value, includes species such as ponyfish, silver pomfret, dark-banded goatfish, hairtail, barracuda, stingray, and spotted butterfish (Indonesian: *Kiper*).

The annual fish catch by trawl net fishermen varies due to several factors, including fishing seasons and weather conditions. These fluctuations impact productivity as a result of the vessel's ability to capture fish annually.

Key factors to consider in mini trawl fisheries include the vessel, crew, and fishing trips, all of which significantly influence the mini trawl fishery business. Effective fisheries management in Cilincing, supported by well-functioning cooperatives, can enhance the fishery's performance and positively impact the ecosystem. However, intense fishing activities without environmental conservation measures can deplete fish resources. Furthermore, mini trawls are categorized as illegal by the Indonesian Ministry of Marine Affairs and Fisheries due to their destructive nature and impact on the ecosystem. Besides that, mini trawl fishermen are also reluctant to substitute their gear with other fishing gears that are more environmentally friendly. Therefore, it is essential to conduct research on mini trawl fisheries in Cilincing. The objectives of this research are: (1) Determining the feasibility of the mini trawl net fishery business in North Jakarta's Cilincing area, and (2) Assessing the productivity of the mini trawl net fishing units and comparing it with bottom gillnet (*rampus*) and blue swimming crab gillnet fisheries which operate mostly in the same fishing grounds.

2 Material and method

2.1 Research location and times

The research has been conducted from January to May 2024 in the waters of Cilincing (Figure 1). The stages of the research were as follows: (1) Preliminary Survey Conducted in October to familiarize the research team with the study location; (2) Primary Data Collection conducted from January to February 2024, focusing on three aspects: economic, social, and institutional; (3) Secondary Data Collection; (4) Data Analysis.

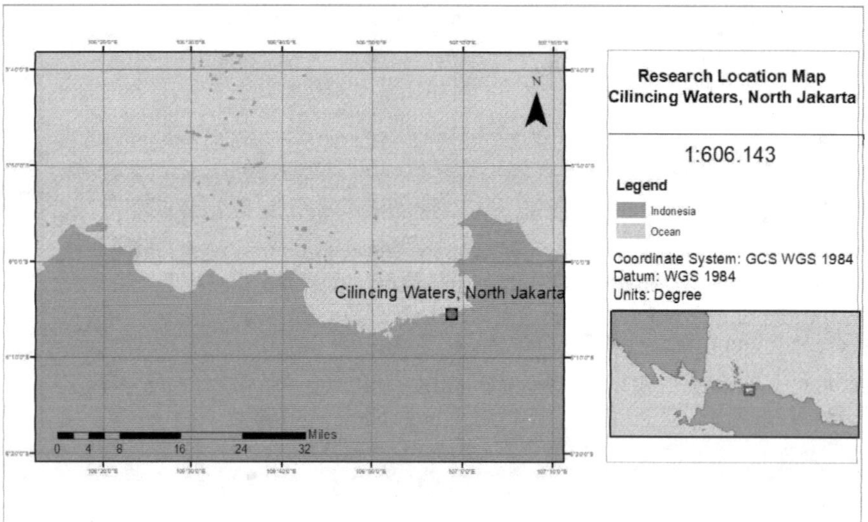

Fig. 1. Map of Cilincing, showing its location in North Jakarta.

2.2 Tools and materials

The tools and materials used in this research are listed in Table 1.

Table 1. Tools.

No	Tools	Function
1	Questionnaire	Contains questions for respondents

2.3 Research methodology

The survey method involved observation or obtaining detailed information on a specific issue. The data used in this study consist of primary and secondary data. Primary data collected include operational costs, fixed costs, variable costs, types of catch, fish prices, vessels' tonnage, number of crew, engine power and fishing trips. Primary data was obtained through interviews with a group of mini trawl net fishermen. Respondents were selected deliberately, considering their positions and roles in their daily activities, using purposive sampling [2]. The respondents included 15 fishermen and collectors. Additionally, secondary data was obtained from government agencies and published research on mini trawl nets, such as the Provincial Fisheries and Marine Affairs Agency and the Ministry of Marine Affairs and Fisheries, as well as scientific publications on mini trawl fisheries.

2.4 Data analysis method

Analysis of data in terms of economic analysis was conducted through quantitative analyses, i.e.:

a) Financial Analysis to determine the feasibility of the business venture. This analysis included feasibility analysis using investment criteria such as Net Present Value (NPV), Net Benefit Cost Ratio (Net B/C), and Internal Rate of Return (IRR) [3]. Long-term analyses such as NPV, Net B/C, and IRR can be used to ascertain the benefits and costs of utilizing fisheries resources. This analysis offers insights into the benefits derived from the exploitation of fisheries resources [4].
b) Productivity analysis of mini trawl nets, bottom gillnets, and blue swimming crab gillnets in Cilincing.

- **NPV (Net Present Value)**

$$NPV = \sum_{t=1}^{n} \frac{(Bt-Ct)}{(1+i)^t} \quad (1)$$

B : Benefit
C : Cost
i : Discount rate
t : Year of operation

In the NPV method, there are three investment assessments [5]:

1) NPV ≥ 0 (feasible): This indicates that financially, the venture is feasible to be undertaken because the benefits outweigh the costs.
2) NPV = 0 (break even): This means financially the venture is at a break-even point because the benefits obtained are equal to the costs incurred.

3) NPV ≤ 0 (not feasible): This indicates that financially, the venture is not feasible to be undertaken. This is because the benefits obtained are less than the costs/incurred, and they are insufficient to cover the expenses.

- **PP (Payback Period)**

The Payback Period (PP) can be determined using the formula:

$$Payback\ Period = \frac{Investment}{Benefit} \quad (2)$$

- **IRR (Internal Rate of Return)**

The Internal Rate of Return (IRR) can be determined using the formula below [6]:

$$IRR = i + \frac{NPV}{NPV - NPV'}(i' - i) \quad (3)$$

Information:

i' : first trial value for the discount rate;
i'' : second trial value for the discount rate;
NPV : first net present value;
NPV'' : second net present value.

- **B/C Ratio (Benefit Cost Ratio)**

The Net Benefit Cost (Net B/C) ratio is a comparison between the present value of benefit flows and the present value of cost streams. This figure indicates the level of additional benefits for each additional unit of cost in terms of currency. The criterion used for selecting the Net B/C ratio measure for project benefits is to choose all projects with a Net B/C ratio of one or more if the benefits are discounted at the opportunity cost rate. However, if the Net B/C ratio is less than 1, the project is deemed not feasible [3]. The formula used is as follows:

$$Net\frac{B}{C} = \frac{\sum_{t=1}^{n} \frac{(Bt-Ct)}{(1+i)^t}}{\sum_{t=1}^{n} \frac{(Ct-Bt)}{(1+i)^t}} \quad (4)$$

Information:
Net B/C : Benefit Cost Ratio
Bt : Benefit year t
Ct : Cost Year t (IDR)
n : Time period (year)
i : Discount rate (%)
t : Period (year)

The general formulas for calculating the productivity of fishing gear are as follows [7]:

$$Productivity\ of\ Fishing\ Vessel = \frac{\sum Catch(tons)}{\sum Gross\ Tonnage(GT)} \quad (5)$$

$$Productivity\ of\ Crew = \frac{\sum Catch(tons)}{\sum Crew(person)} \quad (6)$$

$$Productivity\ of\ Mini\ Trawl's\ Engine = \frac{\sum Catch(tons)}{\sum Engine\ Power(HP)} \quad (7)$$

$$Productivity\ of\ Fishing\ Trip = \frac{\sum Catch(tons)}{\sum Fishing\ Trip(Trip)} \quad (8)$$

3 Results and discussion

Arad fisheries, or mini bottom trawls, are primarily used for catching shrimp. However, they are controversial due to their low selectivity and potential for environmental damage. *Arad* fishing is associated with challenges like bycatch, a high percentage of fish caught being younger than their length at first maturity, and socio-economic issues for fishermen, who often have lower education levels. Regulations have been introduced to ban *arad* nets and encourage the use of more environmentally friendly alternatives, but adoption rates can be low due to factors like the perceived lower value of substitute gear.

3.1 NPV

Net Present Value (Net Present Value = NPV) is the total profit over the life of the business that has been multiplied by a discount factor. The results of the analysis of the feasibility of the mini trawl business show a NPV of 40,843,137 IDR. As the NPV has a positive value, this shows that the mini trawl business is worth continuing. This is in accordance with the statement of Umar [5], who mentioned that a business is said to be feasible if the NPV value > 0.

3.2 Payback Period (PP)

Payback period is the rate of return on capital or the length of time used to cover the original investment costs. The faster the return on investment of a business, the better the business pattern because the smoother the capital turnover [8]. The PP value in *arad* fisheries is 2 years and 6 months. Thus, *arad* business units in Cilincing have a rate of return on investment costs that fall into the fast category, because the *arad* business units in Cilincing waters have a payback period of less than 3 years.

3.3 IRR

The internal rate of return (IRR) is an investment calculation method in which an interest rate that equates the present value of an investment with the present value of future receipts is calculated [4]. The result of the analysis shows that the value of the internal rate of return (IRR) is 38%, which means it provides a profit of 38% per year from all investments invested during the 10-year fishing life.

3.4 B/C ratio

The ratio of receipts and costs (B/C) is a comparison of total receipts with total costs (operational costs and fixed costs). The B/C ratio in *arad* fisheries is 2, meaning that the businesses of *arad* fisheries exceed 0 and have met the requirements for an economically beneficial activity.
The feasibility of the *arad* fishery business is based on favourable NPV, Net B/C and IRR values. The result of the business feasibility analysis is shown in Table 2.

Table 2. Results of the Business Feasibility Analysis of the *arad* fishery

Criterion	Value	Information
NPV	IDR 40,843,137	Worthy, NPV value > 0
Net B/C	2.21	Worthy, net value B/C > 1
IRR	38%	Feasible, IRR value > Interest Rate (8.55%)
PP	2.54 yrs	2 years and 6 months until investment costs are returned

Positive NPV, B/C > 1, and IRR value greater than the interest rate used, which is 10%, show that the *arad* fishery business is financially feasible because the return on capital is faster and the profit is higher than the interest rate. This is in accordance with Pinkerton [9], who mentioned that fishing businesses with a high level of feasibility can support the independent economic development of local fishermen.

3.5 Mini trawl fisheries productivity

Based on the obtained data of catches landed at TPI Cilincing for the 2022-2023 period, we can calculate the productivity of mini trawl nets. There are four productivity calculations using the obtained data, namely; the productivity of fishing vessel, crew, engine and fishing trip.

- **Productivity of fishing vessel**

The value of vessel productivity is determined by landed catch data and vessel tonnage. The size of the trawl net ship is known to be 7 GT. In this way, the productivity value of the Trawl vessel can be calculated. The results showed that one unit of mini trawl vessel has a catch productivity value of 148.58 tons/GT of the vessel.

$$Productivity\ of\ Mini\ Trawl's\ Vessel = \frac{\sum 1.040\ tons}{\sum 7GT} = 148.59\ tons/GT \quad (9)$$

The productivity value of mini trawl net vessels in Cilincing waters is relatively lower when compared to the productivity value of trawl net vessels in Pandeglang waters, namely 1326/7 = 189.42 tons/GT [10]. The productivity value of this trawl fishing gear tends to be far below the minimum limit value [10]. The productivity obtained is the overall productivity where the catch still consists of main and bycatch. The main catch in trawl nets is generally less than the by-catch [13].

- **Productivity of crew**

Based on the obtained data of catches landed at TPI Cilincing for the 2022-2023 period, combined with data on the average number of ship crew members obtained through questionnaire interviews, the productivity value of mini trawl net crew members can be determined. The results of the interviews showed that the average number of crew members on the trawl boats at TPI Cilincing was 3 people. The research results show that the annual productivity value of trawl net crew members is 346.70 tons/person.

$$Productivity\ of\ Mini\ Trawl'sCrew = \frac{\sum 1.040\ tons}{\sum 3\ person} = 346.70\ tons/person \tag{10}$$

This value tends to be lower when compared to the annual productivity of mini trawl net crew members in Pandeglang waters, namely 442 kg/person [10]. Differences in productivity values can be influenced by many things such as the number of fishing trips, capital and number of crew members. The productivity can be determined by energy, the number of workers and the fishing ground [11]. Apart from that, crew skills also play an important role in increasing productivity values. Workers play a central role in defining and achieving decent work [12].

- **Productivity of fishing engines**

Based on the obtained data of catch landed at TPI Cilincing for the 2022-2023 period, combined with data on ship engine power obtained through questionnaire interviews, the productivity value of trawl net machines can be determined. Interview results showed that the average trawl vessel at TPI Cilincing is powered by a 24 HP engine. The results showed that the productivity of trawl fishing engine is 43.34 tons/HP.

$$Productivity\ of\ Mini\ Trawl's\ Engine = \frac{\sum 1.040\ tons}{\sum 24\ HP} = 43.34\ tons/HP \tag{11}$$

Engine power is vital for the mini trawl fishing unit because mini trawls are an active fishing gear in which a net is pulled to sweep the bottom of the sea. Apart from that, auxiliary engines are also needed to pull the net onto the ship during the hauling process. Trawl net boats have two engines where the main engine functions to run the ship and drag the gear during fishing operation, whilst the auxiliary engine is used to pull the net during the hauling process [13]. Without the help of machines, the hauling process will be slower and of course reduce fishing productivity.

- **Productivity of fishing trip**

Based on the obtained data of catches landed at TPI Cilincing for the 2022-2023 period, combined with fishing trip data obtained through questionnaire interviews, the productivity value of mini trawl net fishing trips can be determined. The mini trawl productivity value per trip is 0.32 tons/trip. This value is greater than trawl productivity per trip in Pemalang waters which is only 0.012 tons/trip [13]. This significant difference in mini trawl productivity per trip can be caused by the availability of resources, fishing grounds and fishermen's experience. Low value of mini trawl productivity is thought to be an indication that the shrimp population in the research area is starting to decline [13].

$$Productivity\ of\ Mini\ Trawl'sFishing\ Trip = \frac{\sum 1.040\ tons}{\sum 3.300\ Trip} = 0.32\ ton/trip \quad (12)$$

We also calculated the productivity value of bottom gillnets (*rampus*) and blue swimming crab gillnets as a comparison. Based on the obtained data of catches landed at TPI Cilincing, the productivity values of ships, crew, engines and fishing trips was calculated. The fishing productivity of trawls, bottom gillnets (*rampus*) and blue swimming crab gillnets can be seen in Table 3.

Table 3. Productivity of mini bottom trawls (trawl), bottom gillnets (*rampus*) and blue swimming crab gillnet.

No	Productivity	Productivity Score		
		Trawl	*Rampus*	Blue Swimming Crab gillnet
1	Ship	148.59	26.80	20.73
2	Crew	346.70	134	51.82
3	Engine	43.34	11.17	8.64
4	Trip	0.32	0.08	0.06

The results show that overall mini trawl nets have a much higher productivity value compared to *rampus* nets and blue swimming crab gillnets. This difference in productivity values is thought to be caused by the active nature of mini trawl nets compared to the passive nature of rampus and blue swimming crab gillnets. Trawl nets are a fishing method that actively catches fish at the bottom of the water, while *rampus* and blue swimming crab gillnets tend to be passive and wait for the target to come and be caught in the net. Gillnets are usually operated passively in waters, and have a lower productivity value than seine nets and trawls due to their passive nature [14]. However, this makes gillnets more environmentally friendly than trawls as they do not have an impact on the seabed ecosystem [15].

4 Conclusion

The conclusion of this study are

1. The feasibility of the mini trawl fishery business has been shown based on NPV, Net B/C and IRR values. Positive NPV, B/C greater than one, and an IRR value greater than the interest rate used, which is 10%, imply the mini trawl business is financially feasible and show that the *arad* fishery business is financially suitable because the return on capital is faster and the profit is higher.
2. Mini trawl nets have a much higher productivity value compared to bottom gillnets and blue swimming crab gillnets. This difference in productivity values is thought to be caused by the active nature of trawl nets compared to the passive nature of bottom gillnet and blue swimming crab gillnets.

Acknowledgements

This paper was supported by the Directorate of Research and Innovation, IPB University with National Research (Ri-Na) program activities with the title Seine Net fishing and Its impact in Cilincing waters, Jakarta based on cooperation agreement letter No.537/IT3.D10/PT.01.03/P/B/2023 dated 24th November 2023. The author would also like to thank the Sub-Directorate of Food Security, Maritime Affairs and Agriculture, Cilincing, and the *arad* fishermen in Cilincing, North Jakarta.

Conflict of Interest

The authors declare that there is no conflict of interest.

References

1. H.A. Nugroho, A. Rosyid, A.D.P. Fitri, Analisis indeks keanekaragaman, indeks dominansi dan proporsi hasil tangkapan non target pada jaring arad modifikasi di perairan Kabupaten Kendal [Analysis of diversity index, dominance index, and proportion of non-target catch in modified trawl nets in Kendal Regency waters]. J. Fish. Resour. Util. Manag. Technol. **4**(1), 1–11 (2015).
2. Mujio, L. Adrianto, K. Soewardi, Y. Wardiatno, Analisis potensi konflik pemanfaatan ruang kawasan pesisir: integrasi rencana tata ruang darat dan perairan pesisir [Analysis of potential spatial conflicts at coastal and marine zones: integration of the spatial planning of land and coastal water]. Sodality: Jurnal Sosiologi Pedesaan **4**(2), 139–144 (2016). https://doi.org/10.22500/sodality.v4i2.13379
3. J.P. Gittinger, Project economics analysis (UI Press, Jakarta, 1986)
4. B.O. Nababan, T. Kusumastanto, L. Adrianto, A. Fahrudin, Analisis ekonomi alat penangkapan ikan arad di Pantai Utara Provinsi Jawa Tengah [An economic analysis of 'arad' fishing gear in the North Coast of Central Java Province]. Jurnal Sosial Ekonomi Kelautan dan Perikanan **15**(1), 1-14 (2020). http://dx.doi.org/10.15578/jsekp.v15i1.8492
5. H. Umar, Marketing and consumer behavior research (Gramedia Pustaka Utama, Jakarta, 2003).
6. Kadariah, K. Lien, C. Gray, Introduction to project evaluation (Faculty of Economics, University of Indonesia, Jakarta, 1999).
7. Choliq, R. Wirasasmita, Sofyan, Introduction to project evaluation (Pionir Jaya, Bandung, 1994).
8. Pujianto, H. Boesono, D. Wahyu, Analisis kelayakan usaha aspek finansial penangkapan mini purse seine dengan ukuran jaring yang berbeda di PPI Ujungbatu Kabupaten Jepara [Business feasibility analysis of the financial aspect of catching mini purse seine with different net sizes in PPI Ujungbatu, Jepara Regency]. J. Fish. Resour. Util. Manag. Technol. **2**(2), 124–133 (2013).
9. E. Pinkerton, Attaining better fisheries management through co-management: prospects, problems, and propositions, in Co-operative management of local fisheries: new directions in improved management and community development, ed. E. Pinkerton (University of British Columbia Press, Vancouver, Canada, 1989), pp. 3–33.
10. Yonvitner, E. Yuliana, D.E. Yani, L.E. Setijorini, Nurhasanah, A. Santoso, M. Boer, R. Kurnia, S.G. Akmal, Fishing gear productivity related to fishing intensity and stock vulnerability potency in Sunda Strait. IOP Conf. Ser.: Earth Environ. Sci. **404**, 012066 (2020). https://doi.org/10.1088/1755-1315/404/1/012066
11. M.M. Oliveira, M.B. Gaspar, J.P. Paixão, A.S. Camanho, Productivity change of the artisanal fishing fleet in Portugal: a Malmquist index analysis. Fish. Res. **95**(2-3), 89–197 (2009). https://doi.org/10.1016/j.fishres.2008.08.020

12. A.J.G. Lozano, J.L.D. Sparks, D.P. Durgana, C.M. Farthing, J. Fitzpatrick, B. Krogh-Poulsen, G. McDonald, S. McDonald, Y. Ota, N. Sarto, A. Cisneros-Montemayor, G. Lout, E. Finkbeiner, J.N. Kittinger, Decent work in fisheries: current trends and key considerations for future research and policy. Mar. Pol. **136**, 104922 (2022). https://doi.org/10.1016/j.marpol.2021.104922
13. T. Ernawati, B. Sumiono, Hasil tangkapan dan laju tangkap jaring arad (mini bottom trawl) yang berbasis di TPI Asemdoyong Pemalang [Catch and catch rate of mini bottom trawl in Asem Doyong Landing Place, Pemalang]. Jurnal Penelitian Perikanan Indonesia **16**, 267–274 (2010).
14. A.D.P. Fitri, H. Boesono, B.B. Jayanto, T.D. Hapsari, F. Adiyanto, The productivity of modified mono- and multifilament gill nets operating in Rembang waters. AACL Bioflux **14**(1), 274–281 (2021).
15. M.W. Kraus, J.M. Rucker, J.A. Richeson, Americans misperceive racial economic equality. *Proc. Natl. Acad. Sci. U.S.A.* **114**, 10324–10331 (2017). https://doi.org/10.1073/pnas.1707719114

AUTHOR INDEX